# STONE AGE INSTITUTE PUBLICATION SERIES

*Series Editors Kathy Schick and Nicholas Toth*

Stone Age Institute
Gosport, Indiana

and

Indiana University,
Bloomington, Indiana

---

*Number 1.*
THE OLDOWAN: Case Studies into the Earliest Stone Age
*Nicholas Toth and Kathy Schick, editors*

## *Forthcoming Volumes:*

*Number 3.*
THE CUTTING EDGE: New Approaches to the Archaeology of Human Origins
*Kathy Schick and Nicholas Toth, editors*

*Number 4.*
THE HUMAN BRAIN EVOLVING: Papers in Honor of Ralph L. Holloway
*Douglas Broadfield, Michael Yuan, Kathy Schick, and Nicholas Toth, editors*

---

# The Stone Age Institute Press Publication Series

## Comment by Series Editors

The Stone Age Institute Press has been established to publish critical research into the archaeology of human origins. The Stone Age Institute is a federally-approved non-profit organization whose mission is investigating and understanding the origins and development of human technology and culture throughout the course of human evolution. The ultimate goal of these undertakings is to provide a better understanding of the human species and our place in nature.

An unbroken line of technology and culture extends from our present day back in time at least two-and-a-half million years. The earliest stone tool-makers were upright-walking, small-brained ape-men, and from these primordial origins, the human lineage embarked upon a pathway that is extraordinary and unique in the history of life. This evolutionary pathway has shaped our bodies, our brains, and our way of life, and has had an ever-accelerating impact on the earth and its other organisms. Well over 99% of the time span of this technological evolution occurred in the Stone Age, so it is no exaggeration to say that modern humans have evolved as Stone Age creatures.

The Stone Age Institute's goal is to provide a haven for this research, providing support for scholars from around the world conducting cutting-edge human origins research. Our research team and associates include professional research scientists in long-term positions, postdoctoral research fellows, and visiting scholars. Fieldwork, collecting primary data from important archaeological sites, as well as laboratory studies and experimental archaeological research are all critical components of our operation. Other vital functions of the Institute include organizing workshops - convening top scientists to discuss and work together on special topics–and organizing and sponsoring conferences and lectures to disseminate information to a broader, general public.

An equally critical component of our operations is the publication of research results in journal articles and books. *Breathing Life into Fossils: Taphonomic Studies in Honor of C.K. (Bob) Brain* is the second volume in a new publication series initiated by the Stone Age Institute Press. This volume emanates from a Stone Age Institute conference in honor of the lifework of C.K. (Bob) Brain in natural history and taphonomy. This meeting convened researchers at the Institute for two days in April of 2004 in a workshop conference on taphonomic studies and their impact on human origins research. The Stone Age Institute Press is devoted to the publication of seminal research in the form of high-quality, data-rich volumes, including information and interpretations from original research as well as critical syntheses of research results.

For more information on the Stone Age Institute, including its research programs, lectures, conferences, and publications, we invite you to visit our web site at www.stoneageinstitute.org.

**Nicholas Toth and Kathy Schick**
*Co-Directors, Stone Age Institute*
*Series Editors, Stone Age Institute Press Publication Series*

STONE AGE INSTITUTE PUBLICATION SERIES

*NUMBER 2*

*Series Editors Kathy Schick and Nicholas Toth*

# BREATHING LIFE INTO FOSSILS:

## Taphonomic Studies in Honor of C.K. (Bob) Brain

*Editors*

**Travis Rayne Pickering**
University of Wisconsin, Madison

**Kathy Schick**
Indiana University

**Nicholas Toth**
Indiana University



Stone Age Institute Press · www.stoneageinstitute.org
1392 W. Dittemore Road · Gosport, IN 47433

*COVER CAPTIONS AND CREDITS.*

*Front cover, clockwise from top left.*

*Top left:*
*Artist's reconstruction of the depositional context of Swartkrans Cave, South Africa, with a leopard consuming a hominid carcass in a tree outside the cave: bones would subsequently wash into the cave and be incorporated in the breccia deposits. © 1985 Jay H. Matternes.*

*Top right: The Swartkrans cave deposits in South Africa, where excavations have yielded many hominids and other animal fossils. ©1985 David L. Brill.*

*Bottom right: Reconstruction of a hominid being carried by a leopard. © 1985 Jay H. Matternes.*

*Bottom left: Photograph of a leopard mandible and the skull cap of a hominid from Swartkrans, with the leopard's canines juxtaposed with puncture marks likely produced by a leopard carrying its hominid prey. © 1985 David L. Brill.*

*Center: Photo of Bob Brain holding a cast of a spotted hyena skull signed by all of the taphonomy conference participants. © 2004 Kathy Schick, Stone Age Institute.*

*Back cover credits.*

*Top: © 2004 Stone Age Institute.*

*Bottom left: © 2004 Kathy Schick, Stone Age Institute.*

*Bottom right: © 2005 Greg Murphy.*

Published by the Stone Age Institute.
ISBN-10: 0-9792-2761-5
ISBN-13: 978-0-9792-2761-5

# Contributors

**Peter Andrews** (pjandrews@uwclub.net) Natural History Museum, London, United Kingdom

**Anna K. Behrensmeyer** (behrensa@si.edu) Department of Paleobiology, National Museum of Natural History, Smithsonian Institution, Washington, D.C., USA

**Robert Blumenschine** (rjb@rci.rutgers.edu) Center for Human Evolutionary Studies, Department of Anthropology, New Brunswick, New Jersey, USA

**Bob Brain** (brainnew@iafrica.com) Transvaal Museum, Pretoria, South Africa

**Maia Bukshianidze** Georgian National Museum, Tbilisi, Georgia

**Henry Bunn** (htbunn@wisc.edu) Department of Anthropology, University of Wisconsin-Madison, Madison, Wisconsin, USA

**Salvatore Capaldo** (sdcap24@hotmail.com) Department of Anthropology, Rutgers University, New Brunswick, New Jersey, USA

**Ron Clarke** (Clarker@anatomy.wits.ac.za) School of Anatomical Sciences and Institute for Human Evolution, University of the Witwatersrand, Johannesburg, South Africa

**Naomi E. Cleghorn**, (cleghorn@berkeley.edu) Department of Anthropology, University of California at Berkeley, Berkeley, California, USA

**Manuel Domínguez-Rodrigo** (m.dominguez.rodrigo@gmail.edu) Departmento de Prehistoria, Universidad Complutense, Madrid, Spain

**Charles Egeland** (cegeland@indiana.edu) Department of Anthropology, Indiana University, Bloomington, Indiana, USA

**Reid Ferring** Department of Geography, University of North Texas, Denton, Texas, USA

**Tom Gehling** (tj-gehling@comcast.net) El Cerrito, California, USA

**Gary Haynes** (gahaynes@unr.edu) Department of Anthropology, University of Nevada-Reno, Reno, Nevada, USA

**Gail Krovitz** (gailk@ecollege.com) eCollege, Denver, Colorado, USA

**Kathleen Kuman** (kathleen.kuman@wits.ac.za) School of Geography, Archaeology and Environmental Studies and Institute for Human Evolution, University of the Witwatersrand, Johannesburg, South Africa

**David Lordkipanidze** (dlordkipanidze@museum.ge) Georgian National Museum, Tbilisi, Georgia

**Curtis W. Marean** (curtis.marean@asu.edu) Institute of Human Origins, School of Human Evolution and Social Change, Arizona State University, Tempe, Arizona, USA

**Jackson Njau** (jknjau@yahoo.com) National Natural History Museum, National Museums of Tanzania, Arusha, Tanzania

**Charles Peters** (chaspete@uga.edu) Department of Anthropology and Institute of Ecology, University of Georgia, Athens, Georgia, USA

**Travis Pickering** (tpickering@wisc.edu) Department of Anthropology, University of Wisconsin-Madison, Madison, Wisconsin, USA

**Briana Pobiner** (PobinerB@si.edu) Human Origins Program, Smithsonian Institution, National Museum of Natural History, Washington, D.C., USA

**Kathy Schick** (kaschick@indiana.edu) Stone Age Institute and Indiana University, Bloomington, Indiana, USA

**Pat Shipman** (pls10@psu.edu) Department of Anthropology, Pennsylvania State University, State College, Pennsylvania, USA

**Martha Tappen** (tappe004@umn.edu) Department of Anthropology, University of Minnesota, Minneapolis, Minnesota, USA

**Francis Thackeray** (thack@nfi.org.za) Transvaal Museum, Pretoria, South Africa

**Nicholas Toth** (toth@indiana.edu) Stone Age Institute and Indiana University, Bloomington, Indiana, USA

**Abesalom Vekua,** Georgian National Museum, Tbilisi, Georgia

**Alan Walker** (axw8@psu.edu) Department of Anthropology, Pennsylvania State University, State College, Pennsylvania, USA

**Tim White** (timwhite@berkeley.edu) Human Evolution Research Center and Department of Integrative Biology, University of California at Berkeley, Berkeley, California, USA

# Acknowledgements

The editors would like to thank a number of organizations and offices for their support of the conference that led to this volume. These include the Wenner-Gren Foundation for Anthropological Research, the Stone Age Institute and Friends of CRAFT, and the following offices of Indiana University: Vice President for Research and the University Graduate School, Office of the Chancellor, College of Arts and Sciences, and Office of International Programs. We would also like to extend special thanks to Charles Egeland, Melanie Everett, Leslie Harlacker, Jason Heaton, Anneliese Pickering, and Dietrich Stout for their valuable assistance during the conference. Funds for the production of this volume have been provided by the Stone Age Institute and Friends of CRAFT, Inc.

# BREATHING LIFE INTO FOSSILS: TAPHONOMIC STUDIES IN HONOR OF C.K. (BOB) BRAIN

EDITED BY

TRAVIS RAYNE PICKERING, NICHOLAS TOTH AND KATHY SCHICK

STONE AGE INSTITUTE PRESS, 2007

## CONTENTS

# Foreword: In Appreciation of Bob Brain

Chapter 37 of the Old Testament Book of Ezekiel tells the story of its author being set in the midst of a valley full of dry moldering bones, which are eventually brought to life with God's breath. The secular world of paleontology does not have it so easy, but indeed our goal as historical scientists is to breathe life into long-dead bones. The disconnection between that aim (reconstructing the behavioral and ecological dynamics of prehistoric animals) and the static databanks for doing so (the paleontological and archaeological records) has long been recognized. This disjunction is most successfully overcome through the application of the principle of actualism. Lee Lyman's (1994: 69) comprehensive review of the articulation of actualism and taphonomy concludes that "[a]ctualistic research is presently perceived as the basis for most taphonomic…analysis and interpretation."

Even a cursory review of Bob Brain's research portfolio (see Rubidge, 2000), which spans 50+ years, reveals that this is certainly the case in his career. It is impossible to compartmentalize Bob as a person or as a scientist. An ostensible geologist (his 1957 doctorate was entitled *The Ape-Man-Bearing Cave Deposits of the Transvaal*), it is more appropriate to identify Bob as a consummate naturalist, in the best and most professional sense of that appellation. From research as seemingly disparate as studies on rotifer biology to reconstructions of the geomorphology of Pleistocene caves in the Sterkfontein Valley, all of Bob's work is united in a purpose that is ultimately behavioral.

In Chapter 1 of this volume, Bob recounts Raymond Dart's role in provoking him to enter into a sphere of research in which the fossil and living worlds converge, with observations made in the latter employed to breathe life into the bony residues of the former. As a result, Bob ushered the developing discipline of taphonomy into paleoanthropology, culminating in a new standard of scientific rigor in the field.

On a more specific level, Bob's actualistically grounded analyses of the fossil faunas from Swartkrans, Sterkfontein and Kromdraai caves toppled Dart's (e.g., 1957) hypothesis of australopithecines as "Killer Apes," the cannibalistic alpha predators of Pleistocene Africa. Bob's observations of modern primates being consumed by carnivores combined with his recognition of tooth marks on hominid fossils led him to construct an opposing model of australopithecines as being more commonly prey than predators (e.g., Brain, 1981, 1993a). Thus, began Bob's interest in predation as a major behavioral factor that conditioned not just human evolution, but the evolution of the very first animals, *c*. 600 million years ago.

Grappling productively with an overarching "grand theme"—predation, for instance—is a key reason why many "famous scientists" come to be venerated. But, it is also appropriate to fully understand the body of data generated by the "famous scientist" in service of posing and exploring the myriad of questions relevant to his "grand theme." To his great credit, such an appreciation comes easily in the case of Bob Brain. For example, his meticulous studies of carnivore feeding behavior, of the production of bone "pseudotools" through naturalistic processes, of the modification of actual bone tools through hominid digging and of the effects of burning on bones come readily to mind. Together, they create a much richer view of early hominids then them simply serving as the prey of large carnivores. Certainly *some* australopithecines were hapless victims (Figure 1); ac-

*Figure 1. There is no other example from the fossil record that better documents the link between early hominids and carnivore feeding (and probable predation) than Bob Brain's observation of puncture holes in the juvenile hominid calotte (SK 54) from the Hanging Remnant of Swartkrans. The holes match exactly the spacing of the canine teeth of a fossil leopard mandible from the same deposit. Photograph by David Brill.*

cordingly, predation probably influenced their social organization and perhaps other aspects of their biology. Verifying the causal links predicted on good theoretical grounds between predation pressure and socioecology is vexing even in modern higher primates (e.g., Zuberbühler and Jenny, 2002), which are observable—much-less in the fossil record, in which behavior can only be inferred (Pickering, 2005).

The beauty of Brain's approach to investigating his "grand theme" is that predation never became an exclusive fixation to him. Rather, it always was (and is) just that—a *theme* in the truest sense of the word. Predation is the great anchor of his work as evidenced by his consistent return to it time and time again. But, from that anchor the disparate lines of his work fan out in an interconnected web that, taken together, describe far-ranging aspects of prehistoric behavior.

An example illustrates both the agility with which Bob approaches paleoanthropology and his enviable ability to avoid the myopia that can often blemish a good but lesser scientist's standing. Figure 2 reproduces Bob's estimates of the minimum number of individuals (MNI) for some major taxonomic groups recovered from Members 1–3 of the Swartkrans Formation. As is apparent, the proportional representation of hominids drops markedly between the earliest formed Members 1 and 2 versus Member 3, a more recent Pleistocene deposit. Bob interpreted this shift as behaviorally significant, arguing that while predation had an important impact on the hominids in Members 1 and 2, the data suggest otherwise for the Member 3 hominids. Interestingly, while abundant *Australopithecus robustus* remains have been recovered from all three members, *Homo erectus* remains are much less common and do not occur at all in Member 3. The inference is that the recovered hominids represent, for the most part, victims of large carnivores. Thus, the paucity and eventual complete absence by Member 3 times of *H. erectus*, the presumptive direct ancestor of modern people, suggests the greater success of this species in avoiding predation. In addition to his taxonomically based conclusions, Bob documented the presence of burned bones in Member 3. His actualistic, chemical and histological analyses suggest that these fossils were heated in humanly controlled fires. Putting together these findings, Bob transcended the textbook caricature that each and every South African fauna was created exclusively by the feeding behavior of large carnivores. Instead, he cast a nuanced, sophisticated and (most importantly) testable hypothesis of broad-scale behavioral complexity through time at Swartkrans:

> "Taphonomic reconstructions at Swartkrans have emphasized the importance of feline predation on early hominids, at a stage when human influence on the natural environment was very slight, and when the balance of power lay with the cats. But the Swartkrans record also documents a technological innovation of immense importance: the management of fire, providing a measure of protection from nocturnal predators. In Member 3 at Swartkrans, a hint is discernable of a shift in the power balance towards hominids—the first in a series of technologically based triumphs that have established people as dominant animals on earth" (Brain, 1993b: 264).

In a wonderful display of magnanimity, Bob has encouraged and joined a new group of researchers at Swartkrans in testing this hypothesis through zooarchaeological and taphonomic analyses of Swartkrans fossils he recovered between 1979 and 1986. I am happy to be part of this group and we present some of our results on the Member 3 fauna in Chapter 13 of this volume (analyses of the Member 1 and 2 faunas are forthcoming and will be published elsewhere). This is just one of countless examples in which Bob has taken a genuine interest and tangible action in encouraging the work of others for the service of advancing our knowledge of Swartkrans, particularly, and our understanding of prehistoric life, generally. For this, a simple volume of papers in his honor does not suffice, but still, it does stand as an outward expression of our great gratitude. Bob sparked my interest in taphonomy when I first read his work as a student and has continued to foster it since we have become friends and collaborators. I am just one of a multitude of

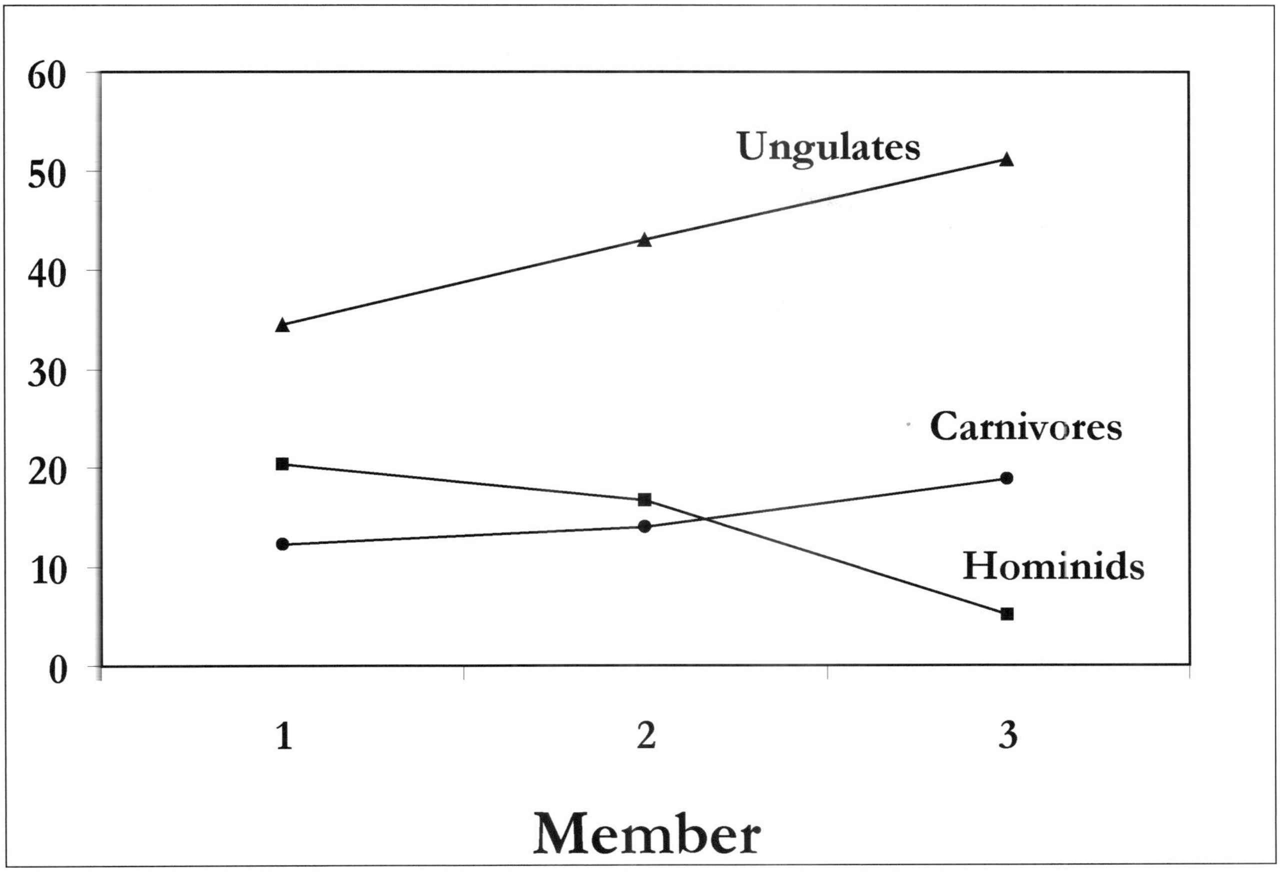


*Figure 2 Percent representation of three major taxa (ungulates, carnivores, and hominids) based on the total minimum number of individuals (MNI) in the faunas of Swartkrans Members 1–3 (after Brain, 1993b: 263, Figure 8). Total MNI estimates for each member: Member 1 (Lower Bank) = 443; Member 2 = 114; Member 3 = 176 (Brain, 1993b).*

researchers whom Bob taught that combining an interest in the living and fossil worlds is not only possible, but is also scientifically productive. We are all indebted.

*31 May 2006*
*Travis Rayne Pickering*
*Madison, Wisconsin, USA*

## References


Brain, C.K. 1981. The Hunters or the Hunted? An Introduction to African Cave Taphonomy. University of Chicago Press, Chicago.

Brain, C.K. (Ed.). 1993a. Swartkrans: A Cave's Chronicle of Early Man. Transvaal Museum, Pretoria.

Brain, C.K. 1993b. A taphonomic overview of the Swartkrans fossil assemblages. In: Brain, C.K. (Ed.), Swartkrans: A Cave's Chronicle of Early Man. Transvaal Museum, Pretoria, pp. 255-264.

Dart, R.A. 1957. The Osteodontokeratic Culture of *Australopithecus prometheus*. Transvaal Museum, Pretoria.

Lyman, R.L. 1994. Vertebrate Taphonomy. Cambridge University Press, Cambridge.

Pickering, T.R. 2005. Underrated prey? Evolutionary Anthropology 14, 159-164.

Rubidge, B.S. 2000. Charles Kimberlin (Bob) Brain—A tribute. Palaeontologia Africana 36, 1-9.

Zuberbühler, K., Jenny, D. 2002. Leopard predation and primate evolution. Journal of Human Evolution 43, 873-886.

# Introduction

**Travis Rayne Pickering, Kathy Schick and Nicholas Toth**

Taphonomy, the study of the processes leading to the fossilization of organic remains, is one of the most important avenues of inquiry in research into human origins. By carefully examining processes and patterns in the modern, observable world (actualistic studies), we are able to gain crucial insights that can be used in the data collection, analysis and interpretation of the prehistoric record. Such actualistic studies have grown tremendously in the past few decades, providing a wealth of information for use in paleoanthropological-research.

The conference *African Taphonomy: A Tribute to the Career of C.K. (Bob) Brain* was convened at the Stone Age Institute in Bloomington from April 29–30, 2004 to discuss the latest research and developments in taphonomy—a field introduced to African prehistorians in large part through the early work of C.K. (Bob) Brain. Brain's (1981) book *The Hunters or the Hunted?*, published after roughly 20 years of accumulated actualistic research, is still heralded as a classic treatise in human evolutionary studies. It was apparent from the conference presentations that Brain's work, as summarized in that book, continues to have a far-reaching and lasting impact. In honor of this legacy, we have collected in this volume 16 papers that emanated from the conference. As with Brain's own work, it is difficult to pigeonhole many of the chapters herein; most cross-cut various types of actualistic work. We have, however, made an effort to arrange the contributions into five broad themes.

The first set of papers includes two essays that discuss the influences on Brain's development as a scientist and his own subsequent impact on paleoanthropology. Chapter 1, by Bob Brain, presents the central storyline of the development of African cave taphonomy. The broad outline of that story will be familiar to many readers, but it is quite illuminating to read Brain's own take on the relevance of its various components. Especially valuable and inspiring are the dual themes of optimism and fun that run through his lively narrative. Science can and should be a pleasurable pursuit, one worthy of a lifetime's devotion, as in the case of Brain. Gary Haynes, in Chapter 2, a sociohistorical account of Brain's influence on the development of taphonomy in Paleoindian studies, stresses another aspect of Brain as a scientist and person. Haynes makes the point that we would all do well to use Brain's approach as a model in our own research:

> "He reviewed others' work, collected data, and spelled out his alternative interpretations with grace and tact… Brain's contribution to Paleoindian research went beyond merely providing examples of taphonomic studies to emulate. To his greatest credit, he also showed us how to stalwartly present a case without alienating colleagues and friends."

The remaining chapters are largely empirically based, but can still be crudely sub-divided. A group of papers by Naomi Cleghorn and Curtis Marean, Francis Thackeray, and Kathy Schick and colleagues deal ostensibly with mammalian carnivores as taphonomic agents—one of the major concentrations of Brain's research. Cleghorn and Marean (Chapter 3) discuss the growth of a general model for bone survival in zooarchaeological assemblages, with a special emphasis on carnivore destruction of skeletal elements. Their model separates bones into a low-survival set (elements that lack thick cortical bone) and a high-survival set (elements comprised predominately of thick cortical bone) and argues that because of their resistance to complete destruction (through processes such a carnivore ravaging), the dense midshaft portions of limb bones are the most useful category of bone for reconstructing early hominid behavior. Thackeray's chapter (Chapter 4) also focuses on carnivore contributions to the formation of paleoanthropological bone assemblages and on limb bone shaft fragments. In particular, he explores the usefulness of mean limb bone shaft lengths and carnivore: ungulate ratios to assess the biotic agent(s) of bone accumulation at the important early hominid sites of Kromdraai, Swartkrans and Sterkfontein (South Africa). Schick and her colleagues (Chapter 5) present data on the bone assemblage from a modern striped hyena den they excavated in Jordan. The presentation is of particular relevance to the South African paleontological record since it is hypothesized that an extinct subspecies of striped hyena (*Hyaena hyaena makapani*) was a likely contributor of the Makapansgat Limeworks Grey Breccia fauna. This is the very fauna upon which Raymond Dart based his notion of the Osteodontokeratic Culture of *Australopithecus prometheus* (now *A. africanus*).

The third set of chapters, by Alan Walker, Martha Tappen and colleagues, Kay Behrensmeyer, Rob Blumenschine and colleagues, and Kathleen Kuman, explore site- or landscape-level issues in taphonomy and paleoanthropological assemblage formation. Walker's (Chapter 6) discussion of two Miocene sites on Rusinga Island (Kenya) deals with the most remote time period covered in the book and some of the most fascinating taphonomic circumstances known in primate paleontology. R114, which yielded the type specimen of *Proconsul heseloni*, is the site of large hollow tree that was ultimately infilled by matrix and bones, while the Kaswanga Primate Site is possibly an infilled carnivore burrow. Chapter 7, by Tappen and colleagues, describes taxonomic and taphonomic aspects of the important Pleistocene fauna from Dmanisi (Republic of Georgia), associated with the earliest securely dated hominids outside of Africa. The site is tantalizing taphonomically, with the authors' concluding preliminarily that it does not conform to the "plutonic ideals of human habitation sites, hyena dens, or mass death sites." Behrensmeyer's (Chapter 8) contribution returns to Africa, with a report on changes in skeletal part survival and bone surface modification in the Amboseli (Kenya) ecosystem over her 30 years of work there. She links many of the changes to a marked increase in the spotted hyena population and the decline of other large predators over the last decade, suggesting that such an inter-specific demographic in the past would have resulted in increased competition for carcasses and minimal opportunities for aggressive scavenging by early hominids. Landscape taphonomy is also the concern of Blumenschine and his colleagues. In Chapter 9, they present data from their Olduvai Landscape Paleoanthropology Project (Tanzania). In order to reconstruct landscape facets existing during Bed I and Lower Bed II times within the Olduvai Basin, the researchers have conducted modern taphonomic surveys in the Serengeti, along Lake Masek and the Lower Grumeti River. Especially fascinating is their study of crocodile taphonomy in these settings and its relevance for determining fine-scale landscape features. Kuman's chapter (Chapter 10) shifts focus from bones to stones and from East to South Africa, as she describes varying land use by hominids in the Stone Age. She concludes that nearly all of the earliest sites in South Africa are secondary deposits within karstic cavities, while actual occupation of caves occurred only much later, after 600,000 years ago. All other sites are open air and usually close to standing water. Kuman also elaborates on her current research on the late Acheulean and Middle Stone Age archaeology of the Mapungubwe National Park, in Limpopo Province (South Africa).

Two other papers, by Ron Clarke and Gail Krovitz and Pat Shipman, comprise the fourth section of this volume. Clarke (Chapter 11) provides taphonomic comparisons of three australopithecine skeletons from Sterkfontein (South Africa). The Little Foot (Stw 573) skeleton, from the Member 2 level at that site, is far more complete than the partial torsos from Member 4, Sts 14 and Stw 431, which each have only one partial limb preserved. Clarke concludes that Stw 573 was apparently mummified and buried before its bones separated, while the two Member 4 skeletons were probably ravaged by the same type of biotic actor, resulting in their similar degree and kind of incompleteness. Krovitz and Shipman (Chapter 12) provide methods for reconstructing the taphonomy of immature hominid crania, of particular relevance to the human fossil record, which has yielded many important specimens of juvenile status. Indeed, the authors then apply their methods to the cases of three such specimens, the Taung Child (*A. africanus*), Mojokerto (*Homo erectus*) and Herto BOU-VP-16/5 (*H. sapiens idaltu*).

The final four chapters are organized together because of their emphasis on hominids as taphonomic agents. Travis Pickering and his colleagues (Chapter 13) elaborate upon Brain's interpretation of early hominid behavior at Swartkrans Member 3 (South Africa) by presenting evidence of 163 fossil specimens bearing newly identified stone tool cutmarks and hammerstone percussion damage. Data presented on tooth marks indicate that carnivores contributed more predominantly than hominids to the formation of the Member 3 fauna, but hominids still appear to have been capable foragers. Based on the anatomical distribution of cutmarks, it is argued that hominids gained access to carcass parts usually defleshed early and entirely by carnivores before that happened. Although mostly a presentation of data on carcass modification by felids, the work of Manuel Domínguez-Rodrigo and his colleagues (Chapter 14) makes the point that hominid-induced bone damage is the preferred class of data upon which inferences of hominid behavior should be made. In response to previous models, they emphasize that the order of carnivore and hominid access to carcasses could be modeled more specifically and usefully with a taxon-specific (felid versus hyenid) consideration of tooth mark frequencies, and present some useful steps in this direction. Henry Bunn's chapter (Chapter 15) is also concerned with inferring early hominid access to carcasses, but concentrates on the utility of his ethnoarchaeological observations of Hadza foragers (Tanzania) for doing this. He emphasizes that Hadza maximize their return rates by transporting to base camps essentially whole carcasses of zebra size and smaller animals. This makes sense when one considers that the sophisticated chopping (metal axes) and boiling technologies of these modern humans assure they can extract nutrients from skeletal parts that must have been difficult or impossible for Oldowan hominids to exploit. Tim White and Nicholas Toth (Chapter 16) close-out the volume by discussing the likelihood that feeding hominids modified bones with their teeth, as well as with stone tools. The chapter will serve as an important caution to taphonomists against automatically attributing every tooth mark observed in a fossil fauna to carnivores. The broader implication is that the usefulness of models of carnivore-hominid interaction that are dependent on tooth mark frequencies might be suspect.

Twenty-five years after the publication of the *Hunters or the Hunted?*, Bob Brain's masterwork, each of the chapters in this volume reflects the continuing and encompassing influence of the man and his work on the field of paleoanthropological taphonomy and its practitioners. Bob is still relevant and the research questions he posed and then so eloquently explored still resonate and inspire advancement in our understanding of human evolution.

# CHAPTER 1

# Fifty Years of Fun with Fossils: Some Cave Taphonomy-Related Ideas and Concepts that Emerged Between 1953 and 2003

C.K. Brain


## Abstract

During the last 50 years, African cave taphonomy has been established as a discipline in its own right, capable of informing us about the lives and deaths of animals whose remains are preserved in the caves concerned. It all started with Raymond Dart who, shortly after his interpretation and description of *Australopithecus africanus* in 1925, began to speculate about what the faunal remains, associated with the child skull, could tell us about the circumstances in which our early ancestors lived and died. Twenty years later, when Dart became interested in fossils from the Makapansgat Limeworks cave in the northern Transvaal, he made a pioneering study of over 7,000 fossil bone pieces that had been laboriously chipped from the rock-hard breccia matrix. He concluded that the vast fossil bone accumulation in this cave had been collected by *Australopithecus* for use as tools and weapons. In his view these hominids were undergoing a "predatory transition from ape to man" and were already mighty hunters, capable of killing the most powerful animals of their time and making use of an "osteodontokeratic culture." He put forward these ideas in a series of 39 publications, making use of dramatic and provocative prose that was intended to stimulate and provoke others to take the issues further. In my case, his provocation and encouragement was so great that I spent 40 years examining southern African caves and their associated fossil assemblages, in particular the Swartkrans one, as well as documenting contemporary taphonomic processes. In the paper that follows, some of the facts, concepts and ideas that emerged from this work are discussed. They are:

1. Understanding skeletal part disproportions in bovid bone assemblages.
2. The complication of bone pseudo-tools.
3. "Head-hunters" and "professional decapitators"?
4. "The myth of the bone-accumulating hyena"?
5. The comparative vulnerability of primate and bovid skeletons to carnivore damage.
6. Stone tools in the South African early hominid bone accumulations.
7. The finding and interpretation of bone tools.
8. The importance of cave-form to taphonomic reconstructions.
9. The effects of progressive Cainozoic cooling on African habitats and fauna.
10. Evidence for the management of fire at Swartkrans.
11. Evidence from Swartkrans for predation on early hominids.
12. The significance of predation to the evolution of intelligence in hominids and much older ancestral animals.
13. The pleasure of seeing taphonomic themes being carried forward from their early beginnings.


## The Early Days of Taphonomy in Africa

The emerging discipline of vertebrate taphonomy,

with particular reference to Africa, received a major boost in July 1976 when the Wenner Gren Foundation of New York, under the enthusiastic research direction of Lita Osmundsen, sponsored a symposium entitled *Taphonomy and Vertebrate Paleoecology, with special reference to the Late Cenozoic of Sub-Saharan Africa.*

*Figure 1. In 1976, Lita Osmundsen was Director of Research at the Wenner Gren Foundation in New York and her enthusiasm was of critical importance in developing the new discipline of taphonomy.*

This was held at the Foundation's Conference Centre, the castle of Burg Wartenstein, in the Austrian Alps and was attended by 17 delegates from a wide variety of disciplines. The meeting was organised by Anna K. Behrensmeyer, Andrew P. Hill, Alan Walker and myself, while the proceedings appeared thereafter in book form, *Fossils in the Making–Vertebrate Taphonomy and Paleoecology*, edited by Behrensmeyer and Hill (1980). This volume served to place vertebrate taphonomy on a secure footing, that was reinforced by other highly significant books that appeared thereafter, such as Pat Shipmans's *Life History of a Fossil. An Introduction to Taphonomy and Paleoecology* (1981), Lewis Binford's *Bones: Ancient Men and Modern Myths* (1981) and Kathy Schick and Nick Toth's *Making silent stones speak: Human evolution and the dawn of technology* (1993).

## Raymond Dart - the Provocative Pioneer of African Cave Taphonomy

The sub-discipline of *African Cave Taphonomy* had its roots a long time ago, although it was initially not designated as such. It started with Raymond Dart who described the child skull from Taung (Dart, 1925a) as *Australopithecus africanus*, the southern ape of Africa, claiming that it showed features intermediate between those of apes and humans. A few years later (Dart, 1929) started to speculate about the significance of all the other fossil bones found in association with the child skull. He wrote:

> "Examination of the bone deposit at Taungs shows that it contains the remains of thousands of bone fragments. It was a cavern lair or kitchen-midden heap of a carnivorous beast. It was not a water-borne deposit and the Taungs remains could not have been washed into the cavern from the surface. The bones are chiefly those of small animals like baboons, bok, tortoises, rodents, rats and birds. Egg shells and crab shells have also been found. This fauna is one that is not characteristic of the lair of a leopard, hyaena or other large carnivore, but is comparable with the cave deposits formed by primitive man. The deposit was, therefore, formed by primitive man or by Australopithecus, an advanced ape with human carnivorous habits. As no human remains have been found there, as no Australopithecus remains have been found elsewhere in known Pleistocend deposits, I am of the opinion that the deposit was formed by the Taungs sub-man himself."

*Figure 2. Delegates at the symposium on Taphonomy and Vertebrate Paleoecology, with Special Emphasis to the Late Cenozoic of Sub-Saharan Africa, held at the Wenner Gren Foundation's Conference Centre of Burg Wartenstein in the Austrian Alps during 1976. It can be said that Taphonomy crystallised as a significant new discipline at this meeting.*

This opinion was strengthened by Dart's observations, published a few years later (Dart, 1934) concerning the damage that baboon skulls from Taung had suffered, in particular the holes and fractures he observed in their brain-cases. He concluded that these had been caused by blows from clubs and stones wielded by the ape-men. But it was only when Dart became preoccupied with the vast fossil deposits at the Makapansgat Limeworks that his ideas on the hunting ability of the ape-men were really formulated. His attention was drawn to the fossil potential of the Makapansgat Limeworks Cave by a local school teacher, Wilfred Eitzman, during the early 1920s. Among the numerous fossils that Eitzman sent to Dart at that time were several blackened bones, enclosed in the calcified cave earth, that Dart suspected had been burnt. He arranged for chemical analyses of the bones to be done and these showed the presence of free carbon, suggesting that the bones had indeed been in a fire. On the strength of this evidence, together with that of the broken bones from a wide variety of animals, Dart (1925b) concluded that Makapansgat had been "a site of early human occupation." Subsequently, following a University of the Witwatersrand student expedition, led by Phillip Tobias in 1945, new fossils were found at the Limeworks that encouraged Dart to visit there the following year. Dart immediately recognised the importance of the cave as a potential early hominid locality and employed James Kitching, Alun Hughes and their helpers to sort the lime-miners' dumps. This resulted in the finding of the first Makapansgat hominid fossils (Dart, 1948) which Dart named *Australopithecus prometheus*, assuming that they had been responsible for the burning of blackened bones found in the deposit. Subsequent research on this topic by Kenneth Oakley (1956) failed to confirm the presence of free carbon in the bones and the conclusion was reached that the blackening was caused by the presence of manganese dioxide. It has been suggested that the carbon initially detected in the first samples may have come from the blasting activities of the lime-miners.

The long-term operation of sorting miners' dumps at Makapansgat also produced very numerous blocks of highly fossiliferous grey breccia that had been blasted from the lower levels of the cave. Dart arranged for many of these blocks to be transported to the University in Johannesburg, where the individual fossil bones were manually chipped from their matrix.

At the Third Pan-African Congress on Prehistory, held at Livingstone in 1955, Dart (1957a) presented the results of his taphonomic investigation of the Makapansgat grey breccia (now termed Member 3) fossil assemblage. His sample consisted of 7,159 pieces of fossil bone of which, 4,560 were found to be sufficiently complete to allow allocation to skeletal part and taxon, while the remaining specimens consisted of bone flakes and fragments. Dart found that 91.7% of the identifiable fossils were of bovid origin, 4.0% came from non-bovid ungulates and the rest were from non-ungulates, such as primates and carnivores. Among the 293 individual antelope represented, 39 were large such as kudu. 126 were medium-sized, 100 were from gazelle-sized antelope and 28 were from small species such as duiker. Non-bovid ungulates were represented by four equids, six chalicotheres, five rhinos, 20 pigs, one hippo and six giraffids. Among the primates, there were remains of 45 baboons and five australopithecines, together with a variety of other animals that included 17 hyaenas, a sabre-toothed cat, porcupines, as well as other small mammals and reptiles, including terrapins. Non-ungulate mammals were typically represented by skull-parts only, but the antelope had contributed a wide variety of skeletal parts which, however, showed striking and unexpected disproportionate representations. Most common of all parts were skull pieces, particularly mandibles; neck vertebrae, especially the atlas and axis, were well represented, but thoracic vertebrae were scarce and those from the tail were absent. Among the limb bones, disproportions were most striking of all: in the case of the humerus, for instance, the distal ends were ten times more common than were the proximal ends.

Dart also made a detailed study of the damage that the fossil bones had suffered. He described how, in his

*Figure 3. Raymond Dart in 1965 with a block of richly fossiliferous grey breccia from the Makapansgat Limeworks Cave. It was on the basis of fossils from this site that he developed his concept of the Osteodontokeratic Culture of* Australopithecus.

opinion, broken antelope cannon bones had been pounded with a pointed object, perhaps a bovid calcaneus, converting them into scoop-like tools. As would be the case in a contemporary taphonomic investigation, Dart speculated on the possible agents of accumulation for the collection of bones in the cave and the behaviour of the animals involved. Since the early conclusions of William Buckland (1822) about the role of hyaenas as bone collectors in the Kirkdale Cave of Yorkshire, it had been customary to implicate these scavengers in the accumulation of fossil assemblages elsewhere. Dart (1956a), however, dismissed this concept in his paper on "the myth of the bone-accumulating hyaena." He concluded instead that the entire accumulation of bones in the grey breccia, running to many hundreds of thousands, had been taken to the cave by hominids, who ate the meat and then used the bones as a variety of tools and weapons. The implication was that australopithecines were powerful hunters as Dart (1956b) made clear:

> "The fossil animals slain by the man-apes at Makapansgat were so big that in 1925 I was misled into believing that only human beings of advanced intelligence could be responsible for such manlike hunting work as the bones revealed ... These Makapansgat protomen, like Nimrod long after them, were mighty hunters."

In his 1957 monograph, Dart elaborated his theory of the "osteodontokeratic" (bone, tooth and horn) culture of *Australopithecus prometheus*. He explained the striking disproportions in skeletal parts apparent in the Makapansgat assemblage in terms of deliberate selection of certain bones in view of their potential as tools and weapons. Parts of antelope skeletons not suitable for these purposes were simply left at the kill sites, hence their absence from the cave. Uses were suggested for virtually all the bones in the fossil assemblage: the tooth-rows of mandibles made good saws, for instance, while the distal ends of humeri served as convenient clubs. Early in the investigation, Dart (1949) had suggested that humeral clubs had been responsible for the depressed fractures he observed on the calvaria of baboons and hominids from the caves of Taung, Sterkfontein and Makapansgat. In the case of fossil animals, such as baboons and carnivores, where only skulls are found in the Limeworks assemblage, Dart suggested that the exclusive presence of these, too, represented deliberate selection, concluding that the hominids had been "head-hunters" and "professional decapitators."

In the course of the 20-year-long duration of his Makapansgat project, Dart published 39 papers, the text of which often contained powerful provocative prose. For instance, in his paper "The predatory transition from ape to man" (1953), he wrote:

> "On this thesis, man's predecessors differed from living apes in being confirmed killers: carnivorous creatures that seized living quarries by violence, battered them to death, tore apart their broken bodies, dismembered them limb from limb, slaking their ravenous thirst with the hot blood of victims and greedily devouring livid writhing flesh."

As this style of writing struck me as unusual in a serious scientific context, I asked Dart what he hoped to achieve by using it. He replied without a moment's hesitation. "That will get 'em talking!" he said, and he certainly succeeded in his objective. His highly provocative ideas and style of presentation sparked heated discussion in scientific circles in many parts of the world, while the American dramatist, Robert Ardrey, was so impressed with Dart's theme of "the blood-bespattered archives of humanity" that he wrote a series of five widely-read books on the subject, starting with *African Genesis* in 1961.

My own imagination was so captivated by Dart's ideas on the behaviour of our early ancestors that I spent 40 years developing the new discipline of Cave Taphonomy, in the hope that fossils in African caves could be interpreted with ever increasing confidence. This included a long-term investigation of the Swartkrans cave in the Sterkfontein valley where Robert Broom and John Robinson had worked between 1948 and 1951, and had recovered numerous fossils of robust australopithecines and demonstrated, for the first time, the co-existence of these hominids with early humans. This co-existence has subsequently been confirmed in many parts of Africa.

But as the principles of cave taphonomy became established, alternative interpretations emerged for almost all the observations that Dart had made on his Makapansgat fossil assemblage. His concepts of the "mighty hunters," the "predatory transition from ape to man" and the "osteodontokeratic culture" all gave way to other ideas which, although less dramatic, are probably more realistic. As each of these ideas surfaced, I made a particular point of telling Dart about them. For instance, when I found that a collection of goat bones from the Namib Desert showed skeletal disproportions very similar to those that he had encountered at Makapansgat (see below), and that such disproportions had a very simple explanation, Laura Brain and I took the collection over to Dart's laboratory in Johannesburg. There I explained that it was no longer necessary to suggest that hominids had deliberately selected certain skeletal parts for tools and weapons and had left others at the kill sites, but that such disproportions were to be expected in any assemblages worked over by carnivores or scavengers. When Dart realised this, he was silent for a few minutes. Then his eyes lit up and he said "This is wonderful—at last we are getting closer to the truth!" A few days later he nominated me for an award! But despite the fact that Dart was always willing to accommodate alternative viewpoints, I doubt if he ever really gave up his conviction that our pre-human ancestors had been bloodthirsty killers. For instance, when we found bone tools with the early hominid remains at Swartkrans, we showed them to Dart before describing them. He was over 90 years old at the time and his eyesight was failing, but he felt the smooth, tapering points with his fingers. Then he said: "Brain, I always told you that *Australopithecus* made bone tools, but you never believed me! What were these used for?" I

*Figure 4. In 1981, Raymond Dart and his wife Marjorie visited the Transvaal Museum to examine bone tools from Swartkrans. They were delighted to see evidence of bones being used as tools by early hominids.*

replied that I thought that they had been used for digging in the ground. Dart slumped back in his chair with a look of total disbelief on his face. "That" he said "is the most unromantic explanation I have heard of in my life!" He than grabbed the longest of the bone points and stuck it into my ribs saying, "Brain, I could run you through with this!"

It was Dart's generosity of spirit that added greatly to the pleasure I experienced from my varied taphonomic studies in the early hominid field that had sprung from Dart's stimulation and provocation. And like Dart, I too made various assumptions in the course of my research that proved to be false. For instance, when I was doing my Ph.D. project 50 years ago, on the cave deposits from which South African hominid fossils came, I spent a lot of time trying to reconstruct past climatic conditions at the times that the cave entrances were admitting surface-derived sediments. I found that the older sediments, that had been laid down at Makapansgat and Sterkfontein, contained significant proportions of aeolean sand and therefore suggested that the climate and environment at that time was more desertic than it was when the later deposits at Kromdraai and Swartkrans accumulated. Some years later, Karl Butzer (1976) pointed out that my assumption that surface-derived sediment, entering an underground cave, was a valid indicator of climatic conditions at that particular time, was false. He said that, during the Pliocene, much of the southern African interior surface consisted of the African Peneplain, on which thick deposits of Kalahari desert sand had been laid down in earlier times. This meant that the presence of aolean sand grains in the cave deposits could not be used to infer desert conditions at the time of their entry to the caves. They could be reflections of much more ancient desiccation. It was only when continental uplift disrupted the old African surface and resulted in the downcutting of streams and rivers, that the residual aeolean sediment was eroded away. By the time of Swartkrans Member 1, for instance, the surface-derived sediment entering the cave appears to have been in equilibrium with conditions at the time. Thus, my deduction that sediments at Makapansgat and Sterkfontein indicated arid conditions, at the time of their deposition, was completely at variance with subsequent evidence from plant and animal fossils that reflected lush riverine vegetation.

What follows now is a brief overview of some facts, ideas and concepts that emerged during the 50 years of personal taphonomic involvement.

## 1. Understanding skeletal disproportions in bovid bone assemblages

Today it is taken for granted that any ancient bone assemblage is likely to contain certain parts of the skeletons of the animals that contributed to it, but not others. But, at the time that Dart did his analysis of the fossil assemblage from the Makapansgat grey breccia this likelihood had not been appreciated. Consequently, when Dart found that some skeletal parts were present to the partial or complete exclusion of others, he came up with his imaginative explanation, implicating the hominids in deliberating selecting and preserving some useful bone pieces, while ignoring others.

Unexpected insights to this question came my way, unexpectedly, in the Namib Desert. At Gobabeb, on the banks of the normally dry Kuiseb River bed, 96 km inland from Walvis Bay in southwestern Africa, the Transvaal Museum built a Desert Ecological Research Station in 1963. Two years later I visited this spot to set up a bone-weathering experiment in an arid environment and noticed that there was an abundance of goat bones, lying on the desert sand in the nearby villages occupied by Nama Hottentot people. Out of sheer curiosity I picked up a sample of these bones and laid them out at the research station as an exercise in osteology. It struck me at once that some skeletal parts were well represented, while others were rare or absent. Distal humeri, for instance, that had been so important in Dart's osteodontokeratic cultural concept, were common but, search as I might, I could not find a single proximal end of a humerus. The explanation was not difficult to find: the bones represented the resistant residue of goat skeletons, able to survive the treatment they had received. But what was this treatment? Inquiries and observations during the following week showed that goats were virtually the only source of meat for these rural people. When a goat was slaughtered, its body was treated in a traditional manner and those parts that the people found inedible were tossed to the dogs. When they, in turn, were finished, the parts unchewable by person or dog were left on the desert surface, where recovery was easy for me, as the sand was devoid of vegetation in the villages. Apart from occasional crows, no other carnivores or scavengers were involved.

After the initial reconnaissance in 1965 (Brain, 1967a), I returned the following year to collect all the available bones and to investigate the circumstances in greater detail. On this occasion, I was accompanied by Trefor Jenkins who undertook a thorough genetic and demographic study of the Nama population there. We found that the total population of the lower Kuiseb valley was 133 people who lived in eight separate villages. Between them they had 40 dogs and 1754 goats, while the spacing of the villages along the riverband was determined by the number of goats kept at each, since grazing could only take place in the riverbed and the extent of a village's pasturage was measured in a linear fashion (Brain and Jenkins, 1967).

The treatment of goat carcasses, slaughtered by these Nama people has been described elsewhere (Brain, 1969, 1981). All meat eaten by the people was cooked before it was eaten, either by boiling in large metal pots or direct roasting over the fire. The horns were broken at their bases from the skulls by sharp blows from an axe and were discarded. The dogs chewed the horn-core bases before rejecting them. The complete head was then boiled for several hours in a pot, standing over a fire. All edible meat was picked from it and eaten, after which the braincase was smashed in the occipital region with a hammerstone for the removal of the brain. The skull and mandibles were passed to the dogs. As eating progressed, all marrow-containing bones were broken by the people, by being held on a rock anvil and hammered with another stone. Neither the anvil nor the hammerstone were artefacts in the usual sense of the word, as they were simply suitable pieces of rock that happened to be lying around. These Hottentots habitually ate while squatting on the ground and, apart from their anvils and hammerstones, their only utensils were pocket knives. Their feeding behaviour seems to be a mixture of long-standing tradition and European influence.

*Figure 5. A Nama Hottentot village in the Namib Desert of Namibia, on the bank of the Kuisib River bed, inland from Walvis Bay. The discarded goat bones from here, shown in the foreground, provided valuable information on the survival and disappearance of skeletal parts in 1965.*

Once discarded by the people, the goat bones were gnawed for many days by the dogs, all of which were about the size of jackals. When lying on the sandy surface, the bones became bleached and de-greased in about three months. The collection made in the villages consisted of 2,373 pieces which included 385 horns and horn-core pieces from an estimated number of 190 individuals, estimated on these elements. However, it was clear that horns survived the weathering effects of the arid desert climate much better than did other bony pieces and in long-deserted villages, these were the only parts to be found. Therefore, in the final estimation of the number of animals involved, horns were excluded in favour of the next most abundant element, which happened to be mandible pieces, as in Dart's Makapansgat sample. I found that the 188 fragments could be divided into 53 left and 64 right half-mandibles, indicating a minimum of 64 individual goats that contributed to the sample. On known tooth-eruption and wear criteria, it transpired that there was one goat in the sample under 6 months of age, 23 between 9 and 12 months, 7 between 15 and 30 months, and 35 more than 30 months in age. The goats had therefore been slaughtered largely when either just under a year in age or when fully mature. The village people confirmed that this was their usual practice, the

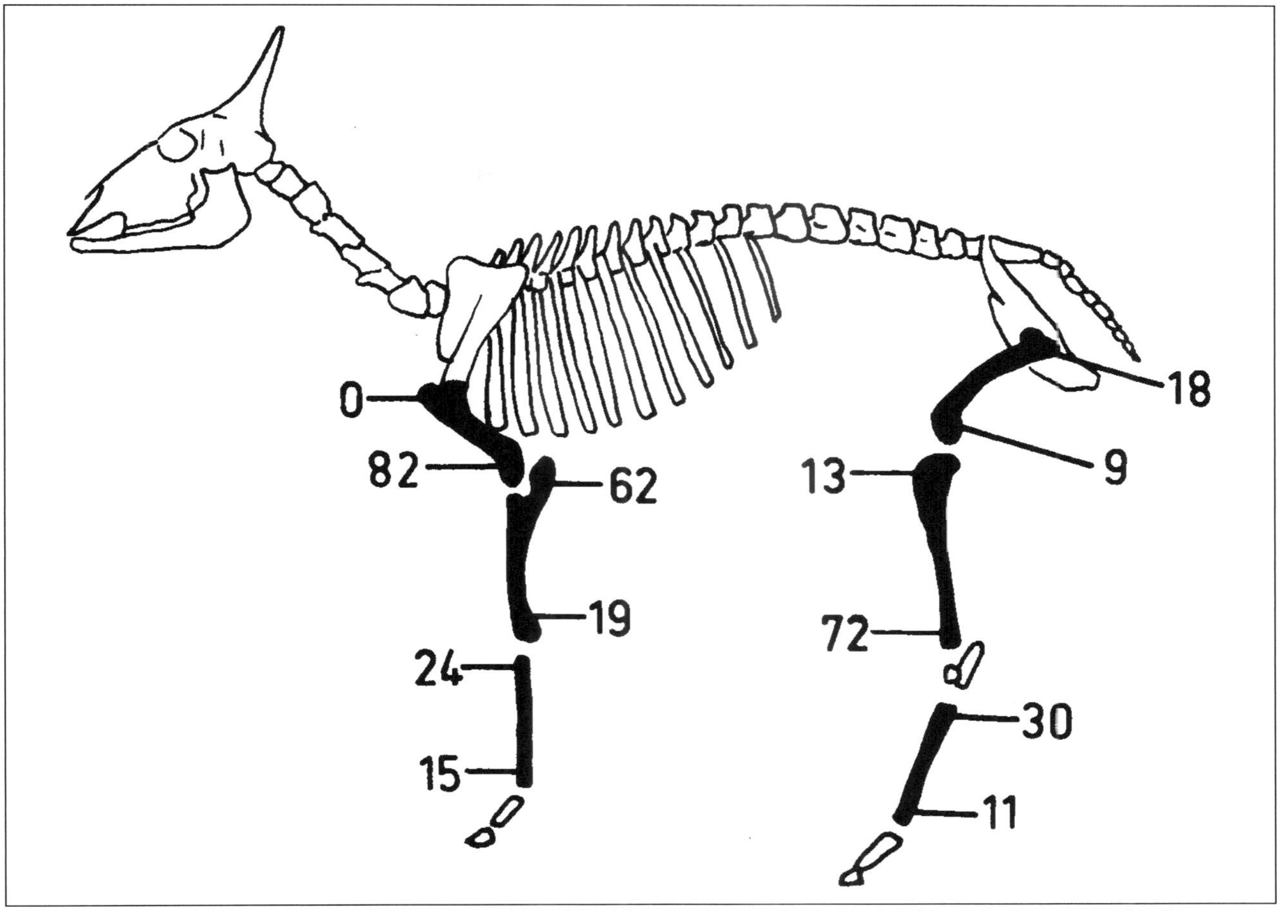


*Figure 6. Diagram of a goat skeleton published in 1967. The numbers refer to ends of long bones present in the sample collected in the Namib Desert.*

yearlings usually being the surplus males.

After mandible pieces, the most commonly preserved skeletal element was the distal humerus, followed by the distal tibia, proximal radius and ulna and so on. Parts absent altogether were caudal vertebrae and proximal humeri. So it became clear that the parts of the goat skeleton that survive best are the unchewable ones. In the course of this study it occurred to me that, in a sample derived essentially from immature animals, the survival of limb-bone ends could be related to the time at which the epiphysis of that bone fused to its shaft. Consider the humerus for instance, in which survival of the proximal end is nil but that of the distal end amounts to 64%. The proximal epiphysis is likely to fuse to the shaft at about 36 months, whereas the distal epiphysis is fully fused by 12 months. An unfused epiphysis is linked to its shaft by a cartilaginous interface that is easily broken, making the two pieces vulnerable to damage. This means that, when a year-old goat is eaten, the distal end of the humerus will be fully ossified and unchewable, while the proximal end remains chewable. But, in addition to fusion times, structural considerations are very important. The proximal end of the humerus is wide, thin-walled and filled with spongy bone; the distal end is comparatively narrow and compact. Such qualities may be expressed in terms of specific gravity of each end of the bone. When I made the necessary measurements, I found that the proximal end of a goat humerus had a specific gravity of about 0.6, while that of the distal end was approximately 1.0. There turned out to be a clear and direct relationship between the specific gravity of the end of a long bone and its percentage survival. In fact, percentage survival of a part is related directly to the specific gravity of that part, but inversely to the fusion time expressed in months. The conclusion to be drawn is simply that survival is not haphazard, but is related to the inherent qualities of the parts (Brain, 1981, p. 21).

In contrast to the goat-bone sample, Dart's Makapansgat sample was made up of remains from 293 antelopes, ranging in size from large species (39 individuals), medium (126 individuals), small (100 individuals) and very small antelopes (28 individuals). Using the total number of 293 individuals, it was possible to calculate the percentage survival of various parts of the skeletons as was done for the Namib goat bones. When listed and plotted in descending order of survival, the Makapansgat bones showed a remarkable similarity to those from the goats, despite the fact that the fossil bones came from such a wide species and size range and that they could have been subjected to a variety of destructive processes. So, what the goat-bone study did show, is that skeletal-part disproportions are extremely likely to occur in an assemblage and that they can inform us on the destructive influences that the contributing skeletons had suffered.

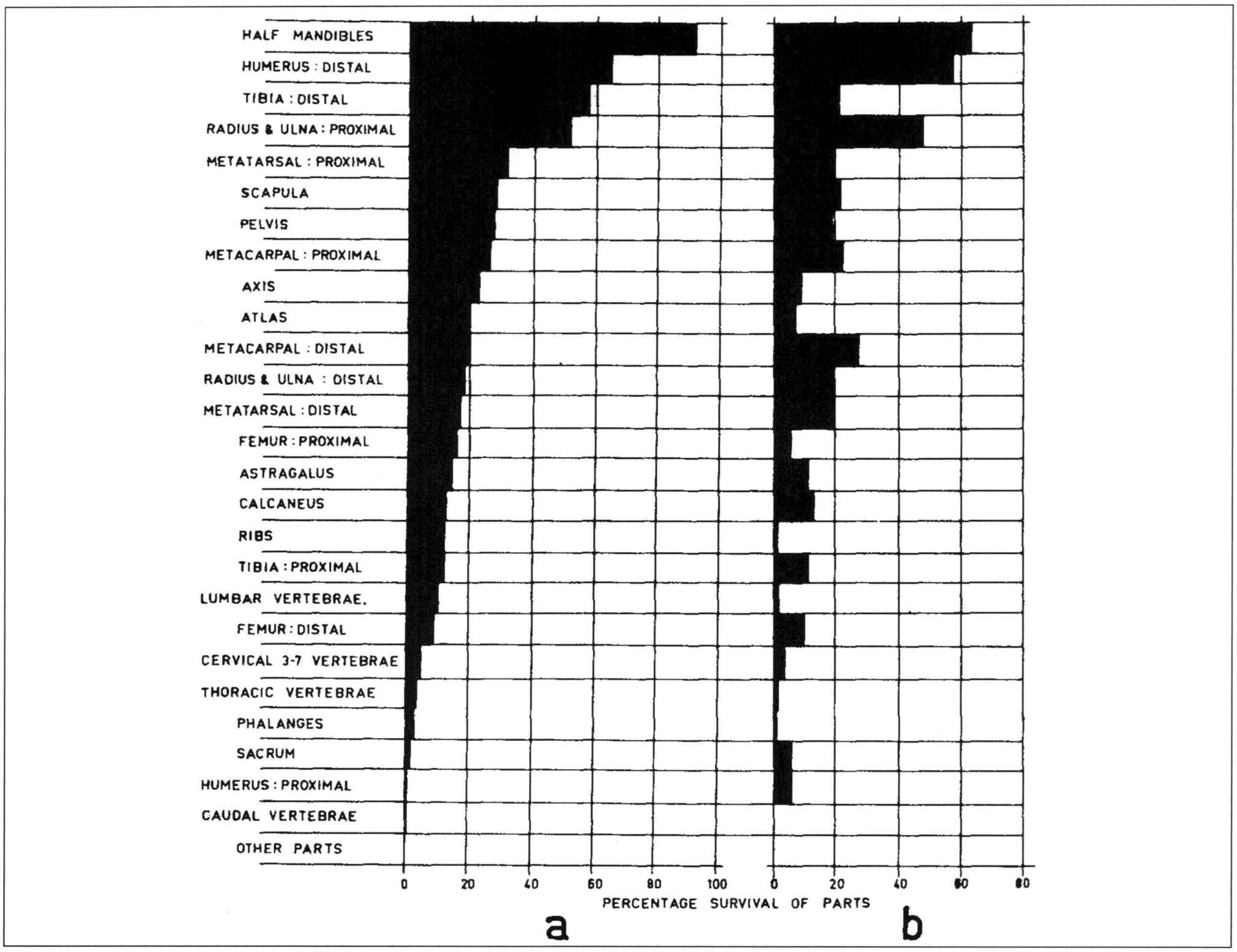


*Figure 7. A diagram published in 1969. (a) Histogram showing percentage survival of parts of goat skeletons from the Kuisib River villages. Calculations are based on a minimum of 64 individuals. (b) Percentage survival of parts of fossil bovid skeletons from Makapansgat, arranged in the same order as for (a).*

## 2. The complication of bone "pseudo-tools"

Another unexpected insight into the interpretation of Dart's Makapansgat fossil assemblage came with the finding of "pseudo-tools," which showed overall wear and polish. Similar specimens from the Makapansgat assemblage were thought by Dart to have acquired their surface modification by hominid handling and use. In connection with the Namib evidence, I wrote (Brain, 1981, p.15):

> "While collecting bone fragments from the vicinity of the Hottentot villages, I was surprised to find many pieces that appeared to be bone tools. They tapered to points and showed wear and polish that had surely resulted from human use. In reply to my queries, the Hottentots denied that they made use of bone tools at all, and I had to find a different explanation for the remarkably suggestive appearance of these 'pseudotools.' Further observation showed that the worn and polished bones were specially abundant in areas regularly used by men and animals, such as around the Ossewater water hole, where 460 goats converge daily to drink, in the immediate vicinity of goat kraals, and along paths used by the Hottentots and their goats in the riverbed. If protected among rocks, for instance, the bones would develop their characteristic chalky surfaces but would lack signs of wear and polish. The mechanism of pseudotool production was therefore clearly related to the disturbance of the sand in which the bones lay by the feet of animals and men (Brain, 1967c). The process may therefore be summarized as follows: bones come to rest on the sand, and their surfaces weather to a chalky consistency. Regular disturbance of the sand by the feet of animals abrades the chalky surface as it forms, leading to bones that are both worn and polished. If the whole piece of bone is lying in the disturbed sand zone, it is likely to acquire wear and polish on all surfaces, but if some part of it is buried deeper this will remain protected, and only a part of its surface will be converted into a pseudotool. Selective abrasion of this kind has been observed on a number of metapodial and other limb-bone pieces that have been buried with their long axes vertical, or at least inclined. This meant that parts of such bones were

buried too deep to be affected by superficial sand movements, so that wear and polish occurred on one end only. Pseudotool production is not restricted to arid environments like that of the Kuiseb River, and the mechanism should be borne in mind when any interpretation of a bone assemblage is undertaken."

With respect to the Makapansgat situation I wrote (Brain, 1981, p.264):

> "Over the years, a picture has developed in my mind of how the Limeworks cave may have looked when the bones were accumulating there. I visualize an extensive amphitheatre that had resulted from a collapse of part of the cavern system's roof, while from this amphitheater openings to the cavern system we know today led downward. I visualize, too, a permanent water hole in the amphitheatre, perhaps at the point where the Makapansgat stream descended into the subterranean chambers. Finally, I visualize large numbers of animals regularly visiting the water hole and some of them being killed there by carnivores that perhaps included hominids. Their bones would lie about in abundance within the catchment area of the cavern's mouth. Some would be modified by australopithecines, all would be worked over by scavengers, and large numbers would be transported to the inner recesses of the cavern by breeding hyaenas and resident porcupines. While lying in the much disturbed sand around the fringes of the water hole, some of the bones would acquire the wear and polish so characteristic of certain specimens in the gray breccia assemblage. Like all other fossil assemblages in caves, the Makapansgat bones could be taphonomically interpreted with assurance only if they were excavated with due regard to subtle detail. I have no doubt that if an in situ deposit of bone-rich gray breccia could be stripped of its overburden and if the individual fossils could then be chipped out as they lay in a carefully controlled grid system, it would be possible to assess with confidence the accumulation pattern that originally operated. Such a task would be difficult, but it would be highly rewarding in the interpretation of a situation that has excited the imagination of paleontologists for years."

Fortunately detailed work is progressing at the site on various fronts, such as that of Latham, Herries and Kuykendall (in press), and the results are anticipated with great interest.

## 3. "Head-hunters" and "professional decapitators?"

As mentioned earlier, Dart was intrigued to find in his Makapansgat fossil assemblage that the remains of non-ungulate animals such as hominids, other primates and carnivores were typically represented by little more than their skulls. He thought that this might have been the result of deliberate selection, for trophy purposes, on the part of hominid hunters.

Early in my taphonomic career I made a detailed study of several contemporary caves that had been used for many years by leopards (Brain, 1981, p. 85-93) as breeding and feeding lairs. These were on the farms Valencia and Portsmut in the Hakos Mountains of Namibia, about 160 km southwest of Windhoek where their owner, Atilla Port, was very enthusiastic about such projects. We found that the leopards frequently fed on *Procavia* hyraxes in these caves, invariably eating the entire body of each, with the exception of the heads, the anterior parts of which accumulated in the lairs. Similar observations were made on a captive leopard that we caught and confined in a cage on Valencia, until it managed to escape one night. Feeding experiments with captive cheetahs there confirmed that they, too, only leave the stomach and the head, when feeding on hyraxes.

But, considering the bone-crushing ability of hyaenas, it came as a complete surprise to find that brown hyaenas (*Parahyaena brunnea*) frequently leave the skulls of their smaller prey animals undamaged. When these hyaenas have cubs in a breeding lair, they tend to kill a variety of small animals, including other carnivores, which they take back to the lair for the young to feed on. The cubs typically leave the skulls of these prey animals and, it seems as if the adults have an inhibition against themselves eating any of the food they provide for their cubs. For instance, food remains collected from brown hyaena breeding lairs in the Kalahari Gemsbok National Park (Mills and Mills, 1977) contained the virtually undamaged skulls of 6 black-backed jackals, 11 bat-eared foxes, 4 caracal, 1 aardwolf and 1 ratel (Brain, 1981, p. 295). Observations in the Kruger National Park have shown that baboons are also killed by brown hyaenas for the feeding of young.

So, if the Makapansgat cave was originally a hyaena lair, as will be discussed shortly, there is no need to suggest, as Dart did, that skulls of non-bovid animals had been collected there by head-hunting hominids.

## 4. "The myth of the bone-accumulating hyena?"

Since Dart published his paper with this title in 1956, important observations have been made by numerous people in various parts of Africa on the bone-collecting behaviour patterns of Spotted, Brown and Striped hyaenas. There is now no question that all three species collect significant numbers of bones at their lairs which, might well be in caves. At Makapansgat, the hyaenas represented in the fossil assemblage are Striped Hyaenas, *Hyaena hyaena*, which do not occur at present in southern Africa, but are found further north on the continent and in the Middle East. Various studies have been made on the bones found in their feeding and breeding lairs, both in East Africa and Israel and there is now no

*Figure 8. The discovery of stone artefacts by Bob Brain at Sterkfontein in May 1956. Shown here in the West Pit, close to the main Type Site quarry, are (left-right) Ben Grobbelaar, Daniel Mosehle and Absalom Lobelo.*

question that this species can accumulate vast numbers of bones in the caves that they use. The current consensus opinion appears to be that, during Member 3 grey breccia times, the low-roofed parts of the cavern served as breeding and feeding lairs for Striped hyaenas, that collected the bones for adult and cub consumption. Additional input was made by porcupines and other carnivores, such as leopards.

Following his detailed taphonomic study of the fossil assemblages from Sterkfontein Member 5 West, Travis Pickering (1999, p. 159) wrote:

> "The weight of taphonomic evidence presented here–bone surface modifications, the presence of juvenile hyena remains, and to a lesser degree, bovid skeletal part ratios–indicates that hyenas probably played a significant role in the accumulation of Member 5 West faunal assemblage. Furthermore, consideration of species-specific behavioral adaptations implicates brown hyenas as the most likely primary collectors of bone at the site. Porcupines also contributed to the accumulated assemblage, in addition to possibly spotted hyenas. The serial use of single den sites by brown hyenas, spotted hyenas and porcupines is well documented in modern habitats (e.g., Mills and Mills, 1977). Considering the presumably much longer accumulation period of the Member 5 West assemblage, alternating occupation of the site would not be surprising."

## 5. The comparative vulnerability of primate and bovid skeletons to carnivore damage

When I analysed the composition of the available fossil assemblages from the various hominid-bearing cave deposits in the 1960s, it became apparent that antelope were represented by a far wider range and abundance of skeletal parts than were primates. In fact, postcranial remains of hominids, baboons and monkeys were rare in comparison with cranial ones, whereas bovid postcranial fossils were comparatively common. The reason for this was not immediately apparent, but some light was thrown on the question when I made a study on the feeding behaviour and food-remains of cheetahs. My reason for studying these topics was that I thought cheetahs might tell us something about the food remains of sabre toothed cats, the dentition of which was specialized for

*Figure 9. May 1956. One of the newly-discovered stone artefacts, made on a quartzite pebble, still embedded in the breccia of the Sterkfontein West Pit.*

the cutting and slicing of meat, as is the case, albeit to a lesser extent, in cheetahs.

On the Valencia Ranch in Namibia, mentioned above, Attila Port caught several wild cheetahs for me and put them in a large enclosure, where they were fed mainly on karakul sheep and the occasional springbok. It was clear that the cheetahs could do little damage to the skeletons of these animals, apart from some chewing on the ventral ends of the ribs, vertebral processes and the blades of the scapulae. But, on one memorable day we were unable to find a springbok so Attila shot, instead, a large male baboon that had been taunting us from the top of a nearby cliff. The baboon was offered to the cheetahs with unexpected results, as is reflected in this extract from my field notes (Brain, 1981, p.24):

> "20 March 1968: The body of an adult male baboon weighing 29.5 kg was placed in the enclosure at 9:05 a.m. It was immediately taken by the two male cheetahs and carried by its arms to the shade of a tree. All three cheetahs started to feed on the ventral surface of the abdomen; the viscera were removed and part of the intestine eaten. The rib cage was quickly chewed away and the vertebral column simply crunched up and swallowed—quite unlike the antelope situation. As the vertebral column was destroyed, the pelvis and both hind limbs were removed by one cheetah and carried a short distance away. The sacrum was eaten so that the femurs, still articulated into the innominates, were separated. One cheetah left the baboon after 1hr. 10 min., the others remained 15 min. longer, then left, but all three returned intermittently throughout the day."

When the remains were removed and photographed the next day we found that the entire vertebral column, from atlas to first caudal, had disappeared, as had most of the ribs. The innominate bones showed damage round the edges, and both knee joints had been disarticulated and chewed. The disappearance of the vertebral column in this baboon carcass came as a complete surprise and suggested that a primate backbone was less resistant to carnivore chewing than its bovid counterpart. To test this suspicion, an adult sheep of almost exactly the same liveweight as the baboon was fed to the cheetahs when they showed equivalent signs of hunger. Once again the vertebral column was left intact, apart from the tail that had been eaten, while slight damage was done to the innominates and ends of the ribs.

Little new information on the reasons for the comparative durability of primate and bovid skeletons came to light in recent years until the research of Travis Pickering and Kristian Carlson was undertaken on "intrinsic qualities of primate bones as predictors of skeletal element representation in modern and fossil carnivore feeding assemblages" (Carlson and Pickering, 2003). This publication was preceded by one (Pickering and Carlson, 2002) in which the question of "bulk bone mineral density" (bulk BMD) was addressed in baboon and bovid skeletal elements on the assumption that such measurements would provide an indication of the durability of the parts involved. They concluded that the differences in bulk BMD between bovid and baboon skeletal parts were not always sufficient to explain the presence and absence of parts in the Swartkrans fossil assemblages, and that "factors other than bone density—such as bone size, length, shape, and/or the relative palatability of surrounding soft tissue on bones" would have to be implicated. This certainly seems to be the case. In the cheetah feeding experiments, for instance, several of the baboons had their hands and feet completely chewed away. This could never have happened in the case of a bovid, with its resistant hooves and lack of palatable meat in those parts.

## 6. Stone tools in the South African early hominid cave deposits

At the time that I started my Ph.D. project on *The Ape-Man-Bearing Cave Deposits* (Brain, 1958), stone artefacts were not known from any of these sites. So, in 1955 I was excited to find a number of dolomite pieces in the Makapansgat Limeworks deposit that, in my opinion, had been artificially chipped. These came from the stony breccia, now known as Member 4, overlying the grey breccia from which Dart's fossil assemblages came. I showed these to Dart and to the well-known archaeologist of that time, C. van Riet Lowe, who said that he had no doubt that they were artefacts. An illustrated report appeared in *Nature* (Brain, van Riet Lowe and Dart, 1955) entitled "Kafuan stone implements in the post-australopithecine breccia at Makapansgat." Subse-

*Figure 10. One of the Swartkrans bone tools, showing how the smooth wear is restricted to the tip.*

*Figure 11. One of many experiments in which bone pieces were used to dig edible plant food from the contemporary hillside at Swartkrans. Here Conrad Brain digs out a tuber of a* Hypoxis *lily, using a bone flake from a wildebeest humerus generated by hyena feeding, while Bob Brain looks on. Photograph by David Brill.*

quent to this, australopithecine fossils were found in this Member 4 breccia also (Dart, 1955), suggesting that the time interval between it and the underlying grey breccia (Member 3) was not very great. A problem was that all these presumed artefacts were made on dolomite pieces and critics argued that they would not be convinced of their authenticity as tools until some were found made of quartzite.

In May 1956, I was writing up my doctoral thesis and, in the Sterkfontein chapter, I had a heading "Cultural material from the deposit," but found that I had nothing to report. Despite the fact that I had spent many weeks working through the Sterkfontein profiles that were available for study at that time, I decided to devote that particular day looking specifically for stone artefacts. To my considerable surprise, when I re-examined the breccia walls of the "West Pit" on the hilltop, a few metres from the west wall of the Type Site, where so many hominid fossils had been found, I was amazed to find a number of unquestionable *in-situ* artefacts, made of pieces of quartzite and diabase that must have been selected from the river gravels in the vicinity. There were also many artefacts in pieces of breccia on the miners' dumps surrounding the West Pit. Several of these were illustrated in my thesis (Brain, 1958, p.72-73). This find prompted John Robinson to re-open excavations in the West Pit area, after Revil Mason and I had shown that the artefact-containing breccia was continuous with that in the west wall of the Type Site, beneath a soil overburden that contained Middle Stone Age artefacts. Robinson's excavation proceeded for two years and produced 58 stone artefacts, one bone tool and several hominid teeth (Robinson and Mason, 1957; Robinson, 1959).

Further excavations in the West Pit area by Alun Hughes, Ron Clarke, and their team revealed stratigraphic complexity that had not previously been suspected, while Kathy Kuman showed that while the artefacts from the West Pit were Early Acheulean, there were also Oldowan artefacts to be found in the lower levels of the Member 5 infill (Clarke, 1994; Kuman, 1994a and b; Kuman and Clarke, 2000; Kuman, 2003).

In my early taphonomic interpretation of the Sterkfontein Member 5 assemblages (Brain, 1981, p. 217) I wrote: "The remarkable density of artefacts in the excavated part of Member 5 strongly suggests that the cave was intensively occupied during this accumulation phase. It would therefore be remarkable if the bone pieces associated with the artefacts did not represent hominid food remains." The later work of Travis Pickering (1999) and Kathy Kuman (2003) does not support this conclusion. They point out that the virtual absence of cut-marks on the bones, and the weathered state of many of the artefacts, imply that the latter were left outside the cave for a considerable time before being washed into it. Human involvement in the bone accumulation process might well have been negligible.

## 7. The finding and interpretation of bone tools

As the Swartkrans excavation proceeded between 1979 and 1986, a total of 68 fossil bones were found that appeared to have been used as tools. Of these, 17 came from the Lower Bank of Member 1, 11 from Member 2 and 40 from Member 3. They typically tapered to smooth points and some of them showed superimposed polish. Most of these specimens were bone flakes, although there were several horn cores and other skeletal parts represented. The wear was very reminiscent of that suffered by the metal screwdrivers that we used when excavating the lightly calcified cave earth in various parts of the Swartkrans cave, so it seemed possible that the bone pieces had been used for digging in the ground. I had often watched baboons digging edible bulbs and tubers from the ground with their hands, about 20km north of Swartkrans, but they were only able to do this in the reasonably soft alluvial soils at the bottom of the valleys. The plants most commonly involved were a lily, *Scilla marginata*, and two species of "African potato," *Hypoxis costata* and *H. rigidula*, all of which were particularly common on the rocky dolomite hillsides, although the baboons could usually not dig them out there. It seemed conceivable that, had the hominids access to digging tools, they would have been able to extract the bulbs and tubers from the rocky situations as well. A series of digging experiments were therefore carried out, using bone flakes from the limb bones of a wildebeest, killed by lions in the Kruger National Park, the bones of which were worked over by spotted hyaenas. These flakes were used for between four and eight hours of digging on the rocky Swartkrans hillside, where we found it possible to extract a *Scilla* bulb or *Hypoxis* corm in 14 to 30 minutes, depending on the stoniness of the ground.

Wear similar to that seen on the fossil bones could be caused in four hours of digging, while subsequent wear proceeded more slowly.

I was familiar with the work that Pat Shipman was doing on worn bones from Olduvai (Shipman and Rose, 1988; Shipman, 1989), so took the collection of Swartkrans specimens, as well as two experimental bone tools, over to the Department of Cell Biology and Anatomy of the Johns Hopkins University Medical School in Baltimore. Here, Pat Shipman replicated selected parts of the specimens, using silicone-based dental impression material to make negative impressions, in conjunction with epoxy resin used to make positive casts, according to the method described by Rose (1983). The surfaces of these were then examined with a scanning electron microscope.

The similarity of wear-patterns on the experimental digging tools and the Swartkrans fossil specimens was striking, leading us to conclude that the latter specimens had, in fact, been used for digging by the hominids about one million years ago. But, in addition to this, three of the fossil specimens showed a polish, superimposed on the characteristic wear and scratch-marks. Our suggested interpretation was that the digging tools had been used also for rubbing a soft substance, presumably animal skin. Of interest in this regard is the description of a bone tool by Robinson (1959) from his excavation of Sterkfontein Member 5. It was made on a bone flake with a natural point and has a well-defined worn facet, showing fine linear scratching and a high polish. Robinson's interpretation was that the bone had been repeatedly rubbed on a soft substance, presumably animal hide.

Another bone tool, from Swartkrans Member 3, is also suggestive of use on animal hides. It is a delicate awl-like artefact (SKX 37052) that consists of a thin flake of bone tapering to a worn point and showing longitudinal and circumferential scratching, together with polish. This tool may have been used for piercing holes in skins or other soft materials, as similar microscopic wear has been documented on experimental awls (Olsen, 1984).

The evidence discussed here suggests that the Swartkrans hominids may have made simple carry bags from animal skins, in which they transported their tools and possibly their gathered food. This could explain the evidence for the apparent use of the same tools over successive days or weeks.

More recently, Lucinda Backwell (2000) completed a Masters project at the University of the Witwatersrand on "A critical assessment of southern African early hominid bone tools." She has since progressed to a Ph.D. on this topic, with her research supervised by Francesco d'Errico of Bordeaux. While they concur that these are genuine bone tools, they conclude, on the basis of image analysis of the marks and striations on the bones, that they were used for digging in termite mounds, rather than in the stony soil of the dolomite hillsides (Backwell and d'Errico, 2001; d'Errico, Backwell and Berger, 2001).

More recently, Backwell and d'Errico (2004) have identified another 16 bone pieces from the Swartkrans assemblage that they regard as being bone tools. Most of these had been looked over earlier by Pat Shipman and myself, but they had not been included in the original sample as they lacked proper stratigraphic documentation, or, in our opinion, could have been pseudo-tools. The possibility that the Swartkrans hominids had augmented their diet with termites, dug out with bone tools is an exciting one.

*Figure 12. Pat Shipman making molds of Swartkrans bone tools in her laboratory at the Johns Hopkins University Medical School in Baltimore during 1987.*

## 8. The importance of original cave-form to taphonomic reconstructions

All of the South African early hominid cave sites have suffered considerable surface erosion, and in some cases mining, since the bones originally found their way into the caves. But, if we are to interpret the bone accumulating agencies with confidence, it is important that the original form of the cave be reconstructed. In the case of the Taung site, for instance, the fossils occur in secondary infillings of cavities in a massive travertine boss that had been laid down by lime-rich water flowing over the edge of a dolomite cliff. However, mining of this boss has seriously complicated the reconstruction of the time that bones accumulated.

In the 1980s, I had the opportunity of examining a series of very similar, but undamaged, travertine deposits, set in the wild and natural environment of the Namib-Naukluft Park of Namibia. Here it is possible to observe travertines forming and eroding, while natural taphonomic processes may be studied in the numerous secondary cavities within the travertine masses themselves. The travertines are currently forming outward over a cliff in a steep-sided valley in Precambrian dolomite, where it is possible to see how closely the formation of carapace travertine is linked to the presence of moss and algae. The steeply-inclined carapace layers are formed where water flows or seeps slowly over living moss, often in the form of hanging curtains. Quite apart from the natural evaporation of this lime-rich water, it is the photosynthesis of the moss and algae that removes carbon dioxide from the water, leading to the precipitation of calcium carbonate. Moss-banks in various stages of calcification may be seen in such places and it is not unusual to find natural cavities behind hanging curtains of moss. These usually have an opening at one end, leading into a chamber perhaps four metres high, five metres wide, and up to 30 metres long. These are probably the sort of places into which the Taung fossils, including the numerous baboon skulls and that of the *Australopithecus* child found their way. Examination of the bones currently accumulating in the Naukluft travertine caves show that the expected taphonomic accumulating processes are at work there. Some bones have been accumulated by porcupines; others show unmistakable signs of leopard feeding activity —this is to be expected as leopards may be observed in the immediate vicinity (Brain, 1985).

Where normal dolomite caves are concerned, as at Sterkfontein, Swartkrans and Makapansgat, it is important to try to reconstruct their form at the time that the fossiliferous sediments were accumulating. For instance, was the entrance to the cave a death-trap situation, meaning that if an animal fell into it, it would not be able to escape? If the cavern was accessible to animals, was the roof high or low? This is important, as hyaenas and porcupines, for instance, definitely favour low-roofed chambers for their feeding- and breeding-dens, whereas baboons prefer more open situations for their sleeping sites. This may have been the case for hominids as well.

## 9. Effects of progressive Cainozoic cooling on African habitats and fauna

In the 1960s, people interested in reconstructing past climatic and environmental conditions in Africa were working with the concept that, during the last one to two million years, there had been three pluvial and three non-pluvial episodes. At that time, Waldo Meester and I speculated as to how these might have served as biological isolating mechanisms (Brain and Meester, 1965, pp. 332-340), with particular reference to small mammals and the vegetation they required in their habitats. But, as I gradually unravelled the complicated stratigraphy of the Swartkrans cave, it became apparent that there had been numerous episodes of deposition of the cave sediment, interspersed with others, during which erosion took place. It seemed inevitable that such episodes were climatically induced, and we were fortunate in that Swartkrans proved to be a sensitive indicator of climatic change. The reason for this was that the cavern had been linked upward to the hillside surface with several openings, but also downwards to lower caverns. The main cave would rapidly fill up with sediment and then, in a successive climatic cycle, parts of this filling would be eroded and carried away to the lower caverns. It seemed that the infilling processes had been relatively rapid, compared to the much longer intervals when erosion took place.

It seemed likely to me that such depositional/erosional cycles must have been climatically induced, but also that changing temperature was more likely to be the primary factor than was rainfall. I was invited by the Geological Society of South Africa to give the 17th Alex L. Du Toit Memorial Lecture in 1979 and, for this, decided to look into the evidence then available for temperature changes during the last few million years. In the published version of the lecture (Brain, 1981) I wrote:

> "The aim of this lecture is a simple one—*to draw the attention of those interested in human evolution to a remarkable record of past global temperatures that has recently become available, and to point out that certain low temperature episodes, reflected in this record, could well have served as stimuli for critical steps in hominoid evolution.* The record of past global temperatures, to be described here, is based largely on isotope compositions of foraminiferal tests preserved in deep-sea sediments. Fluctuations in global temperature are regarded as *primary environmental changes*, which then led to *secondary effects*, such as rainfall and vegetation changes. In the case of many African habitats, the secondary effects could well have been more important as evolutionary stimuli than were the primary temperature fluctuations."

The deep-sea temperature record, for the southern oceans, showed a progressive cooling trend during the last 50 million years, and it seems that this trend, involving about 12 degrees C, could be attributed largely to conti-

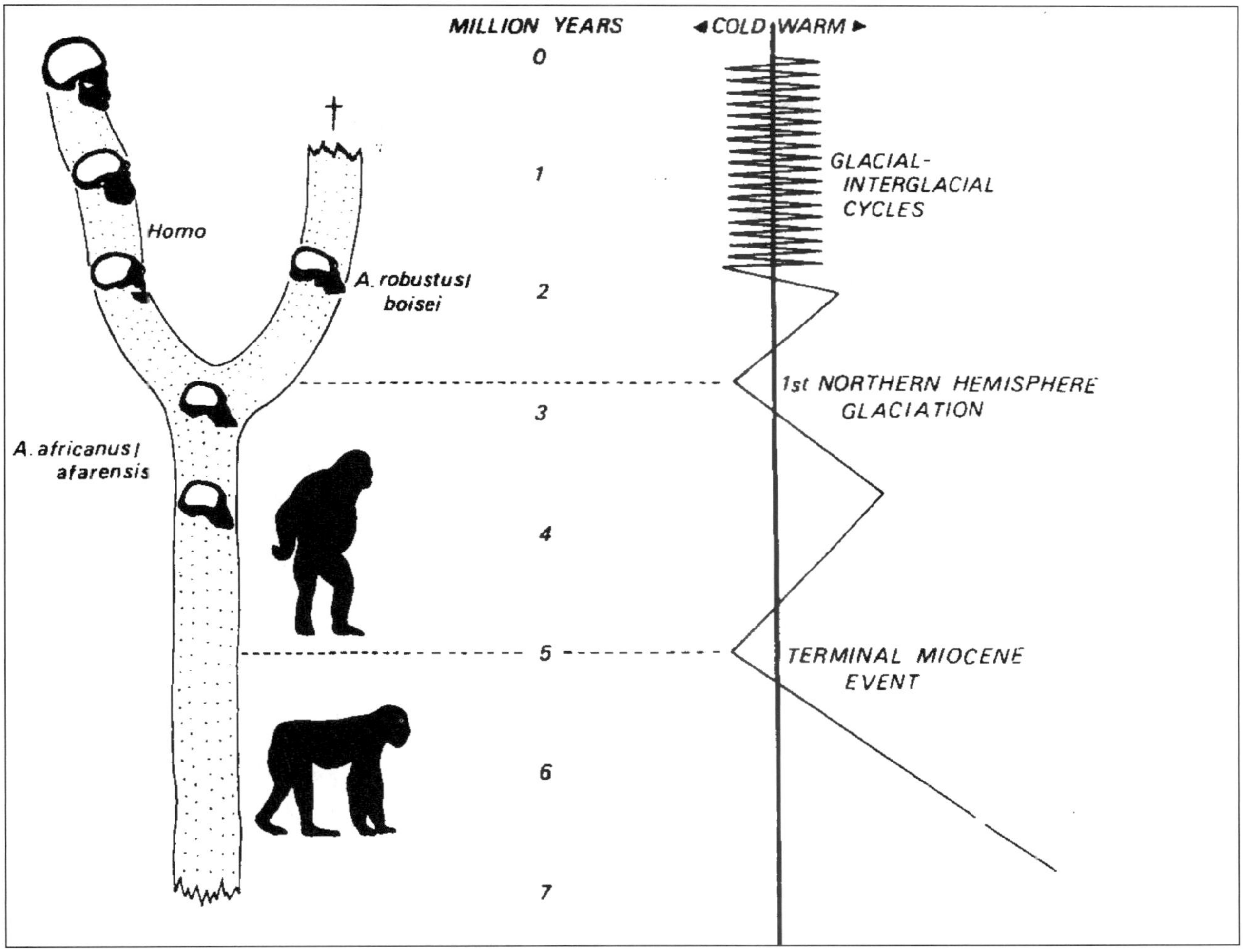


*Figure 13. An early attempt in 1979 (Brain, 1981) to correlate events in hominid evolution with low-temperature global climatic episodes. This was prompted by the realization that the Swartkrans cave filling reflected repeated cycles of deposition and erosion.*

nental drift, following the breakup of the supercontinent Gondwanaland, that initially embraced South America, Africa, Antarctica, and Australia. When Antarctica took up its south polar position, other continents drifted away from it to the north, and an open seaway was created around its periphery. This started the Circum-Antarctic Current, driven by the rotation of the earth and the resultant westerly winds. Once this happened, the thermal isolation and refrigeration of Antarctica began.

As far as African habitats go, the global cooling trend seems to have crossed a critical threshold about 6.5 million years ago with the "Terminal Miocene Event." At about this time, a widespread sea-level drop has been recorded and this, in addition to tectonic movements in the Gibralter area, resulted in the isolation of the Mediterranean from the Atlantic. The seawater in the Mediterranean basin then dried out completely, depositing over one million cubic kilometres of sea salt, which, as Ryan (1973) pointed out, constitutes about 6% of the dissolved salts in the world's oceans; its removal from circulation must have resulted in a significant freshening of sea water that would have facilitated the formation of sea-ice near Antarctica. This *Messinian Salinity Crisis* had striking physical and biological consequences, as was first described by Hsu et al. (1977). For instance, a dry-land connection was opened between Africa and Europe, facilitating the free exchange of fauna and flora, while the cool, arid climate around the desiccated Mediterranean might well have promoted an early expansion of African savannahs (Brain, 1984). The evidence further suggested that the connection between the Atlantic Ocean and the Mediterranean basin was re-established abruptly at about five million years ago, which brought the salinity crisis to an end.

It has long been surmised that the Terminal Miocene Event was precipitated by the sudden establishment of the west Antarctic ice-cap, and its link-up with the long-standing east Antarctic equivalent. While this concept still seems to be current, the situation has been complicated by evidence of tectonic activity, active rifting and the rise of the Trans-Antarctic Mountains (Denton, 1995; Cande et al., 2000).

The next important event in the cooling trend was the onset of the first Northern Hemisphere glaciation, which is currently placed at about 2.54 million years ago (Clark et al., 1999). This trend was accompanied by the regular glacial-interglacial cycles that have been such a feature of more recent times, first with a periodicity of 41,000 years until about one million years ago, followed by the establishment of the 100,000 year cycles, in which we are still currently locked. In Africa, the effects of cooling on habitats was greatly enhanced by volcanic activity and tectonic movement; in East Africa, combined effects of the Rift valley development, associated volcanoes and regional uplift was striking, while, in southern Africa, uplift of between 600 and 900 metres along the eastern regions may have had as much effect itself as the superimposed global cooling trend (Partridge et al., 1995).

So, in 1979, when I tried to correlate some of the features of hominid evolution with global low temperature events, it seemed reasonable to assume that the dramatic breakup of tropical evergreen forest areas at the time of the Terminal Miocene Event, between six and five million years ago, could be related to the acquisition of bipedal posture in our hominid ancestors. Following that, the next major low temperature event, at about 2.5 million years, seems to coincide with a split in the hominid lineage, with one line leading to humans and the other to the robust australopithecines, whose strategy for coping with ever more difficult environmental challenges did not, in the end, succeed. By contrast, adaptations in our own lineage proved to be more viable.

Since then, the regular glacial/interglacial cycles must have been very important in the spread of grasslands, at the expense of woodlands, the shrinking and breakup of tropical lowland forest areas (Hamilton, 1976), and the fragmentation of other habitats. Such effects must surely have promoted allopatric speciation in a variety of animals; in fact, Elisabeth Vrba has documented the first appearance of 37 new African antelope species, many of them open country grazers, between 2.7 and 2.5 million years ago, at the time when the robust australopithecine lineage split from that leading to *Homo*.

In September 1982, Elisabeth Vrba organised a symposium at the Transvaal Museum in Pretoria on "Species and Speciation," at which I again emphasized the significance of *Temperature-induced environmental changes in Africa as evolutionary stimuli* (Brain, 1985, pp. 45-52). The following year an international symposium was held by the South African Society for Quaternary Research in Swaziland, where I again drew attention to *The Terminal Miocene Event: a critical environmental and evolutionary episode* (Brain, 1984, pp. 491-498). At this meeting, Elisabeth Vrba took up the temperature/evolution theme with her usual energy and enthusiasm. She teamed up with George Denton, Tim Partridge, and Lloyd Burckle to organise a series of workshops on palaeoclimate and evolution at the Lamont-Doherty Earth Observatory, Palisades, New York in September 1984; at Sun City, South Africa in February 1985, and again at the Lamont-Doherty Earth Observatory in May 1985. The proceedings were published as extended abstracts in three dedicated issues of the South African Journal of Science (vols 81(5), 1985; 82(2), 1986 and 82(9), 1986). Finally a conference was held in Airlie, Virginia, in May 1993 under the title *Paleoclimate and evolution, with emphasis on human origins* where many important topics were addressed. Here I discussed *The influence of climatic changes on the completeness of the early hominid record in southern African caves, with particular reference to Swartkrans* (Brain, 1995), stressing that the fossiliferous infills of caves such as Swartkrans, probably reflected only the interglacial periods of the last two million years. The much longer glacial episodes were probably not represented at all in the deposits.

## 10. Evidence from Swartkrans for predation on early hominids

Early on in the analysis of the fossil assemblage from Swartkrans, it became apparent that, although the numerous hominid individuals were well represented by cranial fossils, post-cranial bones were very rare, and I was confronted with "the mystery of the missing bodies," as we called it then. In addition to the hominids, four baboon species were represented and we found that, the larger the body-size of these, the more juvenile remains appeared in the assemblage. An obvious conclusion to draw was that we were dealing with the food remains of a predator, which had preference for prey of a particular body-size. Leopards immediately came to mind, as they are well-known to select prey within preferred limits and this suspicion was confirmed by the specific damage that some of the bones had suffered (Newman, 1993). One well-known specimen, the calvaria of a hominid child, was found to have two punctures in its parietal bones and the distance between these was matched by the spacing of the canines of a fossil leopard from the same part of the cave. The reconstruction that I suggested was that the child had been killed by a leopard, perhaps by the usual throat-bite method, and that it had then been picked up by its head, as leopards are inclined to do, and dragged off to a feeding place in the dark recesses of the cave. This carrying behaviour, observed in contemporary leopards with monkey or baboon prey, results in the upper canines gripping the face of the prey, while the lower canines penetrate the back of the skull (Brain, 1969, 1974, 1981).

The detailed taphonomic analysis of the fossil assemblages from Swartkrans Members 1 and 2 suggested that hominids and baboons came to shelter within the entrance area of the cave on cold winter nights and that they were preyed upon there by leopards and sabre-tooth cats. The predators took their victims to the lower parts of the cave and ate them; what scraps survived their attention, and that of scavengers such as hyaenas—whose coprolites in the deposits testify to their visits—contributed to the fossil assemblage (Brain, 1993).

In broad perspective, my impression is that the life of hominids in environments such as that of the Sterkfontein valley, one and a half million years ago, would have been a hazardous one, calling for continual vigilance against a wide variety of predatory threats, day and night. In my opinion, such threats must have represented a significant selective pressure in favour of any advance in intelligence and resulting technology that could have reduced the threat.

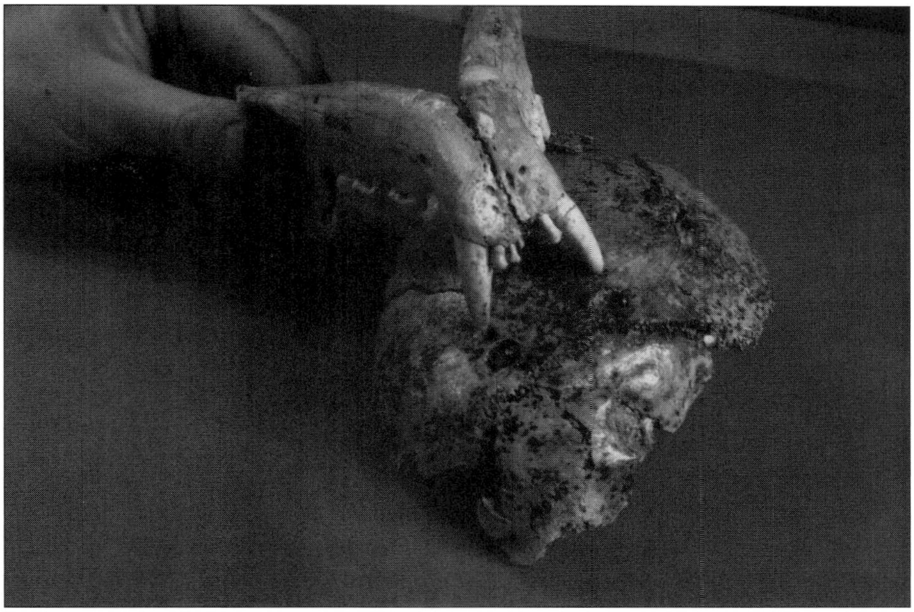

Figure 14. Part of a juvenile australopithecine cranium (SK 54) with two round holes in it, from the Swartkrans Hanging Remnant. The spacing of these holes is matched closely by that of the lower canines of a fossil leopard from the same deposit.

## 11. Evidence for the management of fire at Swartkrans

As mentioned above, there can surely be no doubt that humans eventually established their current dominance in the natural world through intelligence and its product, technology. But were the initial steps along this path also mediated in this way? I have the impression that some of the evidence from the Swartkrans cave confirms this possibility. Excavation revealed that the Member 3 deposit accumulated in a roofed erosional gully, about 20 m long and up to 5 m wide, running between the west wall of the cave and an eroded vertical bank of older sediments on the east side of the gully. Initially I was not aware that the calcified sediment in this gully was different from that further to the east, but when pieces of burnt bone started turning up with regularity, suspicions were aroused and a near-vertical unconformable contact became apparent between the contents of this gully and what surrounded it. The excavation proceeded to a depth of 850 cms and produced 59,488 pieces of fossil bone, including nine fossils of robust australopithecines and 270 pieces of bone that showed signs of having been burnt. Careful chemical analyses were undertaken at the University of Capetown by Andy Sillen, while we measured the temperatures attained in a number of experimental fires, using different kinds of wood. We then made thin sections of bones heated to such temperatures in a kiln fitted with a very precise temperature regulating device. After all this, we were able to estimate the approximate temperatures to which each of the fossil pieces had been heated. (Brain and Sillen, 1988; Sillen and Hoering, 1993; Brain, 1993b)

Figure 15. A reconstruction in the Transvaal Museum of the Swartkrans ape-man child suggesting that it had been killed by a leopard and then carried to a feeding lair in the cave. The lower canines of the leopard are in the back of the child's head, while the upper canines are in its face.

It is to be expected that natural grass fires passing the entrance to a cave should burn any pieces of bone lying around, and that these could later make their way back into the lower parts of the cave. In fact, three pieces of fossilised burnt bone had turned up, among 153,781 other fossils in the Lower Bank of Member 1, and one may assume that the burning could have happened in this way. But when pieces of burnt bone made their appearance in 17 excavation grid squares (1 m × 1 m), and in up to 23 vertical excavation spits (each 10 cms thick) in

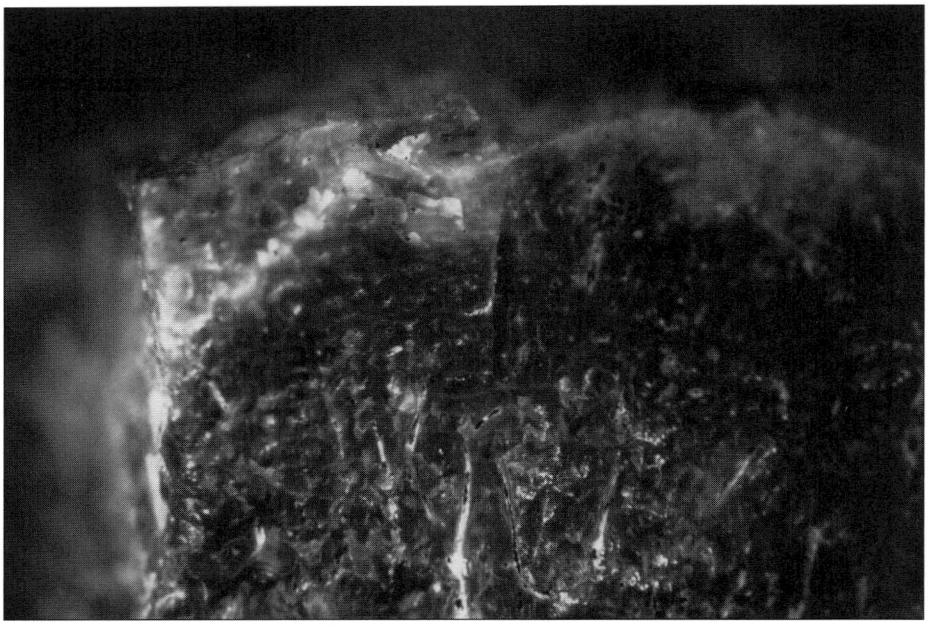

*Figure 16. A piece of fossil burnt bone, one of many found in the Member 3 sediments at Swartkrans cave. The dark interior of this particular bone had been carbonized, while the pale exterior was calcined.*

the newly exposed Member 3 deposit, one was obviously dealing with a different situation. The interpretation that we proposed was that fires had been tended in the entrance area of the Member 3 gully repeatedly during the accumulation period of this sedimentary profile, and that pieces of bone heated in these fires had made their way down the talus slope to their final repository. There is no evidence that people at this time had mastered the technique of *fire making*, but had presumably collected burning branches from natural, lightning-induced grass fires—that are very much a feature of the highveld grassland in early summer—and brought this fire back to their sleeping place. If Member 3 is about one million years old, hundreds of thousands of years would probably have had to pass before the deliberate making of fire became a reality. But whatever the source of the fire, its presence in the cave entrance would have given these early human groups some measure of protection from the ever-present danger of waiting leopards. In addition to the burnt bones, we also recognised 16 pieces in the Member 3 assemblage with cut- and chop-marks on them. Since then, many others have been recognised in the collection through the careful work of Travis Pickering and his colleagues. Such damage has not been seen on any of fossils from Members 1 and 2, suggesting that hominid meat-eating at the cave, presumably round a camp-fire, became a reality in the cold interval between Members 2 and 3. Presumably, without the protection af-

*Figure 17. An experimental camp fire at Swartkrans, using the local* Celtis *wood, in 1985. Here Virginia and Tim Brain measure temperatures within the fire, using a long thermocouple proble, linked to a digital thermometer.*

forded by fire, it would have been too dangerous to bring meat to the cave for fear of attracting the attention of other carnivores. In my view, fire-management of this kind, must have represented a critical early step in human emancipation from subservience to more powerful carnivores that ultimately led to their domination.

As a result of further intelligence-driven technology, humans then went on to become highly effective social hunters and predators in their own right. The selective pressures driving this process were presumably similar to those that had allowed the human emergence from a former subservient role. Among the variety of selective pressures that drove the evolution of the large human brain, it can be argued that the demands of predation, first in surviving its dangers and later in the successful practice, were ever-present and powerful in their effects.

## 12. The significance of predation to the evolution of intelligence in hominids and much older ancestral animals

By any standards, the increase in brain size relative to that of the body in our human ancestors during the last two million years was a remarkable zoological event. When the earliest known members of the *Homo* lineage appeared on the scene, in the form of *H. habilis* or *H. rudolfensis*, their average brain-capacity was about 654 cc; this had risen to about 850 cc in *H. ergaster* and *H. erectus*, and to 1400 cc in archaic *H. sapiens* towards the end of the Middle Pleistocene. As Leslie Aiello and Peter Wheeler (1995) pointed out, this event is all the more remarkable because a brain is built of "expensive tissue"—although a human brain may only make up 2 to 3% of the weight of the whole body, it uses 16-20% of the energy consumed by the resting body. To double the size of the brain, relative to that of the body, would usually mean that the basic metabolic rate of the animal would have to be substantially increased. Oddly enough, this has not been observed in humans, in comparison to related primates, and this prompted Aiello and Wheeler to conclude that human brain expansion occurred at the expense of the size of the gut, that has apparently shrunk during the course of human evolution. To be able to function with a much smaller gut implies that ancestral humans changed to a diet of higher quality, such as one including animal protein, and they would have done this by scavenging and active hunting. There is, in fact, good evidence for this behaviour among the 2.5-million-year-old Bouri hominids of Ethiopia, as Tim White and his colleagues have pointed out (de Heinzelin et al., 1999).

Clearly, a greatly increased brain-size is not a luxury to be acquired lightly. It is something that would only have evolved under strong selective pressure. For many years it has been suggested that brain expansion, and the benefits that it brings to humans, has been linked to the problems of making a living in the changed and more open habitats that characterized Africa during the last two million years. Frequently cited is the need to cope with the more complex foraging strategies than had been the case when ancestral hominids lived in evergreen forests. I have no doubt that this need would have been one of the factors. But, as I have discussed above, I believe that it was also the need to survive the ever-present threat of predation by carnivores in these more open and alien habitats that provided the needed selective pressure (Brain, 2001a).

At the end of my Swartkrans project in the early 1990s, I thought it would be fun to find out more about the ancient roots of predation in the animal lineage. These would have been among ancestral invertebrates, some of which were very small, and I was attracted to them as I have had a long-standing interest in living micro-invertebrates in African fresh waters. These have included testate amoebae, rotifers and planarians, which I had the chance to study intensively while serving as Curator of Invertebrates at the Transvaal Museum between 1991 and 1995. But, to gain any insights on the ancient roots of predation, one has to turn to the fossil record, going back at least 600 million years to Terminal Proterozoic times. Some of the best evidence in this regard can be found in sediments belonging to the Nama Group of Namibia, that were accumulating in a shallow sea on the western edge of the Kalahari craton at the time of the assembly of Gondwanaland (Brain, 1997b). It is to the fossil record from the Nama Group, as well as from the somewhat older Otavi Group on the Congo Craton further to the north, that I have given my attention in the last few years.

Late proterozoic times, when animals first left abundant traces in the fossil record, were preceded by several very severe glacial periods. Glacial deposits from two of these episodes, each with their very distinctive "cap-carbonates," have been recognised in northern Namibia (Hoffmann and Prave, 1996), as well as elsewhere in the world. Based on these, the "snowball earth" scenario has been invoked (Hoffman et al, 1998), addressing the problem of low-latitude glaciations, as are indicated by the Namibian evidence. It proposes that the oceans froze over and that biological productivity collapsed for some millions of years. It was only through the abundant production of carbon dioxide by active volcanoes that a "greenhouse" situation developed, rapidly melting the global ice and swinging the climate to an opposite extreme, as indicated by the cap carbonates, immediately above the glacial sediments.

The period following the last of these glacials, starting at about 580 million years, saw two remarkable radiations of animal life. The first is known as the Ediacaran radiation and involves soft-bodied organisms whose impressions are typically preserved in sandstones. The first evidence of these turned up in the Nama Basin of southern Namibia as early as 1908, and was described by Gürich in 1933. These organisms were typically flat or leaf-like with a very characteristic quilted structure, reminiscent of an air-mattress, but there were also circular medusoid-like structures. A similar fauna came to light during 1946 in the Ediacara Hills of South Australia and it is from this locality that the radiation gained its name. Since then, similar fossils have become known from at least 30 localities on five continents (Narbonne, 1998).

The remarkable structure shown by these organisms prompted Adolf Seilacher (1992) to create a new Kingdom, the Vendobionta, for them, although their affinities have been the subject of vigorous debate during the last 20 years. But, whatever they actually were, it is now clear that this fauna existed for about 55 million years, showing maximum diversity during the last 20 million of their existence. The Ediacaran community was apparently composed of sedentary or very slow-moving individuals, luxuriating in a tranquil "Garden of Ediacara," as Mark McMenamin (1986) has called it. Also attached to the shallow sea's "biomat" of that time were small cone-in-cone structures, originally described by Gerard Germs from the Nama Basin as *Cloudina*. These have since been found in many parts of the world and are regarded as a Terminal Proterozoic index fossil. The cones presumably housed a filter-feeding metazoan, of at least cniderian-grade organisation, with tentacles protruding from the top. Of particular interest is the fact that many fossils from China studied by Bengtson and Zhao (1992), showed evidence of having been bored into by predators. Recently, I came across similarly bored *Cloudina* tubes from the Nama Basin (Brain, 2001b) and the hunt is now on for fossils of the predator that caused the damage. But whatever the zoological affinity of this first predator might be, recent molecular evidence suggests strongly that the animal lineages would have gone back far further in time than currently known fossils might indicate. For this reason, I am working these days on micro-invertebrate fossils from Otavi Group limestones in northern Namibia, that date from between the two snowball earth glacials, 720–590 million years ago. This project is very labour-intensive, necessitating the preparation of hundreds of acetic acid residues of the limestones as well as the cutting and grinding of thin-sections that Laura Brain does on the back verandah of our home. But we have found a limestone succession where preservation of these small, soft-bodied creatures is exceptionally good, even though interpretation of the fossils that we find is fraught with problems (Brain et al., 2001). Some of these interpretive problems are now being overcome (Brain et al., 2003) and future prospects are exciting. So the fun with fossils continues!

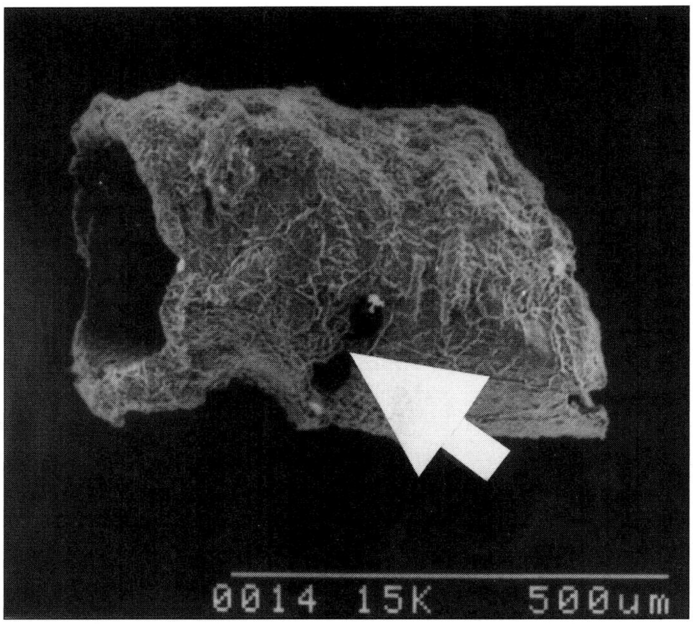


*Figure 18. The calcified tube of an ancestral animal,* Cloudina, *preserved in 550 million year-old Nama limestones of Namibia. It is thought to have been a coelenterate, with tentacles protruding in life from the open upper end of the tube. But in this specimen, a predator has drilled holes (arrow) through the wall of the tube, to gain access to the soft-bodied interior. This provides very early evidence for predatory activity.*

Returning to the theme of the appearance of the first predators, it is clear that this anticipated the demise of the tranquil Garden of Ediacara. The end came with the second major radiation of those times, the "Cambrian explosion of animal life," so well documented in the fossils of the Burgess Shale and other sites such as those at Chengjiang in China. These fossils suggest that within a brief period of geological time, representatives of almost every known phylum of animal life made their appearance in Middle Cambrian seas. Among these animals were, of course, burrowers and grazers, that rapidly destroyed the microbial biomats that had been such a feature of shallow seas for three billion years. But of particular relevance here was the appearance, with the Cambrian radiation, of the first effective predators that the world had seen. By definition, an animal is a multicellular heterotroph–an organism that feeds on other living things or their remains, while predators form but a subset of these. To feed on other true animals will generally require active pursuit and the overpowering of reluctant prey. For this, coordinated mobility and the ability to locate evasive prey are required. In other words, mobility and sense organs are needed.

Predators obviously rely on a variety of senses, but for the present purpose, let us consider just one—that of sight. The largest and most fearsome of Cambrian predators was *Anomalocaris*, growing up to 50 cms long and known now from a variety of localities in different parts of the world. The expanded lateral lobes along the length of the body made it an active swimmer, while the two powerful appendages anterior to the mouth must have been effective in the capture of prey, such as trilobites. These appendages had, in fact, been described (Whiteaves, 1892) many years before the entire animal was recognised and each was thought to represent a phyllocarid crustacean in its own right, although the "heads" of these could never be found, hence the name *Anomalocaris*. Now that the entire animal can be examined in fossil form (Collins, 1996), it is clear that two very large eyes were one of its striking features, each providing information to a central coordinating nervous system.

Eyes that are even more remarkable can be seen in fossils of another Burgess Shale organism of uncertain affinity, known as *Opabinia*, According to the reconstruction and interpretation provided by Briggs *et al.* (1994), *Opabinia* had five large eyes at the front of the head and a long flexible proboscis that ended in an array of grasping spines used to capture prey as the animal swam rapidly over the seafloor, relying on its lateral lobes for propulsion and using its tail as a stabilizer.

Throughout 500 million years of animal evolution, every advance that a predator could make to its effectiveness as a hunter had to be countered by comparable improvements in the survival ability of its prey, if one or the other were to avoid extinction. In this way, sense organs and coordinating neural systems were under constant selective pressure to promote their improvement. Cranial expansion, improved intelligence and the use of technology by hominids, faced with completely new and unprecedented predatory challenges, appears to have been one of the solutions. The fact that robust australopithecines did not follow this route of cranial expansion, may well have contributed to their disappearance from the merciless African savannah.

## 13. The pleasure of seeing taphonomic themes being carried forward from their early beginnings

The overview of my taphonomic career that I have provided here has obviously been retrospective. But writing it has made me realise what I appreciate most about my current situation. It lies in observing how many of the themes with which I have been concerned, are now being carried forward by young and enthusiastic taphonomists, like Travis Pickering and his colleagues, in ways that I could not have imagined (Pickering et al., 2004 a, b; Egeland et al., 2004). They bring new enthusiasms, insights and skills to the quest, and all my good wishes go with them.

I currently serve as Chief Scientific Adviser to the Palaeo-Anthropology Scientific Trust, or PAST, a South African organisation that, for the last ten years has provided financial support to students, researchers and educators, formerly in palaeo-anthropology, but now also in the wider field of palaeontology (Brain and Read, 2002a and b; Brain, 2003). A real pleasure in this is that I am brought into contact with students and others in the palaeo-field, throughout southern Africa, and am able to follow the progress of their interesting projects.

## Acknowledgements

It is with particular gratitude that I plan to participate in this meeting and thank Travis Pickering, Nick Toth and Kathy Schick for their invitation, and for all the hard work that they have put into its organisation. I would like to wish them and their colleagues at CRAFT every success with their work in the wonderful new facilities at Indiana University.

By way of acknowledgement, I also wish to record that much of my early work in this field was supported by the Wenner Gren Foundation and, in particular, by the interest of its then Director, Lita Osmundsen. Subsequently, generous research grants from the National Research Foundation in South Africa made my scientific work possible and this support is gratefully acknowledged. In the topics outlined above I have had the encouragement and support of many colleagues, without which the work could not have been done. For instance, Francis Thackeray started helping me at Swartkrans when he was 16 years old, and I was joined by Ron Clarke and Clark Howell in the description of new *Homo* remains from there, as long ago as 1970. My gratitude goes to many such people.

Fossil projects have provided me with a great deal of fun over the last 50 years, due largely to the presence of Laura Brain who, with our four children, have participated so enthusiastically. Finally, the encouragement and generosity of spirit shown to me by Raymond Dart is something for which I will always be grateful.

## References

Aiello, L. C., Wheeler, P. 1995. The expensive tissue hypothesis: The brain and the digestive system in human and primate evolution. Current Anthropology 36, 199-221.

Ardrey, R. 1961. African Genesis: A Personal Investigation into the Animal Origins and Nature of Man. Collins, London.

Backwell, L. R. 2000. A critical assessment of southern African early hominid bone tools. M Sc Thesis, University of the Witwatersrand.

Backwell, L. R., d'Errico, F. 2001. Evidence of termite foraging by Swartkrans early hominids. Proceedings of the National Acadamy of Sciences, USA 98, 1356-1363.

Backwell, L. R., d'Errico, F. 2004. Additional evidence of early hominid bone tools from Swartkrans. In: Brain, C. K. (Ed.), Swartkrans: A Cave's Chronicle of Early Man, 2nd Edition. Transvaal Museum, Pretoria, pp. 279-295.

Behrensmeyer, A. K., Hill, A. P. (Eds.) 1980. Fossils in the Making: Vertebrate Taphonomy and Palaeoecology. University of Chicago Press, Chicago.

Bengtson, S., Zhao. Y. 1992. Predatorial borings in Late Precambrian mineralized exoskeletons. Science 257, 367-369.

Binford, L. R. 1981. Bones: Ancient Men and Modern Myths. Academic Press, New York.

Brain, C. K. 1958. The Transvaal Ape-Man-Bearing Cave Deposits. Transvaal Museum, Pretoria.

Brain, C. K. 1967a. Hottentot food remains and their bearing on the interpretation of fossil bone assemblages. Scientific Papers of Namib Desert Research Station 32, 1-11.

Brain, C. K. 1967b. Bone-weathering and the problem of bone pseudo-tools. South African Journal of Science 63, 97-99.

Brain, C. K. 1969a. The contribution of Namib Desert Hottentots to the understanding of australopithecine bone accumulations. Scientific Papers of the the Namib Desert Research Station 39, 13-22.

Brain, C.K. 1969b. The probable role of leopards as predators of the Swartkrans australopithecines. South African Archaeological Bulletin 24, 127-143.

Brain, C. K. 1974a. A hominid skull's revealing holes. Natural History 33, 44-45.

Brain, C. K. 1980. Criteria for the recognition of bone-collecting agencies in African caves. In: Behrensmeyer, A. K., Hill, A. P. (Eds.), Fossils in the Making: Vertebrate Taphonomy and Palaeoecology. University of Chicago Press, Chicago, pp. 107-130.

Brain, C. K. 1981a. Hominid evolution and climate change. South African Journal of Science 77, 104-105.

Brain, C. K. 1981b. Presidential address: Taphonomy as an aid to African palaeontology. Palaeontologia Africana 24, 1-4.

Brain, C. K. 1981c. The evolution of man in Africa: was it a consequence of Cainozoic cooling? Proceedings of the Geological Society of Southern Africa, Annexure to Vol. LXXXIV, pp. 1-19. Alex. L. Du Toit Memorial Lecture no. 17.

Brain, C. K. 1981d. The Hunters or the Hunted? An Introduction to African Cave Taphonomy. University of Chicago Press, Chicago.

Brain, C. K. 1982a. Cycles of deposition and erosion in the Swartkrans cave deposit. Palaeoecology of Africa and of the Surrounding Islands and Antarctica 15, 27-29.

Brain, C. K. 1982b. The Swartkrans site: Stratigraphy of the fossil hominids and a reconstruction of the environment of early *Homo*. Proc. Congres International de Paleontologie Humaine, Nice, October 1982, pp. 676-709.

Brain, C. K. 1984. The Terminal Miocene Event: a critical environmental and evolutionary episode? In: Vogel, J. C. (Ed.), Late Cainozoic Palaeoclimates of the Southern Hemisphere. A. A. Balkema, Rotterdam, pp. 491-498.

Brain, C. K. 1985a. New insights into early hominid enviromnents from the Swartkrans cave. In: Coppens, Y. (Ed.), Le Environment des Hominides au Plio-Pleistocene. Fondation Singer-Polignac, Paris, pp. 325-343.

Brain, C. K. 1985b. Temperature-induced environmental changes in Africa as evolutionary stimuli. In: Vrba, E. S. (Ed.), Species and Speciation. Transvaal Museum, Pretoria, pp. 45-52.

Brain, C. K. 1985c. Interpreting early hominid death assemblages: The rise of taphonomy since 1925. In: Tobias, P. V. (Ed.), Hominid Evolution, Past, Present and Future. Alan R. Liss Inc., New York, pp. 41-46.

Brain, C. K. 1986a. Prospects for a southern African continental record of Cainozoic temperature changes. South African Journal of Science 80, 94-96.

Brain, C. K. 1987. Some adaptations for survival by African hominids during the last five million years. South African Journal of Science 83, 610-612.

Brain, C. K. (Ed.) 1993. Swartkrans: A Cave's Chronicle of Early Man. Transvaal Museum, Pretoria.

Brain, C. K. 1993a. Structure and stratigraphy of the Swartkrans cave in the light of the new excavations. In: Brain, C. K. (Ed.), Swartkrans: A Cave's Chronicle of Early Man. Transvaal Museum, Pretoria, pp. 23-33.

Brain, C. K. 1993b. The occurrence of burnt bones at Swartkrans and their implications for the control of fire by early hominids. In: Brain, C. K. (Ed.), Swartkrans: A Cave's Chronicle of Early Man. Transvaal Museum, Pretoria, pp. 229-242.

Brain, C. K. 1993c. A taphonomic overview of the Swartkrans fossil assemblages. In: Brain, C. K. (Ed.), Swartkrans: A Cave's Chronicle of Early Man. Transvaal Museum, Pretoria, pp. 257-264.

Brain, C. K. 1994. The Swartkrans Palaeontological Research Project in perspective: Results and conclusions. South African Journal of Science 90, 220-223.

Brain, C. K. 1995. The importance of predation to the course of human and other animal evolution. South African Archaeological Bulletin 50, 93-97.

Brain, C. K. 1995a. The influence of climatic changes on the completeness of the early hominid record in southern African caves, with particular reference to Swartkrans. In: Vrba, E. S., Denton, G.H., Partridge, T.C., Burckle, L.H., (Eds.) Paleoclimate and Evolution with Emphasis on Human Origins. Yale University Press, New Haven, pp. 451-458.

Brain, C. K. 1995b. The contribution of Raymond Dart to the development of cave taphonomy. Palaeontologia Africana 33, 81-83.

Brain, C. K. 1997. The importance of Nama Group sediments and fossils to the debate about animal origins. Palaeontologia Africana 34, 1-13.

Brain, C. K. 2001a. Do we owe our intelligence to a predatory past? 70th James Arthur Lecture on the Evolution of the Human Brain. American Museum of Natural History, New York, pp. 1-32.

Brain, C. K. 2001b. Observations on *Cloudina*, a Neoproterozoic index fossil from Namibia. Journal of African Geology 33, 475-480.

Brain, C. K. 2003. A perspective on the PAST (Palaeoanthropology Scientific Trust). South African Journal of Science 99, 235-236.

Brain, C. K., Jenkins, T. 1967. The peoples of the lower Kuiseb Valley, South West Africa. Scientific Papers of the Namib Desert Research Station 35, 1-24.

Brain, C. K., Meester, J. A. J. 1965. Past climatic changes as biological isolating mechanisms in southern Africa. In: Davis, D. H. S. (Ed.), Ecological Studies in Southern Africa. Re. W. Junk Publishers, The Hague, pp. 332-340.

Brain, C. K., Read, C. 2002. A future for the PAST. Africa Geographic 10, 66-73.

Brain, C. K., Shipman, P. 1993. The Swartkrans bone tools. In: Brain, C. K. (Ed.) Swartkrans. A Cave's Chronicle of Early Man. Transvaal Museum, Pretoria, pp. 195-215.

Brain, C. K., Sillen, A. 1988. Evidence from the Swartkrans cave for the earliest use of fire. Nature 336, 464-466.

Brain, C. K., Watson, V. 1992. A guide to the Swartkrans early hominid cave site. Annals of the Transvaal Museum 35, 343-365.

Brain, C. K., Clarke, R. J., Howell, F. C. 1970. More evidence for an advanced hominid at Swartkrans. Nature 225, 1219-1222.

Brain, C. K., van Riet Lowe, C., Dart, R. A. 1955. Kafuan stone implements in the post-australopithecine breccia at Makapansgat. Nature 175, 16-20.

Brain, C. K., Churcher, C. S., Clark, J. D., Grine, F. E., Shipman, P., Susman, R. L., Turner, A., Watson, V. 1988. New evidence of early hominids, their culture and environment from the Swartkrans cave. South African Journal of Science 84, 828-835.

Brain, C. K., Hoffmann, K.-H., Prave, A. R., Fallick, A. E, Coetzee J., Botha, A. J. 2001. Interpretive problems in a search for micro-invertebrate fossils from a Neoproterozoic limestone in Namibia. Palaeontologia Africana 37, 1-12.

Brain, C. K., Prave, A. R., Fallick, A. E., Hoffmann, K.-H. 2003. Sponge-like microfossils from Neoproterozoic intertillite limestones of the Otavi Group in northern Namibia. In: Frimmel, H. E. (Ed.), Proceedings of the 111 International Colloquium on the Vendian-Cambrian of West Gondwana. University of Cape Town, Cape Town, pp. 19-22.

Briggs, D. E. G., Erwin, D. H., Collier, F. J. 1994. The Fossils of the Burgess Shale. Smithsonian Institution Press, Washington.

Buckland, W. 1822. Account of an assemblage of fossil teeth and bones of elephant, rhinoceros, hippopotamus, bear, tiger and hyaena and sixteen other animals discovered in a cave at Kirkdale, Yorkshire in the year 1821, with a comparative view of five similar others on the Continent. Philosophical Transactions of the Royal Society of London 112, 171-237.

Butzer, K. W. 1976. Lithostratigraphy of the Swartkrans Formation. South African Journal of Science 72, 136-141.

Cande, S. C. 2000. Cenozoic motion between East and West Antarctica. Nature 404, 145-150.

Carlson, K. J., Pickering, T. R. 2003. Intrinsic qualities of primate bones as predictors of skeletal element representation in modern and fossil carnivore feeding assemblages. Journal of Human Evolution 44, 431-450.

Clark, P. U., Alley, R. B., Pollard, D. 1999. Northern hemisphere ice-sheet influences on global climate change. Science 286, 1104-1111.

Clarke, R. J. 1994. On some new interpretations of Sterkfontein stratigraphy. South African Journal of Science 90, 212-214.

Collins, D. 1996. The "evolution" of *Anomalocaris* and its classification in the Arthropod Class Dinocarida (Nov.) and Order Radiodonta (Nov.). Journal of Paleontology 70, 280-293.

Dart, R. A. 1925a. *Australopithecus africanus*: the man-ape of South Africa. Nature 115, 195-199.

Dart, R. A. 1925b. A note on Makapansgat: a site of early human occupation. South African Journal of Science 26, 454-455.

Dart, R. A. 1929. A note on the Taungs skull. South African Journal of Science 26, 648-658.

Dart, R. A. 1934. The dentition of *Australopithecus prometheus*. Folia. Anatomic Japon. 12, 207-221.

Dart, R. A. 1948. The Makapansgat proto-human *Australopithecus prometheus*. American Journal of Physical Anthropology 6, 259-284.

Dart, R. A. 1949. The bone-bludgeon hunting technique of *Australopithecus*. South African Journal of Science 2, 150-152.

Dart, R. A. 1953. The predatory transition from ape to man. International Anthropological and Linguistic Review 1, 201-218.

Dart, R. A. 1955. The first australopithecine fragment from the Makapansgat pebble culture stratum. Nature 176, 170.

Dart, R. A. 1956a. The myth of the bone-acumulating hyena. American Anthropologist 58, 40-62.

Dart, R. A. 1956b. Cultural status of the South African manapes. Smithsonian Report No. 4240, 317-318.

Dart, R. A. 1957a. The Makapansgat australopithecine osteodontokeratic culture. Proceedings of the Third Pan-African Congress on Prehistory (Livingstone, 1955). Chatto and Windus, London.

Dart, R. A. 1957b. The Osteodontokeratic Culture of *Australopithecus Prometheus*. Transvaal Museum, Pretoria.

de Heinzelin, J., Clark, J. D., White, T., Hart, W., Renne, P., WoldeGabriel, G., Beyene, Y., Vrba, E. 1999. Environment and behavior of 2.5-million-year-old Bouri hominids. Science 284, 625-635.

Denton, G. H. 1995. The problem of Pliocene paleoclimate and ice-sheet evolution in Antarctica. In: Vrba, E. S., Denton, G.H., Partridge, T.C., Burckle, L.H. (Eds.), Paleoclimate and Evolution with Emphasis on Human Origins. Yale University Press, New Haven, pp. 213-229.

d'Errico, F., Backwell, L. R., Berger, L. R. 2001. Bone tool use in termite foraging by early hominids and its impact on our understanding of early hominid behaviour. South African Journal of Science 97, 71-75.

Egeland, C. P., Pickering, T. R., Dominguez-Rodrigo, M., Brain, C. K. 2004. Disentangling Early Stone Age palimpsests: A method to determine the functional independence of hominid- and carnivore-derived portions of archaeofaunas. Journal of Human Evolution 47, 343-357.

Gürich, G. 1933. Die Kuibis-Fossiliender Nama Formation von Südwest Afrika. Palaeontologia Zoologica 15, 137-154.

Hamilton, A. 1976. The significance of patterns of distribution shown by forest plants and animals in tropical Africa for the reconstruction of Upper Pleistocene palaeoenvironments: a review. Palaeoecology of Africa and of the Surrounding Islands and Antarctica 9, 63-97.

Hoffmann, K.-H., Prave, A. R. 1996. A preliminary note on a revised subdivision and regional correlation of the Otavi Group based on glaciogenic diamictites and associated cap dolostones. Communications of the Geological Survey of Namibia 11, 77-82.

Hoffman, P. F., Kaufman, A. L., Halverson, G. P., Schrag, D. P. 1998. A Neoproterozoic snowball earth. Science 281, 1342-1346.

Hsu, E.T. 1977. History of the Mediterranean salinity crisis. Nature 267, 399-403.

Kuman, K. 1994a. The archaeology of Sterkfontein: Preliminary findings on site formation and cultural change. South African Journal of Science 90, 215-219.

Kuman, K. 1994b. The archaeology of Sterkfontein—Past and present. Journal of Human Evolution 27, 471-495.

Kuman, K. 2003. Site formation in the early South African Stone Age sites and its influence on the archaeological record. South African Journal of Science 99, 251-254.

Kuman, K., Clarke, R. J. 2000. Stratigraphy, artefact industries and hominid associations for Sterkfontein Member 5. Journal of Human Evolution 38, 827-847.

Latham, A. G., Herries, I. R., Kuykendall, K. in press. The formation and sedimentary infilling of the Limeworks Cave, Makapansgat, South Africa. Palaeontologia Africana.

McMenamin, M. 1986. The garden of Ediacara. PALAIOS 1, 178-182.

Mills, M. G. L., Mills, M. E. J. 1977. An analysis of bones collected at hyaena breeding dens in the Gemsbok National Park. Annals of the Transvaal Museum 30, 145-155.

Narbonne, G. M. 1998. The Ediacaran biota: A Terminal Proterozoic experiment in the evolution of life. GSA Today 8, 1-6.

Newman, R. A. 1993. The incidence of damage marks on Swartkrans fossil bones from the 1979-1986 excavations. In: Brain, C. K. (Ed.), Swartkrans: A Cave's Chronicle of Early Man. Transvaal Museum, Pretoria, pp. 217-228.

Oakley, K. P. 1956. The earliest fire-makers. Antiquity 30, 102-107.

Olsen, S. L. 1984. Analytical approaches to the manufacture of bone artefacts in prehistory. Ph.D. Dissertation, University of London.

Partridge, T. C., Wood, B. A., de Menocal, P. B. 1995. The influence of global climatic change and regional uplift on large mammalian evolution in East and southern Africa. In: Vrba, E. S., Denton, G.H., Partridge, T.C., Burckle, L.H. (Eds.), Paleoclimate and Evolution with Emphasis on Human Origins. Yale University Press, New Haven, pp. 331-355.

Pickering, T. R. 1999. Taphonomic interpretations of the Sterkfontein early hominid site (Gauteng, South Africa) reconsidered in light of recent evidence. Ph.D. Dissertation, University of Wisconsin-Madison.

Pickering, T. R., Carlson, K. J. 2002. Baboon bone mineral densities: Implications for the taphonomy of primate skeletons in South African cave sites. Journal of Archaeological Science 29, 883-896.

Pickering, T. R., Dominguez-Rodrigo, M., Egeland, C. P., Brain, C. K. 2004a. New data and ideas on the foraging behaviour of Early Stone Age hominids at Swartkrans Cave. South African Journal of Science 100, 215-219.

Pickering, T. R., Dominguez-Rodrigo, M., Egeland, C. P., Brain, C. K. 2004b. Beyond leopards: Tooth marks and the contribution of multiple carnivore taxa to the accumulation of the Swartkrans Member 3 fossil assemblage. Journal of Human Evolution 46, 595-604.

Robinson, J. T. 1959. A bone tool from Sterkfontein. Nature 184, 583-585.

Robinson, J. T., Mason, R. J. 1957. Occurrence of stone artefacts with *Australopithecus* at Sterkfontein. Nature 180, 521-524.

Rose, J. J. 1983. A replication technique for scanning electron microscopy: Applications for anthropologists. American Journal of Physical Anthropology 62, 255-261.

Ryan, W. B. F. 1973. Geodynamic implications of the Messinian crisis of salinity. In: Drooger, C. W. (Ed.), Messinian Events in the Mediterranean. North Holland Publishing Co., Amsterdam, pp. 26-38.

Seilacher, A. 1992. Vendobionta and Psammocorralia: Lost constructions in Precambrian evolution. Journal of the Geological Society of London 149, 609-613.

Schick, K. D., Toth, N. 1993. Making Silent Stones Speak: Human Evolution and the Dawn of Technology. Touchstone, New York.

Sillen, A., Hoering, T. 1993. Chemical characterization of burnt bones from Swartkrans. In: Brain, C.K. (Ed.), Swartkrans: A Cave's Chronicle of Early Man. Transvaal Museum, Pretoria, pp. 243-250.

Shipman, P. 1981. Life History of a Fossil: An Introduction to Taphonomy and Paleoecology. Harvard University Press, Cambridge.

Shipman, P. 1989. Altered bones from Olduvai Gorge, Tanzania: Techniques, problems and implications of their recognition. In: Bonnichsen, R., Sorg, M. H. (Eds.), Bone Modification. Center for the Study of the First Americans, Orono, (ME), pp. 317-334.

Shipman, P., Rose, J. J. 1988. Bone tools: An experimental approach. In: Olsen, S. L. (Ed.), Scanning Electron Microscopy in Archaeology. British Archaeological Reports International Series, Oxford 442, 303-335.

Whiteaves, J. F. 1892. Description of a new genus and species of phyllocarid crustacean from the Middle Cambrian of Mount Stephen, British Columbia. Canadian Record of Science 5, 205-208.

# CHAPTER 2

# Rather Odd Detective Stories: A View of Some Actualistic and Taphonomic Trends in Paleoindian Studies

Gary Haynes

## Abstract

During the last three decades of American Paleoindian research, some taphonomists played a mug's game while others knew all about the game's ambiguous rules. After Paleoindianists discovered a string of influential 1970s publications by researchers working mainly in Africa, they changed their attitude towards taphonomy. But many Paleoindianists idiosyncratically used taphonomy to create support for unusual propositions or to lend plausibility to off-beat theses such as an unexpectedly early human presence in the Americas, instead of testing hypotheses through taphonomic analysis. After the 1980s, taphonomic research has greatly advanced in allowing clear and definite interpretations of Paleoindian bone assemblages, but stubborn personalities and the tendency to "brand" certain sites continue to discourage the most rigorous skeptical inquiry that is taphonomy. The process of explaining archaeological contexts through taphonomy is a make-or-break step that must be applied to the earliest sites.

## Introduction

This is a detective story, but a rather odd one.
C. K. Brain (1981)

America's deep prehistory is a very foreign country, and clever detectives are needed to uncover how people did things then.[1] Some Paleoindianists have been able to show us through taphonomic research what the world of foragers was like in the distant past, but not all the detectives have been equally canny. Years ago C. K. Brain said something that helped me recognize how the detective work can bedevil our imperfect minds.

I first met C. K. Brain in 1982 when I went to Africa to find field sites and agreeable governments willing to issue research permits for a planned study of elephant bones. My research plans came together in Zimbabwe instead of South Africa, so I did not see Bob Brain again until 1984, when we met in Carson City, Nevada, at a conference about animal-bone modifications. One day Dr. Brain, while eating lunch with Kate Scott and me at an A&W Root Beer stand across the street from the soon-to-be-bankrupt casino where the conference took place, warned us that "Taphonomy is a mug's game." Here's what I think he meant: Too many taphonomists were duping themselves into serving causes instead of seeking the more complex truths about site-formation processes. In fact, maybe ultimate complex truths were unattainable, which is an insight reached by other conscientious taphonomists. In spite of years of study of all the variables, no single predictor could tell us how to explain every example of bone settling, bone survival, bone subtraction from assemblages, bone marking by human butchers and feeding carnivores, or the other end-effects of taphonomic processes.

In this paper, I offer a personal view of the last three decades of American Paleoindian research, some of which was carried out by taphonomists who may never have realized they were playing a mug's game, and some of which was done by *au courant* researchers who knew all about the game's ambiguous rules. Readers will soon understand that because I was a participant in this recent period of history, my viewpoint has affected how I inter-

[1] Apologies to readers of L. P. Hartley's 1953 novel *The Go-Between* (Hartley, 2002).

pret the scientific activity and opinions of the time. But having convictions is an unavoidable part of the practice of science.

## First Flush Of Ambition

I take it you are in the first flush of ambition, and just beginning to make yourself disagreeable. You think (do you not?) that you have only to state a reasonable case, and people must listen to reason and act upon at once. It is just this conviction that makes you so unpleasant.

F. M. Cornford (1908)

The starting point of my review is the middle of the 1970s, when I was a graduate student learning about the first native cultures in eastern North America. The themes that occupied my Paleoindianist colleagues and teachers were primarily (1) the timing of the earliest human colonization and (2) the technology and subsistence of the first peoples. Both themes were ripe for the application of taphonomic principles.

Contemporary developments in hominid paleontology were barely given notice in the papers and publications written by Paleoindianists of those days. Yet by 1976, thanks to the Wenner-Gren Conference and the resulting *Fossils in the Making* book, which "crystallize[d] the new science of taphonomy and [helped] to chart its future course" (Brain, 1981: ix; Brain, 1980:73), Paleoindianists discovered a string of influential publications by Andrew Hill, Kay Behrensmeyer, Pat Shipman, and others. To me in my Paleoindian program, which in those days had the status of a déclassé suburb far from the bustling metropolis of hominid evolutionary studies, the taphonomists seemed to be training through one of two axes—Harvard's or rival Berkeley's. Without mentoring or training and without peers sharing interests within my own Paleoindian program, I ingenuously entered the arena with a few observational papers about carnivore-gnawing, based on a series of studies of Pleistocene fossil bone collections, a zoo-animal-feeding study, and some actualistic work in American wildlands. These papers were often given harsh treatment by Paleoindianist referees and paleontologists trying to prevent them from being published. I look back on the taphonomists of those days as avatars of the maverick Hollywood detectives whose mulish supervisors stall their homicide investigations.

The 1970s saw the appearance of several of Brain's taphonomy papers preceding the *Hunters or the Hunted* book, as well as Rob Bonnichsen's *Pleistocene Bone Technology* monograph, in which he set out his propositions about bone-flaking in prehistory and its relationship to the pre-Clovis stage of America's colonization. In 1978, tying in neatly with Bonnichsen's proposals, the Owl Cave site in Idaho was described by a paleontologist and geologist (Miller and Dort, 1978) as an example of how prehistoric people deliberately flaked mammoth bones into tools. While the process of flaking bone was replicated and thus plausible, it still needed actualistic testing to show whether noncultural processes could be eliminated as potential causes of the same results.

Also in the 1970s, thanks to Professor C. Vance Haynes (who is not related to me) and then-graduate student Jeff Saunders, both at the University of Arizona, Paleoindianists could clearly see how relevant age-profiling can be in explaining the possible agencies that contributed animal bones to fossil sites. Saunders (1977) thought the mass mammoth site of Lehner, AZ, contained the remains of a herd of related animals killed together, because the age distribution was so similar to what is seen in modern African elephant herds. When I went to Africa a few years later to study elephant biology and behavior, I was determined to see how age profiles could vary in elephant bone assemblages when the causes of mortality varied. After a few years of fieldwork in Africa, I suggested (Haynes, 1987) that mammoth age-profiles in Clovis sites such as Lehner might reflect climatic stresses on the populations rather than mass-hunting by humans, an interpretation at odds with Saunders' (1980).[2]

A hinge point in Paleoindianists' changing attitude towards taphonomy developed during the key years 1978-1985. Within this span, perhaps 1981 was most critical: C. K. Brain's *The Hunters or the Hunted* book appeared in the same year as Lewis Binford's *Bones* book, Pat Shipman's book *Life History of a Fossil*, a *Science* paper by Stanford and colleagues about an elephant they butchered to create bone flakes—thus in their opinion proving that pre-Clovis people flaked mammoth bones in North America—and the hiring of A. K. Behrensmeyer at the Smithsonian Institution's Natural History Museum, which slimly avoided a federal hiring-freeze (Harrison, 1981). My own doctoral dissertation was completed that same year, to far less effect than the other publications. Other products of the year were the first announcement of a Clovis-associated mastodont killsite in eastern North America (Graham et al., 1981), and the first description of the Lamb Spring site in Colorado (Stanford, Wedel, and Scott, 1981) that had yielded a component of stacked and flaked mammoth bones. I was co-author of a second Lamb Springs paper the next year (Rancier et al., 1982), which added fuel to the debate about bone-flaking and the possible existence of a pre-Clovis human presence in North America.

The main use of taphonomy in Paleoindian publications—the word taphonomy being loosely and implicitly defined as bone-modification analysis—was to serve a very narrow cause, namely finding support for unusual propositions or for lending plausibility to off-beat theses such as the evidence for an unexpectedly early human presence in the Americas, based on flaked bone speci-

[2] It is worth noting that one later study of some mass mammoth Clovis sites now may indicate that the dead animals came from different source ranges and were not all related family members (Hoppe 2004).

mens. These were not really examples of detective work, as Brain had called his own taphonomic studies—they were one-sided editorials with taphonomy added to increase the plausibility. Many papers with a taphonomic bent seemed to be polemical rather than truth-seeking, intended to advance opinions without addressing the strengths and weaknesses of competing hypotheses.

Some developments in Paleoindian taphonomic work were considered pivotal at the time, but in fact they might have deflected the flow of research, like the investigative work of an obsessed but blindered detective. An example is small-scale elephant-butchering, which nearly became a cottage industry in actualistic research. The refereed *Science* paper by Stanford, Morlan, and Bonnichsen (1981) summarizing the Ginsberg experiment (also see Callahan, 1994) could not elevate the elephant-butchering projects (for example, Matyukhin, 1984; Rippeteau, 1979) from makeshift or impromptu happenings to replicative science. None of the experiments was ever written up adequately. These events achieved an almost folkloric presence as background in some of the ensuing literature about butchering marks to be found on megamammal elements and the expectable ways that prehistoric people must have sectioned huge prey carcasses. Yet these and other individual bone-modifying experiments were too easily transformable into lawlike generalizations about human behavior (as in Bonnichsen,, 1979). All too often, as shown in these examples, and following the precocious post-processual trend of the times, taphonomic studies involved a novel but reckless form of induction. Referring to the observable traces created by an individual's unmatched acts (such as Bonnichsen's bone-breaking or other archaeologists' attempts to butcher carcasses and produce cutmarks), these studies then proposed universalities about butchering practices in the past.

Of course not all work was driven by scholars trying to advance unyielding points of view. A very interesting and less slanted literature was also being produced in this period. For example, Dinah Crader in (1983) and (1984) described Bisa elephant butchering—very pertinent for Paleoindianists trying to understand mammoth-butchering—and the resulting traces of carcass sectioning and bone-processing created by people having a real economic interest in the meat and bones. When Hill (1976, 1984) described the testing of competing hypotheses about fossil animal-bone accumulations, he showed the process to be extremely challenging and requiring a rigor not seen often enough in the scientific literature. A flow was not yet underway of taphonomic writings closely relevant to Paleoindian studies, but nevertheless the 1970s and early 1980s did see a turning point in awareness of how such studies could relate to emerging interpretations.

One major emphasis in that decade was on skeletal disarticulation sequences in small and large mammals when different agencies affected the carcasses. Hill (1979) devised a statistical technique for describing the African topi sequence and modeled how the elements become scattered. Hill and Behrensmeyer (1984) soon found the disarticulation sequences to be consistent in a wide range of African mammals. A year later Hill and Behrensmeyer (1985) looked at the sequence of American bison disarticulation at the Olsen-Chubbock late Paleoindian site, and suggested that a few differences from the natural sequences they had recorded for African mammal skeletons might reflect human actions at the archaeological site. Overall, however, most human and nonhuman processes were recognized as producing very similar sequences of separation.

Influenced by this desire to know how animal skeletons are altered by different agencies in nature, and following the lead of both Brain and P. R. K. Richardson (1980), I sought data to produce flowcharts that combined information on how the grey wolf in North America damaged skeletal elements of American bison, moose, and deer and how the bones naturally disarticulated (Haynes, 1980, 1982). As far as I can tell these papers have very rarely been referenced by taphonomists and archaeologists.

## Bad News

In some ways taphonomy is 'bad news' to archaeology. It shows us just how much we don't know about the archaeological record...

Sarah Colley (1990)

In 1984, a bone-modification conference was held in Carson City, Nevada, hosted by the Nevada State Museum and partly funded by Rob Bonnichsen's Center for the Study of Early Man, then located at the University of Maine, Orono. The conference scheduled the actualistic and taphonomic papers early in the program, to be followed by several half-days of presentations by people who, it appeared to me, had paid inadequate attention to the taphonomists. Even in the book that eventually resulted from the conference (Bonnichsen and Sorg 1989) some authors displayed this same selective forgetfulness. For example, early in the book (as at the conference) Oliver (1989) discussed bones showing noncultural impact marks, surface incisions, and other effects of natural processes, as did Behrensmeyer, Gordon, and Yanagi (1989), yet in the book's later articles similarly modified specimens from other sites were said to be affected by human actions only, and figures such as of tooth-marking on bones were interpreted as cultural in origin. When I mentioned the lapses in a journal review of the book, I was made to realize that I was facing lasting hostility from participants and interested parties who had staked their careers on the interpretations. Most Paleoindian researchers intended to be very selective in trying to apply the taphonomists' results for many more years to come.

At another conference held two years later at Lubbock Lake, Texas, I overheard a remark from a prominent archaeologist who advocated mammoth-bone-flaking as proof of a pre-Clovis presence in the Americas. When I approached the podium he said "Here comes a taphonomist to tell us everything we say is wrong." I realized this was the prevailing attitude among the other participants as well—taphonomy was spoiling their stories. I was nettled at the time, but now I can point out that much of what was said really *was* dead wrong.

I remained exasperated through the mid-1980s, especially after unsuccessfully applying for NSF money to support a project to study elephant-bone-flaking. This happened in 1984.[3] An NSF proposal I submitted was returned unfunded, along with comments from anonymous reviewers who did not understand the specialized vocabulary of taphonomy or who speculated that maybe I was making up some of my results. Taphonomic studies were indeed bad news for many of these people, whose neat stories about mammoths and whose attempts to reinvent Paleoindian culture-histories were weakening.

## Unhelpful And Unpleasant

...nothing is ever done until every one is convinced that it ought to be done, and has been convinced for so long that it is now time to do something else.

F. M. Cornford (1908)

In 1986, Paleoindian archaeologists George Frison and Larry Todd published a very short but influential book about a mammoth killsite at Colby, Wyoming. In one chapter Frison and Todd described an experiment with elephant bones, reminiscent of Voorhies' (1969) and Hanson's (1980) experiments to measure the extent to which moving water can displace different elements of mammal skeletons. Frison was a true leader in applying taphonomy to Paleoindian studies. He had come from a ranching and hunting background, and he confided to students and friends that he thought 99% of archaeologists didn't know near enough about animal behavior to interpret human hunting behavior. He made sure his students began learning about the animals that prehistoric people hunted. He entered the taphonomic business with a flourish, becoming an ever-present voice in Paleoindian research, encouraging students and colleagues to devise methods for determining how much the animal bones in High Plains sites had been affected by human versus nonhuman processes.

Frison was (and still is) right about how little archaeologists know of animal biology and behavior, as seen in much Paleoindianist literature. Astonishingly, some archaeologists still believe that prehistoric people butchered large mammal carcasses any way they wanted to, depending on ethnic or cultural preferences presumably, without regard for efficiency or basic anatomical limitations; an example is Storck and Holland (2003: 299, 300) who suggest that even "illogical and unrealistically extravagant" proboscidean-butchering interpretations are acceptable, and that criticisms of such outlandish stories are merely "culturally relative" judgments and therefore not valid. Frison had learned from personal experience and from his intellectual control of the ethnographic literature that human butchering practices were rational, patterned, and understandable.

Frison is an example of a Paleoindianist who wisely and early paid attention to the taphonomists, even when they spoiled some of his stories. In earlier publications by Great Plains archaeologists (such as Frison, 1974) writing about prehistoric bison sites, cultural causes frequently had been assigned to bone modifications that were more likely carnivore-caused. But Frison's experiments and his unusual curiosity opened his eyes and those of his students to the varied end effects of noncultural processes. Frison made two trips to Zimbabwe to throw spears at culled elephant carcasses and take part in large-scale elephant-butchering at the time I was doing my fieldwork there.

At one point in his writings, George Frison tried to introduce a word—"taphonomics"—which could have given a convenient name to the chapter every book should contain discussing the origins of fossil bone modifications—but it was never adopted by other authors. Frison's experiments in bone-floating and spear-throwing produced a limited set of unreplicated data, but the work is still valuable. Thanks to Frison, we know something important about elephant-bone buoyancy, spear penetration, and especially the expected lack of cutmarks on elephant bones when they are butchered by experts.

Larry Todd also continued the taphonomic work by painstakingly documenting patterns in bison bone damage and element attrition, thus helping to clarify the prehistoric cultural and noncultural processes at work on the American High Plains (Todd, 1987; Todd and Rapson, 1988, 1999).

In 1986, Johnson and Shipman published a short description of a study that many Paleoindianist readers hoped would provide a guide for distinguishing incised bone surfaces cut by butchers from specimens cut by noncultural agents. This study was part of a family of other valuable SEM studies of the time (such as Shipman and Rose, 1983a, b, 1984). Paleoindianists made use of these works, but eventually began wondering how many hours of searching under the microscope or how many marks were examined to find the clearest matches between fossil marks and experimentally produced cuts illustrated in these guides. Paleoindianists also wondered about how the documented cutmarks had been created (were they deliberate attempts to mark bone, or were they by-products of economy-based butchering?). More discussion was needed about the range of variability in both true cuts and the fossil marks. Only the best matches and sharpest differences were featured in the widely used guides, so the ambiguity was downplayed. Paleoindianists learned to be a little more cautious over the next

[3] It was not long afterwards that I began writing sometimes biting book reviews for the journals *American Antiquity* and *North American Archaeologist*.

decade when they found that no taphonomic guide was infallible.

The 1980s and 1990s were notable for the increasing volume of publications about essential taphonomic issues. By the end of the 1980s, taphonomic research had greatly advanced in allowing clear and definite interpretations of Paleoindian bone assemblages. By then, instead of anecdotes and conjecture, we had on hand multiple empirically documented records about bone representation at different kinds of sites, bone subtraction due to scavenging carnivores, and so forth. It had begun to seem that archaeologists and paleontologists regularly applied these studies in their own research and that the research to that point had made a start in defining (even if not clearing up) the important ambiguity in fossil bone assemblages. The overlooked classics of the older literature (such as Weigelt, 1989 [original 1927 in German]) were revived in print as interest exploded in taphonomy. Solid and well reported actualistic studies inspired some Old World researchers to go to war with adversaries over the deeper meanings of early hominid bone assemblages (such as Lewis Binford and Rob Blumenschine versus Henry Bunn over Plio-Pleistocene hominid scavenging behavior [Binford, 1986; Blumenschine, 1986; Bunn and Kroll, 1986], or Curtis Marean versus Mary Stiner over Neanderthal diet [Marean, 1998; Stiner, 1994]). Cautionary tales stressed the remaining ambiguities—such as equifinality in bone survival or element distribution or surface-marking—but these were often roughly treated by critics: see the probable career-stalling responses to Rob Gargett's (1989a, b, 1999) rethinking of Neanderthal burial and Nicola Stern's (1993, 1994) cautions about assemblage structure at Olduvai Gorge. Paleoindianists also went to battle over the interpretations of assemblages such as the pre-Clovis broken bones from Old Crow, Yukon, and Lost Chicken Creek, Alaska, but these debates possessed a lower international profile because cautionary tales were often ignored or dismissed behind an authoritative sniff rather than attacked head-on with data and strong arguments.

Celebrity dissidents in Paleoindian studies seem to have developed broadly similar careerist strategies. Aggressive self-promoters rely more on a strong, thick-skinned personality able to stay the course while recounting unorthodox claims, and less on a willingness and ability to carry out adequate actualistic/taphonomic research—sustained detective work—which would uphold their questionable interpretations of the past. When asked about the finer details of their fieldwork practices or the replicability of their interpretive standards, celebrity dissidents in Paleoindian studies often may not respond, perhaps implying that they are infallible. Whenever they do claim to be responding, they mainly attack their inquisitor, which in Paleoindian studies frequently turned out to be Stuart Fiedel.[4]

Fiedel's valid querying of the Monte Verde site's ambiguities was publicly brushed aside (Dillehay et al., 1999a, b), as had been my own private questions. When I commented on a pre-publication chapter ("Zooarchaeological Remains") meant for the now well-known second volume about the site (Dillehay, 1997:661-758), the author wrote me a testy letter rejecting my questioning, and proclaimed that "after 30 years [of experience as an archaeologist] I can recognize an artifact." A peremptory dismissal of a taphonomist's caution is a common reaction, but an archaeologist's proclamation of personal skills does not obviate the need to test the reliability of interpretations.

Monte Verde's huge second volume (Dillehay, 1997) contains an impressive amount of data and interpretations, but it is more imperfect than the first volume in many ways (see Fiedel, 1999, 2000). The book is frustrating because of the frequent impossibility of figuring out where certain key items were found (such as the seemingly unmapped handful of indisputable lithic artifacts) or ambiguity about the specific items that were radiometrically dated, although it is a very large book full of outstanding information. The site yielded hundreds of minimally modified stream-rounded stones, about 400 animal bones or fragments looking like noncultural debris, diverse plant remains including wood fragments with a decidedly wave-washed look, "structures" made of what appear to be strewn wood pieces (for example, Dillehay, 1997: 775), and possible mastodont skin fragments. Overlying the materials interpreted as cultural is a peat layer that preserved the organic remains. The site is fascinating and unique, but "bizarre" would also be an appropriate word.

Paleoindianists are tough fighters when it comes to changing other people's paradigms and defending their own. Yet too often the sampling of taphonomic literature offered to support one set of interpretations is overly selective. The Monte Verde report contained a limited literature review as well as descriptions (Dillehay, 1997:695-703) of neotaphonomic experiments and observations to uphold some of the interpretations of broken bones as being culturally produced. But such experiments must be coldly evaluated, because they can be narrow and faulty if they consist of limited bone set-ups to test possibilities for bone movement and noncultural modifications.

Roosevelt (2000) observed that the discontinuous strata at Monte Verde were complex and contained possible contaminants along with the questionable "artifacts." Thus not only are the cultural materials doubted by a number of New World archaeologists, but Monte Verde's dating procedures themselves are now being questioned. Monte Verde is not the only possibly pre-Clovis site with lingering contextual problems. The Meadowcroft

[4] Unpleasant disagreements between archaeologists with different interpretations has a long history. For example, when J. L. Lorenzo claimed Irwin-Williams's field crew had fraudulently planted artifacts at Valsequillo, Mexico, in the 1960s, Irwin-Williams (n.d.:12) accused Lorenzo of "distorted personal animosity and irrational inability to change an opinion."

rockshelter site in Pennsylvania has been known for decades as a stratified site containing unquestioned lithic artifacts associated with pre-Clovis radiometric dates. Excavator James Adovasio is the site's long-suffering archaeologist who has had to face down the skeptics for nearly 30 years. He spent the 1970s and 1980s becoming a well-known dissident in Paleoindian studies, due to his advocacy of Meadowcroft's early dates, and over time his self-defense has been fine-tuned into near-churlish responses to critics (see Adovasio and Page, 2002).

The problems in this case are centered not around determining if modified bones are genuine artifacts, but around the dating itself. Two published reviews (Flannery, 2003; Roosevelt, 2000) of Adovasio's co-authored book (Adovasio and Page, 2002) about the site and its surrounding controversies alert readers to the fact that naturally occurring coal might have contaminated some of the dated materials at Meadowcroft. The process of analyzing the samples used in radiometric dating is part of a make-or-break contextual study that hasn't been done in either the Meadowcroft or Monte Verde cases. Yet while directly dating the lowermost Meadowcroft and Monte Verde organics, such as plant fiber, wood, or nutshells, might serve to test the possible contamination of sediments with dead carbon, these materials also must be proven to have cultural associations, and that requires much more taphonomic detective work.

The prevailing strategy in Paleoindian debates is mainly trying to strip opponents of respect instead of objectively answering the criticisms about dubious interpretations. Also favored is accusing critics of misunderstanding or distorting one's views, although distortions and misrepresentations are rhetorical sins that all parties commit. In the words of C. Hitchens (2004:28), writing in reply to a similar kind of response to his criticisms, "When a man thinks any stick will do, he tends to pick up a boomerang."

## Brandscapes

> This, like other species of patriotism, consists in a sincere belief that the institution to which you belong is better than an institution to which other people belong.
>
> F. M. Cornford (1908)

Such personality-driven debates in place of collaborative detective work possibly arise from the unconscious process of archaeological "brandscaping," a term I borrow from modern marketing and cultural studies. This word usually means the marketing of an object by creating special spaces, designs, and associated products that consumers can identify with it. The word as I use it here refers to the transformation of what should be merely an archaeological interpretation into a career-centered cause. When an archaeological interpretation/scenario/discovery is introduced by one or a few sources, it may strike a chord with archaeologists and become a widespread belief instead of a testable (and in-need-of-testing) possibility, even though we all know that archaeological stories are unproven. Eventually, the belief comes to be shared by people who know little to nothing of the original evidence behind the interpretation. In the case of the South American site called Monte Verde, the belief that the site is reliably interpreted rests firmly in many people who may not have read the two big volumes about it. The site becomes part of a mass belief system, in which certain key concepts are always linked and firmly accepted.

Mass belief systems may not be "very deep or long lasting," but they are superpotent (Twitchell, 2003:vii). The population of believers may not share wider interests or even specific knowledge, but they understand each other because they share a branded thing, such as a Monte Verde point of view about American prehistory (*viz.*, pre-Clovis populations spread across the New World with minimal visibility and little ecological impact, speaking different languages and having different geographic origins, etc.). An attack on Monte Verde is an attack on a global brand name having a huge list of consumers. To consume the Monte Verde story is perhaps to feel part of a new cognoscenti, a special class of archaeologists, a fresh generation of prehistorians who feel entitled to believe in a site where almost everything is unique, unreplicated elsewhere, different from all other sites.

This is a safe way to consume the brand's rejection of the status quo and is thus a downstream form of iconoclasty (Twitchell, 2003), or a secure way for brand adherents to feel in the know about the past's most cryptic evidence. Monte Verde is an example of a Paleoindian brandscape—a nesting collection of ideas, attitudes, and scenarios that are identifiable and coherent, and most importantly are an ensemble. Perhaps some people have decided to inhabit the Monte Verde brandscape because it is a trend—many of them do not actually calculate its strengths, only its mass appeal. Monte Verde is an object of aggressive marketing, and now it is considered unarguable truth by many people.

## Good News

> If we knew what we were doing, it wouldn't be called research, would it?
>
> Attributed to Albert Einstein.

> Basic research is what I'm doing when I don't know what I'm doing.
>
> Attributed to Wernher Von Braun.

Although I've described examples of the biased adoption of taphonomic work in one-sided support of Paleoindian interpretive causes, nonetheless many other examples of enduring and important taphonomic studies have been done specifically for Paleoindian research. As mentioned already, Lawrence Todd in particular has quietly taken Paleoindian taphonomy to a higher level, along with a few other students who trained with George Frison or Lewis Binford. Their work has provided Paleoindianists with necessary guides to taxon-specific pattern-recognition principles, such as Burgett's (1990) study of coyote (*Canis latrans*) scavenging on bison (*Bison bison*) and elk (*Cervus canadensis*) carcasses, Matthew Hill's (2001) part-taphonomic analysis of Paleoindian diet and subsistence, or Todd's (e.g., 1983, 1987; Todd and Rapson, 1988, 1999) series of papers about quantification and precise data-recording standards. My own early publications—where I described general shapes and unquantified central frequencies of bison bones affected by a variety of noncultural processes—appear inadequate today when seen in the light of the work done by these colleagues. The detective business in Paleoindian studies hasn't always been faultless, but I think it has moved along towards maturity.

My elephant-taphonomy studies, which are ongoing to this day, are useful, I hope, but I had to self-fund much of the fieldwork after the NSF review process proved so bigoted in the mid-1980s. The experts who ignored or disliked the work in the 1980s still do, apparently, but I don't think they have read very much of it, judging from their unwillingness to cite the publications, even if only to disagree with them. To quote essayist C. Hitchens (2004:28) again: "After allowing me to shoulder my way, with many a sigh, through all [their] scurvy pages, [they] will not deign to glance in return at what I wrote."

Like scholars-in-disagreement, scholars of a later age are rarely generous towards the output of earlier generations. Many of the taphonomic publications of 25 years earlier suffer criticisms (then and now) for not addressing contemporary keynotes. Yet in spite of the sore points, many hindsight evaluations are also useful (for example, Lyman and Fox (1989) on variability in bone weathering) and do add a new, valuable dimension to the pioneering publications. I am at best a peripheral player in taphonomic dramas, but I too have learned how it can be both ego-feeding and bruising to see one's works dissected and critiqued by sharp graduate students (Graves, 2002) or colleagues (Domínguez-Rodrigo and Piqueras, 2003), but it is the price one pays for publishing in science.

More than two decades after Brain's *The Hunters or the Hunted* book, several of the 1970s-era taphonomists who set the pace for Paleoindian researchers have moved away from fundamental actualistic fieldwork. One example is Pat Shipman, now a successful science writer. Others did not so much leave the field as step laterally to be bigger-picture interpreters, a natural evolution of career trajectories. For example, Andrew Hill, who had plunged into the young and burgeoning field of taphonomy in the late 1960s, has since become a leader in hominid ecology and taxonomy. Others who were Brain's compañeros in the 1960s and 1970s taphonomic research, such as Kay Behrensmeyer (e.g., 1975), never really left the actualistic work behind but expanded upon it to show how ecosystem reconstructions can be improved through taphonomic analyses. The trend towards moving on from taphonomy has also depleted the ranks of Paleoindianists who once ran actualistic or neotaphonomic projects. In my 22 years of full-time teaching, I've had only one doctoral student in taphonomy, but I'm still doing my own taphonomic fieldwork, even if I can't interest anybody else.

The global community of taphonomic researchers has grown well beyond Paleoindianists and the specialists working in Africa. Researchers from South America and Europe in the past years have carried out their own new actualistic or taphonomic studies (for example, Mondini, 1995, 2000, 2001a,b; Wojtal, 2001; Wojtal and Sobczyk, 2003), aimed at correcting earlier studies' errors or shaping the research towards local conditions, thus keeping the field alive, to use an ironic modifier. Each new study reveals the temporal and geographical variability in taphonomic processes, thus potentially adding to Paleoindian bone assemblages a wider array of new sources of proxy information about the past.

The most encouraging trend in Paleoindian taphonomic research has been the awareness (still not fully emerged) that controversy is actually good and skepticism is even better. Archaeologist Charles Keally, comparing the nature of America's Paleoindian debates with the controversy about the nonexistence of an Early Paleolithic stage in Japan, pointed out that the Paleoindian debate has become interdisciplinary, scientific, and academic, and while criticisms are often heated, "conferences and publications purposefully include contributors from both (all?) sides" (Keally, 2001). In Keally's view, controversy is exciting and useful, most ideas are openly admitted to be only speculation, questioning is and should be common, solid scientific evidence is required [to address problems], vigorous and public debate is normal, people should be encouraged to change their minds after hearing new evidence or arguments, and scholars should enjoy having their ideas criticized. While some Paleoindianists (Adovasio and Page, 2002; Dillehay, 2000) may not appear to agree with these precepts, the current generation of taphonomists must have gotten used to them by now.

## Brain's Sway

> ...small things [can be used to] discover great [things]...better than great can discover the small.
>
> Francis Bacon (1973; orig. 1605)

Is the post-"Hunters or Hunted" period a case of Paleoindianists behaving as Feyerabend (1975) suggested they might want to do during a period of changing interpretations (anything goes—anarchy and intellectual dishonesty are acceptable and valid when exploring the unknown), or as Bourdieu (1977) theorized they usually would (careerism is as important in shaping scientific trends as any so-called objective search for truths), or as Kuhn (1962) had generalized (periods of paradigm shift are full of programmatic confusion and leadership struggles)?

The last quarter-century of Paleoindian studies had these and other socio-political processes taking place, but the model C. K. Brain had established was available as a filter to pass the three kinds of pettiness through. Brain had asked a question in the title of his major book —*The Hunters or the Hunted*?—and that willingness to query was intended to encourage readers to weigh evidence and seek answers, rather than to decide those answers in advance and to merely invent *ad hoc* models to support hardened opinions.

Practically speaking, Brain's book showed him engaged in multiple modes of taphonomic study. He had done feeding experiments with large carnivores, he had carried out actualistic studies, he had learned the patterns of fossil bone modifications, and he had carefully read the growing literature about other people's taphonomic research. He thus put together his case step-by-step.

Rubidge (2000:5) has pointed out that Brain enjoyed the day by day process of just doing science. He looked for answers creatively and often in the same ways the old fashioned naturalists did it, by allowing himself to veer off intellectually in many different directions when it seemed to be needed. He was not a project-driven careerist obsessed with achieving prominence in his field, which I think set him apart from many Paleoindianists.

He was inspired to do the taphonomic work first by his knowledge (and eventually doubts) about Raymond Dart's hunting-ape hypotheses (Brain, 1997), and second by his first-hand knowledge of Plio-Pleistocene fossils—accumulated through decades of "hard labor at Swartkrans" (Brain, 1973, 1974, 1976a). He was also inspired by the new ideas emerging from meetings with other researchers who had similar puzzles to solve. Especially catalytic was the Wenner-Gren conference of 1976 (Brain 1976b). Yet Brain was a true all-around naturalist —he worked as a geological scientist, a paleontologist, a lower-vertebrate zoologist, the director of a major natural history museum, a historian of science, a biographer of scientists—in short, he had no end to the shifting problems he wanted to address.

He patiently kept at the taphonomic work for over two decades—never expecting to solve the problems in a single field season or a single research process. The specifically taphonomic set of problems did not completely monopolize his attention from the late 1960s through the 1990s, but it came close to doing so.

He had the benefit of living on a continent where the most directly relevant taphonomic fieldwork could be done (such as seen in Hill, 1975; Maguire et al., 1980; Richardson, 1980). Meanwhile, Paleoindian taphonomists trying to work within North America faced a shortage of landscapes where they could study noncultural processes such as carcass-feeding by the same free-roaming carnivores that would have been present before the colonial era.

He designed and carried out many different and related projects—examining the effects of bone weathering (Brain, 1967b), collecting ethnographic data (Jenkins and Brain, 1967), observing patterns in humanly butchered remains (Brain, 1967a, 1969), experimentally feeding animal carcasses to carnivores (Brain, 1981), collecting animal bones from wild carnivore lairs (Brain, 1981), collecting bones from owl roosts in caves, and so on.

These projects were examples of actualism, neotaphonomy, and classical taphonomy (defined here as the laboratory interpretation of fossil bone histories). They formed the basis for his "rather odd detective story" about Plio-Pleistocene hominids in South Africa. He was comfortable carrying forward his line of reasoning one small maneuver at a time. He reviewed others' work, collected data, and spelled out his alternative interpretations with grace and tact.

Did Paleoindian taphonomists follow suit? I don't think we ever really did, but once in a while some scholars came close. My own body of work is incomplete and unbalanced compared to Brain's. Other Paleoindianists' work of the 1970s and 1980s also seems unfulfilled or provisional, but several classic references will never lose their usefulness. Yet, Paleoindian studies in general have stayed at an unfledged stage because American prehistorians often seek "brands" of interpretations instead of facing the complexities and ambiguities that a long-term commitment to taphonomy reveals.

I end this paper by acknowledging that Brain's contribution to Paleoindian research went beyond merely providing examples of taphonomic studies to emulate. To his greatest credit, he also showed us how to stalwartly present a case without alienating colleagues and friends.

## References

Adovasio, J. M., Page, J. 2002. The First Americans: In Pursuit of Archaeology's Greatest Mystery. Random House, New York.

Bacon, F. 1973 (original, 1605). The Advancement of Learning, J.M. Dent and Son, London.

Behrensmeyer, A.K. 1975. The taphonomy and paleoecology of Plio-Pleistocene vertebrate assemblages east of Lake Rudolf, Kenya. Harvard University Museum of Comparative Zoology Bulletin 146, 473-578.

Behrensmeyer, A.K., Gordon, K.D., Yanagi, G.T. 1989. Nonhuman bone modification in Miocene fossils from Pakistan. In: Bonnichsen, R., Sorg, M.H. (Eds.), Bone Modifications. Center for the Study of the First Americans, Orono (ME), pp. 99-120.

Binford, L.R. 1981. Bones: Ancient Men and Modern Myths. Academic Press, New York.

Binford, L. R. 1986. Comment on "Systematic butchery by Plio/Pleistocene hominids at Olduvai Gorge" by H.T. Bunn and E. Kroll. Current Anthropology 27, 444-446.

Blumenschine, R.J. 1986. Comment on "Systematic butchery by Plio/Pleistocene hominids at Olduvai Gorge" by H.T. Bunn and E. Kroll. Current Anthropology 27, 446.

Bonnichsen, R. 1979. Pleistocene bone technology in the Beringian Refugium. National Museum of Man [Canada] Mercury Series, Archaeological Survey of Canada Paper No. 89.

Bonnichsen, R., Sorg, M.H. (Eds.). 1989. Bone Modifications. Center for the Study of the First Americans, Orono (ME).

Bourdieu, P. 1977. Outline of a Theory of Practice. Cambridge University Press, Cambridge.

Brain, C. K. 1967a. Bone weathering and the problem of pseudo-tools. The South African Journal of Science 63, 97-99.

Brain, C. K. 1967b. Hottentot food remains and their bearing on the interpretation of fossil bone assemblages. Scientific Papers of the Namib Desert Research Station 32, 1-11.

Brain, C. K. 1969. The contribution of Namib Desert Hottentots to an understanding of australopithecine bone accumulations. Scientific Papers of the Namib Desert Research Station 39, 13-22.

Brain, C. K. 1973. Seven years' hard labour at Swartkrans. Bulletin of the Transvaal Museum 14, 5-6.

Brain, C. K. 1974. Some suggested procedures in the analysis of bone accumulations from southern African Quaternary sites. Annals of the Transvaal Museum 29, 1-8.

Brain, C. K. 1976a. Some principles in the interpretation of bone accumulations associated with man. In: Isaac, G.Ll., McCown, E.R. (Eds.), Human Origins: Louis Leakey and the East African Evidence. W. A. Benjamin, Mnelo Park (CA), pp. 97-116.

Brain, C. K. 1976b. Some criteria for the recognition of bone-collecting agencies in African caves. Paper prepared in advance for participants in Burg Wartenstein Symposium No. 69, "Taphonomy and Vertebrate Paleoecology: With Special Reference to the Late Cenozoic of Sub-Saharan Africa," July 2-11, 1976. Wenner-Gren Foundation for Anthropological Research.

Brain, C. K. 1980. Swartkrans as a case study in African cave taphonomy. Palaeontologia Africana 23, 73-74.

Brain, C. K. 1981. The Hunters or the Hunted? An Introduction to African Cave Taphonomy. University of Chicago Press, Chicago.

Brain, C. K. 1997. The contribution of Raymond Dart to the development of cave taphonomy. Palaeontologia Africana 33, 81-83.

Bunn, H.T., Kroll, E.M. 1986. Systematic butchery by Plio/Pleistocene hominids at Olduvai Gorge, Tanzania. Current Anthropology 27, 431-452.

Burgett, G. R. 1990. The Bones of the Beast: Resolving Questions of Faunal Asemblage Formation Processes Through Actualistic Research. Ph.D. Dissertation, University of New Mexico.

Callahan, E. 1994. A mammoth undertaking. Bulletin of Primitive Technology 1, 23-39.

Colley, S.M. 1990. Humans as taphonomic agents. In: Solomon, S., Davidson, I., Watson, D. (Eds.), Problem Solving in Taphonomy: Archaeological and Palaeontological Studies From Europe, Africa, and Oceania." Tempus 2, 50-64.

Cornford, F.M. 1908. Microcosmographia Academica, Being a Guide For the Young Academic Politician. Bowes and Bowes Publishers Ltd., Cambridge.

Crader, D. C. 1983. Recent single-carcass bone scatters and the problem of "butchery" sites in the archaeological record. In: Clutton-Brock, J., Grigson, C. (Eds.), Animals and Archaeology. Volume 1: Hunters and Their Prey. Oxford, British Archaeological Reports International Series No. 163, pp. 107-141.

Crader, D. C. 1984. Elephant butchery techniques of the Bisa of the Luangwa Valley, Zambia. Paper presented at the 49th Annual Meeting of the Society for American Archaeology, 11-14 April, Portland, Oregon.

Dillehay, T.D. 2000. The Settlement of the Americas: A New Prehistory. Basic Books, New York.

Dillehay, T. D. 1997. Monte Verde: A Late Pleistocene Settlement in Chile, Volume 2: The Archaeological Context and Interpretations. Smithsonian Institution Press, Washington, D.C.

Dillehay, T.D., Collins, M.B., Pino, M., Rossen, J., Adovasio, J., Ocampo, C., Navarro, X., Rivas, P., Pollack, D., Henderson, A.G., Saavedra, J., Sanzana, P., Shipman, P., Kay, M., Munoz, G., Karathanasis, A., Ugent, D., Cibull, M., Geisler, M.. 1999a. On Monte Verde: Fiedel's confusions and misrepresentations. Accessed November 2000 on website http://www.uky.edu/Projects/MonteVerde/monteverde.pdf

Dillehay, T.D., Pino, M., Rossen, J., Ocampo, C., Rivas, P., Pollack, D., Henderson, G. 1999b. Reply to Fiedel, Part I. Scientific American Discovering Archaeology [1(6)] Special Report: Monte Verde Revisited, pp. 12-14.

Domínguez-Rodrigo, M., Piqueras, A. 2003. The use of tooth pits to identify carnivore taxa in tooth-marked archaeofaunas and their relevance to reconstruct hominid carcass processing behaviours. Journal of Archaeological Science 30, 1385-1391.

Feyerabend, P. 1975. Against Method: Outline of an Anarchistic Theory of Knowledge. NLB, London.

Fiedel, S.J. 1999. Artifact provenience at Monte Verde: Confusion and contradictions. Scientific American Discovering Archaeology [1(6)] Special Report: Monte Verde Revisited, pp. 1-14.

Fiedel, S. J. 2000. Response to Dillehay. Published electronically on "About Archaeology" website. Accessed December 2000 on website http://archaeology.about.com/science/archaeology/library/weekly/aa070200c.htm

Flannery, T. 2003. Who came first? [Review of "The First Americans" by J. M. Adovasio with J. Page, and "America Before the European Invasions" by A. B. Kehoe] The New York Review of Books 50(10), 51-53.

Frison, G.C. 1974. The Casper Site: A Hell Gap Bison Kill on the High Plains. Academic Press, New York.

Frison, G.C., Todd, L.C. 1986. The Colby Mammoth Site: Taphonomy and Archaeology of a Clovis Kill in Northern Wyoming. University of New Mexico Press, Albuquerque.

Gargett, R.H. 1989a. Grave shortcomings: The evidence for Neandertal burial. Current Anthropology 30, 157-190.

Gargett, R.H. 1989b. Reply to comments on "Grave shortcomings: The evidence for Neandertal burial". Current Anthropology 30, 326-329.

Gargett, R.H. 1999. Middle Palaeolithic burial is not a dead issue: the view from Qafzeh, Saint-Césaire, Kebara, Amud and Dederiyeh. Journal of Human Evolution 37, 27-90.

Graham, R.W., Haynes, C.V., Johnson, D.L., Kay, M.. 1981. Kimmswick: A Clovis-mastodon association in eastern Missouri. Science 213, 1115-1117.

Graves, A. 2002. Unpublished MA Dissertation, University of Wyoming.

Hanson, R.B. 1980. Fluvial taphonomic processes: models and experiments. In: Behrensmeyer A.K., Hill, A. (Eds.), Fossils in the Making: Vertebrate Taphonomy and Paleoecology. University of Chicago Press, Chicago, pp. 156-181.

Harrison, J. 1981. National Museum of Natural History [News From Members]. Society of Vertebrate Paleontology News Bulletin 122, 16.

Hartley, L. P. 2002 (original 1953). The Go-Between. New York Review of Books, New York.

Haynes, G. 1980. Prey bones and predators: potential ecologic information from analyses of bone sites. Ossa 7, 75-97.

Haynes, G. 1982. Utilization and skeletal disturbances of North American prey carcasses. Arctic 35, 266-281.

Haynes, G. 1987. Proboscidean die-offs and die-outs: Age profiles in fossil collections. Journal of Archaeological Science 14, 659-668.

Hill, A. 1975. Taphonomy of Contemporary and Late Cenozoic East African Vertebrates. Ph.D. Dissertation, University of London.

Hill, A. 1976. On carnivore and weathering damage to bone. Current Anthropology 17, 335-336.

Hill, A. 1979. Disarticulation and scattering of mammal skeletons. Paleobiology 5, 261-274.

Hill, A. 1984. Hyaenas and hominids: Taphonomy and hypothesis testing. In: Foley, R. (Ed.), Hominid Evolution and Community Ecology. Academic Press, London, pp. 111-128.

Hill, A., Behrensmeyer, A.K.. 1984. Disarticulation patterns of some modern East African mammals. Paleobiology 10, 366-376.

Hill, A., Behrensmeyer, A.K.. 1985. Natural disarticulation and bison butchery. American Antiquity 50, 141-145.

Hill, M.G. 2001. Paleoindian Diet and Subsistence Behavior on the Northwestern Great Plains of North America. Ph.D. Dissertation, University of Wisconsin-Madison.

Hitchens, C. 2004. [Reply to a letter to the editor]. The Atlantic Monthly 293 (1), 27-28.

Hoppe, K. 2004. Late Pleistocene mammoth herd structure, migration patterns, and Clovis hunting strategies inferred from isotope analyses of multiple death assemblages. Paleobiology 30, 129-145.

Irwin-Williams, C. n.d. Irwin-William's Rejoinder to Lorenzo. (Typescript, photocopy in possession of author).

Jenkins, T., Brain, C.K. 1967. The peoples of the Lower Kuiseb Valley, South West Africa. Scientific Papers of the Namib Desert Research Station 35, 1-24.

Johnson, E., Shipman, P. 1986. Scanning electron microscope studies of bone modification. Current Research in the Pleistocene 3, 47-48.

Keally, C.T. 2001. America's Clovis vs pre-Clovis controversy and Japan's Early Palaeolithic controversy: A comparison in approaches. Accessed January 2004 on website <http://www.tnet.ne.jp/~keally/Hoax/clovis.html>

Kuhn, T. 1962. The Structure of Scientific Revolutions. University of Chicago Press, Chicago.

Lyman, R. L., Fox. G.L. 1989. A critical evaluation of bone weathering as an indication of bone assemblage formation. Journal of Archaeological Science 16, 293-317.

Maguire, J. M., Pemberton, D., Collett, M.H.. 1980. The Makapansgat Limeworks grey brecccia: Hominids, hyaenas, hystricids or hillwash? Palaeontologia Africana 23, 75-98.

Marean, C.W. 1998. A critique of the evidence for scavenging by Neanderthals and early modern humans: New data from Kobeh Cave (Zagros Mountains, Iran) and Die Kelders Cave 1 Layer 10 (South Africa). Journal of Human Evolution 35, 111-136.

Matyukhin, A. E. 1984. Opyty po razdelke tush krupnykh zhivotnykh orudimi paleoliticheskogo oblika. Sovetskaya Arkheologiya 4, 5-24.

Miller, S.J., Wakefield, D. 1978. Early man at Owl Cave: Current investigations at the Wasden site, eastern Snake River Plain, Idaho. In: Bryan, A.L. (Ed.), Early Man in America From a Circum-Pacific Perspective. University of Alberta Department of Anthropology Occasional Paper No. 1, pp. 129-139

Mondini, M. 1995. Artiodactyl prey transport by foxes in Puna rock shelters. Current Anthropology 36, 520-524.

Mondini, M. 2000. Tafonomia de abrigos rocosos de la Puna: Formación de conjuntos escatológicos por zorros y sus implicaciones arqueológicas. Archaeofauna 9, 151-164.

Mondini, M. 2001a. Taphonomic action of foxes in Puna rock shelters: A case study in Antofagasta de la Sierra (Province of Catamarca, Argentina). In: Kuznar, N.L.A. (Ed.), Ethnoarchaeology of Andean South America: Contributions to Archaeological Method and Theory. International Monographs in Prehistory, Ann Arbor, pp. 266-296.

Mondini, M. 2001b. Carnivore taphonomy in early archaeological sites in the southern Puna. Paper prepared for the 66th Annual Meeting of the Society for American Archaeology, New Orleans, Louisiana.

Oliver, J.S. 1989. Analogues and site context: Bone damages from Shield Trap Cave (24CB91), Carbon County, Montana, U.S.A. In: Bonnichsen R., Sorg, M.H. (Eds.), Bone Modification. Center for the Study of the First Americans, Orono (ME), pp. 73-98.

Rancier, J., Haynes, G., Stanford, D. 1982. 1981 excavations at Lamb Spring. Southwestern Lore 8, 1-17.

Richardson, P. R. K. 1980. Carnivore damage to antelope bones and its archaeological implications. Palaeontologia Africana 23, 109-125.

Rippeteau, B. (Ed.). 1979. Megafauna Punchers' Review 1.

Roosevelt, A.C. 2000. Who's on first? There's no end to the controversy over when and how humans populated the New World (review of "The Settlement of the Americas" by T. Dillehay and "Bones, Boats, and Bison" by E. J. Dixon). Natural History 109(6), 76-79.

Rubidge, B.S. 2000. Charles Kimberlin (Bob) Brain – A tribute. Palaeontologia Africana 36, 1-9.

Saunders, J.J. 1977. Lehner Ranch revisited. In: Johnson, E. (Ed.), Paleoindian Lifeways. The Museum Journal 17, West Texas Museum Association, pp. 48-64.

Saunders, J.J. 1980. Lehner Ranch revisited. In: Johnson, E. (Ed.), Paleoindian Lifeways. The Museum Journal 17, West Texas Museum Association, pp. 48-64.

Shipman, P. 1981. Life History of a Fossil: An Introduction to Taphonomy and Paleoecology. Harvard University Press, Cambridge.

Shipman, P., Rose, J. 1983a. Early hominid hunting, butchering, and carcass-processing behaviors: Approaches to the fossil record. Journal of Anthropological Archaeology 2, 57-98.

Shipman, P., Rose, J. 1983b. Evidence of butchery and hominid activities at Torralba and Ambrona: An evaluation using microscopic techniques. Journal of Archaeological Science 10, 465-474.

Shipman, P., Rose, J. 1984. Cutmark mimics on modern fossil bovid bones. Current Anthropology 25, 116-117.

Stanford, D., Bonnichsen, R., Morlan, R.E. 1981. The Ginsberg experiment: Modern and prehistoric evidence of a bone flaking technology. Science 212, 438-440.

Stanford, D., Wedel, W.R., Scott, G.R. 1981. Archaeological investigations of the Lamb Spring site. Southwestern Lore 47, 14-27.

Stern, N. 1993. The structure of the Lower Pleistocene archaeological record: A case study from the Koobi Fora Formation. Current Anthropology 34, 201-225.

Stern, N. 1994. The implications of time-averaging for reconstructing the land-use patterns of early tool-using hominids. Journal of Human Evolution 27, 89-105.

Stiner, M.C. 1994. Honor Among Thieves: A Zooarchaeological Study of Neandertal Ecology. Princeton University Press, Princeton.

Storck, P.L., Holland, J.D. 2003. From text to context: Hiscock in the Paleoindian world. In: Laub, R.S. (Ed.), The Hiscock Site: Late Pleistocene and Holocene Paleoecology and Archaeology of Western New York State. Bulletin of the Buffalo Society of Natural Sciences 37, pp. 281-300.

Todd, L.C. 1983. The Horner Site: Taphonomy of an Early Holocene Bison Bonebed. Ph.D. Dissertation, University of New Mexico.

Todd, L.C. 1987. Analysis of kill-butchery bonebeds and interpretation of Paleoindian hunting. In: Nitecki, M. H., Nitecki, D.V. (Eds.), The Evolution of Human Hunting. Plenum Press, New York, pp. 105-110.

Todd, L.C., Rapson, D.J. 1988. Long bone fragmentation and interpretation of faunal assemblages: approaches to comparative analysis. Journal of Archaeological Science 15, 307-325.

Todd, L.C., Rapson, D.J. 1999. Formational analysis of bison bonebeds and interpretation of Paleoindian subsistence. In: Brugal, J-Ph., David, F., Enloe, G., Jaubert, J. (Eds.), Le Bison: Gibier et Moyen de Subsistence des Hommes du Paléolithique aux Paléoindiens des Grandes Plaines (Actes du colloque international, Toulouse 1995). Éditions APDCA, Antibes, pp. 479-499.

Twitchell, J.B. 2003. Living It Up: America's Love Affair with Luxury. Simon and Schuster, New York.

Voorhies, M. 1969. Taphonomy and population dynamics of an Early Pliocene vertebrate fauna, Knox County, Nebraska. University of Wyoming Contributions to Geology Special Paper No. 1.

Weigelt, J. 1989 (original 1927 in German). Recent Vertebrate Carcasses and Their Paleobiological Implications. University of Chicago Press, Chicago.

Wojtal, P. 2001. The woolly mammoth (*Mammuthus primigenius*) remains from the Upper Palaeolithic site Kraków Spadzista Street (B). In: Cavarretta, G., Gioia, P., Mussi, M., Palombo M.R. (Eds.) Le Terra Degli Elefanti—The World Of Elephants: Proceedings Of The 1st International Congress. Rome, CNR, Rome, pp. 367-372.

Wojtal, P., Sobczyk, P. 2003. Taphonomy of the Gravettian site Kraków Spadzista Street (B). In: Reumer, J.W.F., De Vos, J., Mol, D. (Eds.), Advances in Mammoth Research (Proceedings of the Second International Mammoth Conference, Rotterdam, May 16-20 1999). Deinsea 9, 557-561.

# CHAPTER 3

# THE DESTRUCTION OF SKELETAL ELEMENTS BY CARNIVORES: THE GROWTH OF A GENERAL MODEL FOR SKELETAL ELEMENT DESTRUCTION AND SURVIVAL IN ZOOARCHAEOLOGICAL ASSEMBLAGES

NAOMI CLEGHORN AND CURTIS W. MAREAN

## ABSTRACT

In the 1960s, Brain published on a series of taphonomic studies in which he observed the destruction of goat bones by pastoralists and domestic dogs. Those studies were notable and novel for a variety of reasons: 1) the attempt to control for complex parameters through the use of what we now recognize as experimental and naturalistic actualism, 2) documentation of the destructive impact on skeletal element abundance by secondary carnivore consumers, and 3) the attempt to understand the mechanical aspects of this process, and thus establish the foundation for justifiable uniformitarianism. This work set the stage for a proliferation of research, and today the differential destruction of bone by secondary carnivore consumers is considered a significant, perhaps the most important, determinant of zooarchaeological patterning. This process selectively removes less dense portions of bones (the articular ends, in the case of long bones), and therefore demands a methodological shift away from the easily identified articular fragments to the more challenging shaft portions. Carnivore ravaging also destroys greasy and less dense elements such as axial bones disproportionately, resulting in different survival potentials between elements. This paper reviews the long accrual of knowledge initiated by Brain, evaluates what is known and unknown, re-examines the relationship between mechanical properties (density) and skeletal element survival, develops a general model of archaeological bone survival, and concludes with a methodological roadmap for zooarchaeology's future studies of skeletal element abundance.

## INTRODUCTION

It is now widely recognized that taphonomic analysis founded on actualistic research is fundamental to all zooarchaelogical interpretation. Two books published in 1981—Brain's *The Hunters or the Hunted* and Binford's *Bones: Ancient Men and Modern Myths*—set off a flurry of research leading to the widespread acceptance of taphonomy in zooarchaeology. Despite the equally revolutionary nature of both books, Binford's was easily the more controversial of the two. Writing in a pointed, sometimes bombastic style, Binford sought to educate the archaeological community as to how proper research should be done. He was harshly critical in his treatment of some established ideas about early hominid behavior and evolution, as well as the people who put them forth. Although *Bones* presented very little new data, it overflowed with analyses of research originally presented elsewhere by Binford (1978) and others. Chapter two of *Bones* remains one of the clearest articulations of the essential nature of taphonomy and actualistic studies (along with Gifford-Gonzalez, 1991). Binford could have been pointing to Brain's work as an example.

By contrast, Brain (1981) writes in a subdued, balanced, and understated manner, giving careful thought to caveats and competing hypotheses (as in the "leopard hypothesis" of chapter 14). Brain's book is also more empirical. It is full of direct observations and data, many of these packed into an appendix that has been mined and re-used by a multitude of researchers. Although lacking explicit discussions of epistemology and never taking on the tone of a lecture, the text is replete with lessons by example. The message and significance of the book is subtle but transformational once realized. It is a lesson plan for studying the past grounded in more than 14 years of research. The book was an epiphany for many young taphonomists starting out in the early 1980s, and a model of research for many participating in this symposium.

Together, Brain and Binford provided the following guidelines for actualistic research:

1. Make the uniformitarian assumption that physical and biological systems operated in the past as they do today.
2. Ground inferences about the past in an understanding of processes operating in the present.
3. Justify the relevance of these modern processes to the past, insuring that cause and effect are well understood and would likely apply to the past.
4. Isolate all the processes that can produce the physical traces and patterns that comprise archaeological evidence.
5. Study these processes in the modern world to develop criteria for their recognition in an archaeological context. This is what Binford meant by the search for "signature patterns."
6. Insure that there is little or no doubt about the link between the agent and the resulting traces.

Gifford-Gonzalez (1991) provided a much-needed re-statement and terminological clarification of these principles along with a research protocol for actualistic studies seeking to develop the uniformitarian assumptions referred to above. In the Gifford-Gonzalez system, a *trace* is a visible attribute displayed by a bone that has undergone a taphonomic process. The *causal agency* is the immediate physical cause producing a trace, such as a tooth skidding across a bone. The *effector* is the item or material that effects the modification of a bone, such as the tooth. The *actor* is the source of the force or energy that creates the trace, such as a hyena biting on to a bone. This analytical construct provides a protocol for evaluating the strength of bridging arguments linking actor to trace (cause to effect), and thus a procedure to evaluate the robustness of uniformitarian propositions.

Over the 20 plus years since Brain (1981) and Binford (1981) laid the foundations, the taphonomic approach has been embraced by many zooarchaeologists, ignored by others, and argued against by still others. In 1995, Marean noted that the field had largely split into two camps: one practicing "actualistic taphonomy," and the other relying on "comparative taphonomy." The former examines processes in the modern world to make inferences about these in the past, while the latter makes these inferences by comparing patterns from the past.

Actualistic taphonomy covers a spectrum of methodology ranging from naturalistic to experimental, in which the analyst observes the relationship between the trace, causal agent, effector, and actor. In a fully naturalistic context, the analyst only observes and records without intentionally manipulating the parameters of the process. In experimental studies, the analyst actively controls the parameters of the observed process to better understand the relations linking the actor to the trace. There are also intermediate approaches in which the analyst controls some parameters, but only observes the actors. Ideally, there is a dynamic relationship between naturalistic and experimental studies, as there is between analogous fields of endeavor such as field and laboratory primatology. Naturalistic studies often set the research agenda while experimental studies refine one's knowledge of the bridging arguments.

Proponents of comparative taphonomy cite two main arguments in its favor. The first is that the range of modern processes is insufficient for understanding the past because there were conditions in the past that do not exist today. For example, since sabertooth cats are now extinct, we cannot observe their bone chewing behavior, and we must therefore rely on studies of fossil bone assemblages where we are reasonably confident that sabertooths were the agent of accumulation and destruction. The second argument for comparative taphonomy is that there are geological processes acting on the fossil record that are impossible to model on anything less than geologic time scales. These can only be understood by trying to control for their effects through the comparison of fossil assemblages. Perhaps the strongest advocates and applicants of the comparative approach are Klein (Klein et al., 1999) and Stiner (1994, 2002).

While it is certainly true that adaptively unique taxa were present in the past, and that geologic processes operate at time scales that cannot be directly replicated, neither proposition provides compelling support for the comparative method, nor overturns the primacy of the actualistic method. Interpretations based on comparative taphonomy fail Gifford-Gonzalez's (1991) litmus test for method: there is no direct control over the links between trace, causal agent, effector, and actor. Thus, the results are bridging arguments reliant on circumstantial evidence. Comparative studies can only produce unverified models that, if left unverified, have a good probability of producing a Type II error—that is, the erroneous acceptance of the hypothesis that similar ancient patterns are indicative of similar processes. What then is the proper role of comparative studies in taphonomy? These studies form a fertile source of hypotheses that can guide research design and testing using actualistic methods; first in a naturalistic context, then further refined by experimental studies. The result is a robust bridging argument that can then be effectively applied to the interpretation of traces and their patterns in the fossil record.

Modern approaches to skeletal element analysis (SEA) developed out of this heuristic process and have been strongly influenced by actualistic studies. As early as the 1950's, White (1954, 1955) recognized that element profiles, although a potentially rich source of information on human behavior, were skewed by taphonomic processes related to both marrow processing and carnivore ravaging. Brain (1967, 1969) brought this idea into the realm of actualistic research in his now classic studies with goat remains at Hottentot camps. It has since become apparent that many factors affect the SEA, including prey/carcass availability, transport decisions, butchery destruction, carnivore ravaging, sedimentary processes, and excavation/curation procedures. It is no wonder that the interpretation of SEA remains contentious (Stiner, 2002; Pickering et al., 2003).

In this paper, we review and analyze the history of research examining the impact of carnivore destruction on the SEA. In particular, we will do the following: 1) review the evidence for the impact of carnivore ravaging on bone, 2) focus on data from actualistic research, 3) re-examine the idea that bone density is the primary mediator of SEA, 4) develop a general model for bone survival and destruction, and 5) suggest a course for future research.

## Cause and Effect: The Role of Carnivores in the SEA

Discarded food remains have been magnets attracting carnivores to hominin locales (i.e., FLK *Zinj*) for at least 1.7 million years. Despite the debate over the interpretation of the Plio-Pleistocene faunal assemblages, all would agree that there is excellent evidence that at least some carcass remains discarded by hominins were ravaged by carnivores. Actualistic studies demonstrate that defleshed carcasses discarded in natural habitats are rapidly discovered and ravaged by carnivores (Blumenschine, 1988; Capaldo, 1995). By the Late Pleistocene, sites in Eurasia and Africa document intense carnivore ravaging of hominin-discarded carcass parts (Marean and Kim, 1998; Marean et al., 2000). It is likely that this attraction to human refuse eventually put canids on the path to domestication. Once canids were domesticated around 12,000 BP, the discarded carcass parts became one of their primary sources of nutrition, and carnivore ravaging probably became even more regular and intense. Despite clear evidence for carnivore damage to faunal assemblages, the extent to which ravaging could affect element representation was not always appreciated.

Brain (1967, 1969) was one of the first to recognize and study this process, and he did so with the goal of explaining a pattern of element representation observed by Dart (1957) at Makapansgat. In his 1967 paper, Brain addressed two patterns identified by Dart. The first was the abundance of cranial fragments in the Makapansgat assemblage. Dart had argued that the abundance of heads resulted from intentional collection habits of the early hominins, resulting from their desire to have mandibles as saws. The second pattern was the differential survival of the ends of limb bones—particularly the great distinction between proximal and distal humeri. Dart (1957) also noted that lower limb bones (specifically metapodials) occurred in much larger numbers than expected.

Skeletal patterns such as this typically result in negative curvilinear correlations between element representation and food utility (the so-called "reverse utility curves", Marean and Frey, 1997). This head-and-foot, head-dominated, and reverse utility pattern is the most widespread skeletal element distribution in zooarchaeology. Although it occurs in paleontological, Paleolithic, and complex society sites, it is still commonly attributed to selective transport (by humans) when found in archaeological contexts. For example, head-dominated Middle Paleolithic assemblages in Italy are interpreted as an indication that Neanderthals were scavenging the heads of carcasses and bringing them back to caves (Stiner, 1991). The one thing these assemblages share is a collection or quantification procedure that excludes isolated shaft fragments from analysis (Marean and Kim, 1998; Bartram and Marean, 1999), a practice shown to result in this typical element distribution (Marean and Frey, 1997; Marean, 1998). Although these methodological issues were raised much later, Brain's (1969) work showed beyond a doubt that post-depositional process could significantly alter the distribution of skeletal elements.

Brain (1969:13) presciently noted "The reconstruction of events from the remote past is always an indirect process" and "In such circumstances it is enormously helpful when contemporary situations can be found in which comparable events are taking place." The contemporary situation to which he referred was the butchery and discard of goat bones by Khoi-khoi pastoralists (Kuiseb River, Namibia), and the subsequent carnivore (canid) ravaging. Brain collected the discarded bones from meals for which he had reliable estimates of the original number of carcasses entering the taphonomic system. He also procured an entire goat and observed its butchery and consumption, isolating the carcass from dogs so he could see the difference in destruction. The study illustrates the dual use of the naturalistic and experimental models. The canid ravaging of the discarded goat bones produced an assemblage that mimicked many of the patterns present in the Makapansgat assemblage, and Brain (1969:22) concluded, "The bones preserved would have been those best able to survive the destructive treatment to which they had been subjected." Thus, the resulting skeletal element pattern required no special appeal to hominin behavior.

In 1969, Brain expanded his 1967 study in two exciting ways. First, he calculated a percent survival of skeletal elements. In the case of limb bones, Brain (1969: 19, table III) calculated survival for both proximal and distal portions. Calculating a percent survival was reasonable because Brain could legitimately assume that the entire goat carcass entered the deposit, and he could estimate the original minimum number of carcasses. Second, he estimated the resistance of skeletal elements to destruction by calculating a specific gravity per element portion, which is a gross estimate of density as a proxy for resistance to destruction. He concluded (1969:20): "survival is not haphazard, but is determined by inherent qualities of the parts." Here, as early as the 1960s, we see a search for cause and effect between skeletal element survival and a contemporary process. That work set in motion two parallel research trajectories that often intersected: the study of carnivore ravaging's impact on the skeletal element profile, and the attempt to estimate bone's structural properties as a tool for understanding bone survival. We now turn to a critical review of both.

## Carnivore impact on skeletal element survival

Skeletal element survival can be studied in at least two dimensions: intra-bone survival (variation in the survival of different element portions) and inter-bone survival (variation in the survival of different elements). The two are intimately related. The relationship between intra-bone survival and the element portions coded for analysis will affect the analytical results of inter-bone survival. In other words, if limb bone ends and shafts survive at different rates, then inter-bone survival will appear distinct depending on whether one codes ends, shafts, or both. Table 1 provides a summary of the published record on carnivore destruction of bone in which data are relevant to either intra- or inter-bone survival.

Actualistic studies of how carnivores destroy bone range from experimental to naturalistic approaches. In Table 1 we have made a distinction between procedure ("type of study") and context in indicating whether or not a study is experimental or naturalistic. The study type is experimental if the analyst presented the carcass or carcass parts for study, and naturalistic if the analyst simply observed behavior that was occurring naturally. The context of a study was also classified as experimental if it took place in some type of controlled environment, such as an enclosure or zoo, and naturalistic if out in an area where wildlife occur free-ranging (preserve, park, or ranch).

Studies of carnivore impact on skeletal element survival have focused on patterning at dens, kill sites, and the scavenged remains from human butchery. These contexts can be anticipated to have widely varying results. In the first two situations, carnivores encounter a carcass (through either hunting or scavenging), conduct at least some on-site consumption, and then may elect to transport a small selection of carcass parts back to a den. Importantly, the den researcher only sees the outcome of this process, and cannot determine the original number of carcasses or elements. In the last context, carnivores encounter skeletal elements or carcasses that humans may have modified in any of the following ways: selective transport, defleshing, demarrowing, cooking, and then discard. The differences in the initial carcass conditions affect carnivore behavior and will likely have a major impact on what elements survive (Blumenschine and Marean, 1993; Lupo, 1995). However, all the processes of the latter two contexts can be observed, and in some cases controlled, by the researcher. Our review will segregate the literature based on this distinction between den recovery and behavioral observation.

### *Bone collection at carnivore dens*

Many of the first carnivore taphonomy studies were based on den research (see Table 1, "Context" column), and most of these focused on the frequency of taxa and skeletal elements. Den studies are distinct from other actualistic research in that they do not allow direct observation of the process of destruction. Many of these studies were done, we believe, with the hope that there would emerge a carnivore specific skeletal element pattern that could then be used to identify carnivore accumulations (Brain, 1981). Stiner has argued that it is unlikely that skeletal element abundance can be used to diagnose agents of collection, but that this instead "reflects the predominant foraging strategy employed" (i.e. scavenging or hunting; Stiner, 1991: 169).

However, the data from these studies are of limited use for pattern identification for several reasons. First, in most (but not all) of the studies the thoroughness of the collection method is unclear. Was there an excavation below the sediment surface, were the sediments sieved, and was everything picked up? Two studies employed archaeological recovery methods (Kerbis-Peterhans and Horwitz, 1992; Lam, 1992). This is particularly significant given the fact that hyenas fragment bones, often consuming articular ends and other greasy parts, and thus produce shaft fragments. Small shaft fragments quickly penetrate the sediment surface (Gifford-Gonzalez et al., 1985), protecting them from surface collection. Unexcavated assemblages would likely be biased toward larger fragments that, for one reason or another, have survived ravaging.

Second, the methods used in many of the den studies to quantify skeletal element abundance are unclear, and almost certainly do not meet the standards set today in zooarchaeology. This is further complicated by the fact that some of the analysts are not zooarchaeologists (i.e. Bearder, 1977 and Skinner et al., 1980) and thus would not be expected to have robust methods for estimating skeletal element abundance from highly fragmented specimens, while others (e.g., Bunn, 1983) are zooarchaeologists and would likely have better developed zooarchaeological methods.

Third, the data are reported in ways that vary both in the measure being reported and the grouping of skeletal elements and portions. For example, Henschel et al. (1979) provide a table listing bone abundance, but it is unclear if this is the number of individual specimens (NISPs) or minimum number of elements (MNEs). Some bones are listed twice (e.g., "Tibia, complete" and "Tibia, distal") and are not collated into a final measure of abundance. Lam (1992) provides data consistent with modern usage, including NISPs and MNEs on all bones and individual portions. Data of these two types are not comparable.

Despite these problems, some valuable information has emerged from these studies. One of the initial questions asked of den assemblages was simple: do hy-

Table 1. Summary of actualistic studies documenting or discussing carnivore destruction of bone

| Source | Type of study | Context | Consumer taxon/taxa | Original condition of elements when encountered by carnivore | Type of Intra-Element Presentation | Intra-Element Data NISP or MNE/ MNI | Shaft Portions or Fragments Included in Quant Analysis | Type of Inter-Element Presentation | Inter-Element Survival Control | Inter-Element Data NISP, MNE, MNI, or MAU | Consumed taxon/taxa | Locality |
|---|---|---|---|---|---|---|---|---|---|---|---|---|
| Brain (1967, 1969) | Natur. | Natur. (village refuse) | Human, domestic dog | complete carcass | quantitative | NISP and MNI | In NISP analysis only | quantitative | yes | NISP and MNI | Size class 1 – 2 bovids | Namibia |
| Brain (1970) | Exper. | Exper. (feeding in enclosure) | Cheetah | complete carcass | narrative description | nr | nr | brief comments | yes | nr | Baboon | Namibia |
| Sutcliffe (1970) | Natur. | Natur. (dens) | Spotted hyena | unknown | narrative description | nr | nr | none | nr | nr | Size class 1 – 3 mammals | Kenya |
| Lyon (1970) | Natur. | Natur. (house refuse) | Human, domestic dog | unknown | narrative description | nr | nr | none | nr | nr | Size class 1 – 2 mammals | Peru |
| Binford & Bertram (1977) | Natur. | Natur. (camp refuse) | Domestic dog | Cooked, uncooked, defleshed | quantitative | MNI | **no** | quantitative | yes | MNI | Size class 1-3 caribou and sheep | USA |
| Mills & Mills (1977) | Natur. | Natur. (dens) | Spotted hyena | unknown | narrative description | nr | nr | none | nr | nr | Size class 1-4 | South Africa |
| Bearder (1977) | Natur. | Natur. (dens) | Spotted hyena | unknown | narrative description | nr | nr | quantitative | **no** | NISP | Size class 1-4 | South Africa |
| Henschel et al. (1979) | Natur. | Natur. (dens) | Spotted hyena | unknown | narrative description | nr | nr | none | nr | nr | Size class 1 – 3 animals | Namibia |
| Richardson (1980) | Natur. and Exper. | Natur. (ranches and wildlife reserves) | Feral dog, jackal, spotted hyena, brown hyena, lion | complete carcass | quantitative | MNI | no | quantitative | yes | MNI | Size class 1 – 5 bovids | South Africa, Zimbabwe |
| Haynes (1980, 1981, 1983a, b) | Natur. | Natur. (wilderness areas) | Wolf, bear | various (complete carcass and unknown) | narrative description | nr | nr | none | nr | nr | Size class 1-4 | North America |

"Natur." = Naturalistic, "Exper." = Experimental, nr = not relevant, NISP = number individual specimens, MNI = minimum number individuals.

*Table 1. Summary of actualistic studies documenting or discussing carnivore destruction of bone (continued)*

| Source | Type of study | Context | Consumer taxon/taxa | Original condition of elements when encountered by carnivore | Type of Intra-Element Presentation | Intra-Element Data NISP or MNE/MNI | Shaft Portions or Fragments Included in Quant Analysis | Type of Inter-Element Presentation | Inter-Element Survival Control | Inter-Element Data NISP, MNE, MNI, or MAU | Consumed taxon/taxa | Locality |
|---|---|---|---|---|---|---|---|---|---|---|---|---|
| Haynes (1980, 1981, 1983a, b) | Exper. | Exper. (zoos) | Wolf, bear, large cats, hyenas | defleshed tibia and femora | narrative description | nr | nr | none | nr | nr | Size class 4 (*Bos taurus*) | North America |
| Skinner et al. (1980) | Natur. | Natur. (dens) and Exper. | Striped hyena | unknown for den, complete carcass for feedings | quantitative | NISP | yes | quantitative | no for den, yes for feedings | NISP | Size class 1-4 | Israel |
| Binford (1981) | Natur. | Natur. (den and kill sites) | Wolf | unknown for den, complete carcass for kills | quantitative | MNI | no | quantitative | no for den, yes for kill sites | MNI | Size class 3 (*Rangifer tarandus*) | Alaska |
| Hill (1981, 1984), Potts (1988) | Natur. | Natur. (dens) | Spotted hyena | unknown | narrative description | nr | unclear | quantitative | no | MNE | Size class 1-4 mammals | Kenya |
| Bunn (1983) | Natur. | Natur. (dens) | Spotted hyena | unknown | narrative description | nr | yes | quantitative | no | unclear | Size class 1-4 mammals | Kenya |
| Stallibrass (1984) | Exper. | Natur. (farm) | Foxes | complete carcass | quantitative | MNI | no | quantitative | yes | MNI | Size class 1-2 sheep | England |
| Payne & Munson (1985) | Exper. | Exper. (enclosure) | Domestic dog | Cooked, uncooked, defleshed | none | nr | no | quantitative | yes, but not precise | MNE | Size class 2 bovids, small mammals | USA |
| Snyder (Snyder, 1988; Klippel et al., 1987) | Exper. | Exper. (enclosure) | Gray wolf | complete carcass | quantitative | MNI | no | quantitative | MNI | MNI | Size class 2 cervids | USA |

*"Natur." = Naturalistic, "Exper." = Experimental, nr = not relevant, NISP = number individual specimens, MNI = minimum number individuals.*

*Table 1. Summary of actualistic studies documenting or discussing carnivore destruction of bone (continued)*

| Source | Type of study | Context | Consumer taxon/taxa | Original condition of elements when encountered by carnivore | Type of Intra-Element Presentation | Intra-Element Data NISP or MNE/ MNI | Shaft Portions or Fragments Included in Quant Analysis | Type of Inter-Element Presentation | Inter-Element Survival Control | Inter-Element Data NISP, MNE, MNI, or MAU | Consumed taxon/taxa | Locality |
|---|---|---|---|---|---|---|---|---|---|---|---|---|
| Binford et al. (1988) | Exper. | Natur. (wildlife reserve) | Spotted hyena | defleshed limb bones | narrative description | nr | nr | none | nr | nr | Size class 4 bovids | South Africa |
| Blumenschine (1988) | Exper. | Natur. (wildlife reserve) | Spotted hyena, lion | defleshed limb bones | quantitative | NISP | yes | none | nr | NISP | Size class 1 – 3 bovids | Tanzania |
| Horwitz (Horwitz & Smith, 1988) | Natur. | Natur. (den) | Striped hyena | unknown | narrative description | nr | nr | narrative description | nr | nr | human | Israel |
| Marean (Marean & Spencer, 1991) | Exper. | Exper. (enclosure) | Spotted hyena | defleshed limb bones | quantitative | MNE | yes | quantitative | MNE | yes | Size class 1-2 sheep | USA |
| Kerbis-Peterhans and Horwitz (1992), Horwitz (1998) | Natur. | Natur. (den) | Striped hyena | unknown | narrative description | nr | no | quantitative | no | MNI | various mammals and reptiles | Israel |
| Lam (1992) | Natur. | Natur. (den) | Spotted hyena | unknown | quantitative | NISP/ MNE | yes | quantitative | no | NISP/ MNE | Size class 1–2 bovids | Kenya |
| Bartram (1993; Bartram & Marean, 1999) | Natur. | Natur. (abandoned human sites) | Human, Brown hyena, jackal, | unknown | quantitative | NISP/ MNE | yes | quantitative | no | NISP/ MNE | Size class 1–5 mammals | Botswana |
| Bunn (1993) | Natur. | Natur. (abandoned human sites) | Human, spotted hyena, jackal, | unknown | none | nr | yes | quantitative | no | MAU | Size class 1–4 bovids | Tanzania |

*"Natur." = Naturalistic, "Exper." = Experimental, nr = not relevant, NISP = number individual specimens, MNI = minimum number individuals.*

*Table 1. Summary of actualistic studies documenting or discussing carnivore destruction of bone (continued)*

| Source | Type of study | Context | Consumer taxon/taxa | Original condition of elements when encountered by carnivore | Type of Intra-Element Presentation | Intra-Element Data NISP or MNE/ MNI | Shaft Portions or Fragments Included in Quant Analysis | Type of Inter-Element Presentation | Inter-Element Survival Control | Inter-Element Data NISP, MNE, MNI, or MAU | Consumed taxon/taxa | Locality |
|---|---|---|---|---|---|---|---|---|---|---|---|---|
| Hudson (1993) | Natur. | Natur. (occupied campsites) | Human, domestic dog | defleshed and some hammer-stone broken | quantitative | MNE | yes | quantitative | yes | MNE | Size class 1–2 mammals | Central African Republic |
| Selvaggio (1994, 1998) | Natur. | Natur. (wilderness areas) | Large felids, spotted hyena, jackal, human | complete carcass | none | nr | nr | quantitative | yes | NISP | Size class 1 and larger | Tanzania |
| Lupo (1995) | Natur. | Natur. (abandoned human sites) | Human, unknown East African carnivores | unknown | quantitative | NISP, some MNE | yes | quantitative | some | NISP, some MNE | Size class 2-3 bovids, Size class 3 Equid | Tanzania |
| Capaldo (1995) | Exper. | Natur. (simulated abandoned human sites) | Unknown East African carnivores | defleshed bones | quantitative | NISP | yes | quantitative | yes | NISP | Size class 1 – 3 bovids | Tanzania |
| Domínguez-Rodrigo (Domínguez-Rodrigo, 1997; Pickering et al., 2003) | Exper. | Natur. (simulated abandoned human sites) | Unknown East African carnivores | defleshed bones | quantitative | NISP | yes | none | nr | nr | Size class 3 bovids | Kenya |
| Pickering (Pickering et al., 2003) | Exper. | Exper. (enclosure) | Spotted hyena, leopard | complete carcass | quantitative | MNE | yes | none | nr | nr | Size Class 2 primates | South Africa |

*"Natur." = Naturalistic, "Exper." = Experimental, nr = not relevant, NISP = number individual specimens, MNI = minimum number individuals.*

enas accumulate bones at dens? Dart (1956) argued no, but these studies overwhelmingly provide an answer in the affirmative, and thus accumulation by carnivores in caves and rock shelters is a potentially serious problem for zooarchaeologists. The more complicated question addresses the issue of patterning in skeletal element representation and whether this can be used to diagnose an accumulator or identify a particular type of predatory niche.

Stiner (1991) argues that there is a pattern in the collections. Her analysis draws on a selected series of carnivore dens, including several unpublished dens from data provided by Gary Haynes and Lewis Binford. She indexes skeletal element completeness and finds that striped and brown hyenas (obligate scavengers) tend to have horn- or head- dominated assemblages, wolves tend to have assemblages dominated by the more meaty parts of the skeleton, and spotted hyenas vary between the two extremes.

The complicating factor here of course is that skeletal element abundance at dens minimally reflects both transport behavior and bone consumption capabilities. Hyenas are far more capable bone consumers than wolves (Ewer, 1973), and it is likely that their assemblages will display more attrition of the trabecular portions of postcranial elements than those of wolves. The result would be a greater relative representation of head, horn, and limb bone cortical fragments at hyena accumulations. This means that any (particularly hyena) study that fails to include shaft fragments in the analysis (most of these studies do not include shafts) will inevitably underestimate the number of limb bones present. This is clearly displayed in Lam's (1992) data (not included in Stiner's 1994 survey), which shows a strong representation of limb bones and a low frequency of horn and head parts. Another problem with comparing these carnivores is that cervids (the main prey of the wolves) only have antlers for part of the year (and these are in a soft state for some of that time). By contrast, bovids (the main prey of the hyenas) have horns (which survive well) all year round (Brain, 1967, 1969). Thus, it is inevitable that hyena assemblages will have more horns than wolf assemblages will have antlers without any appeal to differences in the carcass portions transported.

Several of these studies have noted that carnivores, in the process of accessing within-bone nutrients, chew away and consume the softer, greasy, cancellous portions of bones, and leave behind the harder cortical bone fragments. Sutcliffe (1970) studied several hyena dens in East Africa, leading him to develop a list of four types of hyena damage to bones. "Type 1" damage is typified by a regular pattern of relative bone destruction—robust bone portions survive, while cancellous portions are commonly destroyed. Sutcliffe goes on to note that when hyenas feed on human remains, the ends of limb bones are destroyed, but the shafts survive. Most of the den reports make similar observations on various prey taxa (Henschel et al., 1979; Lam, 1992; Kerbis-Peterhans and Horwitz, 1992).

Juvenile hyenas break bones far less effectively than do adults, and their gnawing creates multiple striations perpendicular to the long axis of the bone (Sutcliffe's "Type 2" damage), sometimes even wearing a hole. Various authors have noted that den assemblages typically have reasonable frequencies of "bone cylinders"—limb bone shafts that retain the complete circumference (Potts, 1988). However, they are rare to absent in assemblages where adult hyenas have been presented with complete long bones. One of us (CWM) has observed that juvenile hyenas, through this regular bone gnawing, produce cylinders because they cannot reduce the more resistant shafts. Thus, gnawed bone cylinders implicate juveniles when hyenas are the agent, and perhaps smaller adult carnivores of taxa less capable of crushing bone.

One pattern that has arisen from the den studies, but has been widely ignored in zooarchaeological literature, is the presence of regurgitations and their potential as an indicator of the persistence of hyena activity in caves and rockshelters. Bearder (1977) notes that spotted hyenas regurgitate in and around their dens on a regular basis (see also Brain, 1981). Although we lack precise data on transit times, we noted that regurgitations among the captive hyenas at Berkeley occurred at least several hours after consumption. Hyena den assemblages should therefore display fairly high frequencies of regurgitated bone. Along with tooth mark frequencies, this is likely to provide a useful measure of the contribution of hyenas to fossil bone assemblages.

In summary, we have gained a modest amount of knowledge from the currently published studies of den assemblages but these have had a limited impact on our interpretation of the SEA. Their importance in identifying a skeletal profile unique to dens could be revitalized by renewed research using strictly controlled collection and analysis procedures. However, unless done in very controlled situations, these will always lack the critical link between observer and initial conditions.

### *Bone ravaging at kills and areas of human discard*

Studies examining how carnivores ravage bones discarded by humans have been done in at least three contexts: 1) studies of people living in traditional economic pursuits (ethnoarchaeology), 2) studies of carnivore kills, and 3) simulations of human butchery and discard. The first two contexts have necessarily less control over the process, but they at least have a more naturalistic context. We have lumped carnivore kill studies together with hominid-first research for two reasons. First, the process of observation can be done completely—that is, all skeletal inputs to the system can be known. Second, these studies are done in order to identify basic parameters of carnivore destruction and to put hominid-first destruction into perspective.

Ethnoarchaeological contexts have included both residential and butchery sites of hunter-gatherers or pastoralists. Some researchers observed behavior at occupied sites, making it easier to estimate the original number of elements entering the system (Brain, 1967, 1969; Hudson, 1993). However, when studying sites abandoned for varying lengths of time, researchers (e.g. Bartram, 1993) sometimes had little knowledge of the original number of carcasses.

Experimental studies provide a direct measure of the impact of carnivore feeding and ravaging through direct observation (securing the link between trace and actor), controlled input of skeletal elements, and thorough collection of bone fragment output. With such control over the parameters of the experiment, one can confidently estimate a percentage survival (Brain, 1967) or percentage change (Marean and Spencer, 1991). Without these statistics, the cause and effect relationship between carnivores and element representation is obscured by other factors that could affect representation (such as variation in human transport decisions) and thus diminishes the predictive power of the model. We have indicated in Table 1 whether this control applies in each study in the column titled "Inter-Element Survival Control."

Several studies maintained control over input parameters (in both naturalistic and experimental contexts) by providing carnivores with complete fleshed carcasses, observing the process of destruction, and then collecting the remains. More recent studies (such as those of Capaldo, Selvaggio, Domínguez-Rodrigo, and Marean) have focused on gaining even greater control through the use of a "simulated site" approach as first illustrated by Binford et al. (1988) and Blumenschine (1988). These studies typically model several scenarios of carcass part access by hominins and carnivores, preparing carcass parts accordingly. For example, the hominid-first (a.k.a. "hominid to carnivore") scenario involves defleshing bones, processing them for marrow, and then allowing carnivores to ravage them. These have since been permutated into a variety of sequences (e.g. carnivore only, carnivore to hominid, carnivore to hominid to carnivore, etc.).

Brain (1967, 1969) conducted one of the earliest studies of sequential carcass access, and documented both intra-bone and inter-bone survival. He found that human food preparation, followed by dog ravaging of the discarded bones, typically resulted in the elimination of limb bone epiphyses (except for those epiphyses with greater "specific gravity" and earlier fusion). His inter-bone analysis showed that the mandible survived the best (nearly 100% survival), while vertebrae, pelves, scapulae, ribs, small compact bones, and phalanges survived at rates less than 30%. Without a percentage survival based on shafts, limb bone portions other than the distal humerus survived poorly.

Brain (1970) also fed baboons and size 1 and 2 bovids to cheetahs, a felid with a rather flesh-specialized dentition (Ewer, 1973; see Brain, 1981: 9 for size class definitions). He found that cheetahs minimally damaged the bones of even small bovids such as springbok, but did far more damage to the skeletons of similarly sized baboons. The cheetahs consumed the entire baboon vertebral column, hands, and feet, and did significant damage to the ends of the limb bones. Brain (1970) argued that baboon skeletons were less dense than those of bovids, even those of the same body size, and thus were more susceptible to carnivore destruction. There are no quantitative data on survival presented in the study.

Lyon (1970: 214) studied the Wachipaeri of eastern Peru and observed the ravaging of size class 1 and 2 mammal bones that were discarded after human processing. These studies were done in the context of village sites with domestic dogs. Lyon notes "dogs generally chewed off all the articulations and occasionally completely consumed the long bones." The dogs totally consumed all the bones of small animals including fish, birds, and small mammals, but only damaged most of the bones of the larger mammals. There are no quantitative data presented in the study.

Binford and Bertram (1977) provided data and analyses of Nunamiut and dog destruction of caribou bones, and Navaho and dog destruction of sheep bones. Importantly, in one Nunamiut study the authors had nearly perfect control over the percentage survival (Binford and Bertram, 1977: 81, table 3.2, last column) and in two Navaho studies they had good, though not perfect, control. They did not present any data on shaft portions, but did document differential inter-bone survival that they argued was due to differences in density. Axial parts such as vertebrae, ribs, and pelves did not survive well, nor did small compact bones and phalanges.

Binford (1981) also reports on the inter-bone survival from 24 individual caribou killed by wolves. As in the Nunamiut study (Binford and Bertram, 1977), inter-bone survival does not appear to be density dependent, and axial elements have high rates of survival. Binford (1981) also provides useful information on the relative representation of intra-long bone survival, although not for individual elements. These data show that shaft splinters are nearly eight times more numerous than articular ends. Of course, this does not tell us that MNE would have been higher if calculated on shafts, but it suggests this possibility. Bone cylinders are also reported, but it is unclear whether they might have been included in the calculation of MNI per long bone end. Binford (1981: 210-217) compares the wolf kill data with a couple of dens and another suspected kill site, but he expresses some doubts about the agents of accumulation in these latter contexts.

Between 1977 and 1981, Haynes (1980, 1981, 1982, 1983a, 1983b) observed carcass destruction in natural contexts and bone destruction in zoo feeding experiments. In the former, he observed the effects of wolf (*Canis lupus*) and bear (*Ursus sp.*) ravaging of primarily large and medium ungulates (e.g., *Bison bison, Odocoileus virginianus, Alces alces*). These carcasses were examined seasonally for progressive change. Fieldwork was complemented by experimental research in which tibia and femora of *Bos taurus* were fed to several carnivores (bear, wolf, large cats, and hyenas). Haynes (1980, 1982, 1983a, 1983b) published detailed narrative descriptions of the resulting damage (including fracture type, surface marks, and general extent of destruction). In addition to the morphological distinctions between damage caused by different carnivores, Haynes noted that the hyenas were generally much more destructive than the wolves, and both generally did more damage to bone than the other carnivores surveyed. Even so, Haynes (1982) documented at least one instance in which wolves reduced a white-tailed deer carcass to a handful of isolated limb shaft fragments—very similar in appearance to the remains from Marean's (Marean and Spencer, 1991) hyena feeding experiments.

Richardson's (1980) study of the damage inflicted by various carnivores on 89 bovid, equid, and giraffe carcasses (size classes 1–5) shows differential destruction of limb bone ends relative to shafts across all body sizes. Richardson had control over the number of carcasses set out, and therefore provided a true percentage for inter-element survival. He found remarkable similarity in bone survival when comparing carnivore ravagers, but hyenas inflicted far more damage than other taxa. The idea that hyenas can have such a significant impact on large mammal skeletons has recently come under attack by Klein et al. (1999), who argue that hyenas are unable to chew the ends off size 4 and 5 mammals. As evidence, they provide a drawing of a hyena skull at maximum gape attacking the proximal tibia of *Pelorovis* just as a human would eat a hot dog. Both Richardson (1980: 113, figure 3b) and Brain (1981: 71, figure 63) illustrate giraffe proximal humeri that were completely removed by hyena chewing. Blumenschine (1988) and Capaldo (1995) both report destruction of size 4 and 5 limb bone ends in their hyena observations. One of us (CWM) has fed whole size 4 bones to hyenas, and found that hyenas have a standard approach to reducing large mammal bones. Rather than attacking the bone like a hot dog, hyenas grip protuberances in their teeth and exert leverage with their powerful neck muscles. They snap off chunks (some of which are swallowed) and thus produce craggy areas that can be gripped and attacked again, eventually opening the medullary cavity. These multiple actualistic observations, including Richardson's (1980), clearly falsify the Klein et al. (1999) hypothesis that hyenas are unable to destroy bones that exceed their maximum gape.

Stallibrass (1984) observed the impact of scavenging foxes and birds on 18 complete sheep carcasses with no human processing, and reported percentage element survival. She found significant variation in survival between elements as well as density related differences in intra-bone survival (in tibia and humerus). Small elements such as phalanges and compact bones survived poorly, and ribs and vertebrae survived the worst. Overall, limb bones with very dense epiphyses survived best. She does not provide survival data on limb bone shafts.

Payne and Munson (1985) fed the bones of squirrels, rabbits, and goats to a dog. They found that teeth survive well, as do early-fusing long bone ends. Survival is moderate for foot bones and low for late-fusing long bone ends, scapulae, pelves, and phalanges. They had good control over the original number of bones fed to the dog and were able to calculate percentage survival, but their data do not include shaft fragments, and some of the survival estimates are given as ranges.

Snyder (Klippel et al., 1987; Snyder, 1988) observed the feeding behavior of gray wolves on fleshed white-tailed deer carcasses. She found that limb bone end survival ranged from 87.5% (distal humerus) to 0% (distal radius and distal metacarpal). Axial remains such as vertebrae, pelves, and ribs survived very poorly, as did smaller bones like phalanges and carpals. Snyder had excellent control of percentage survival, but did not report the survival of limb bone shafts.

Binford and colleagues (1988) simulated hominid bone discard followed by hyena ravaging in a park in South Africa in which size 4 bovid bones (African buffalo) were defleshed and broken open with hammerstones. The researchers observed, "the bone elements remaining in their original positions were either long-bone splinters or impact chips, and none of them had been gnawed by hyenas" (Binford et al., 1988: 125). The focus was on limb bones and there are no data presented on inter-bone survival.

Blumenschine conducted similar, but more extensive, studies in northern Tanzania using size 1–3 mammals and presents NISP data on limb bone portion survival (including mid-shaft survival; Blumenschine, 1988: 488, table 2). He concludes:

> "The most conspicuous effect [of scavenger disturbance] is the virtually complete deletion or on-site destruction of hammerstone-produced epiphyseal fragments, a pattern that mirrors carnivore consumption of whole bone…At the same time, midshaft fragments produced by hammerstone breakage seem to be largely if not totally ignored by scavengers and to bear features distinctive of hammerstone breakage only" (Blumenschine, 1988: 495-496).

Marean and Spencer (1991) reported on the destruction of defleshed sheep limb bones, offered to hyenas as either unbroken bones or hammerstone broken fragments. They documented survival across five portions of each limb bone. They had excellent control of both the number of bones presented and of recovery, and their results showed significant destruction of end portions and rather complete survival of shafts. A subsequent paper (Marean et al., 1992) examined inter-bone survival and provided data on both the sequence of skeletal element choice by the hyenas and on ultimate levels of destruction. Axial bones tended to be chosen first and ravaged more intensely. The combined results of both papers indicate that only the middle shaft portions of long bones regularly withstand hyena attack.

Bunn (1993) described carnivore ravaging at Hadza

base camps, noting that long bone ends were preferentially removed, but that this bias was less pronounced when sites were occupied for prolonged periods. He noted similar destruction of ends at San sites in the Kalahari. He did not include quantitative data on intra-element survival at the Hadza camps, but he did publish some graphs showing relative element representation. No information was given indicating the original number of skeletal elements that entered the system.

Hudson (1993) detailed the destruction of size 1 and 2 mammal skeletal elements by domestic dogs in Aka pygmy camps. Her observations with regard to the impact of dog ravaging are clear and "can be used to suggest a baseline of expected survival frequencies on an ordinal scale: the preferential survival of heads, the loss of the articular ends of limb bones, and of carpals, tarsals, phalanges, and the under representation of vertebrae and ribs" (Hudson, 1993: 320). Hudson also presents excellent control data on percentage of inter-element survival. The results of the study are somewhat limited, however, in that excavation occurred after Aka camps were abandoned and the assemblage included multiple taxa. In fragmented assemblages, taxonomic variability will naturally result in a higher proportion of unidentifiable bone specimens. Indeed, Hudson (1993: 305) reported a 48% loss of the original minimum number of individuals (MNI).

Bartram (1993; Bartram and Marean, 1999) conducted research among Kua San hunter-gatherers of Botswana, in which he regularly monitored the bone refuse from their kills of sizes 1 to 5 mammals. His (1993) data, presented in detail in a series of appendices and summarized in a clear discussion, show that carnivores at Kua camps regularly deleted the ends of limb bones while leaving shafts undisturbed. Furthermore, Bartram's results show that the deletion of ends over shafts is elevated in size 3 and 4 versus size 1 and 2 mammals. Bartram (1993) did not have control over the number of skeletal elements that entered the system, so percentage survival data are not available, but his published quantification, including NISP and MNE on all long bone portions is unmatched in completeness.

Selvaggio (1994, 1998) observed various carnivores consuming prey taxa in naturalistic contexts in Tanzania. For 19 of the 32 carcasses studied, she was able to document the process beginning with the hunt. She then collected long bones from the carcasses and marrow-processed all but 35 of these, simulating a "carnivore to hominid" assemblage. Carnivores were allowed to continue ravaging nineteen carcasses from this sample, creating a "carnivore to hominid to carnivore" assemblage. Although her published research focused on the resulting bone surface modification, she provided some information on bone portion (epiphysis, near-epiphysis, and shaft) representation after marrow processing. For all long bones combined, there was little epiphyseal destruction within her carnivore to hominid sample. However, an additional period of ravaging dramatically reduced epiphyseal survival (from 42% to 11%) while increasing midshaft representation (46% to 74%; Selvaggio, 1998: 196, table 4). Because she did not segregate her data by element, it is not possible to determine whether density influenced the initial survival of bone ends. The fact that bones were collected shortly following the initial episode of carnivore consumption may have also affected bone survival data. It is possible that there is more intra-bone survival information within her original data.

Capaldo (1995) simulated hominid discard assemblages ravaged by carnivores in semi-naturalistic contexts in northern Tanzania. He had excellent control over the number of bones that entered the system. To date Capaldo has reported the survival potential of skeletal elements and portions in NISP, but not in MNE, limiting the value of the results. These data together with data collected in a similar study by Domínguez-Rodrigo (Pickering et al., 2003) document severe preferential destruction of limb bone ends of size 1–3 bovids. Pickering (Ibid.) reports the same pattern for a study of carnivore ravaged baboon limb bones.

### *Summary of skeletal element survival*

There are several clear patterns documented in the literature reviewed above, and we will review these both qualitatively and quantitatively. First, carnivores have a dramatic impact on intra-bone survival. The literature shows time and again that carnivores selectively destroy limb bone ends in preference to limb shafts. Pickering et al. (2003) provide a quantitative analysis of these data showing that, when presented by portion, limb shafts nearly always provide higher MNE estimates than ends. The critical implication of this pattern is that long bone counts that include isolated shaft portions are more accurate than counts based only on long bone ends. There is less discussion in these studies, and certainly no quantitative description, of intra-bone survival among other bones, such as axial elements. The literature agrees that when axial elements survive at all, their protuberances are most likely to be destroyed. It might be useful in future studies to examine this pattern in more detail.

Carnivores also have a dramatic impact on inter-bone survival. The question "how do different skeletal elements survive carnivore destruction?" is best answered by examining the rate of survival, not the skeletal element pattern (i.e. head and foot) that results from the process. The latter may be partly an outcome of selective transport and not only a measure of destruction. We must therefore first identify those studies in which we can directly measure rate of survival—that is, those studies with accurate quantification of element input and output.

Because the quantification of long bone shafts is so essential to accurate reporting of percentage survival (as discussed above), we have divided the analyses into those studies where shaft portions of limb bones were included, and those where they were not. For the majority of the studies surveyed here, data from limb shafts were

*Table 2. Percentage survival of skeletal elements in studies that **did not** report long bone shaft survival*

| | A. | | B. | | C. | | D. | | E. | | F. | | G. | All Studies | | |
|---|---|---|---|---|---|---|---|---|---|---|---|---|---|---|---|---|
| | MNI | % Survival | MNI | % Survival | MNI | % Survival | MNI | % Survival | MNE | % Survival | MNE | % Survival | % Survival MNE | Mean % Survival | Min % Survival | Max % Survival |
| Skull | 11.0 | 47.8 | 16.5 | 76.7 | 11.0 | 61.1 | ni | ni | ni | ni | 1 | 25.0 | ni | 52.7 | 25.0 | 76.7 |
| Mandible | 18.5 | 80.4 | 21.5 | 100.0 | 14.0 | 77.8 | ni | ni | 117 | 91.4 | 8 | 100.0 | 47.0 | 82.8 | 47.0 | 100.0 |
| Atlas | 12.0 | 52.2 | 4.0 | 18.6 | 5.0 | 27.8 | 3.0 | 75.0 | 12 | 18.8 | 2 | 50.0 | 33.0 | 39.3 | 18.6 | 75.0 |
| Axis | 17.0 | 73.9 | 4.0 | 18.6 | 5.0 | 27.8 | 2.0 | 50.0 | 14 | 21.9 | 1 | 25.0 | 28.0 | 35.0 | 18.6 | 73.9 |
| Cervical | 12.8 | 55.7 | 3.7 | 17.2 | 8.3 | 46.1 | 3.0 | 37.5 | 12 | 3.8 | 3 | 15.0 | 40.0 | 30.8 | 3.8 | 55.7 |
| Thoracic | 12.2 | 52.8 | 3.2 | 14.9 | 3.0 | 16.7 | 3.2 | 35.0 | 21 | 2.5 | 2 | 3.8 | 23.0 | 21.2 | 2.5 | 52.8 |
| Lumbar | 15.6 | 67.8 | 4.5 | 20.9 | 5.6 | 31.1 | 6.7 | 47.8 | 31 | 8.1 | 7 | 29.2 | 30.0 | 33.6 | 8.1 | 67.8 |
| Sacral | ni | ni | ni | ni | ni | ni | ni | ni | 1 | 1.6 | ni | ni | 28.0 | 14.8 | 1.6 | 28.0 |
| Caudal | ni | ni | ni | ni | ni | ni | ni | ni | 0 | 0.0 | ni | ni | ni | 0.0 | 0.0 | 0.0 |
| Pelvis | 23.0 | 100.0 | 9.5 | 44.2 | 13.0 | 72.2 | 9.0 | 81.8 | 34 | 26.6 | 7 | 87.5 | 60.0 | 67.5 | 26.6 | 100.0 |
| Rib | 10.8 | 47.0 | 1.9 | 8.8 | 2.4 | 13.3 | 2.6 | 22.7 | 170 | 10.2 | 1 | 1.0 | 18.0 | 17.3 | 1.0 | 47.0 |
| Scapula | 12.0 | 52.2 | 6.0 | 27.9 | 14.5 | 80.6 | 2.5 | 50.0 | 35 | 27.4 | 4 | 50.0 | 47.0 | 47.9 | 27.4 | 80.6 |
| Humerus Prx. | 2.5 | 10.9 | 1.5 | 7.0 | 2.5 | 13.9 | ni | ni | 0 | 0.0 | 2 | 25.0 | 19.0 | 12.6 | 0.0 | 25.0 |
| Humerus Dis. | 9.5 | 41.3 | 7.0 | 32.6 | 18.0 | 100.0 | ni | ni | 82 | 64.0 | 7 | 87.5 | 58.0 | 63.9 | 32.6 | 100.0 |
| Rad.-Ulna Prx. | 8.0 | 34.8 | 7.5 | 34.9 | 12.5 | 69.4 | ni | ni | 65 | 50.8 | 6 | 75.0 | 71.0 | 56.0 | 34.8 | 75.0 |
| Rad.-Ulna Dis. | 4.5 | 19.6 | 2.5 | 11.6 | 10.0 | 55.6 | ni | ni | 22 | 17.2 | 0 | 0.0 | 67.0 | 28.5 | 0.0 | 67.0 |
| Carpals | 3.5 | 15.2 | 2.7 | 12.6 | 1.8 | 10.0 | ni | ni | ni | ni | 2 | 4.2 | ni | 10.5 | 4.2 | 15.2 |
| Metacarpal Prx. | 5.5 | 23.9 | 2.5 | 11.6 | 5.0 | 27.8 | 0.5 | 100.0 | 32 | 25.0 | 6 | 75.0 | 35.0 | 42.6 | 11.6 | 100.0 |
| Metacarpal Dis. | 5.5 | 23.9 | 2.5 | 11.6 | 3.5 | 19.4 | 0.0 | 0.0 | 23 | 18.0 | 0 | 0.0 | 33.0 | 15.1 | 0.0 | 33.0 |
| Femur Prx. | 7.0 | 30.4 | 1.5 | 7.0 | 7.0 | 38.9 | 1.0 | 50.0 | 18 | 14.1 | 2 | 25.0 | 36.0 | 28.8 | 7.0 | 50.0 |
| Femur Dis. | 3.5 | 15.2 | 2.5 | 11.6 | 3.5 | 19.4 | 0.5 | 25.0 | 9 | 7.0 | 3 | 37.5 | 31.0 | 21.0 | 7.0 | 37.5 |
| Tibia Prx. | 5.0 | 21.7 | 4.5 | 20.9 | 8.0 | 44.4 | 1.0 | 40.0 | 13 | 10.1 | 1 | 12.5 | 58.0 | 29.7 | 10.1 | 58.0 |
| Tibia Dis. | 11.0 | 47.8 | 7.5 | 34.9 | 13.0 | 72.2 | 1.5 | 60.0 | 72 | 56.3 | 4 | 50.0 | 80.0 | 57.3 | 34.9 | 80.0 |
| Tarsals | 5.0 | 21.7 | 3.6 | 16.7 | 4.0 | 22.2 | 0.5 | 25.0 | ni | ni | 18 | 45.0 | ni | 26.1 | 16.7 | 45.0 |
| Astragalus | 5.5 | 23.9 | 3.0 | 14.0 | 5.0 | 27.8 | 1.0 | 50.0 | 16 | 12.5 | 5 | 62.5 | 19.0 | 29.9 | 12.5 | 62.5 |
| Calcaneum | 5.5 | 23.9 | 2.5 | 11.6 | 5.0 | 27.8 | 0.5 | 25.0 | 14 | 10.9 | 4 | 50.0 | 7.0 | 22.3 | 7.0 | 50.0 |
| Metatarsal Prx. | 8.5 | 37.0 | 4.0 | 18.6 | 4.5 | 25.0 | 1.0 | 40.0 | 39 | 30.4 | 7 | 87.5 | 36.0 | 39.2 | 18.6 | 87.5 |
| Metatarsal Dis. | 4.5 | 19.6 | 2.5 | 11.6 | 5.5 | 30.6 | 0.5 | 25.0 | 20 | 15.6 | 1 | 12.5 | 33.0 | 21.1 | 11.6 | 33.0 |
| Phalanx 1 | 2.6 | 11.4 | 0.9 | 4.0 | 0.9 | 4.8 | 0.4 | 18.5 | ni | ni | 1 | 3.1 | ni | 8.4 | 3.1 | 18.5 |
| Phalanx 2 | 1.9 | 8.1 | 0.6 | 2.9 | 0.6 | 3.4 | 0.3 | 12.5 | ni | ni | 0 | 0.0 | ni | 5.4 | 0.0 | 12.5 |
| Phalanx 3 | 1.5 | 6.5 | 0.4 | 1.7 | 0.4 | 2.1 | 0.1 | 6.0 | ni | ni | 0 | 0.0 | ni | 3.3 | 0.0 | 6.5 |
| Phalanges | 2.6 | 11.4 | 0.9 | 4.0 | 0.9 | 4.8 | 0.4 | 18.5 | 21 | 2.7 | 1 | 0.0 | 9.0 | 7.2 | 0.0 | 18.5 |

*Raw values such as surviving MNIs or MNEs are listed if provided in the original study. ni = element not included in study. A. Binford (1981: 211-213, Table 5.01, total wolf kills). B. Binford and Bertram (1977: 101, Table 3.5, winter sheep). C. Binford and Bertram (1977: 101, Table 3.5, summer sheep). D. Binford and Bertram (1977: 81, Table 3.2, record B). E. Brain (1967: 109, Table 3). F. Klippel et al. (1987: 158, Table 1). G. Stallibrass (1984: Figure 1, data extracted by digitizing bar graph).*

not included (Table 2). Quantifying only cancellous bone portions may provide a good measure of the impact of carnivores on those portions, but not on skeletal element abundance as a whole.

Since carnivores are the secondary or only agents of destruction in all of these studies, we hypothesized that skeletal elements would show similar levels of destruction across the different studies. To test this, we ran a correlation analysis on each pair of samples. We first standardized each pair of samples to the range of elements present. For instance, if one study provided percentage survival for the caudal vertebrae and the other did not, this element was removed from the analysis and the ranking procedure that followed. The remaining elements were then ranked in abundance, and a bootstrap correlation and probability statistic were generated from 1000 permutations of the datasets. The null hypothesis in this case was that the paired samples were not correlated. Table 3 shows that 19 out of 21 pairs of samples are strongly correlated ($p < .05$), and one of the remaining pair are nearly correlated at this level. We therefore reject the null and conclude that these five studies show us that carnivore destruction of human discarded bone results in very similar inter-element levels of destruction when shafts are not included in analysis.

Given this concordance, we calculated a mean percentage survival for each skeletal element in order to show a general pattern (Table 2, last column; Figure 1). Head elements survive best (particularly the mandible), while scapulae and pelves survive reasonably well. The limb bone ends show wide variation (both between elements and between studies). Distal humeri, proximal radius-ulnae, distal tibiae, and proximal metapodials survive well, while all other limb bone ends have mean survival rates below 30%. Vertebral portions, ribs, carpals, tarsals, and phalanges survive quite poorly (all means below 31%). Among these non-shaft portions, there is generally a close relationship between the average rate of survival and the range of variation in survival. That is, those element portions that survived well on average vary more in survival between studies. The fact that an element portion survives well in this grouping is therefore not an indication of consistent high survival.

We conducted the same analysis on those few studies that include shaft portions in their calculations (see Table 4 for the raw data). These were the small and medium duiker samples from Hudson's (1993) Aka study, and Marean's Berkeley hyena colony study (Marean and Spencer, 1991; Marean et al., 1992). The Berkeley hyena MNE's have been recalculated using an updated zooarchaeological coding system, and the numbers differ slightly from those in Marean et al. 1992. The correlation between Hudson's duikers was very low, as was the correlation between Marean's data (either whole bone or hammerstone broken) and the medium duikers (Table 5). However, Marean's data were highly correlated with the small duiker sample. The medium duikers appear to have undergone some taphonomic process different from either Marean's sheep or other duikers in the same study. Although we do not have the contextual data that could clarify this issue, we believe we can show a mechanistic difference when we examine the relationship of these elements to density. We will return to the problem below.

In Figure 2, we compare percentage survival in Hudson's small duiker sample to that of Marean's sheep. In Marean's study, negative values indicate elements not included in the research (no upper limbs were used), while zero values represent elements that were initially present but completely destroyed. Overall, survival is lower in Hudson's study, due in part to some loss during bone recovery (as discussed above). However, the pattern of destruction is quite similar in the two studies, despite differences in carnivores, prey, and fragment collection. Vertebrae survive poorly relative to other post-

*Table 3. Paired rank correlations of skeletal element survival in studies **without** long bone shaft data*

| Study and carnivore taxon | A. Wolf | B. Domestic dog | C. Domestic dog | D. Domestic dog | E. Domestic dog | F. Wolf |
|---|---|---|---|---|---|---|
| B. Domestic dog | 0.737 (.001) | | | | | |
| C. Domestic dog | 0.632 (.001) | 0.819 (.001) | | | | |
| D. Domestic dog | 0.618 (.002) | 0.714 (.001) | 0.783 (.001) | | | |
| E. Domestic dog | 0.408 (.033) | 0.694 (.001) | 0.672 (.002) | 0.620 (.003) | | |
| F. Wolf | 0.498 (.008) | 0.659 (.001) | 0.633 (.001) | 0.795 (.001) | 0.681 (.001) | |
| G. Fox | *0.246 (.138)* | 0.673 (.001) | 0.794 (.001) | 0.496 (.010) | 0.675 (.001) | *0.330 (.057)* |

*Correlation coefficient and p value (in parentheses) are given for paired samples. Insignificant results are indicated in italics. A. Binford (1981: 211-213, Table 5.01, total wolf kills). B. Binford and Bertram (1977: 101, Table 3.5, winter sample). C. Binford and Bertram (1977: 101, Table 3.5, summer sample). D. Binford and Bertram (1977: 81, Table 3.2, record B). E. Brain (1967: 109, Table 3). F. Klippel et al. (1987: 158, Table 1). G. Stallibrass (1984: Figure 1).*

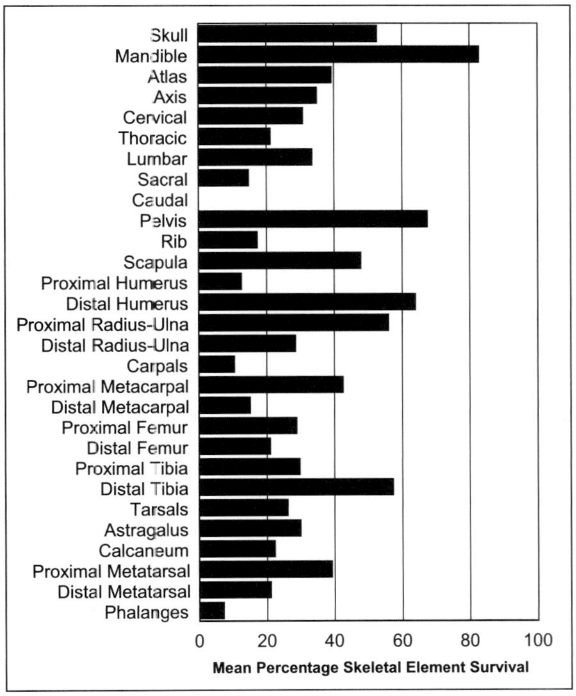


*Figure 1. Mean percentage skeletal element survival from studies that* ***did not*** *report long bone shaft survival (see Table 2).*

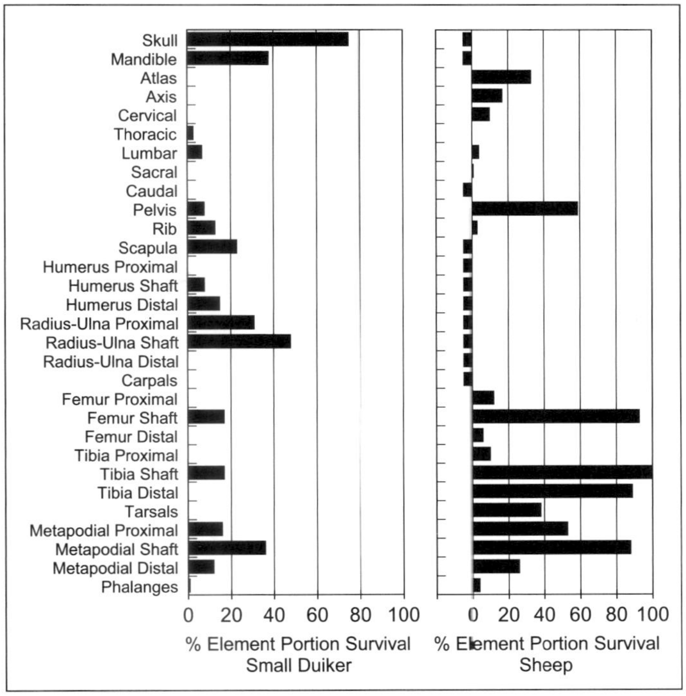


*Figure 2. Percentage element survival from two studies that report long bone shaft survival. Small duiker sample is from Hudson (1993: 316, Table 17.4). Sheep sample is from Marean's hyena feeding experiment (described in Marean and Spencer, 1991). Negative values indicate elements not included in the study. Zero values indicate elements were originally present but completely destroyed.*

cranial elements, and limb shafts generally survive well and give the highest MNE per element (with the exception of small duiker humeri). We now turn to the question of why carnivore ravaging has such an unequal effect on different skeletal portions, and thus wreaks havoc with the skeletal element analysis.

## Bone Density as a Mediator of Skeletal Element Survival

Structural properties have long been recognized as important to bone survival (White, 1953, 1954; Brain, 1969; Binford and Bertram, 1977; Lyman, 1984). The two most important of these properties are bone mineral density and within-bone nutrient distribution—two intimately linked factors. The latter can be determined by observing the distribution of trabecular (grease-bearing) bone, a task that can be accomplished using either comparative or archaeological assemblages. The techniques for estimating bone mineral density are far more complicated, but a consensus is emerging (Lam et al., 2003).

Brain (1967, 1969) made some of the earliest attempts to accurately estimate the structural properties of bone and compare it to skeletal element survival. He used the low-tech, but remarkably effective technique of water displacement to determine "specific gravity" (g/cm$^3$) in goat bones. Lyman (1984), however, pointed out that this technique tended to measure bulk density rather than true density because it inadequately accounted for pore space, particularly within trabecular bone. Recognizing the importance of developing a reliable, widely reproducible method for measuring density, Lyman (1984) turned to photon densitometry. This technique measures the attenuation of a photon beam as it passes through an object. The greater the mineral content of the object, the greater the attenuation of the beam, resulting in a true measure of bone mineralization. Numerous researchers (Kreutzer, 1992; Stahl, 1999; Pickering and Carlson, 2002) saw the practicality of this approach and began producing density value measurements for various species. However, in order to derive density from the mineralization value, it is necessary to know the area of bone over which the beam has passed—that is, the cross-sectional area. Researchers have measured this cross-sectional area in different ways with varying degrees of accuracy, a fact that has created methodological incompatibilities among studies (Lam et al., 2003).

The problem of inaccurate shape estimation introduces considerable error when calculating the density of long bone shafts, fragments of which are common in the archaeological record. Unlike the trabecular portions of skeletal elements, medullary shafts have a dense area of bone around a large empty canal. If cross-sectional area of the bone is calculated based on the external dimensions of the shaft, the mineral content of the cortex is effectively smeared out over the empty medullary cavity. This

*Table 4. Percentage survival of skeletal elements in studies that reported long bone shaft survival*

| | Hudson's (1993) dog ravaging study | | Marean's hyena feeding experiments* | | All Studies |
|---|---|---|---|---|---|
| | Medium duiker % survival | Small duiker % survival | Original MNE | % Survival Sheep | Mean % Survival |
| Skull | 67 | 75 | ni | ni | 71 |
| Mandible | 50 | 38 | ni | ni | 44 |
| Atlas | 50 | 0 | 6 | 33 | 28 |
| Axis | 50 | 0 | 6 | 17 | 22 |
| Cervical | 40 | 0 | 30 | 10 | 17 |
| Thoracic | 21 | 3 | 18 | 0 | 8 |
| Lumbar | 15 | 7 | 56 | 4 | 9 |
| Sacral | 33 | 0 | 140 | 1 | 11 |
| Caudal | 0 | 0 | ni | ni | 0 |
| Pelvis | 40 | 8 | 49 | 59 | 36 |
| Rib | 74 | 13 | 36 | 3 | 30 |
| Scapula | 50 | 23 | ni | ni | 37 |
| Humerus Proximal | 0 | 0 | ni | ni | 0 |
| Humerus Shaft | 67 | 8 | ni | ni | 38 |
| Humerus Distal | 47 | 15 | ni | ni | 31 |
| Radius-Ulna Proximal | 33 | 31 | ni | ni | 32 |
| Radius-Ulna Shaft | 17 | 48 | ni | ni | 33 |
| Radius-Ulna Distal | 0 | 0 | ni | ni | 0 |
| Carpals | 0 | 0 | ni | ni | 0 |
| Femur Proximal | 0 | 0 | 50 | 12 | 4 |
| Femur Shaft | 0 | 17 | 50 | 93 | 37 |
| Femur Distal | 0 | 0 | 50 | 6 | 2 |
| Tibia Proximal | 0 | 0 | 50 | 10 | 4 |
| Tibia Shaft | 40 | 17 | 50 | 101 | 53 |
| Tibia Distal | 0 | 0 | 50 | 89 | 30 |
| Tarsals | 3 | 0 | 150 | 38 | 14 |
| Astragalus | ni | ni | 50 | 36 | 36 |
| Calcaneum | ni | ni | 50 | 28 | 28 |
| Metapodial Proximal | 18 | 16 | 50 | 53 | 29 |
| Metapodial Shaft | 27 | 36 | 50 | 88 | 50 |
| Metapodial Distal | 0 | 12 | 50 | 26 | 13 |
| Phalanges | 6 | 1 | 100 | 4 | 4 |

*Hudson (1993: 316, Table 17.4). *Marean's hyena data (combined hammerstone broken and whole bone feeding experiments) recalculated from original hyena data (experiment details given in Marean and Spencer, 1991). ni = element not included in study.*

*Table 5. Paired rank correlations of skeletal element survival in studies* ***with*** *long bone shaft data (insignificant results indicated with italics)*

| Study, carnivore, and carcass type | A. Domestic dog consuming medium duiker | B. Domestic dog consuming small duiker | C. Spotted hyena consuming sheep (broken bone)* |
|---|---|---|---|
| B. Domestic dog consuming small duiker | *.341 (.063)* | | |
| C. Spotted hyena consuming sheep (broken bone)* | *-.170 (.703)* | .500 (.039) | |
| D. Spotted hyena consuming sheep (whole bone)* | *.182 (.324)* | .758 (.009) | .968 (.001) |

*A. Hudson (1993: 316, Table 17.4, medium duiker). B. Hudson (1993: 316, Table 17.4, small duiker). C. Marean's hyena study (hammerstone broken feeding experiments)*. D. Marean's hyena study (whole bone feeding experiments)*. *Recalculated from original hyena data (experiment details given in Marean and Spencer, 1991).*

calculation significantly underestimates the true density of the shaft. Lam et al. (1998,1999, 2003) proposed computed tomography as an alternative method that would return simultaneous accurate estimates of cross-sectional areas and mineral density. These researchers point out that photon densitometry can provide accurate values for long bone shafts if used in conjunction with a technique that accurately estimates shape (i.e. radiograph or water displacement), but that these have not been widely employed.

## Early studies of the effect of density on the SEA

Although earlier researchers had guessed that bone mineral density probably had an effect on representation, Brain was one of the first to discuss (1967) and show (1969) an actual correlation between these values. Lyman (1984, 1985) and Grayson (1989) recognized the serious implications this would have for faunal analysis. They pointed out that density partially correlated with the bone utility indices, making it difficult to determine whether *in situ* destruction or selective transport were responsible for shaping the final element profile. Both researchers thought that the reverse utility curves found at numerous sites were possibly the result of differential preservation rather than human selectivity. Lyman (1991, 1993, 1994) tested this idea by performing correlation tests between his photon densitometry data and element representation in 143 published archaeological assemblages. Of these, 53% had a positive, significant correlation between density and representation. Furthermore, 71% of the sites (n=38) that had a reverse-utility curve also had a correlation with density (Lyman, 1994: 264). Thus, Lyman showed that density-mediated destruction was widespread in the archaeological record, and would severely limit the application of utility indices. If representation was unrelated to density in any given assemblage, Grayson (1989) had suggested the analyst might avoid the difficulty of disentangling selective transport from *in situ* attrition. Unfortunately, Lyman's study showed that this might obtain in less than half of the record. Marean and Frey (1997) showed that the reverse utility curve in the long bone set collapses when shafts are included, and thus argued that non-cortical bone portions drive much of the patterning in the relation between survivorship and density.

Earlier studies of the relationship between density and element representation (Lyman, 1991, 1993, 1994) were based on photon densitometry data without shape correction (Lam et al., 1998). As discussed above, these values significantly underestimate the density of long bone shafts. In addition, the assemblages used in Lyman's literature survey provided minimum number of elements (MNEs) based on long bone end counts, and did not include shaft portions (Lyman, 1993: 326). At that time, few researchers were publishing derived estimates (such as the MNE) for shaft portions of long bones. Actualistic studies in the 1980s (Blumenschine, 1986, 1988) and early 1990s (Marean and Spencer, 1991, Marean et al., 1992) demonstrated that carnivore ravaging (a significant agent of density-mediated destruction) preferentially deletes long bone ends. As a result, MNEs calculated solely on the basis of ends are almost certain to severely underestimate long bone representation. Pickering et al. (2003) demonstrated this bias in both archaeological and ethnographic sites. Thus, Lyman's (1993) study used data that underestimated both long bone representation and the density of long bone shaft portions. It now seems prudent to revisit the relationship between density and skeletal element survival for the following reasons. We now have more accurate density values. We now have several archaeological studies for which shaft portions are included in limb bone MNE estimates (thus, we have more accurate estimates on those skeletal elements). Finally, we now have some data on skeletal element survival (with shaft portions) following carnivore destruction (see above).

## A new analysis of the relationship between density and skeletal element survival

Based on a survey of actualistic data, we have thus far established that carnivores systematically damage skeletal elements in a way that is consistent across predator and carcass type—that is, trabecular portions are variably affected by the intensity of ravaging, while certain non-trabecular portions are more consistently preserved. We will now re-test the hypothesis that density moderates this pattern of destruction. If supported, the result would present an equifinality problem for skeletal element analyses. That is, in assemblages impacted by carnivores, the archaeologist would not be able to simply differentiate an element pattern shaped by hominid transport decisions from one shaped by carnivore ravaging.

### *Density and carnivore ravaging*

Our first question here is straightforward, and follows Brain's original (1967, 1969) search for the cause of differential skeletal element survival: is carnivore destruction of skeletal elements mediated by density? In order to develop a reliable model for interpretation, we again turn to the actualistic research to secure the link between agent and taphonomic pattern. Using the datasets listed in Table 2, we test the hypothesis that there is a significant ($p \leq .05$) positive correlation between density and percentage survival. Because these studies quantify long bone portions on the basis of ends, we are strictly testing the effect of density on cancellous bone portions. The advantage of this is that we can make inferences about the role density has played in earlier zooarchaeological reports that lack shafts in the analysis.

Because the studies listed in Table 2 used bovids or cervids as prey carcasses, we compare representation to Lam et al.'s (1999) density values for wildebeest (*Connochaetes taurinus*) and reindeer (*Rangifer tarandus*), but not for *Equus*. We are primarily interested here in inter-bone survival, so we use only one density value (the highest) per element. This also precludes any bias that might have resulted from the variability in the number of scan sites and landmarks per element.

*Table 6. Correlation between percentage element survival and bone mineral density (BMD) (insignificant results indicated with italics)*

| **6.1. Studies *without* long bone shaft data** | **R** | **p** |
|---|---|---|
| A. Wolf consuming caribou | 0.313 | *0.064* |
| B. Dog consuming sheep | 0.384 | 0.024 |
| C. Dog consuming sheep | 0.408 | 0.012 |
| D. Dog consuming reindeer | *0.270* | *0.101* |
| E. Dog consuming goat | 0.578 | 0.006 |
| F. Wolf consuming whitetail deer | 0.363 | 0.026 |
| G. Fox consuming sheep | *0.186* | *0.189* |
| H. Spotted hyena consuming size 3 bovid | 0.346 | 0.045 |

*A. Binford (1981: 211-213, Table 5.01, total wolf kills). B. Binford and Bertram (1977: 101, Table 3.5, winter sample). C. Binford and Bertram (1977: 101, Table 3.5, summer sample). D. Binford and Bertram (1977: 81, Table 3.2, record B). E. Brain (1967: 109, Table 3). F. Klippel et al. (1987: 158, Table 1). G. Stallibrass (1984: Figure 1). H. Richardson (1980: 116-117, Figures 8 and 10, spotted hyena data). Rank order from Richardson's study is extrapolated from published bar graphs (no quantitative data provided). Bone mineral density values per element from Lam et al. (1999: Table 1, wildebeest column).*

| | **Wildebeest BMD** | | **Reindeer BMD** | |
|---|---|---|---|---|
| **6.2. Studies *with* long bone shaft data** | **R** | **p** | **R** | **p** |
| A. Dog consuming medium duiker | *0.314* | *0.080* | *0.254* | *0.116* |
| B. Dog consuming small duiker | 0.790 | 0.001 | 0.806 | 0.001 |
| C. Spotted hyena consuming sheep (broken bone)* | 0.712 | 0.009 | 0.816 | 0.002 |
| D. Spotted hyena consuming sheep (whole bone)* | 0.674 | 0.001 | 0.680 | 0.003 |

*A. Hudson (1993: 316, Table 17.4, medium duiker) . B. Hudson (1993: 316, Table 17.4, small duiker). C. Marean's hyena study (hammerstone broken feeding experiments)*. D. Marean's hyena study (whole bone feeding experiments)*. *Recalculated from original hyena data (experiment details given in Marean and Spencer, 1991). Bone mineral density values per element from Lam et al. (1999: Table 1).*

Table 6.1 shows the results: the null hypothesis can be rejected in 5 of the 8 samples, and in the 3 that do not meet the .05 limit, the probability of attaining that correlation through a random process is less than 20%. These results support our hypothesis that carnivore destruction, whenever it occurs, has a significant density-dependent effect on representation.

Ideally, we should be able to test whether the hypothesis holds for the overall skeletal element profile—not just the cancellous portions. Unfortunately, there are only two published actualistic datasets that incorporate shafts into the quantification of percentage survival: Hudson's (1993) Aka research and Marean's hyena colony study (Table 4). In Hudson's (1993) data, density and percentage survival correlate insignificantly among medium duikers, but are very highly correlated among small duikers (Table 6.2). Marean's data (Table 6.2) show a significant correlation between density and survival, comparable to that seen in Hudson's small duikers. As discussed above, element representation correlates well between Marean's study and the small duikers, while the medium duikers appear to be taphonomically distinct. These combined results strongly suggest that the medium duiker component was subjected to relatively less ravaging than the other two datasets.

The critical difference between these actualistic studies (Tables 2 and 4) and archaeological assemblages is, of course, time. A host of taphonomic processes contributes to the formation of an archaeofaunal assemblage (including variable transport, and biogenic and geologic destruction), and actualistic models are not meant to replicate the full range of possible events. Instead, the purpose of these experiments is to develop reliable inferences about parts of the system. The above results demonstrate that carnivores can significantly and systematically modify skeletal element representation even when they are not the primary agents of accumulation. As noted above, this creates a problem of equifinality for the archaeologist.

### *High and low survival elements*

Our review of the record shows that carnivores, when presented with either fleshed or defleshed skeletal remains, will consume and fragment some skeletal portions in preference to others. Our understanding of this process, combined with our understanding of bone density, led us to propose a general model of skeletal element survival in archaeological sites (Marean and Cleghorn, 2003). To explain that model, we will distinguish between nutritive and non-nutritive processes of bone destruction (Blumenschine, 1986, 1988; Capaldo, 1997).

Nutritive processes of destruction are those resulting from attempts to extract nutrition, particularly from bone portions where nutrients and bone are not easily separated. Nutrients include marrow within the cortical portions of long bones and mandibles, grease stored in cancellous bone, and brain matter. Importantly, marrow is separable from cortical bone before consumption, and carnivores typically crack, spit out, ignore, or avoid the surrounding cortical bone portions (Bunn and Kroll, 1986; Blumenschine, 1988; Binford et al., 1988; Marean and Spencer, 1991; Blumenschine and Marean, 1993). Bone grease is not mechanically separable from cancellous bone by non-human animals. As our review above documents, carnivores adapt to this problem by chewing and swallowing the cancellous portions and allowing the digestive tract to render out the grease. To survive these nutritive processes of destruction and thus be countable by the zooarchaeologist, a bone must have a substantial portion of thick cortical bone free of cancellous bone (Figure 3). Any bone portion with associated cancellous bone is more likely to be destroyed or deleted by carnivores scavenging from human meals, and our review above documents that this process is geographically and environmentally widespread. It is now safe to say that nutritive attrition can be considered a law of site formation process that must guide all zooarchaeological analyses where carnivore involvement has been verified.

Non-nutritive bone destruction includes those processes that are not the result of animals attempting to derive nutrients. These include trampling, sediment compaction, chemical leaching, burning, and any other chemical or mechanical process that destroys bone. It is widely believed that these processes are density mediat-

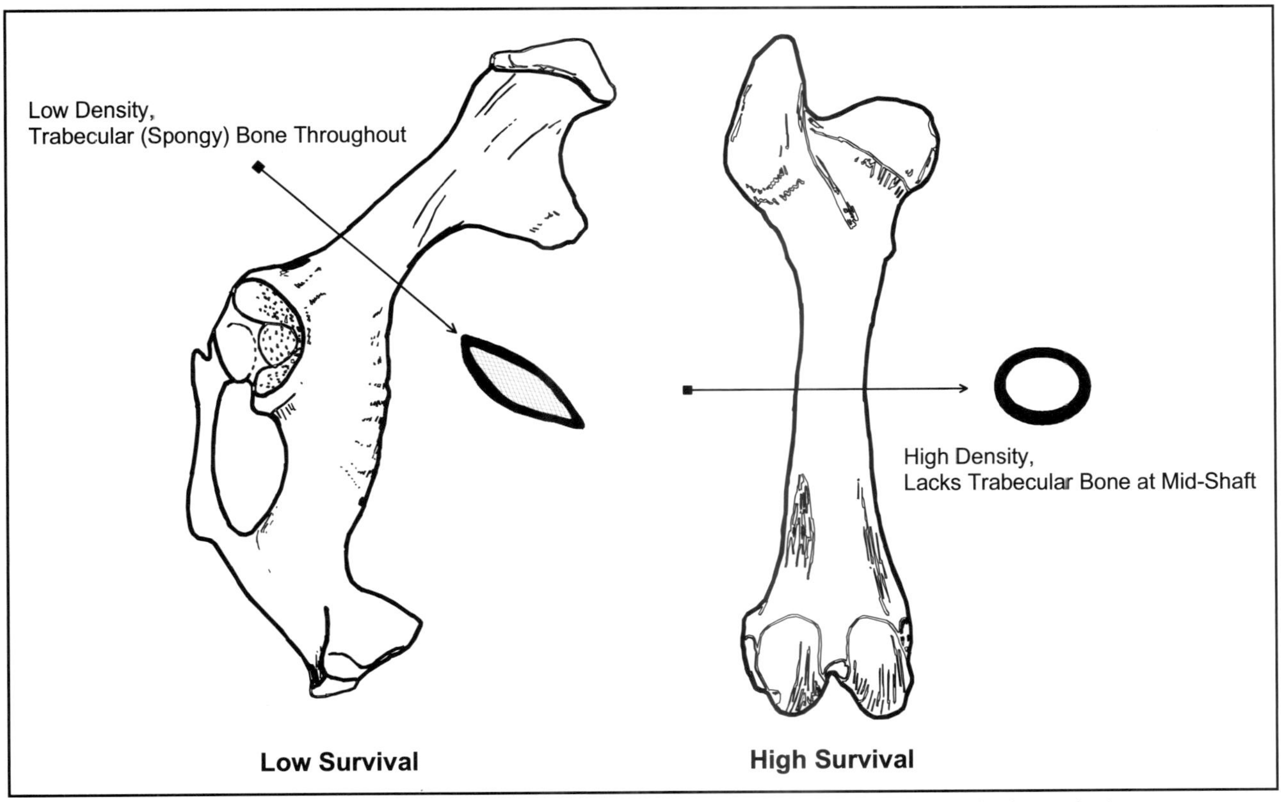


*Figure 3. Trabecular content distinction between high and low survival elements. Although the innominate may sometimes survive well (as noted in the text), this survival is highly variable (as seen in Table 2 and Figure 2).*

ed, meaning that the potential for non-nutritive destruction correlates inversely with bone mineral density (Lyman, 1984, 1985; Grayson, 1989; Lyman, 1992). There is still little experimental research documenting this relationship. If true, however, then there are two important propositions that arise. First, skeletal elements that lack at least some reasonably dense portion will have a lower frequency of survival (in an identifiable state). Bone density studies have shown that the densest parts of bovid and cervid skeletons are the thick cortical portions of long bones, the petrosal, and the teeth (Lyman, 1984; Kreutzer, 1992; Lam et al., 1998). Second, the only skeletal elements that will record relative abundances that reflect their original discard abundance are those that have similar high-density cortical portions free of cancellous bone.

We believe these nutritive and non-nutritive processes of bone destruction divide the skeleton into ***high survival*** and ***low survival*** elements. High survival elements are those that accurately represent their frequency (relative to each other) as it was before carnivore ravaging and other density-mediated destruction. These bones may be consistently useful when investigating hominid behavior through skeletal element analysis. The following three criteria distinguish the high survival set:

1. Each skeletal element has a substantial portion of thick cortical bone, lacking cancellous bone (Figure 3).
2. The density throughout the cortical portion is high and relatively homogeneous among the elements chosen for analysis.
3. The cortical portion is identifiable to skeletal element, and zooarchaeologists can identify and quantify it accurately.

High survival elements include all of the limb bones (excluding the carpals and the phalanges of size 1 and 2 animals), mandibles (which have dense cortical bone and an open medullary cavity similar to long bones), and crania (due to the presence of teeth and the petrosal).

By contrast, the relative representation of low survival elements will reflect their ability to survive the variety of processes that affect the assemblage after transport and discard. These bones include all vertebrae, ribs, pelves, scapulae (which have thick cortical bone but may be difficult to identify and quantify when fragmented), tarsals, carpals, and the phalanges of size 1 and 2 animals since these tend to be swallowed by carnivores (Marean, 1991; see our discussion above and Figure 1). These elements may be useful for evaluating the level of destruction to which the assemblage has been subjected, but their variation may also be the result of differential transport in an archaeological assemblage. The difficulty in distinguishing between the effects of transport and *in situ* destruction make these elements unreliable indicators of either taphonomy or behavior.

It is important to emphasize that the high and low survival elements are not distinguished based on an arbitrary bisection of the density scale. In fact, the primary evidence suggesting the presence of only two real survival sets comes from the data on nutrient-based destruction by carnivores, as discussed above. However, when comparing the density values of the two sets using Lam et al.'s (1999) shape-corrected data, two important differences emerge. First, the highest value per element is generally greater in the high survival set. The exception is the ulna. In bovids and cervids, however, the ulna has a quantifiable landmark (the mid-shaft radius articulation) that often fuses with the cortical portion of the radius and is then preserved along with the radius shaft. The second and more important difference is that there is much less variation in bone mineral density within the high survival set when compared to either the low survival set (Table 8) or to the full spectrum of bone density.

We can now examine the usefulness of the high-low survival dichotomy by testing the following three hypotheses:

*Table 7. Results of correlations between percent survival and bone mineral density (BMD) by high and low survival sets (insignificant results indicated with italics)*

| Hudson's (1993) dog ravaging study | Wildebeest BMD R | Wildebeest BMD p | Reindeer BMD R | Reindeer BMD p |
|---|---|---|---|---|
| medium duiker | | | | |
| High Survival | *0.120* | *0.380* | *0.140* | *0.340* |
| Low Survival | *0.330* | *0.100* | *0.190* | *0.240* |
| small duiker | | | | |
| High Survival | *0.078* | *0.428* | *-0.109* | *0.589* |
| Low Survival | *0.393* | *0.076* | 0.492 | 0.048 |

*Bone mineral density values per element from Lam et al. (1999: Table 1).*

*Table 8. Summary statistics for bone mineral density values*

| | High Survival | | | Low Survival | | |
|---|---|---|---|---|---|---|
| | **Mean** | **sd** | **CV** | **CV** | **sd** | **Mean** |
| Zebra | 1.05 | 0.15 | 14% | 32% | 0.20 | 0.64 |
| Wildebeest | 1.11 | 0.11 | 10% | 30% | 0.21 | 0.70 |
| Reindeer | 1.10 | 0.12 | 11% | 27% | 0.18 | 0.68 |

*Highest bone mineral density values per element from Lam et al. (1999: Table 1). sd = standard deviation, CV = coefficient of variation.*

*Table 9. Archaeofaunal assemblages used in the evaluation of the high and low element survival sets*

| Site | Taxa | Site Type | Location | Zooarchaeologist |
|---|---|---|---|---|
| Ain Dara | Size 1 & 2 (mostly Sheep & Goat) | Iron Age Tell | Syria | C. Frey |
| Ain Dara | Bovid/Cervid Size 3 & 4 | Iron Age Tell | Syria | C. Frey |
| Mezmaiskaya, MP levels | Size 2 (mostly Sheep & Goat) | Mousterian Cave | Caucasus Mtns., Russia | N. Cleghorn |
| Mezmaiskaya, MP levels | Bovid/Cervid Size 3 & 4 | Mousterian Cave | Caucasus Mtns., Russia | N. Cleghorn |
| Die Kelders I | Bovid Size 2 | MSA Cave | South Africa | C. Marean & Students |
| Die Kelders I | Bovid Size 3 & 4 | MSA Cave | South Africa | C. Marean & Students |
| Kunji | Size 1 & 2 (mostly Sheep & Goat) | Mousterian Cave | Zagros Mtns, Iran | C. Marean & Students |
| Kobeh | Size 1 & 2 (mostly Sheep & Goat) | Mousterian Cave | Zagros Mtns, Iran | C. Marean & Students |
| Porc Epic | Bovid Size 2 | MSA Cave | Ethiopia | Z. Assefa |
| Agate Basin, Folsom Comp. | Bison | Open-air Kill | Plains, USA | M. Hill |
| Agate Basin, Folsom Comp. | Pronghorn | Open-air Kill | Plains, USA | M. Hill |
| Agate Basin, Hell Gap Comp. | Bison | Open-air Kill | Plains, USA | M. Hill |
| Agate Basin, Agate Basin Comp. | Bison | Open-air Kill | Plains, USA | M. Hill |
| Clary Ranch | Bison | Open-air Kill | Plains, USA | M. Hill |

*MP = Middle Paleolithic, MSA = Middle Stone Age*

H1 there is a significant ($p \leq .05$) positive correlation between density and representation in the low survival set.

H2 this correlation, while not significant, is greater in the low than in the high survival set.

H3 there is a significant positive correlation between density and representation among the high survival set.

Of these hypotheses, H1 and H2 are mutually exclusive but H3 may co-occur with H1. A high correlation in both high and low survival sets indicates a degree of density-mediated attrition high enough to differentiate even elements that have relatively similar density values. In this case, we can still examine the relative correlations between high and low survival sets.

Ideally, we would like to first test these hypotheses using a large sample of actualistic data. Unfortunately, only two published studies, Marean and Spencer (1991) and Hudson (1993) so far meet the standards of MNE quantification required (i.e., they include long bone shafts). Marean's hyena research is precluded from this analysis by the limited range of carcass elements studied. We compared percentage survival of small and medium duikers (Hudson, 1993) to both wildebeest and reindeer density values (Lam et al., 1999) using a rank correlation test as described above.

The results (Table 7) were not significant. Within the medium duiker set, the correlation is slightly higher within the low survival set, a result supporting our secondary hypothesis (H2). The small duiker results are mixed. This test of this particular dataset suffers from the possible disadvantage of comparing percentage survival with the density values of much larger taxa. Our analysis of the limited actualistic data is therefore inconclusive. A re-analysis of the original collections of Binford and Bertram (1977), Binford (1978), Brain (1969), Stallibrass (1984), and Snyder (Klippel et al., 1987; Snyder, 1988), including a recalculation of percentage survival with limb bone shafts, could usefully expand this investigation.

*Table 10. Correlations between element representation and bone mineral density (BMD)*

| 10.1 Correlations using minimum number of element (MNE) representation | | | | | | | | |
|---|---|---|---|---|---|---|---|---|
| | Highest BMD | | | | Representative BMD | | | |
| | High Survival | | Low Survival | | High Survival | | Low Survival | |
| Archaeofaunal Assemblage | R | p | R | p | R | p | R | p |
| Ain Dara size 1 & 2 | -0.145 | 0.659 | 0.647 | 0.005 | 0.000 | 0.499 | 0.663 | 0.009 |
| Ain Dara size 3 & 4 | 0.033 | 0.479 | 0.631 | 0.009 | 0.620 | 0.058 | 0.803 | 0.006 |
| Kobeh size 1 & 2 | 0.018 | 0.474 | 0.625 | 0.004 | 0.656 | 0.047 | 0.529 | 0.027 |
| Kunji size 1 & 2 | 0.018 | 0.470 | 0.538 | 0.022 | -0.171 | 0.685 | 0.243 | 0.184 |
| Mezmaiskaya MP size 2 | -0.127 | 0.620 | 0.526 | 0.021 | -0.356 | 0.752 | 0.366 | 0.142 |
| Mezmaiskaya MP size 3 & 4 | 0.052 | 0.439 | 0.500 | 0.034 | -0.342 | 0.794 | 0.358 | 0.091 |
| Porc Epic size 2 | -0.018 | 0.503 | 0.499 | 0.020 | -0.151 | 0.649 | 0.499 | 0.020 |
| Agate Basin bison | 0.431 | 0.119 | 0.575 | 0.017 | -0.131 | 0.573 | 0.410 | 0.069 |
| Clary Ranch bison | -0.519 | 0.905 | 0.501 | 0.030 | -0.305 | 0.706 | 0.345 | 0.112 |
| Hell Gap bison | -0.430 | 0.845 | 0.362 | 0.101 | 0.368 | 0.202 | 0.427 | 0.082 |
| Agate Basin, Folsom component bison | -0.142 | 0.614 | 0.379 | 0.103 | -0.084 | 0.586 | -0.062 | 0.572 |
| Agate Basin, Folsom component pronghorn | 0.050 | 0.041 | 0.464 | 0.047 | 0.384 | 0.181 | 0.290 | 0.186 |
| Die Kelders I size 2 | 0.824 | 0.012 | 0.316 | 0.109 | 0.678 | 0.035 | -0.102 | 0.649 |
| Die Kelders I size 3 & 4 | 0.013 | 0.501 | 0.286 | 0.140 | 0.021 | 0.461 | 0.177 | 0.278 |

*Bone mineral density values are from wildebeest (Lam et al., 1999, Table 1).*

| 10.2 Correlations using minimum animal unit (MAU) representation | | | | | | | | |
|---|---|---|---|---|---|---|---|---|
| | Highest BMD | | | | Representative BMD | | | |
| | High Survival | | Low Survival | | High Survival | | Low Survival | |
| Archaeofaunal Assemblage | R | p | R | p | R | p | R | p |
| Ain Dara size 1 & 2 | -0.178 | 0.667 | 0.770 | 0.001 | -0.071 | 0.552 | 0.631 | 0.018 |
| Ain Dara size 3 & 4 | 0.033 | 0.479 | 0.636 | 0.012 | 0.620 | 0.058 | 0.512 | 0.073 |
| Kobeh size 1 & 2 | 0.018 | 0.474 | 0.847 | 0.001 | 0.656 | 0.047 | 0.742 | 0.003 |
| Kunji size 1 & 2 | -0.021 | 0.522 | 0.545 | 0.019 | -0.239 | 0.750 | 0.418 | 0.065 |
| Mezmaiskaya MP size 2 | -0.127 | 0.620 | 0.510 | 0.032 | -0.356 | 0.752 | 0.175 | 0.312 |
| Mezmaiskaya MP size 3 & 4 | -0.053 | 0.571 | 0.431 | 0.049 | -0.379 | 0.794 | 0.386 | 0.109 |
| Porc Epic size 2 | -0.018 | 0.503 | 0.423 | 0.056 | -0.151 | 0.649 | 0.423 | 0.056 |
| Agate Basin bison | 0.431 | 0.119 | 0.504 | 0.038 | -0.131 | 0.573 | 0.713 | 0.003 |
| Clary Ranch bison | -0.519 | 0.905 | 0.394 | 0.073 | -0.305 | 0.706 | 0.255 | 0.183 |
| Hell Gap bison | -0.430 | 0.845 | 0.415 | 0.074 | 0.368 | 0.202 | 0.597 | 0.022 |
| Agate Basin, Folsom component bison | 0.050 | 0.419 | 0.285 | 0.147 | -0.084 | 0.586 | 0.231 | 0.263 |
| Agate Basin, Folsom component pronghorn | 0.606 | 0.041 | 0.015 | 0.476 | 0.384 | 0.181 | 0.112 | 0.378 |
| Die Kelders I size 2 | 0.824 | 0.012 | 0.372 | 0.098 | 0.678 | 0.035 | 0.306 | 0.145 |
| Die Kelders I size 3 & 4 | 0.036 | 0.452 | 0.444 | 0.052 | -0.013 | 0.529 | 0.384 | 0.107 |

*Bone mineral density values are from wildebeest (Lam et al., 1999, Table 1).*

*Table 11. Summary of hypothesis support among archaeofaunal components*

| | Highest Density | | | | Representative Density | | | |
|---|---|---|---|---|---|---|---|---|
| | Wildebeest BMD | | Reindeer BMD | | Wildebeest BMD | | Reindeer BMD | |
| **Archaeofaunal Assemblage** | **MNE** | **MAU** | **MNE** | **MAU** | **MNE** | **MAU** | **MNE** | **MAU** |
| Ain Dara size 1 & 2 | H1 | H1 | H1 | H1 | H1 | H1 | H1 | H2 |
| Ain Dara size 3 & 4 | H1 | H1 | H1 | H1 | H1 | - | (H1)/H3 | H3 |
| Kobeh size 1 & 2 | H1 | H1 | H1 | H1 | H1/H3 | H1/H3 | H3 | H1/H3 |
| Kunji size 1 & 2 | H1 | H1 | H1 | H1 | H2 | H2 | H2 | H1 |
| Mezmaiskaya MP size 2 | H1 | H1 | H1 | H1 | H2 | H2 | - | - |
| Mezmaiskaya MP size 3 & 4 | H1 | H1 | H1 | H1 | H2 | H2 | H2 | H1 |
| Porc Epic size 2 | H1 | H2 | H1 | H2 | H1 | H2 | H1 | H2 |
| Agate Basin bison | H1 | H1 | H2 | H2 | H2 | H1 | H2 | H2 |
| Clary Ranch bison | H1 | H2 | H2 | H2 | H2 | H2 | H2 | H2 |
| Hell Gap bison | H2 | H2 | H2 | H2 | H2 | H1 | H2 | H2 |
| Agate Basin, Folsom component bison | H2 | H2 | - | - | - | H2 | - | H2 |
| Agate Basin, Folsom component pronghorn | (H1)/H3 | H3 | H3 | H3 | - | - | - | - |
| Die Kelders I size 2 | H3 | H3 | H3 | H3 | H3 | H3 | H3 | H3 |
| Die Kelders I size 3 & 4 | H2 | H2 | H2 | H1 | H2 | H2 | H2 | H1 |

*H1: there is a significant ($p \leq .05$) positive correlation between density and representation in the low survival set. H2: this correlation, while not significant, is greater in the low than in the high survival set. H3: there is a significant positive correlation between density and representation among the high survival set. Parentheses indicate the relationship is weaker in the low survival set than in the high survival set.*

*Table 12. Percentage support for each hypothesis and the high-low survival model*

| | Highest BMD | | | | Representative BMD | | | |
|---|---|---|---|---|---|---|---|---|
| | Wildebeest | | Reindeer | | Wildebeest | | Reindeer | |
| | **MNE** | **MAU** | **MNE** | **MAU** | **MNE** | **MAU** | **MNE** | **MAU** |
| H1 | 64% | 50% | 50% | 43% | 29% | 29% | 14% | 29% |
| H2 | 21% | 36% | 29% | 29% | 50% | 50% | 43% | 43% |
| H3 | 14% | 14% | 14% | 14% | 14% | 14% | 21% | 21% |
| **Combined positive (H1 and H2) versus negative (H3) support for the high-low survival model:** | | | | | | | | |
| | **Highest BMD** | | | | **Representative BMD** | | | |
| | **Wildebeest** | | **Reindeer** | | **Wildebeest** | | **Reindeer** | |
| | **MNE** | **MAU** | **MNE** | **MAU** | **MNE** | **MAU** | **MNE** | **MAU** |
| H1 and H2 | 86% | 86% | 79% | 71% | 79% | 79% | 57% | 71% |
| H3 | 14% | 14% | 14% | 14 % | 14% | 14% | 21% | 21% |

*H1: there is a significant ($p \leq .05$) positive correlation between density and representation in the low survival set. H2: this correlation, while not significant, is greater in the low than in the high survival set. H3: there is a significant positive correlation between density and representation among the high survival set.*

## *Density and the final shape of the skeletal element profile*

We can test the power of the high and low survival model within the archaeological record. This may be an appropriate test, since archaeological assemblages have typically undergone both nutritive and post-nutritive processes of destruction. However, this is admittedly a comparative taphonomy approach, resorted to because of the scarcity of appropriate actualistic datasets. We therefore view our results as a guide for future research, and note that the consistency and strength of these results make our conclusions even more compelling.

Table 9 lists the 14 archaeological samples used to test the relationship between density and the final representation of skeletal elements. These samples fulfill two necessary criteria: 1) the analysts present MNE estimates on both shaft and end portions for the limb bones, and 2) we are certain that the procedure used to estimate MNEs provides an adequate estimate for both the shaft and end portions. Our analysis of these assemblages tested whether human, carnivore, and other depositional and post-depositional processes acting on an unknown number of skeletal elements could produce the hypothetical patterns (H1, H2, and H3) as described above. To insure that quantification and data formats minimally affected our test, we tested correlations using several different parameters. For representation, we used both the minimum number of elements (MNE) and the minimum animal units (MAU). Focusing our analyses on the combined bovid/cervid portion of the test assemblages (segregating size 1 and 2 from size 3 and 4 animals), we again used Lam et al's (1999: table 1) wildebeest and reindeer bone mineral density (BMD) values. We used "$BMD_2$" values (as defined in Lam et al., 1999) whenever given, as these have the most accurate area calculation for portions with a medullary cavity. Finally, we ran our correlation tests using two different selection criteria for density. In the first (*highest BMD*), we paired the highest representation of a bone with the highest density value for that element. In this case, we considered that the highest density on a bone represented that element's best chance at entering the archaeological record. In the second comparison (*representative BMD*), we examined each element to find the portion that gave the highest representation, and then compared this to the specific density value from that scan site. Because we were primarily interested in inter-bone survival, we used only one density value per element. This also precluded any bias that might have resulted from variability in the number of scan sites and landmarks per element. Table 10 lists the results of rank correlation tests, and Table 11 provides a summary of hypothesis support.

*Table 13. Hypothesis support and assemblage size*

| Archaeofaunal Assemblage | H1 | H2 | H3 | Maximum MAU |
|---|---|---|---|---|
| Porc Epic size 2 | √ | √ | | 59.4 |
| Kobeh size 1 & 2 | √ | √ | √ | 57.0 |
| Kunji size 1 & 2 | √ | √ | | 45.5 |
| Agate Basin bison | √ | √ | | 39.0 |
| Ain Dara, size 1 & 2 | √ | √ | | 31.5 |
| Mezmaiskaya MP size 2 | √ | √ | | 25.6 |
| Clary Ranch bison | √ | √ | | 20.0 |
| Die Kelders I size 3 & 4 | √ | √ | | 13.5 |
| Mezmaiskaya MP size 3 & 4 | √ | √ | | 13.5 |
| Hell Gap bison | √ | √ | | 11.0 |
| Agate Basin, Folsom comp. bison | | √ | | 7.0 |
| Ain Dara size 3 & 4 | √ | √ | √ | 7.0 |
| Die Kelders I size 2 | | | √ | 5.0 |
| Agate Basin, Folsom comp. pronghorn | | | √ | 3.0 |

*H1: there is a significant ($p \leq .05$) positive correlation between density and representation in the low survival set. H2: this correlation, while not significant, is greater in the low than in the high survival set. H3: there is a significant positive correlation between density and representation among the high survival set.*

Eleven out of the 14 archaeological components we examined supported our primary hypothesis (H1) in at least one (but usually more) of the correlation tests (Figures 4 and 5). Among these assemblages, almost all the tests that did not support H1 supported H2. An additional assemblage (Agate Basin, Main Folsom Component, pronghorn) supported H1 only in a test where H3 was also supported. Another assemblage (Agate Basin, Main Folsom Component, bison) provided some support for only H2. Finally, Die Kelders Cave I (body size 2) did not support either of our main hypotheses, but in fact supported only H3. These last three assemblages, however, appear to be the exceptions to a pattern of significant support for our main hypothesis within the archaeological datasets (Table 11). In seven tests, we found both H1 and H3 to be true. In five of those, the correlation was still much stronger in the low survival set. In the other two, it was not, and we did not count these as supportive.

Table 12 compiles the percentages of support for each hypothesis. This gives us a way to grossly assess the various types of comparisons we made, and show the overall strength of our results when approached with different datasets. Our main hypothesis, H1, was best supported in tests where we compared representation to the maximum density of a given element. Using the best represented density lessened the significance, but did not alter the main trend. Element representation in the low

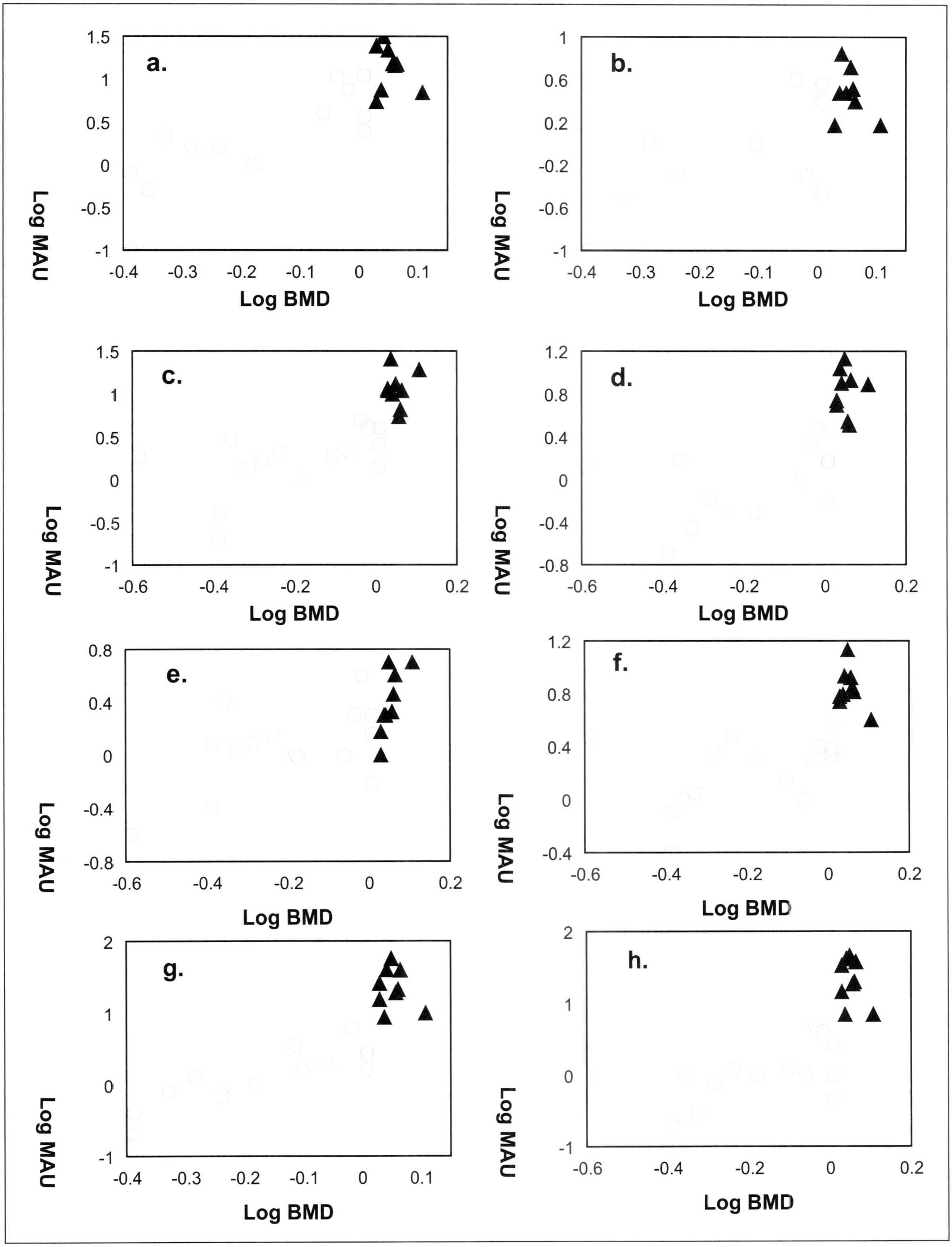


*Figure 4. The relationship between bone mineral density (BMD) and minimum animal unit (MAU) within the archaeological components discussed in the text. High survival elements are shown with triangles (▲), low survival elements are shown with open circles (○). BMD values are from Lam and colleagues (1999, Table 1, wildebeest column). a) Ain Dara, Size Class 1-2. b) Ain Dara, Size Class 3-4. c) Mezmaiskaya, Size Class 2. d) Mezmaiskaya, Size Class 3-4. e) Die Kelders I, Size Class 1-2. f) Die Kelders I, Size Class 3-4. g) Kobeh, Size Class 1-2. h) Kunji, Size Class 1-2.*

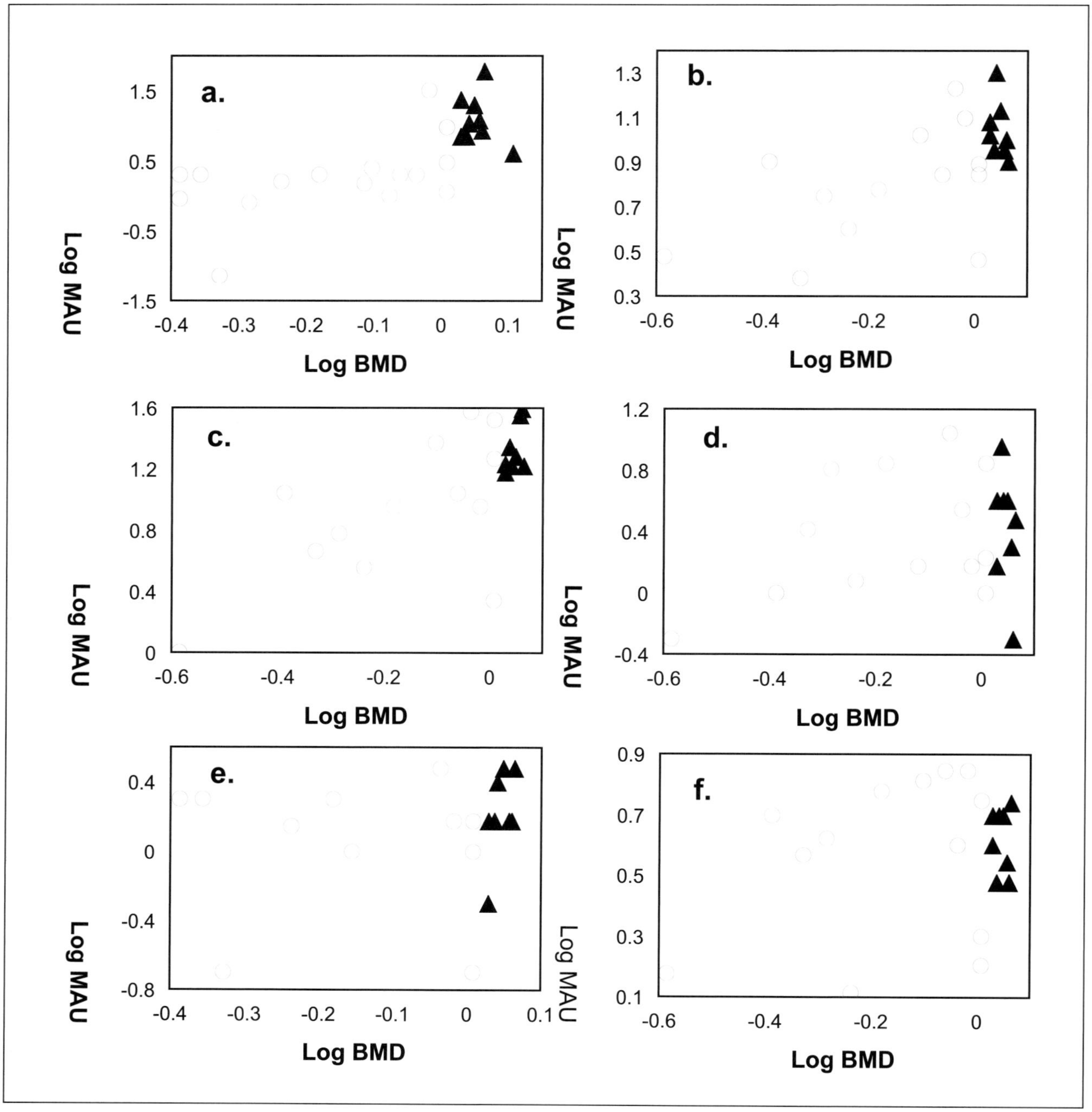


*Figure 5. The relationship between bone mineral density (BMD) and minimum animal unit (MAU) within the archaeological components discussed in the text. High survival elements are shown with triangles (▲), low survival elements shown with open circles (O). BMD values are from Lam et al. (1999, Table 1, wildebeest column). a) Porc Epic, size 2 (from Assefa 2003). b) Clary Ranch, Size Class 4. c) Agate Basin, Size Class 4. d) Hell Gap, Size Class 4. e) Agate Basin, Main Folsom Component, Size Class 4 (bison). f) Agate Basin, Main Folsom Component, (pronghorn). MAU data for graphs b through f from Hill (2001).*

survival set was still positively correlated with density to a much greater extent than in the high survival set.

The choice between the two ways of comparing density (highest per bone or best represented) did not have much of an effect on the rate of our third hypothesis. This is not surprising, since the highest MNE often coincided with the densest scan site per element.

Finally, we return to the question of why our model fits some sites but not others. In Table 13, we show the assemblages sorted by size (maximum MAU) and with the hypotheses they support. Once the assemblages get above a certain size (in this case an MAU of 11), both of our main hypotheses are supported in all assemblages. Lack of support for our model is concentrated in the smaller samples, and could therefore represent a simple sampling error.

## Discussion and Conclusions

About 40 years ago Brain initiated a field of research into actualistic taphonomy and stimulated two parallel research trajectories that regularly intersected: 1) studies of the impact of multiple carnivore agents on skeletal element survival, and 2) studies of bone density with the goal of understanding the mechanical causes for differential bone survival. In reviewing several of these studies of carnivore ravaging, we found investigations set in the following contexts: naturally occurring dens, kill sites with complete prey carcasses, and areas of human refuse involving various states of butchery (from whole defleshed to hammerstone-broken bone). These studies vary widely in usefulness, as can be expected of research in its formative stage. Our review highlights some areas where taphonomists need to revisit old issues.

We have learned some valuable things from studies of dens, but the research return has been rather modest due to highly variable, and sometimes inadequate, methods of assemblage collection and description. The existence of species-specific skeletal element patterns in dens, whether resulting from transport or destruction, is still unresolved. We need new studies with comprehensive fragment collection (i.e., excavation), capturing even very small fragments. We also need studies that apply zooarchaeological standards to quantification methods in order to estimate skeletal element abundance using all bone portions. However, unless the agents of collection are monitored, the researcher will never fully understand the system parameters. For this reason, we remain skeptical that den studies can be used to usefully interpret skeletal element profiles in archaeological contexts.

The results from observations of carcass destruction have been more rewarding. Our review of the literature, of both qualitative observations and quantitative presentations, shows that there is a regular intra- and inter-element pattern of carnivore destruction of skeletal elements. Carnivores of all types preferentially destroy the cancellous portions of bones, though the extent of destruction varies as a function of several factors (e.g., carnivore's bone-crushing abilities, hunger, carnivore group size). Even small domestic dogs follow this generalization. This means that portions of hard cortical bone will survive better when lacking any attached cancellous bone. The implication is that zooarchaeological methods must be able to accurately estimate skeletal element abundance from both types of bone portions. Cortical portions will provide the most accurate estimates of skeletal element abundance as it was before carnivore ravaging, while the relative representation of cortical to non-cortical portions may provide a measure of the intensity of carnivore destruction. It would be useful in future studies to document this pattern in more detail.

These findings have implications for relative skeletal element survival. Our analysis of the percentage survival data from the studies lacking shaft portion estimates closely resembled the pattern Brain (1967) originally described. Head parts, metapodials, and distal humeri survive best. Pelves, scapulae, and the denser long bone ends survive moderately well. Ribs and vertebrae rarely survive. Small bones like phalanges, carpals, and tarsals survive poorly. The survival rate of this sample correlates tightly with density, though it is likely that variation in survival not explained by density is a function of the size and greasiness of bones and bone portions, and the agent of destruction.

When isolated shaft fragments are included in estimates of abundance, long bones and other elements with dense, cortical portions lacking trabeculae generally increase in relative abundance (a function of resistance to carnivore attack and a low nutrient value). This results in a more accurate estimate of element survival. It would therefore be extremely useful to re-study the samples in Table 2 and include shafts in the analyses.

We conducted an updated analysis of the relationship between bone density and skeletal element survival in archaeological faunal assemblages. This was done to overcome two deficiencies in prior analyses (the lack of limb bone shaft portions in quantification and the use of bone density estimates that were not shape-corrected) and to test the explanatory power of our high and low survival set model. We found that the standard analytical practice of setting up regression or rank correlation tests between density and skeletal element abundance using the entire skeleton masks the divergent patterning in these two sets of elements. Within the archaeological assemblages, we found that element representation in the low survival set is largely dictated by the density gradient. This is not the case in the high survival set. Thus, the destruction of cancellous portions of the skeleton is highly variable and subject to a variety of factors related to the identity and condition of the ravaging carnivore. By contrast, high survival portions tend to survive well regardless of these factors.

We conclude that a skeletal element analysis of archaeofaunal assemblages including the low survival set will not generally aid our interpretation of human behavior. Further, the low and high survival sets cannot be compared in order to characterize transport or butchery decisions. The good news is that the high survival set may yet provide answers to some questions of human behavior, with some important caveats. We reiterate that if MNEs are calculated based only on long bone ends or if density values are not shape-corrected, then real distinctions between the high and low survival sets will be lost.

Our conclusions regarding the high and low survival dichotomy, however, derive largely from a comparative taphonomic analysis. As we have noted above, this approach is best used as a means to develop hypotheses to be further tested by actualistic research. Although the published data re-confirm the general relationship between carnivore ravaging and element attrition, a significant expansion of actualistic research is needed to appreciate the subtler (and perhaps more constructive) patterns within this relationship.

## Acknowledgements

CWM thanks Travis Pickering, Kathy Schick, and Nick Toth for hosting a wonderful conference. The analysis of the Kunji and Kobeh faunal collections was funded by NSF grant SBR-9727668 to Marean, and the analysis of the Die Kelders Cave 1 faunal collection was funded by NSF grant SBR 9727491 to Marean. The analysis of Mezmaiskaya was funded by an NSF graduate fellowship, a Fulbright Fellowship, and Wenner Gren grant 6744 to Cleghorn. NC thanks L.V. Golovanova for the opportunity to work with the Mezmaiskaya faunal assemblage. Both authors thank Charles Lockwood for developing and sharing the bootstrap program used in the statistical analysis, Gary Haynes for offering thoughtful advice on the preparation of the text, and Bob Brain for inventing the field of study reviewed by this paper.

## References

Assefa, Z. 2002. Investigations of Faunal Remains from Porc-Epic, A Middle Stone Age Site in Southeastern Ethiopia. Ph.D. Dissertation, State University of New York, Stony Brook.

Bartram, L.E. 1993. An Ethnoarchaeological Analysis of Kua San (Botswana) Bone Food Refuse. Ph.D. Dissertation, University of Wisconsin-Madison.

Bartram, L.E., Marean, C.W. 1999. Explaining the "Klasies Pattern:" Kua ethnoarchaeology, the Die Kelders Middle Stone Age archaeofauna, long bone fragmentation and carnivore ravaging. Journal of Archaeological Science 26, 9-29.

Bearder, S.K. 1977. Feeding habits of spotted hyaenas in a woodland habitat. East African Wildlife Journal 5, 263-280.

Binford, L.R. 1978. Nunamiut Ethnoarchaeology. Academic Press, New York.

Binford, L.R. 1981. Bones: Ancient Men and Modern Myths. Academic Press, New York.

Binford, L.R., Bertram, J.B. 1977. Bone frequencies and attritional processes. In: Binford, L.R. (Ed.), For Theory Building in Archaeology. Academic Press, New York, pp. 77-153.

Binford, L.R., Mills, L.G. L., Stone, N.M. 1988. Hyena scavenging behavior and its implications for the interpretation of faunal assemblages from FLK 22 (the Zinj floor) at Olduvai Gorge. Journal of Anthropological Archaeology 7, 99-135.

Blumenschine, R.J. 1986. Carcass consumption sequences and the archaeological distinction of scavenging and hunting. Journal of Human Evolution 15, 639-659.

Blumenschine, R.J. 1988. An experimental model of the timing of hominid and carnivore influence on archaeological bone assemblages. Journal of Archaeological Science 15, 483-502.

Blumenschine, R.J., Marean, C.W. 1993. A carnivore's view of archaeological bone assemblages. In: Hudson, J. (Ed.), From Bones to Behavior. Center for Archaeological Investigations, Carbondale (IL), pp. 273-301.

Brain, C.K. 1967. Hottentot food remains and their bearing on the interpretation of fossil bone assemblages. Scientific Papers of the Namib Desert Research Station 32, 1-11.

Brain, C.K. 1969. The contributions of Namib Desert Hottentots to an understanding of australopithecine bone accumulations. Scientific Papers of the Namib Desert Research Station 39, 13-22.

Brain, C.K. 1970. New finds at the Swartkrans australopithecine site. Nature 225, 1112-1119.

Brain, C.K. 1981. The Hunters or the Hunted? An Introduction to African Cave Taphonomy University of Chicago Press, Chicago.

Bunn, H.T. 1983. Comparative analysis of modern bone assemblages from a San hunter-gatherer camp in the Kalahari Desert, Botswana, and from a spotted hyena den near Nairobi, Kenya. In: Clutton-Brock, J., Grigson, C. (Eds.), Animals and Archaeology, Vol. 1: Hunters and their Prey. British Archaeological Reports, International Series 163, Oxford, pp.143-148.

Bunn, H.T. 1993. Bone assemblages at base camps: a further consideration of carcass transport and bone destruction by the Hadza. In: Hudson, J. (Ed.), From Bones to Behavior. Center for Archaeological Investigations, Carbondale (IL), pp. 156-167.

Bunn, H.T., Kroll, E.M. (1986). Systematic butchery by Plio-Pleistocene hominids at Olduvai Gorge, Tanzania. Current Anthropology 27, 431-452.

Capaldo, S.D. 1995. Inferring Hominid and Carnivore Behavior from Dual-Patterned Archaeofaunal Assemblages. Ph.D. Dissertation, Rutgers University.

Capaldo, S.D. 1997. Experimental determinations of carcass processing by Plio-Pleistocene hominids and carnivores at FLK 22 (Zinjanthropus), Olduvai Gorge, Tanzania. Journal of Human Evolution 33, 555-597.

Dart, R. 1956. The myth of the bone-accumulating hyena. American Anthropologist 58, 40-62.

Dart, R. (1957). The Osteodontokeratic Culture of *Australopithecus prometheus*. Transvaal Museum, Pretoria.

Domínguez-Rodrigo, M. 1997. Meat-eating by early hominids at the FLK 22 Zinjanthropus site, Olduvai Gorge (Tanzania): an experimental approach using cut-mark data. Journal of Human Evolution 33, 669-690.

Ewer, R.F. 1973. The Carnivores. Cornell University Press, Ithaca.

Gifford-Gonzalez, D. 1991. Bones are not enough: analogues, knowledge, and interpretive strategies in zooarchaeology. Journal of Anthropological Archaeology 10, 215-254.

Gifford-Gonzalez, D., Damrosch, D.B., Damrosch, D.R., Pryor, J., Thunen, R.L., Reinhard, K.J., Ambler, J.R., McGuffie, M. 1985. The third dimension in site structure: an experiment in trampling and vertical dispersal. America Antiquity 50, 803-818.

Grayson, D.K. 1989. Bone transport, bone destruction, and reverse utility curves. Journal of Archaeological Science 16, 643-652.

Haynes, G. 1980. Evidence of carnivore gnawing on Pleistocene and recent mammalian bones. Paleobiology 6, 341-351.

Haynes, G. 1981. Bone Modifications and Skeletal Disturbances by Natural Agencies: Studies in North America. Ph.D. Dissertation, Catholic University of America, Washington, D.C.

Haynes, G. 1982. Utilization and skeletal disturbance of North American prey carcasses. Arctic 35, 266-281.

Haynes, G. 1983a. Frequencies of spiral and green-bone fractures on ungulate limb bones in modern surface assemblages. American Antiquity 48, 102-114.

Haynes, G. 1983b. A guide for differentiating mammalian carnivore taxa responsible for gnaw damage to herbivore limb bones. Paleobiology 9, 164-172.

Henschel, J.R., Tilson, R., von Blottnitz, F. 1979. Implications of a spotted hyaena bone assemblage in the Namib Desert. South African Archaeological Bulletin 34, 127-131.

Hill, A. 1981. Why study paleoecology? Nature 293, 340.

Hill, A. 1984. Hyaenas and hominids: taphonomy and hypothesis testing. In: Foley, R. (Ed.) Hominid Evolution and Community Ecology: Prehistoric Human Adaptation in Biological Perspective. Academic Press, London, pp. 111-128.

Hill, M.G. 2001. Paleoindian Diet and Subsistence Behavior on the Northwestern Great Plains of North America. Ph.D. Dissertation, University of Wisconsin-Madison.

Horwitz, L.K. 1998. The influence of prey body size on patterns of bone distribution and representation in a striped hyaena den. Economie Prehistorique: Les Comportements de Subsistence au Paleolithique. Rencontres Internationales d'Archeologie et d'Histoire d'Antibes, pp. 31-42.

Horwitz, L.K., Smith, P. 1988. The effects of striped hyaena activity on human remains. Journal of Archaeological Science15. 471-481.

Hudson, J. 1993. The impacts of domestic dogs on bone in forager camps. In: Hudson, J. (Ed.), From Bones to Behavior. Center for Archaeological Investigations, Carbondale (IL), pp. 301-323.

Kerbis-Peterhans, J.C., Horwitz, L.K. (1992). A bone assemblage from a striped hyaena (*Hyaena hyaena*) den in the Negev desert, Israel. Israel Journal of Zoology 37, 225-245.

Klein, R.G., Cruz-Uribe, K., Milo, R.G. 1999. Skeletal part representation in archaeofaunas: comments on "Explaining the 'Klasies Pattern': Kua ethnoarchaeology, the Die Kelders Middle Stone Age archaeofauna, long bone fragmentation and carnivore ravaging" by Bartram and Marean. Journal of Archaeological Science 26, 1225-1234.

Klippel, W.E., Snyder, L.M., Parmalee, P.W. 1987. Taphonomy and archaeologically recovered mammal bone from southeast Missouri. Journal of Ethnobiology 7, 155-169.

Kreutzer, L.A. 1992. Bison and deer bone mineral densities: comparisons and implications for the interpretation of archaeological fauna. Journal of Archaeological Science 19, 271-294.

Lam, Y.M. 1992. Variability in the behaviour of spotted hyaenas as taphonomic agents. Journal of Archaeological Science 19, 389-406.

Lam, Y.M., Chen, X., Marean, C.W., Frey, C.J. 1998. Bone density and long bone representation in archaeological faunas: comparing results from CT and photon densitometry. Journal of Archaeological Science 25, 559-570.

Lam, Y.M., Chen, X., Pearson, O.M. 1999. Intertaxonomic variability in patterns of bone denisty and the differential representation of bovid, cervid, and equid elements in the archaeological record. America Antiquity 64, 343-362.

Lam, Y.M., Pearson, O.M., Marean, C.W., Chen, X. 2003. Bone density studies in zooarchaeology. Journal of Archaeological Science 30, 1701-1708.

Lupo, K.D. 1995. Hadza bone assemblages and hyena attrition: an ethnographic example of the influence of cooking and mode of discard on the intensity of scavenger ravaging. Journal of Anthropological Archaeology 14, 288-314.

Lyman, R.L. 1984. Bone density and differential survivorship of fossil classes. Journal of Anthropological Archaeology 3, 259-299.

Lyman, R.L. 1985. Bone frequencies: Differential transport, in situ destruction, and the MGUI. Journal of Archaeological Science 12, 221-236.

Lyman, R.L. 1991. Taphonomic problems with archaeological analyses of animal carcass utilization and transport. In: Purdue, J.R., Klippel, W.E., Styles, B.W. (Eds.), Beamers, Bobwhites, and Blue-Points: Tributes to the Career of Paul W. Parmalee. Illinois State Museum Scientific Papers, Vol. 23, Springfield, pp. 125-138.

Lyman, R. L. 1992. Anatomical considerations of utility curves in zooarchaeology. Journal of Archaeological Science 9, 7-22.

Lyman, R.L. 1993. Density-mediated attrition of bone assemblages: new insights. In: Hudson, J. (Ed.), From Bones to Behavior. Center for Archaeological Investigations, Carbondale (IL), pp. 324-341.

Lyman, R.L. 1994. Vertebrate Taphonomy. Cambridge University Press, New York.

Lyon, P. 1970. Differential bone destruction: an ethnographic example. American Antiquity 35, 213-215.

Marean, C.W. 1991. Measuring the post-depostional destruction of bone archaeological assemblages. Journal of Archaeological Science 18, 677-694.

Marean, C.W. 1995. Of taphonomy and zooarchaeology. Evolutionary Anthropology 4, 64-72.

Marean, C.W., Abe, Y., Frey, C.J., Randall, R.C. 2000. Zooarchaeological and taphonomic analysis of the Die Kelders Cave 1 Layers 10 and 11 Middle Stone Age larger mammal fauna. Journal of Human Evolution 38, 197-233.

Marean, C.W., Cleghorn, N. 2003. Large mammal skeletal element transport: applying foraging theory in a complex taphonomic system. Journal of Taphonomy 1, 15-42.

Marean, C.W., Domínguez-Rodrigo, M., Pickering, T.R. 2004. Skeletal element equifinality in zooarchaeology begins with method: the evolution and status of the "shaft critique". Journal of Taphonomy, 2, 69-98.

Marean, C.W., Frey, C.J. 1997. The animal bones from caves to cities: reverse utility curves as methodological artifacts. American Antiquity 62, 698-711.

Marean, C.W., Kim, S.Y. 1998. Mousterian faunal remains from Kobeh cave (Zagros Mountains, Iran): behavioral implications for Neanderthals and early modern humans. Current Anthropology 39, S79-S114.

Marean, C.W., Spencer, L.M. 1991. Impact of carnivore ravaging on zooarchaeological measures of element abundance. American Antiquity 56, 645-658.

Marean, C.W., Spencer, L.M., Blumenschine, R.J., Capaldo, S.D. 1992. Captive hyena bone choice and destruction, the schlepp effect, and Olduvai archaeofaunas. Journal of Archaeological Science 19, 101-121.

Mills, M.G.L., Mills, M.E.J. 1977. An analysis of bone collected at hyaena breeding dens in the Gemsbok National Parks (Mammalia: Carnivora). Annals of the Transvaal Museum 30, 145-155.

Payne, S., Munson, P.J. 1985. Ruby and how many squirrels? The destruction of bones by dogs. In: Fieller, N.R.J., Gilbertson, D.D., Ralph, N.G.A. (Eds.), Palaeobiological Investigations: Research Design, Methods and Data Analysis. Oxford: British Archaeological Reports, pp. 31-40.

Pickering, T.R., Carlson, K.J. 2002. Baboon bone mineral densities: implications for the taphonomy of primate skeletons in South African Cave Sites. Journal of Archaeological Science 29, 883-896.

Pickering, T.R., Marean, C.W., Domínguez-Rodrigo, M. 2003. Importance of limb bone shaft fragments in zooarchaeology: a response to "On in situ attrition and vertebrate body part profiles" (2002), by M.C. Stiner. Journal of Archaeological Science 30, 1469-1482.

Potts, R. 1988. Early Hominid Activities at Olduvai. Aldine de Gruyter, New York.

Richardson, P.R.K. 1980. Carnivore damage to antelope bones and its archaeological implications. Palaeontologia Africana 23, 109-125.

Selvaggio, M.M. 1994. Carnivore tooth marks and stone tool butchery marks on scavenged bones: archaeological implications. Journal of Human Evolution 27, 215-228.

Selvaggio, M.M. 1998. Evidence for a three-stage sequence of hominid and carnivore involvement with long bones at FLK Zinjanthropus, Olduvai Gorge, Tanzania. Journal of Archaeological Science 25, 191-202.

Skinner, J.D., Davis, S., Ilani, G. 1980. Bone collecting by striped hyaenas, *Hyaena hyaena*, in Israel. Palaeontologia Africana 23, 99-104.

Snyder, L. 1988. A controlled feeding study involving gray wolf (*Canis lupus*) and white-tailed deer (*Odocoileus virginianus*). Tennessee Anthropological Association Newsletter 13, 1-9.

Stahl, P.W. 1999. Structural density of domesticated South American camelid skeletal elements and the archaeological investigation of prehistoric Andean Ch'arki. Journal of Archaeological Science 26, 1347-1368.

Stallibrass, S. 1984. The distinction between the effects of small carnivores and humans on post-glacial faunal assemblages. In: Grigson, C., Clutton-Brock, J. (Eds.), Animals and Archaeology, Vol. 4: Husbandry in Europe. British Archaeological Reports, International Series 227, Oxford, pp. 259-269.

Stiner, M.C. 1991. Food procurement and transport by human and non-human predators. Journal of Archaeological Science 18, 455-482.

Stiner, M.C. 1994. Honor Among Thieves: A Zooarchaeological Study of Neandertal Ecology. Princeton University Press, Princeton.

Stiner, M.C. 2002. On in situ attrition and vertebrate body part profiles. Journal of Archaeological Science 29, 979-991.

Sutcliffe, A.J. 1970. Spotted hyaena: crusher, gnawer, digester and collector of bones. Nature 227, 1110-1113.

White, T.E. 1953. Observations on the butchering technique of some aboriginal peoples, no. 2. American Antiquity 9, 160-164.

White, T.E. 1955. Observations on the butchering techniques of some aboriginal peoples, nos. 7, 8 and 9. American Antiquity 21, 170-178.

# CHAPTER 4

# Hominids and Carnivores at Kromdraai and Other Quaternary Sites in Southern Africa

J. Francis Thackeray

## Abstract

This study serves to examine carnivore:ungulate ratios and mean bone flake lengths of fauna from South African Plio-Pleistocene assemblages from Kromdraai A and B, compared to other assemblages from sites in the Sterkfontein Valley, including Swartkrans (Members 1, 2 and 3) and Sterkfontein (Members 2 and 5). Comparisons are made with Middle and Later Stone Age assemblages from southern Africa, and faunal assemblages accumulated by carnivores (brown hyena, spotted hyena and leopard). Use is made of carnivore: ungulate ratios as well as mean flake lengths of ungulate bones to facilitate the identification of agents of accumulation, and to assess the extent to which one or other agent was responsible for accumulating any given assemblage. The data are used together to establish a general model whereby scavenging activities by hominids may be distinguished from activities associated with longer-term occupation. Although Kromdraai A has previously been referred to as a "non-hominid site," the discovery of stone artefacts indicates a hominid presence, although no hominid fossils have as yet been found at this site, contrasting with Kromdraai B where stone artefacts and hominids (both *Paranthropus* and early *Homo*) have been found. It is suggested that polyhedral core tools (choppers) associated with Early Acheulian/Developed Oldowan assemblages at Kromdraai A may have been used opportunistically to obtain bone marrow from sites which were also used at least temporarily by large carnivores such as sabre-tooth cats, which preyed primarily on juvenile alcelaphines.

## Introduction

Kromdraai is situated within 2 kilometres east of Sterkfontein, and consists of two localities in close proximity to each other, Kromdraai A (KA) and Kromdraai B (KB). The latter yielded the type specimen of *Paranthropus* (*Australopithecus*) *robustus* (TM 1517), described by Robert Broom (1938). The Olduvai Event is represented in the KB deposits (Thackeray et al., 2002), and a minimum date of 1.95 million years ago (Mya) has been indicated for the type specimen of *P. robustus* from Kromdraai B. This is similar in age to the type specimens of *P. boisei* (OH 5) and early *Homo* (OH 7) at Olduvai Bed I, dated circa 1.8 Mya.

Until recently it was thought that all of the hominids at KB represented robust australopithecines (Thackeray et al., 2001), and it appeared anomalous that stone tools from Kromdraai B were present in the apparent absence of early *Homo* at this locality. However, at least one specimen (KB 5223) has been identified as early *Homo* (Braga and Thackeray, 2003). It is thus evident that stone tools, including polyhedral cores, were used at KB where both early *Homo* and *P. robustus* were present, as at Swartkrans (Members 1 and 2), Sterkfontein (Member 5), Olduvai (Bed I) and various other early Pleistocene sites in Africa.

No hominid fossils have as yet been discovered at KA, but stone tools associated with Developed Oldowan/ Early Acheulian technology indicate a hominid presence (Kuman et al., 1997; Thackeray, 1998). The stone tool assemblages at KA include polyhedral cores of the kind which have been found at KB and at Olduvai Bed I.

KB has been considered to have been a potential death trap, but carnivores may have played a role as accumulators of faunal assemblages at both KA and KB (Brain, 1981; Thackeray, 1999; Vrba, 1976).

Although no hominid fossils have as yet been discovered at KA, several questions require attention, regarding agents of accumulation:

1. To what extent, if at all, were hominids responsible for the accumulation of faunal assemblages at Kromdraai A and B?
2. To what extent were carnivores responsible for the accumulation of KA and KB faunal assemblages?
3. In what ways, if at all, do faunal assemblages from KA and KB differ from each other?
4. In what manner do the KA and KB assemblages resemble faunal assemblages from Swartkrans (Members 1,2 and 3) and Sterkfontein (Member 5)?
5. In what manner do early Pleistocene assemblages from Sterkfontein, Swartkrans and Kromdraai assemblages differ from late Quaternary assemblages associated with Middle Stone Age (MSA) and Later Stone Age (LSA) hominids, representing *H. sapiens*?

## Method

In this study I use two indices to facilitate the identification of agents of faunal accumulation. Firstly, carnivore:ungulate ratios can be cautiously used to infer the degree to which carnivores might have contributed to faunal accumulations (Klein, 1975), under the assumption that carnivores tend to consume ungulates as well as other carnivores. Secondly, mean flake lengths (MFL) of long bones (primarily those of ungulates) can reflect hominid activity under the assumption that a high degree of fragmentation is associated with the breakage of bone shafts for the extraction of marrow (Brain, 1981).

The term "bone flake" as used here has been defined by Brain (1974), referring to shaft fragments of ungulate long bones which have been broken through the shaft such that less than half the circumference of the shaft section is intact. Variability in the degree of fragmentation can be assessed not only in terms of patterns in the size distribution of bone flakes, using histograms to illustrate frequencies of bone flakes measured in size classes (Brain, 1981), but also in terms of mean flake length and associated standard deviations.

For purposes of this study, carnivore ungulate ratios (CUR) are calculated by means of the following formula:

$$\text{CUR} = \frac{\text{MNI carnivores}}{\text{MNI ungulates}} \times 100$$

Log-transformed CUR and MFL values have been used to facilitate comparison of these indices, within and between sites. Further, attempts have been made to assess activities of one or more agents of accumulation, by comparing CUR and MFL values against each other, similar to the approach adopted by Isaac (1976) when comparing the relative frequencies of bone and stone artefacts associated with a range of hominid activities. In this case, reference is made not only to hominids but also to carnivores as potential agents of accumulation. The expectation is that high carnivore ungulate ratios (CUR), and a low degree of bone fragmentation (associated with high MFL values) would generally reflect a high degree of carnivore activity at a site. Conversely, low CUR and low MFL values would be expected to be associated with a high probability of having been associated with hominid activity, assuming that hominids generally do not consume carnivores, and assuming that a high degree of fragmentation of ungulate long bones is associated with the tendency to break bones for marrow extraction.

The approach adopted here is an example of how CUR and MFL indices may be used together to explore patterning in their relationship, in an attempt to identify agents of accumulation. However, it is recognized that additional taphonomic approaches, including the analysis of tooth-marking and percussion damage, need to be adopted, and potential post-depositional factors should also be considered. The primary objective in this study is to examine CUR and MFL indices in order to generate hypotheses which may be tested on the basis of other taphonomic studies.

## Materials

The materials used in this study are based on published and unpublished studies of faunal assemblages from the following South African sites: Bushman Rock Shelter in the Mpumalanga Province and Wilton Rock Shelter and in the Eastern Cape Province (Brain, 1981); Nelson Bay Cave, Klasies River Mouth and Swartklip situated in the Western Cape Province (Klein, 1972, 1975, 1976). Also included are faunal assemblages from Fackeltrager (Brain, 1981) and Apollo 11 (Thackeray, 1979) in Namibia , and Pomongwe, a site in Zimbabwe (Brain, 1981) All of these samples date to the Late Quaternary. Plio-Pleistocene assemblages from the Gauteng Province include samples from Sterkfontein (Brain, 1981, Pickering et al., 2004), Swartkrans (Brain, 1981; Watson, 1991) and Kromdraai (Brain, 1981; Vrba, 1976; Thackeray, 1999).

In most cases, both MFL and CUR values could be calculated for assemblages from these sites. In other instances, only one or other index could be determined from available data.

## Results

The results of this study are presented in Table 1, listing carnivore:ungulate ratios and mean flake lengths with associated standard deviations and sample sizes. The distributions of CUR and MFL are presented graphically in Figure 1.

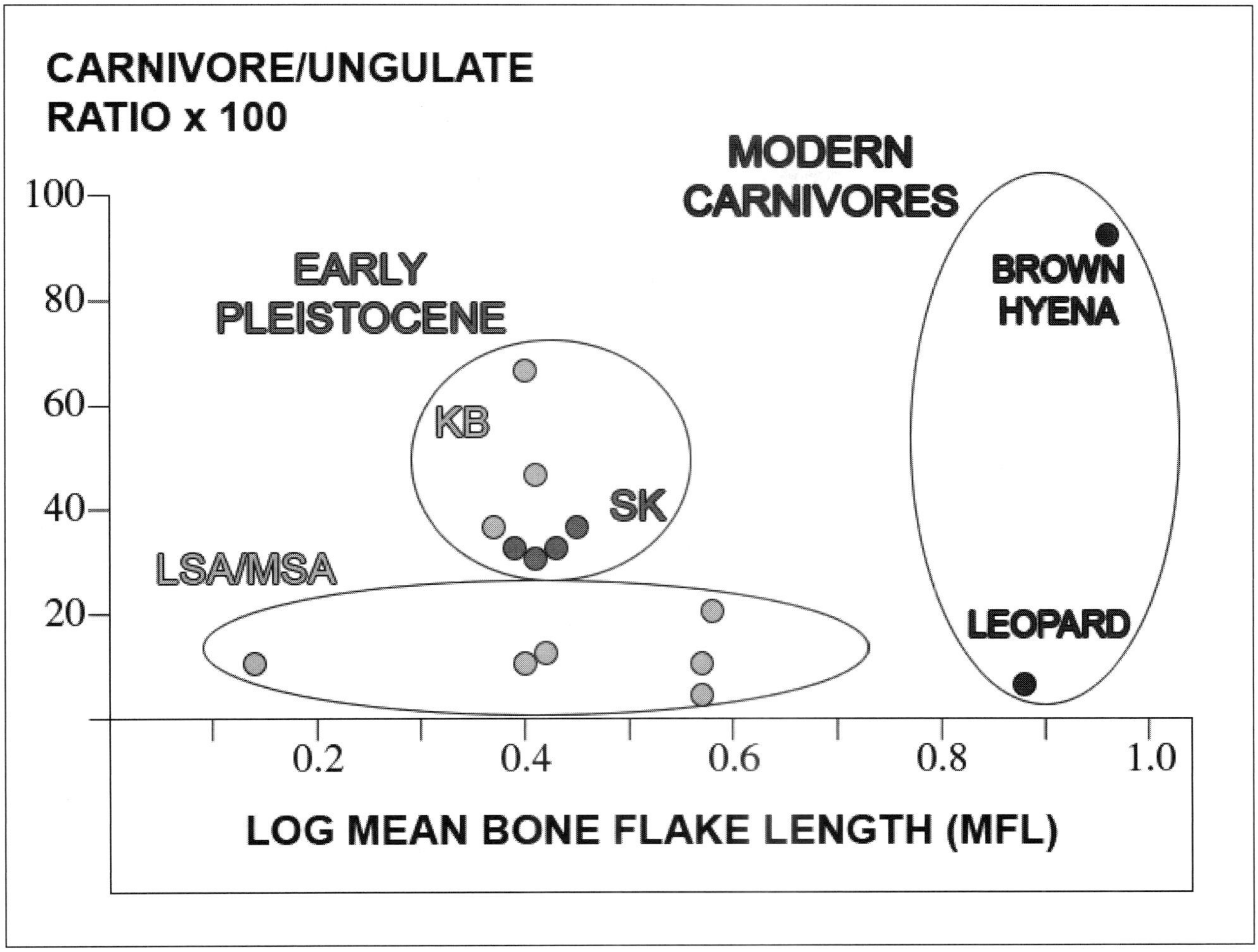


*Figure 1. Distribution of MFL (mean flake length) and CUR values (carnivore:ungulate ratios) for various faunal assemblages from southern Africa. The Later Stone Age (LSA) and Middle Stone Age (MSA) assemblages date to the Late Quaternary.*

## Discussion

Mean flake lengths (MFL) for Late Quaternary assemblages are consistently low, ranging between 1.39 +/- 0.14 cm (n=33,891) at Wilton Rock Shelter, to 3.75 +/- 0.57 cm (n=2723) at Bushman Rock Shelter. Intermediate MFL values have been obtained for the Apollo 11 assemblages (MFL = 2.40 +/- 0.38 cm, n = 10,854). As noted by Brain (1981) and Thackeray (1979), the high degree of fragmentation can be associated with a high probability of hominid activity, related to the breakage of long bones for marrow.

By contrast, the mean bone flake length for assemblages accumulated by brown hyenas (MFL = 11.05 +/- 1.04 cm, n=2887) is significantly higher (p=0.05; Students *t*-tests) than MFL values calculated for Late Quaternary faunal assemblages associated with human activity, notably those from Wilton Rock Shelter, Bushman Rock Shelter, Apollo 11, Fackeltrager and Pomongwe (Table 1) Similarly, the mean bone flake length measured from an assemblage accumulated by spotted hyenas (MFL = 7.20 +/- 0.86 cm, n = 220) is high, significantly higher than the mean MFL values obtained for Late Quaternary hominid sites (p=0.05; Students *t*-tests).

Surprisingly, the mean flake length of bone flakes from Kromdraai A is relatively low (MFL = 2.52 +/- 1.01 cm, n=3016), similar to mean values for Late Quaternary assemblages from sites such as Fackeltrager (MFL = 2.50 +/- 0.40 cm, n=757) and Apollo 11 (MFL = 2.40 +/- 0.38 cm, n=10,854). Noticeable differences are however recognisable in terms of the standard deviations associated with mean values of flake length. The low standard deviations associated with low MFL values of Late Quaternary assemblages reflect a consistently high degree of fragmentation, whereas the KA assemblage is associated with a higher degree of variability in bone flake length. The KA sample include relatively large flakes. The situation at Kromdraai is different in the sense that the assemblage may have included some degree of carnivore activity, although the average flake length is small, similar to that of some Late Quaternary assemblages.

The three assemblages from Kromdraai B show the same pattern displayed at Kromdraai A. The mean flake lengths range from 2.48 - 2.73 cm, and the associated standard deviation for one sample (KB 1) is relatively high, again suggesting some degree of carnivore activity. The standard deviations associated with

*Table 1. Mean flake lengths (MFL) and carnivore: ungulate ratios (C/U x 100) for faunal assemblages from Kromdraai, Swartkrans, Sterkfontein and other hominid sites in southern Africa, and for assemblages accumulated by spotted hyena, brown hyena and leopards*

| | MFL | s.d. | n. | log MFL | C/Ux100 | log (C/Ux100) |
|---|---|---|---|---|---|---|
| **Kromdraai** | | | | | | |
| KA (D13) | 2.52 | 1.01 | 3016 | 0.401 | 34.5 | 1.537 |
| KB 1 | 2.48 | 1.16 | 1512 | 0.394 | 35.7 | 1.533 |
| KB 2 | 2.73 | 0.44 | 627 | 0.436 | 66.7 | 1.824 |
| KB 3 | 2.79 | 0.45 | 748 | 0.446 | 46.2 | 1.665 |
| **Swartkrans** | | | | | | |
| SK M1 (HR & LB) | 2.87 | 0.46 | 6171 | 0.458 | 35.3 | 1.547 |
| SK M2 | 2.74 | 0.44 | 6835 | 0.438 | 32.6 | 1.513 |
| SK M3 | 3.00 | 0.48 | 8923 | 0.477 | 36.7 | 1.565 |
| SK M5 | 3.90 | 0.59 | 757 | 0.591 | 18.5 | 1.267 |
| Sterkfontein M5 | 5.47 | 3.12 | 89 | 0.738 | 16.6 | 1.220 |
| Sterkfontein M2 | - | - | - | - | 250 | 2.39 |
| Wilton | 1.39 | 0.14 | 33891 | 0.143 | 11.1 | 1.045 |
| Bushman Rock | 3.75 | 0.57 | 2723 | 0.574 | 1.7 | 0.230 |
| Apollo 11 | 2.40 | 0.38 | 10854 | 0.380 | 10.4 | 1.017 |
| Fackeltrager | 2.50 | 0.40 | 757 | 0.397 | 11.1 | 1.045 |
| Pomongwe | 3.73 | 0.57 | 9539 | 0.572 | 5.3 | 0.724 |
| **Carnivore-accumulated assemblages** | | | | | | |
| Spotted hyena | 7.20 | 0.86 | 220 | 0.857 | 1.15 | 0.060 |
| Brown hyena | 11.05 | 1.04 | 2887 | 1.043 | 97.2 | 1.987 |
| Swartklip | | | | | 28.9 | 1.461 |
| Leopards (Kafue) | - | - | - | | 3.6 | 0.556 |
| Leopard (Serengeti) | - | - | - | | 5.1 | 0.707 |
| Leopard (Serengeti) | - | - | - | | 3.9 | 0.591 |

MFL values of two other assemblages (KB2 and KB3) indicate a lower degree of variation in flake length, closer to that found in Late Quaternary hominid sites. The implication is that both carnivores and hominids are likely to have contributed to the accumulation of KB faunal assemblages.

MFL values for Swartkrans samples range between 2.87 and 3.00 cm, corresponding closely to values obtained for Kromdraai A and B assemblages. The degree of variation in flake length in the Swartkrans assemblages corresponds closely to values obtained for Kromdraai samples KB2 and KB 3, again reflecting similarities with Late Quaternary assemblages associated with some degree of hominid activity.

Flake length variability needs to be assessed in the context of carnivore:ungulate ratios which are presented in Table 1.

Remarkably, the distribution of MFL and CUR

values for Kromdraai A, Kromdraai B and Swartkrans Members 1, 2 and 3 are all tightly clustered together (Figure 1), and are separated from late Quaternary assemblages by virtue of having higher carnivore:ungulate ratios. Clearly the Swartkrans and Kromdraai assemblages are associated with some degree of carnivore activity. By comparison with CUR values of modern carnivores, it would seem that CUR values for KA and KB correspond most closely to those of assemblages accumulated by brown hyena, but other carnivores may certainly have been involved, including large carnivores such as *Dinofelis*, which have previously been identified as a potential agent of accumulation at KA (Brain, 1981; Vrba, 1976; Thackeray and von Leuvan-Smith, 2001). The degree of fragmentation of bone flakes at KA, KB and at Swartkrans (Members 1 2 and 3), combined with relatively high CUR values, suggest that while carnivores contributed to some extent to the accumulation of faunal remains, hominids may have contributed to the fragmentation of long bone shafts.

Polyhedral cores and chopper artefacts are present at KA, KB, and Members 1, 2, and 3 at Swartkrans. As a working hypothesis, one might suggest that these stone artefacts were used partly to break open long bones of animals that had been killed by carnivores. This hypothesis can be tested by chemical analysis of working edges of polyhedral cores. The intention of current and future research is to explore the working edges of such tools to determine whether or not bone apatite residues are present in the interstices of stone tools where apatite may have been deposited, if the tools had been used to break open long bones for marrow. Several stone artefacts from Kromdraai have been examined using X-ray diffraction at the NECSA facility in Pretoria. Preliminary results indicate the presence of apatite on the working edge of a polyhedral core from Kromdraai A (KA 2776), consistent with the possibility that such artifacts were used to break open bone shafts (Thackeray et al., 2005).

The assemblages from Sterkfontein considered in this study, namely those from Members 2 and 5, are associated with contrasting scenarios. In the case of Member 2, the carnivore:ungulate ratio based on data recently published by Pickering et al. (2004) is high (Table 1). The Silberberg Grotto is likely to have been a deathtrap into which "Little Foot" (a complete skeleton of *Australopithecus*) and other mammals fell (Clarke, 1998), including both carnivores as well as primates (Pickering et al., 2004). Carnivore:ungulate ratios may be high due to a bias towards carnivores, associated with an infrequency of ungulates falling into the death trap.

A small sample of ungulate long-bones from Sterkfontein Member 5 is available for analysis. The mean flake length is 5.47 +/- 3.12 cm (n=89), which is low in comparison with MFL values for carnivore-accumulated assemblages. In addition, Sterkfontein Member 5 is characterised by a relatively low carnivore-ungulate ratio, lower than those obtained from Kromdraai and Swartkrans, placing the values for Sterkfontein Member 5 closer to the distribution of MFL and CUR values of Late Quaternary sites, associated primarily with hominid activity. However, some degree of carnivore activity is likely to have been associated with the Sterkfontein Member 5 sample which has a CUR value that is slightly higher than those of Late Quaternary hominid-accumulated assemblages.

## Conclusion

It is probable that faunal assemblages from Kromdraai A, Kromdraai B, and Swartkrans (Members 1, 2, and 3) were accumulated in part by carnivores, as has been previously suggested (Brain, 1981; Vrba, 1976; Thackeray, 1999; Thackeray and von Leuvan-Smith, 2001), but hominids could have played a significant role in contributing to the fragmentation of bone. At Kromdraai A, hominids may have been using the site opportunistically to access bone marrow, notably at times when the site was not occupied by carnivores. It would seem probable that KA was more frequently used as a lair by carnivores, and that hominids did not make regular use of the site as an occupation site. When they were present at the site, it would seemed probable that they did so temporarily, and discarded artefacts at the site without necessarily manufacturing them there. Stone flake debitage is uncommon relative to core tools. In a study of KA stone tool assemblages, only 24 flakes were found compared to 41 polyhedral cores, chopper cores, edge cores, bipolar cores or other core tools (Kuman et al., 1997).

The hominid species most frequently represented at Kromdraai B, *P. robustus*, may have been victims of carnivore predation, and as at Swartkrans Members 1 and 2, the relatively low frequency of early *Homo* (generally less than 15%) may have been associated with the ability of these hominids to ward off carnivores, or at least to keep their distance from them.

The apparent absence of hominid fossils from KA could be interpreted to mean either that hominids such as early *Homo* were smart enough to avoid falling prey to carnivores (as at Swartkrans Member 3, pene-contemporaneous with KA), or that hominids at that time were simply avoiding the site which was used primarily by large carnivores, except at times when hominids with polyhedral cores attempted to access bone marrow from long bones of animals killed by those carnivores.

As yet there is no evidence for the use of fire at KA, contrasting with the situation at Swartkrans Member 3. Despite this difference, it is surprising that the degree of fragmentation (as reflected by MFL values) is relatively high at both Swartkrans (Members 1, 2, and 3) and at Kromdraai (KA and KB), and the carnivore-ungulate ratios for the same samples are so similar. The possibility that hominids were occasionally making use of these sites cannot be excluded. It would seem probable that hominids and carnivores may have competed for access to the caves.

Whereas the faunal sample from Sterkfontein Member 2 may have been associated primarily with a deathtrap, accounting for the accumulation of many if not all of the mammalian remains, Member 5 is likely to have been associated with hominid activity.

As expected, carnivore:ungulate ratios for Late Quaternary assemblages are relatively low, confirming that samples from sites such Wilton Rock Shelter, Bushman Rock Shelter, Apollo 11, Fackeltrager and Pomongwe were accumulated primarily by hominids (*Homo sapiens*) who consistently broke long bones for marrow as a source of protein. The high degree of fragmentation of long bone flakes as reflected by low mean flake lengths, and the relatively low standard deviations, are indicators of habitual bone breakage by Stone Age people. By contrast, the high carnivore:ungulate ratio obtained for the Swartklip assemblage at a site near Cape Town probably relates to hyenas as agents of accumulation (Klein, 1975).

This study indicates that mean flake lengths and carnivore:ungulate ratios can be used together cautiously to infer aspects of hominid behaviour, as part of the "big picture" which palaeoanthropologists and taphonomists attempt to portray from fragmentary remains. However, it would be necessary to include analysis of features such as percussion damage and tooth marking to test inferences based on CUR and MFL values. Unfortunately, in the case of some sites, it is difficult to examine bone damage in detail if the fossilized material in breccia has been prepared mechanically. The MFL and CUR values presented in this study serve primarily as a preliminary set of data of the kind which can be used to establish working hypotheses, which can themselves be tested by other taphonomic approaches.

As a working hypothesis it can be suggested that the cluster of data associated with Early Pleistocene assemblages from Kromdraai and Swartkrans (Figure 1) relates to scavenging by hominids as well as to carnivore activity. By contrast, the late Quaternary (LSA and MSA) data relate primarily to human behavior, associated with hunting and breakage of bone for marrow.

## Acknowledgements

I would like to thank Travis Pickering, Kathy Schick, and Nick Toth for the opportunity to attend the conference in honour of Bob Brain, to whom I would like to express my gratitude and appreciation for wonderful opportunities, encouragement and fun within a period of more than 30 years. This research has been funded in part by grants from the National Research Foundation (GUN 2047162) and the French Ministry of Foreign Affairs administered by the Northern Flagship Institution.

## References

Braga, J., Thackeray, J.F. 2003. Early *Homo* at Kromdraai B: Probabilistic and morphological analysis of the lower dentition. Comptes Rendus Palevol 2, 269-279.

Brain, C.K. 1974. Some suggested procedures in the analysis of bone accumulations for southern Africa. Annals of the Transvaal Museum 29, 1-8.

Brain, C.K. 1981. The Hunters or the Hunted? An Introduction to African Cave Taphonomy. University of Chicago, Press.

Clarke, R.J.1998. First ever discovery of a well-preserved skull and associated skeleton of *Australopithecus*. South African Journal of Science 94, 460-463.

Isaac, G.Ll., 1976. The activities of early African hominids: A review of the archaeological evidence for the time span two and a half to one million years ago. In: Isaac, G.Ll., McCown, E.R. (Eds.), Human Origins: Louis Leakey and the East African Evidence. Staples Press, Menlo Park (CA), pp. 482-514.

Klein, R.G. 1972. The late Quaternary mammalian fauna of Nelson Bay Cave (Cape Province, South Africa): Its implications for megafaunal extinctions and environmental change and culture change. Quaternary Research 2, 135-142.

Klein, R.G. 1975. Palaeoanthropological implications of the nonarchaeological bone assemblage from Swartklip 1, south-western Cape Province, South Africa. Quaternary Research 5, 275-288.

Klein, R.G. 1976. The mammalian fauna of the Klasies River Mouth sites, southern Cape Province, South Africa. South African Archaeological Bulletin 31, 75-98.

Kuman, K., Field, A.S., Thackeray, J.F. 1997. Discovery of new artefacts at Kromdraai. South African Journal of Science 93, 187-193

Lee Thorp, J.A., van der Merwe, N.J., Thackeray, J.F. 2000. The hunters and the hunted revisited. Journal of Human Evolution 39, 565-576.

Pickering, T.R., Clarke, R.J., Heaton, J.L. 2004. The context of Stw 573, an early hominid skull and skeleton from Sterkfontein Member 2: Taphonomy and palaeoenvironment. Journal of Human Evolution 46, 279-297.

Thackeray, J.F. 1979. An analysis of faunal remains from archaeological sites in southern South West Africa (Namibia). South African Archaeological Bulletin 34, 18-33.

Thackeray, J.F. 1990. Carnivore activity at Klasies River Mouth: A response to Binford. Palaeontologia Africana 27, 101-109.

Thackeray, J.F. 1999. Kromdraai: A summary of excavations since 1938. Excursion Guide, International Quaternary Association Congress (INQUA).

Thackeray, J.F., von Leuvan-Smith. 2001. Implications of crown height measurements of alcelaphine molars from Kromdraai A, South Africa. Annals of the Transvaal Museum 38, 9-12.

Thackeray, J.F., de Ruiter, D.J., Berger, L.R., van der Merwe, N.J. 2001. Hominid fossils from Kromdraai: A revised list of specimens discovered since 1938. Annals of the Transvaal Museum 38, 43-56.

Thackeray, J.F., Kirschvink, J. 2002. Palaeomagnetic analysis of calcified deposits from the Plio-Pleistocene hominid site of Kromdraaai, South Africa. South African Journal of Science 98, 537-540.

Thackeray, J.F., McBride, V.A., Segonyane, S.P., Franklyn, C.B. 2003. Trace element analysis of breccia associated with the type specimen of *Australopithecus* (*Paranthropus*) *robustus* from Kromdraai. Annals of the Transvaal Museum 40, 147-150.

Thackeray, J.F, Venter, A.M., Nothnagel, G. 2005. Bone apatite residues and early Pleistocene stone tools: associations with bone breakage. South African Journal of Science 101, 17-18.

Vrba, E.S. 1976. The Fossil Bovidae of Sterkfontein, Swartkrans and Kromdraai. Transvaal Museum Memoir 21. Transvaal Museum, Pretoria.

Watson, V. 1993. Composition of the Swartkrans bone accumulations, in terms of skeletal parts and animals represented. In: Brain, C.K. (Ed.), Swartkrans: A Cave's Chronicle of Early Man. Transvaal Museum, Pretoria, pp. 35-73.

# CHAPTER 5

# TAPHONOMIC ANALYSIS OF AN EXCAVATED STRIPED HYENA DEN FROM THE EASTERN DESERT OF JORDAN

KATHY SCHICK, NICHOLAS TOTH, THOMAS GEHLING AND TRAVIS RAYNE PICKERING

## ABSTRACT

A recent striped hyena den was excavated in the eastern desert of Jordan to examine taphonomic patterning in the bone assemblage. A total of 4,847 specimens of bones and teeth was recovered from a 16 $m^2$ excavation, with the majority of these (94.7%) buried to a depth of up to 20 cm. While large and even complete bones dominated the surface assemblage, the buried assemblage also contained very large numbers of smaller bones and bone fragments. Taxonomically, the assemblage is composed predominantly of camel, dog, sheep/goat, and gazelle, but also contains donkey, human, horse, fox, stork, hare, hedgehog, other bird, oryx, hyena, honey badger, and snake remains. A minimum number of 54 individuals was represented at the site (26% of which are carnivores), with 510 specimens identifiable to both element and taxonomic group. It is likely that many of the smaller animals could have been introduced to the den as more complete carcasses, while larger animals were likely transported as body parts such as limbs and skulls. The assemblage and its modification (toothmarks and breakage patterns) are consistent with hyena accumulation and consumption, with very little evidence of human or rodent modification. Also examined were the effects of differential bone weathering on toothmark frequencies and fracture patterning. Of special interest is the high degree of fragmentation of limb bones, similar to the patterns seen at many Plio-Pleistocene archaeological sites. This analysis adds to the comparative database of hyena bone accumulations and their taphonomic patterning to aid in interpreting prehistoric faunal assemblages.

## INTRODUCTION

Bob Brain's contributions to taphonomy, human origins studies, and natural history have inspired a generation of researchers, including ourselves. His seminal book, *The Hunters or the Hunted* (1981) was the catalyst that made two of us (K.S. and N.T.) go to Jordan and excavate a recent striped hyena den in the eastern desert. We modeled our methodology after Brain's analysis of the taphonomic patterns at Sterkfontein, Swartkrans and Kromdraai, as well as his studies of modern bone accumulations at locales such as brown hyena dens and Hottentot camps.

Over the past few decades, a sizeable database has been compiled by researchers investigating patterns of bone accumulation and modification by modern and recent hyenas, as well as considering the possible role of hyenas in collecting and modifying bone assemblages in the prehistoric past (e.g., Binford, 1981; Brain, 1981; Bunn, 1982, 1983; Cruz-Uribe, 1991; Henschel et al., 1979; Hill, 1989; Horwitz and Smith, 1988; Hughes, 1954; Kerbis-Peterhans and Horwitz, 1992; Klein, 1975; Kuhn, 2005; Lacruz and Maude, 2005; Leakey et al., 1999; Maguire et al., 1980; Mills and Mills, 1977; Owens and Owens, 1979; Pickering, 2002; Scott and Klein, 1981; Skinner et al., 1980; Skinner et al., 1986; Skinner et al., 1998; Stiner, 1991; Sutcliffe, 1970). Bone collecting activities have now been well documented among all three living species. Although there are some notable differences among hyena species, particularly between, on the one hand, the striped hyena (*Hyaena hyaena*) and the brown hyena (*Parahyaena brunea*), and, on the other hand, the spotted hyena (*Crocuta crocuta*), as to their bone transporting activities and the nature of the bone accumulations they produce, all three species have now been well documented as veritable bone collectors, particularly in situations involving provisioning of young.

Hyenas are of particular interest to the archaeologist since they are one of a select group of species, including humans and porcupines, which sometimes collect large quantities of bones at specific locations over time. The striped hyena (Figures 1 and 2), is especially interesting in view of its possible role in the formation of the Plio-Pleistocene bone assemblages at Makapansgat in South Africa (Brain, 1981). A fossil subspecies of this form *H. hyaena makapani* is known from the Makapansgat site (Toerian, 1952) and has been implicated as a likely source of the faunal accumulations there (Maquire et al., 1980).

This study provides detailed information about a large bone accumulation at a striped hyena den in Jordan that was excavated in order to retrieve detailed taphonomic and comparative information for use in paleoanthropological studies. Surface collection, excavation, and screening of sediment from a substantial portion of the den provided a large bone sample subsequently subjected to detailed taxonomic and taphonomic analysis.

The faunal sample retrieved from this hyena den is especially valuable for paleoanthropological purposes in that a substantial portion of the den was excavated and sieved. Thus, the excavated materials include a good proportion of bone fragments often not retrieved in high numbers in surface collections. In addition, differential weathering (and thus potential for fossilization and modification traces) was observable in the surface versus the buried bone sample, and the buried bone in particular provides a sample well-suited for study of bone fragmentation and surface modification.

## The Umari Den

*Figure 1. A nineteenth century representation of a striped hyena. ("Striped Hyena," aquatint by W. Daniell, from Wood, 1807).*

*Figure 2. A modern striped hyena during a zoo bone-feeding experiment.*

### History

The Umari hyena den was discovered in 1984 by a paleoanthropolocial reconnaissance team searching for fossiliferous sediments. The den is located in the eastern desert of Jordan near the Saudi Arabian border, approximately 45 km southeast of the town of Azraq and six km east of the village of Umari (Figure 3). The team (and their affiliations at the time) included archaeologist Mujahed Meheissen (Yarmouk University), paleoanthropologists Donald Johanson (Institute of Human Origins) and Tim White (University of California at Berkeley), and geologist Robert Drake (University of California at Berkeley). The following year, two of the authors (N.T. and K.S.) undertook excavations of the den site and obtained an extensive excavated sample of faunal material from the den accumulation. Subsequent analysis has provided information regarding the faunal remains represented at the site.

The den was no longer active by the time of its discovery. Of environmental importance in the area is a permanent spring located at Azraq 45 km to the northwest. Historically, recent wild fauna of the region include oryx, gazelle, ass, ostrich, lion, leopard, cheetah, hyena, wolf, jackal, and fox. Although hyenas have reportedly been quite widespread over the Arabian peninsula (except in the interior deserts) in the past, and apparently plentiful as recently as the 1960s, they, along with other wild animals such as oryx and gazelle, appear to have become rare or absent in the

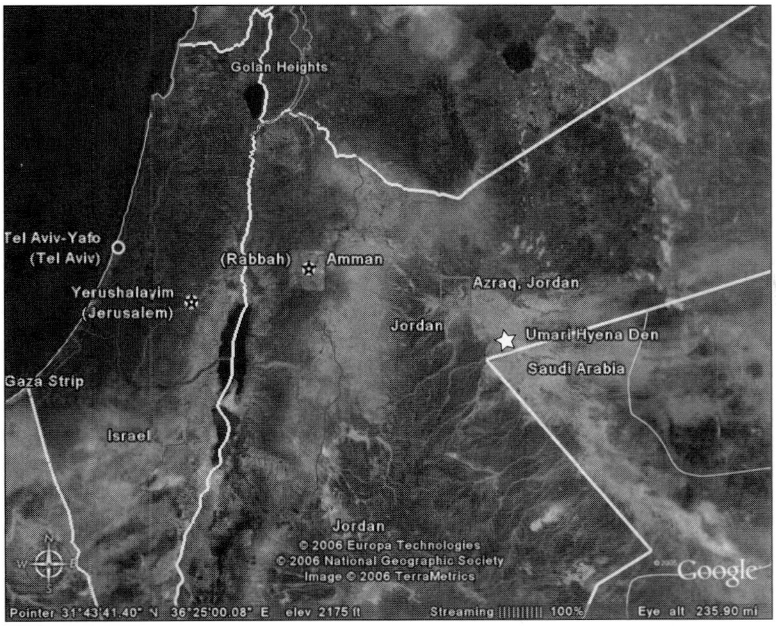


*Figure 3. Map showing the location of the Umari striped hyena den in the eastern desert of Jordan, near the Saudi border.*

eastern desert in recent times (although the oryx has been reintroduced in the region during the past two decades).

## Location

The area in which the Umari den is located is today a sparsely vegetated desert, characterized by flat, flint-paved surfaces and badlands topography where erosion has exposed Miocene marine sediments, creating a network of small escarpments and gullies. The faunal remains were recovered from the surface and buried within up to 20 cm of soft sediment derived from limestone weathering. Due to the unconsolidated nature of the sediment, the lack of identifiable strata, and the visible bioturbation from small animal burrows, this assemblage was considered as one horizon.

The den is situated on a small ridge overlooking a wadi cutting through Miocene limestones (Figures 4–11). A rock layer at the base of the ridge is composed of a less consolidated limestone which has been undercut in many areas along the outcrop, leaving overhangs of the harder, more consolidated limestone. Several narrow, shallow tunnels, too small for most human adults to crawl through, have also been cut or dissolved into the softer limestone layer at the base of the outcrop. Some of these interconnect and some cut all the way through the outcrop to exit on the opposite side of the ridge, steeply overlooking a small wadi. This system of tunnels and overhangs, as well as the gently sloping terrace in front of the den, provides shade, a degree of safety, and an excellent view of the desert plains to the south. The bones had accumulated primarily on the terrace directly in front of the tunnel openings, with very few present within the small tunnels themselves.

The sediment in which the bones were found constitutes erosional residues from the limestones forming the den's rock overhangs and tunnels. This sediment, essentially limestone "flour," was not hard and would be unlikely to produce natural striations on bones that might be interpreted as toothmarks or cutmarks.

## Excavation and Recovery

Bones were subaerially exposed over a total area of approximately 80 m$^2$ but were concentrated especially on the terrace in front of the rock outcrop and, secondarily, on the slope leading down to the nearby wadi. A 16 m$^2$ area was gridded in 2 m × 2 m squares within the region of densest accumulation directly in front of the rock outcrop. Approximately 10% of the den accumulation was excavated in terms of total extent of surface bone, but as the excavation selectively sampled the denser areas of the deposit, it retrieved an estimated 30% of the total surface deposit (in terms of counts of surface bone and bone fragments). As bone burial appeared to be concentrated within the excavation in close proximity to the outcrop face and near the major tunnel entrances, and this area was fully excavated, the excavated sample is estimated to represent minimally 30%, and likely nearly 50% or more, of the buried bone.

*Figure 4. The location of the Umari hyena den from a distance. The arrow shows the location of the den within residual sedimentary outcrops in the Jordanian desert.*

Figure 5. The limestone outcrop at the den location. Bones were concentrated on the apron in front of the outcrop (in the foreground), with highest densities near the openings of the tunnels.

Figure 6. The limestone outcrop with bones scattered across the apron and some extending down the slope toward the adjacent wadi.

Figure 7. Surface scatter of bones in front of the limestone outcrop.

Within each 2 m × 2 m grid unit, the surface bone was plotted and picked up. The underlying sediment was then excavated to the maximum extent of bone burial (up to 20 cm below the surface), the buried bone retrieved, and the sedimentary matrix passed through a 5 mm mesh screen, retrieving small bone and tooth fragments, scat, and other organic materials. (Curiously, scattered over a wide area of the den outside of the excavation were a number of complete and even mummified dog skulls; these were collected but are not part of the formal analysis here of the assemblage from the excavated area.)

The highest density of bones and hyena droppings was not directly under the rock overhangs or in the tunnels, but rather on the broad, flat terrace in front of the limestone outcrop and above the nearby wadi (Figure 11). The density of faunal remains visible on the surface varied along the terrace, with highest densities somewhat closer to the outcrop and to the main tunnel entrances, but as there was no other obvious spatial patterning of surface materials, the excavated sample is here considered as representative of the overall composition of the den assemblage.

## The Bone Sample

A sample of 4,847 bones, bone fragments, and isolated teeth/tooth fragments was recovered from the 16 m$^2$ excavated area. Of these, only 189 specimens were exposed on the surface (3.9 % of the sample), while the remaining 4,658 specimens (96.1%) were buried within the sediment. The surface and buried faunal materials from the 16 m$^2$ excavated area were examined for the purpose of taxonomic and element identification, as well as to identify patterns of modification, including tooth marks, breakage, and weathering stages. The

*Figure 8. View from top of the outcrop showing distribution of exposed bones.*

analysis presented here includes all of the faunal materials from the excavated area, both the small surface sample and the large proportion of buried remains.

This analysis was designed to compare and contrast the Umari hyena den with the taphonomic patterns presented by C.K. Brain in *The Hunters or the Hunted* volume published in 1981. For this reason, levels of element identifiability were comparable to those Brain employed in that classic study.

## Weathering and element fragmentation

It should be noted that the surface and excavated samples from the excavated area show some very important differences. The surface materials overall are more extensive weathered than the buried sample, and consist of relatively larger, even many complete, bones. The excavated sample is overall more highly fragmented and less readily assigned to taxonomic group. While over half (50.3%) of the surface bones are identified to taxon, only a small portion (8.9%) of the buried sample is identifiable. It is probably that larger bones have a greater tendency to "ride high" as animals moving back and forth would be prone to kick them up and help them escape burial, while smaller and more fragmentary bones and teeth would tend to be incorporated into the sediment more readily.

The degree of weathering evident varies greatly among the bones and appears to be a function of the amount of time the bones were exposed to sunlight and the elements. In effect there were two weathering gradients: bones further from the rock outcrop tend to be more heavily weathered, and, as noted above, the surface sample is more heavily weathered than the buried

*Figure 9. Close-up of surface bones. Note the prominence of large limb bones and the advanced degree of weathering of many of the surface materials.*

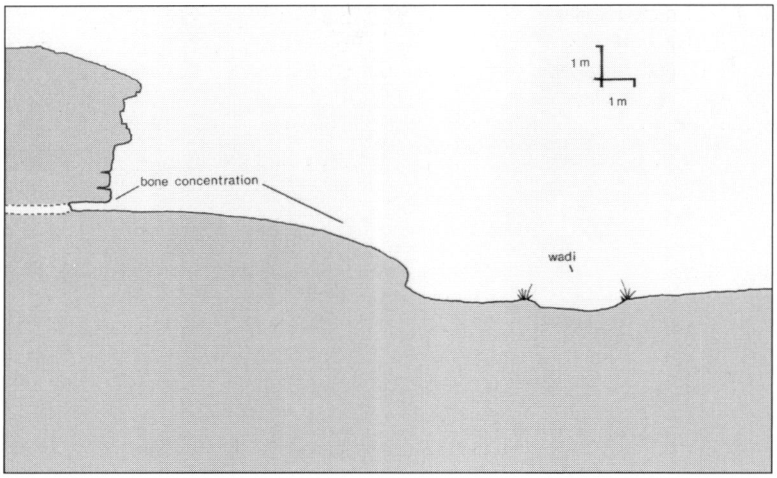


*Figure 10. Cross-section of the major topographic features at the hyena den. The major concentration of bones stretched from in front of the rock outcrop several meters toward the erosion slope above the wadi, with the bones closer to the wadi more highly weathered.*

sample. Bones under the rock overhang in the tunnels are best preserved, sometimes with dried tissue still attached.

The differential weathering of the surface and buried samples likely reflects the differences in rapidity of burial and relative exposure to the elements: the larger bones were more identifiable to taxon, less likely to be buried, and thus more prone to weathering; the smaller bones and bone fragments were less identifiable but more readily buried and hence less vulnerable to weathering processes. As a result, a final assemblage that might become fossilized over time in such circumstances would likely be the more heavily fragmented and less identifiable portion of the faunal assemblage originally present.

## Taxonomic composition

At least 16 different taxonomic groups are represented, with a minimum number of 54 individuals (Table 1 and Figure 12). Taxa present include both wild and domestic animals of the region from modern or recent

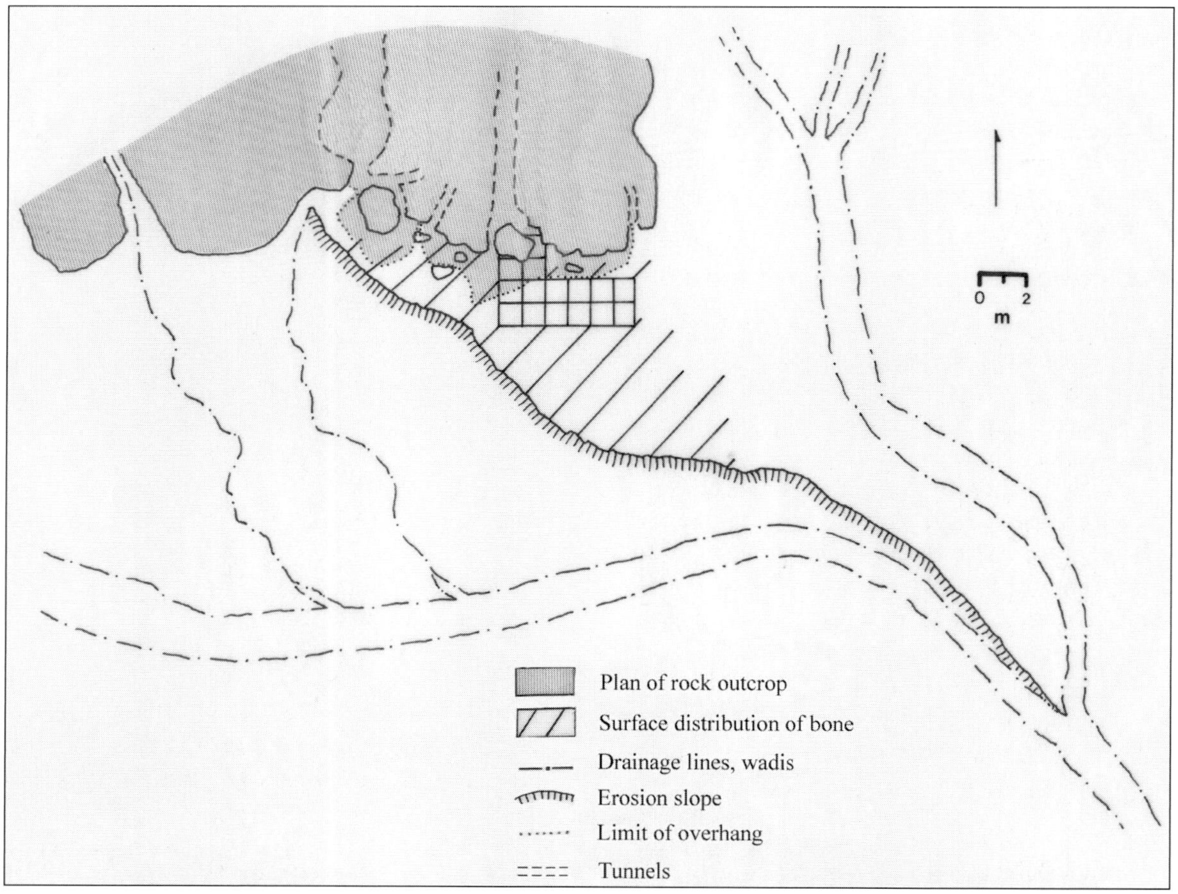


*Figure 11. Plan view of the hyena den showing the extent of surface distribution of bones and the 16-m² excavation grid (extending under the rock overhang in its northwest corner). Bone recovery was complete from within this 16 m² area, with the majority of the recovered bone (96.1%) buried and only 3.9% exposed on the surface.*

*Table 1. Taxa represented at the Umari striped hyena den, showing NISP (number of identifiable specimens), MNI (minimum number of individuals), and the NISP/MNI ratio for each taxonomic group*

| Taxon | NISP | | MNI | | NISP/MNI |
|---|---|---|---|---|---|
| | n | % | n | % | |
| Camel *(Camelus dromedarius)* | 201 | 39.4 | 10 | 18.5 | 20.1 |
| Horse *(Equus caballus)* | 4 | 0.8 | 2 | 3.7 | 2.0 |
| Donkey *(Equus asinus)* | 20 | 3.9 | 3 | 5.6 | 6.7 |
| Oryx *(Oryx leucoryx)* | 1 | 0.2 | 1 | 1.9 | 1.0 |
| Gazelle *(Gazella dorcas)* | 52 | 10.2 | 6 | 11.1 | 8.7 |
| Human *(Homo sapiens)* | 13 | 2.5 | 3 | 5.6 | 4.0 |
| Hyena *(Hyaena hyaena)* | 1 | 0.2 | 1 | 1.9 | 1.0 |
| Goat/sheep *(Capra hircus/Ovis aries)* | 41 | 8.0 | 6 | 11.1 | 6.8 |
| Dog *(Canis familiaris)* | 151 | 29.6 | 10 | 18.5 | 15.1 |
| Honey Badger *(Mellivora capensis)* | 1 | 0.2 | 1 | 1.9 | 1.0 |
| Fox *(Vulpes sp.)* | 7 | 1.4 | 2 | 3.7 | 3.5 |
| Stork *(Ciconia* sp.) | 5 | 1.0 | 2 | 3.7 | 2.5 |
| Hare (Lagomorpha) | 8 | 1.6 | 2 | 3.7 | 4.0 |
| Hedgehog (Erinaceidae) | 2 | 0.4 | 2 | 3.7 | 1.0 |
| Bird indet. (Aves indet.) | 2 | 0.4 | 2 | 3.7 | 1.0 |
| Snake (Reptilia indet.) | 1 | 0.2 | 1 | 1.9 | 1.0 |
| **TOTAL** | 510 | 100 | 54 | 100 | |

times. They likely include remains scavenged by hyenas from death sites, as well as smaller animals that may have been preyed upon by the hyenas, and subsequently transported by hyenas to the den location. The most abundant taxa are camel, dog, goat/sheep, and gazelle. Other taxa include donkey, human, horse, fox, stork, hare, hedgehog, other birds, hyena, honey badger, and snake.

It is likely that animals of Group Size 1 (less than 50 pounds) could have been transported to the den as complete carcasses, while the larger animals were likely transported in as body parts (limb portions, crania and mandibles). (Two fossil Miocene shark teeth were recovered in the excavation, clearly eroded out of the limestone bedrock are not contemporaneous with the rest of the faunal sample, and are not considered in this study.)

The MNI (minimum number of individuals) of each taxonomic group in the excavated den assemblage, by descending body size (for animal size groups, see Brain, 1981; Bunn, 1982), are:

1. Animal Size Group 4 (750-2000 lbs)
   a) Camel (MNI=10)
   b) Horse (MNI=2)
2. Animal Size Group 3 (250-750 lbs)
   a) Donkey (MNI=3)
   b) Oryx (MNI=1)
3. Animal Size Group 2 (50-250 lbs)
   a) Human (MNI=3)
   b) Gazelle (MNI=6)
   c) Hyena (MNI=1)
   d) Goat/sheep (MNI=6)
4. Animal Size Group 1 (less than 50 lbs)
   a) Dog (MNI=10)
   b) Honey badger (MNI=1)
   c) Fox (MNI=2)
   d) Stork (MNI=2)
   e) Hare (MNI=2)
   f) Hedgehog (MNI=2)
   g) Other bird (MNI=2)
   h) Snake (MNI=1)

Camels and dogs have by far the best MNI representation (at least 10 individuals each), while camels are best represented in terms of number of identifiable specimens (NISP=201). Moderately high numbers of bovids (six goats/sheep and six gazelle) and equids (two horses and three donkeys) are also present in the sample. Carnivores overall are relatively well represented, with a minimum of four individuals present in addition to the dogs, including two foxes, a hyena, and a honey badger. The overall taxonomic composition of the Umari den is

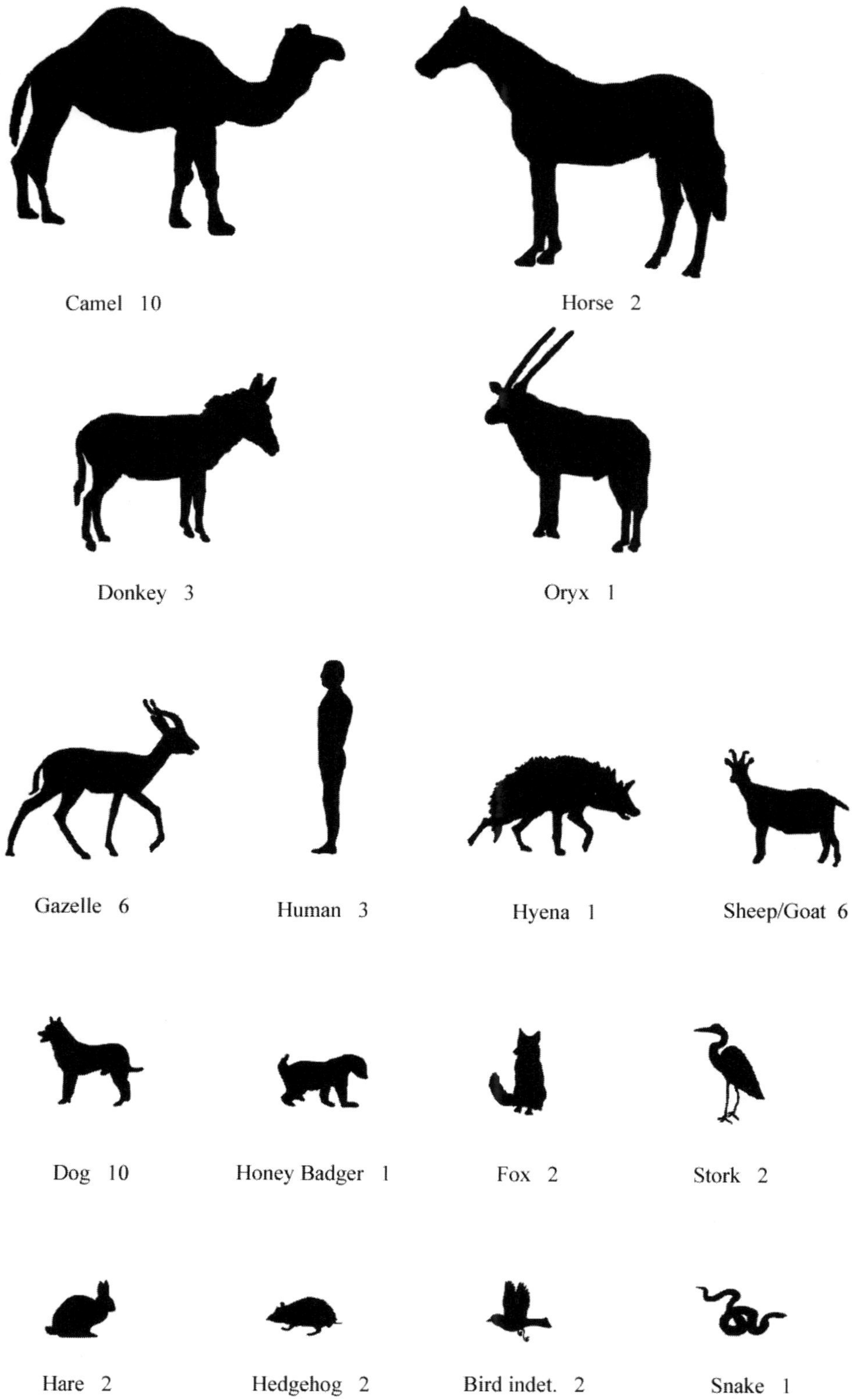


*Figure 12. The sixteen taxonomic groups represented in the Umari den excavation, showing MNI for each. The dominant taxa in terms of MNI and NISP are camel and dog.*

broadly similar to the striped hyena dens elsewhere in the region reported by Skinner et al. (1980), Kerbis-Peterhans and Horwitz (1992), and Kuhn (2005), although the proportion of wild animals included varies somewhat, probably according to differing ecological conditions and effects of human settlement and activities in each locale at the time of active den formation.

*Table 2. The Umari hyena den faunal assemblage broken down by general body part, showing the relative proportions of each body part that were identifiable to taxon and nonidentifiable to taxon*

| Body Part | Total (ID and non-ID to taxon) | | Non-ID to taxon | | ID to taxon | | | |
|---|---|---|---|---|---|---|---|---|
| | n | % assem-blage | n | % assem-blage | n | % body part | % ID | % assem-blage |
| Skull parts (inc. isolated teeth and tooth fragments)[1] | 1665 | 34.4 | 1454 | 30.0 | 211 | 12.7 | 41.4 | 4.4 |
| Vertebrae | 46 | 0.9 | 18 | 0.4 | 28 | 60.9 | 5.5 | 0.6 |
| Ribs | 152 | 3.1 | 140 | 2.9 | 12 | 7.9 | 2.4 | 0.2 |
| Pelves/Scapulae | 21 | 0.4 | 0 | 0.0 | 21 | 100.0 | 4.1 | 0.4 |
| Complete Limbs[2] | 26 | 0.5 | 0 | 0.0 | 26 | 100.0 | 5.1 | 0.5 |
| Limb Ends[2] | 83 | 1.7 | 0 | 0.0 | 83 | 100.0 | 16.2 | 1.7 |
| Limb Shafts[2] | 1404 | 29.0 | 1359 | 28.0 | 45 | 3.2 | 8.8 | 0.9 |
| Manus/Pes | 92 | 1.9 | 8 | 0.2 | 84 | 91.3 | 16.5 | 1.7 |
| Subtotal | 3489 | 72.0 | 2979 | 61.4 | 510 | | 100.0 | 10.5 |
| Other Fragments | 1358 | 28.0 | 1358 | 28.0 | 0 | | 0.0 | 0.0 |
| Total | 4847 | 100.0 | 4337 | 89.5 | 510 | | 100.0 | 10.5 |

[1]Includes 1247 tooth fragments
[2]Limb counts include ungulate metapodials

## General body part representation

The overall breakdown of the Umari den faunal assemblage by general body part is presented in Table 2. Of the 510 elements that were identified to taxon (10.5% of the entire assemblage), the greatest proportion of these consisted of cranial elements and teeth (41.5% of the identified specimens), with limb epiphyses and foot bones (manus and pes) also well represented among the taxonomically identifiable specimens. The specimens not taxonomically identifiable consisted largely of cranial and tooth fragments (30% of the assemblage), limb shaft fragments (28% of the assemblage), and other bone fragments (28% of the assemblage).

## Element representation

Table 3 and Figure 13 show the elements represented for the three major animal groups: camel, small bovid (sheep/goat and gazelle combined), and dog. For camel, the best represented elements are the tibia, mandible, metacarpal, and calcaneus. For the small bovids, the best represented elements are the mandible, maxilla, cranium/horn core, and tibia. For dog, the best represented elements are the mandible, maxilla, and cranium. Element representation (relative to MNI) is, of course, a function of both hyena transport and preferential destruction/survival of elements. The smaller animals are especially well-represented by head elements (mandible, maxilla, and identifiable cranial fragments), while the camel has good representation not only of some head elements (especially mandibles) but also of many of the larger and/or denser limb elements such as the tibia, metacarpal, calcaneus, metatarsal, astragalus and radius-ulna.

A camel death site was discovered a few kilometers from the hyena den (Figure 14) that had apparently been ravaged by carnivores, probably hyenas, with some parts of the body removed or destroyed. The remaining skeleton was dominated by axial elements (cranium, vertebrae, ribs, pelvis), but only one limb was present (presumably this limb had been underneath the carcass and harder for carnivores to access). Interestingly, the elements represented at this ravaged death site (with the exception of this one forelimb) generally had an inverse relation to the camel elements that were present at the hyena den (Figures 15 and 16). Figures 15 and 17 show the preferential element representation for the three major taxonomic groups at the den: camels, dogs, and small bovids.

## Cranial/postcranial ratios

Table 4 shows the cranial/postcranial ratios for each of the taxonomic groups found at the Urami hyena den. As can be seen, overall small animals have a much higher cranial/postcranial ratio (0.68) than do larger animals (0.19). Since it is more likely that smaller animals could have been transported by hyenas as complete carcasses to the den, while large animals were probably transported especially as disarticulated body parts, especially limbs and mandibles, the lower proportion of postcrania among small animals is probably partially due to greater destruction of identifiable limb ends and more extensive breakage of small animal limb shafts into unidentifiable fragments.

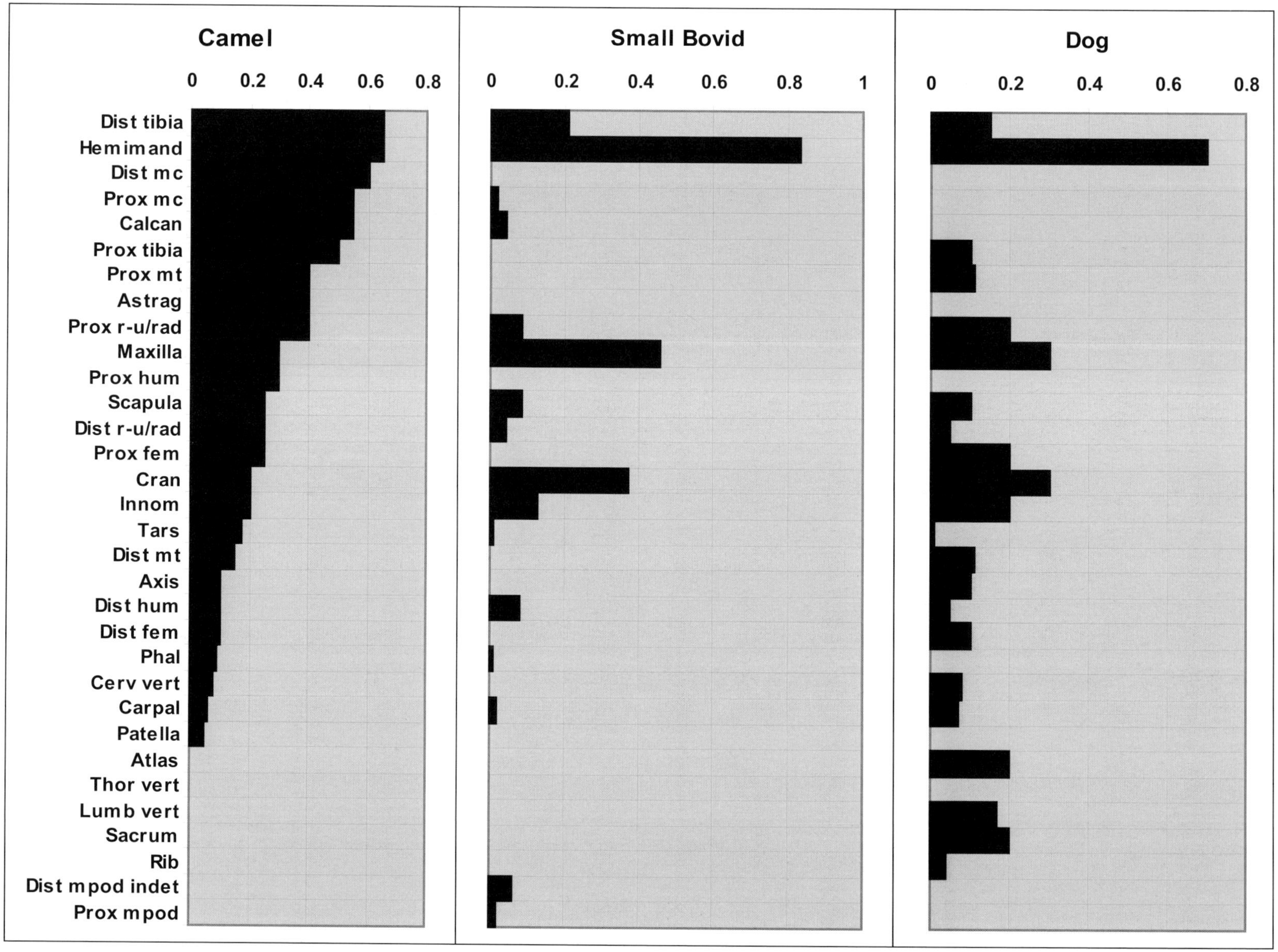


*Figure 13. Comparison of relative element frequencies of camel, bovid and dog. Camels exhibit high percentage of the hemimandibles as well as many limb elements (distal tibia, metacarpal, calcaneus, proximal tibia, proximal metatarsal, astragalus, and proximal radius-ulna), but low frequencies or absence of many other, especially axial, elements (including ribs, sacrum, lumbar and thoracic vertebrae, atlas, patella, carpals, caudal vertebrae, phalanges, distal femur, distal humerus, axis vertebrae, and distal metatarsals). Note that both the smaller-bodied bovids (goat, sheep and gazelle combined) and dogs are best represented by the hemimandible, maxilla and other cranial parts, but dogs also have relatively high frequencies of various other axial as well as appendicular elements, including the proximal femur, innominate, proximal tibia, atlas vertebra, sacrum, and lumbar vertebra.*

## Axial/appendicular ratios

The frequencies of axial and appendicular elements for the major taxa at the Umari den are presented in Table 5. Large animals tend to have a much lower axial/appendicular ratio (0.17) than small animals (1.37). Again, this is likely a result of hyenas having transported more complete carcasses of the smaller animals, so that proportionally more axial elements were likely introduced to the den site, and probably also having preferentially destroyed or heavily comminuted the smaller mammal appendicular elements relative to those of the larger mammals.

## Forelimb/hindlimb ratios

Table 6 shows the forelimb/hindlimb ratios for the major taxa. The overall ratios for the large and small animals tend to be similar, and in both cases identifiable hindlimbs outnumber the forelimbs. Whether these differences are due to differential transport or differential destruction of identifiable elements is not clear.

## Limb fragmentation

Figure 18 shows limb fragmentation for a range of prehistoric and modern bone accumulations, including a number of Plio-Pleistocene sites, hyena dens, a porcupine lair, and a recent hunter-gatherer camp. As can be seen, the Umari hyena den shows a very high degree of fragmentation, with the vast majority of the limbs represented by shafts and with complete limbs and epiphyses present in very small proportions relative to the shafts. Interestingly, the Umari den clusters with many of the Plio-Pleistocene sites in East Africa (Koobi Fora and Olduvai Gorge). The high proportion of limb shafts is likely a function of this hyena den assemblage having

*Table 3. Element representation among the three most abundant taxonomic categories at the Umari hyena den. (small bovids=sheep/goats and gazelle; MNEP=minimum number of elements present; ENE=expected number of elements based on MNI; %=percentage present for the expected number of elements for MNI for that taxonomic group)*

| ELEMENT | CAMEL (MNI=10) | | SMALL BOVIDS (MNI=12) | | DOG (MNI=10) | |
|---|---|---|---|---|---|---|
| | MNEP/ENE | % | MNEP/ENE | % | MNEP/ENE | % |
| Distal tibia | 13/20 | 65.0 | 5/24 | 20.8 | 3/20 | 15.0 |
| Hemi-mandible | 13/20 | 65.0 | 20/24 | 83.3 | 14/20 | 70.0 |
| Distal metacarpal | 12/20 | 60.0 | 1/24 | 4.2 | 0/100 | 0.0 |
| Proximal metacarpal | 11/20 | 55.0 | 1/24 | 4.2 | 0/100 | 0.0 |
| Calcaneus | 11/20 | 55.0 | 1/24 | 4.2 | 0/20 | 0.0 |
| Proximal tibia | 10/20 | 50.0 | 0/24 | 0.0 | 2/20 | 10.0 |
| Proximal metatarsal | 8/20 | 40.0 | 0/24 | 0.0 | 11/100 | 11.0 |
| Astragalus | 8/20 | 40.0 | 0/24 | 0.0 | 0/20 | 0.0 |
| Proximal rad-ulna/rad. | 8/20 | 40.0 | 2/24 | 8.3 | 4/20 | 20.0 |
| Maxilla | 3/10 | 30.0 | 11/24 | 45.8 | 3/10 | 30.0 |
| Proximal humerus | 6/20 | 30.0 | 0/24 | 0.0 | 0/20 | 0.0 |
| Scapula | 5/20 | 25.0 | 2/24 | 8.3 | 2/20 | 10.0 |
| Distal rad-ulna/rad. | 5/20 | 25.0 | 1/24 | 4.2 | 1/20 | 5.0 |
| Proximal femur | 5/20 | 25.0 | 0/24 | 0.0 | 4/20 | 20.0 |
| Cranium/horn core | 2/10 | 20.0 | 9/24 | 37.5 | 3/10 | 30.0 |
| Innominate | 4/20 | 20.0 | 3/24 | 12.5 | 4/20 | 20.0 |
| Tarsal | 14/80 | 17.5 | 1/132 | 0.7 | 1/120 | 0.8 |
| Distal metatarsal | 3/20 | 15.0 | 0/24 | 0.0 | 11/100 | 11.0 |
| Axis | 1/10 | 10.0 | 0/12 | 0.0 | 1/10 | 10.0 |
| Distal humerus | 2/20 | 10.0 | 2/24 | 8.3 | 1/20 | 5.0 |
| Distal femur | 2/20 | 10.0 | 0/24 | 0.0 | 2/20 | 10.0 |
| Phalanx | 11/120 | 9.2 | 2/288 | 0.7 | 0 | 0.0 |
| Cervical vertebra | 4/50 | 8.0 | 0/60 | 0.0 | 4/50 | 8.0 |
| Carpal | 7/120 | 5.8 | 2/132 | 1.5 | 1/140 | 0.7 |
| Patella | 1/20 | 5.0 | 0/24 | 0.0 | 0/20 | 0.0 |
| Atlas vertebra | 0 | 0.0 | 0 | 0.0 | 2/10 | 20.0 |
| Thoracic vertebra | 0 | 0.0 | 0 | 0.0 | 0 | 0.0 |
| Lumbar vertebra | 0 | 0.0 | 0 | 0.0 | 12/70 | 17.0 |
| Sacrum | 0 | 0.0 | 0 | 0.0 | 2/10 | 20.0 |
| Rib | 0 | 0.0 | 1/312 | 0.3 | 10/260 | 3.8 |
| Dist. metapod. indet. | 0 | 0.0 | 3/48 | 6.3 | 1/200 | 0.5 |
| Prox. metapod. indet. | 0 | 0.0 | 1/48 | 2.1 | 0 | 0.0 |

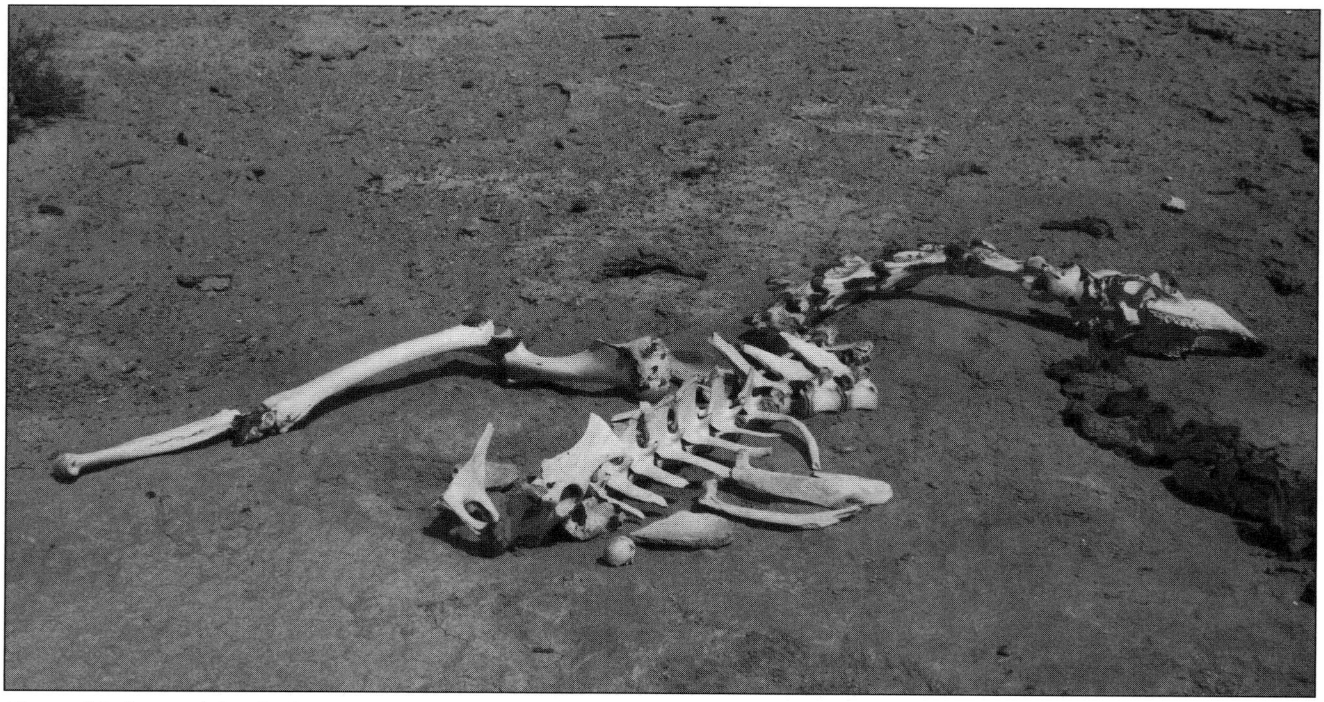

*Figure 14. A camel death site several kilometers from the Umari den. Note the dominance of axial skeletal elements (which are relatively poorly represented at the Umari den) but the near absence of appendicular elements, with the exception of one forelimb (that likely was under the carcass at time of death and inaccessible to scavengers).*

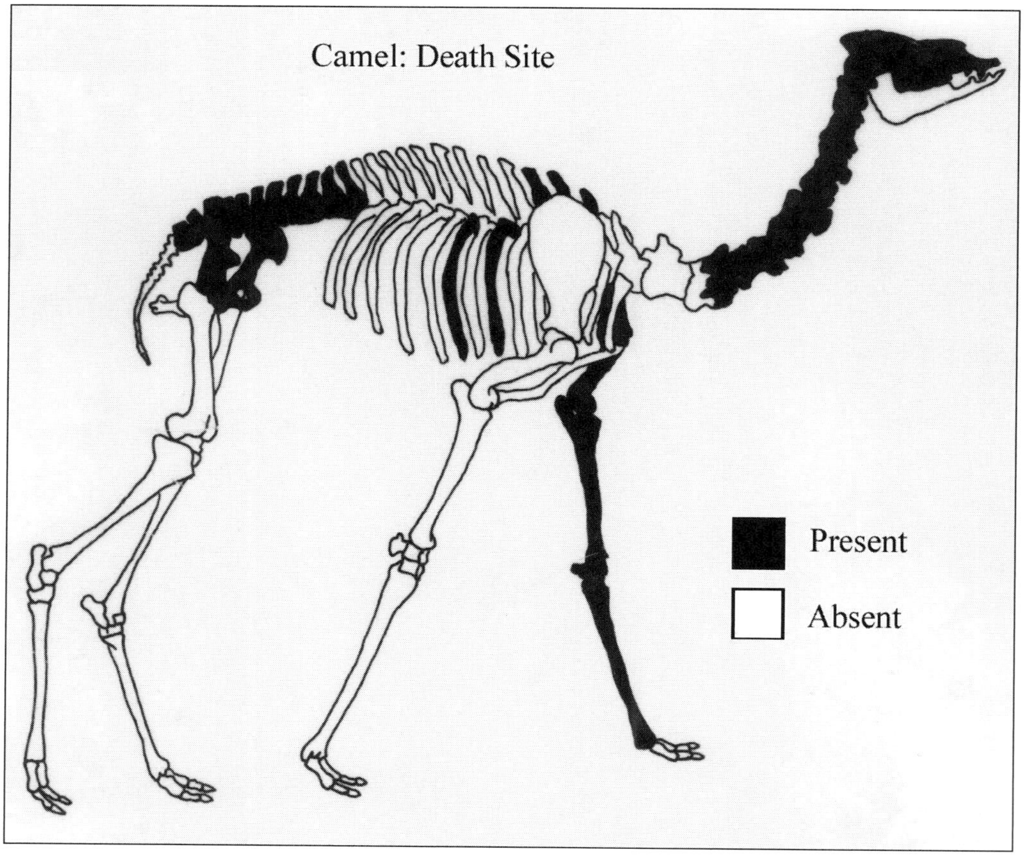


*Figure 15. Element representation at the camel death site. Bones present are represented in black. As mentioned in Figure 12, primarily axial elements are represented, with the exception of one forelimb.*

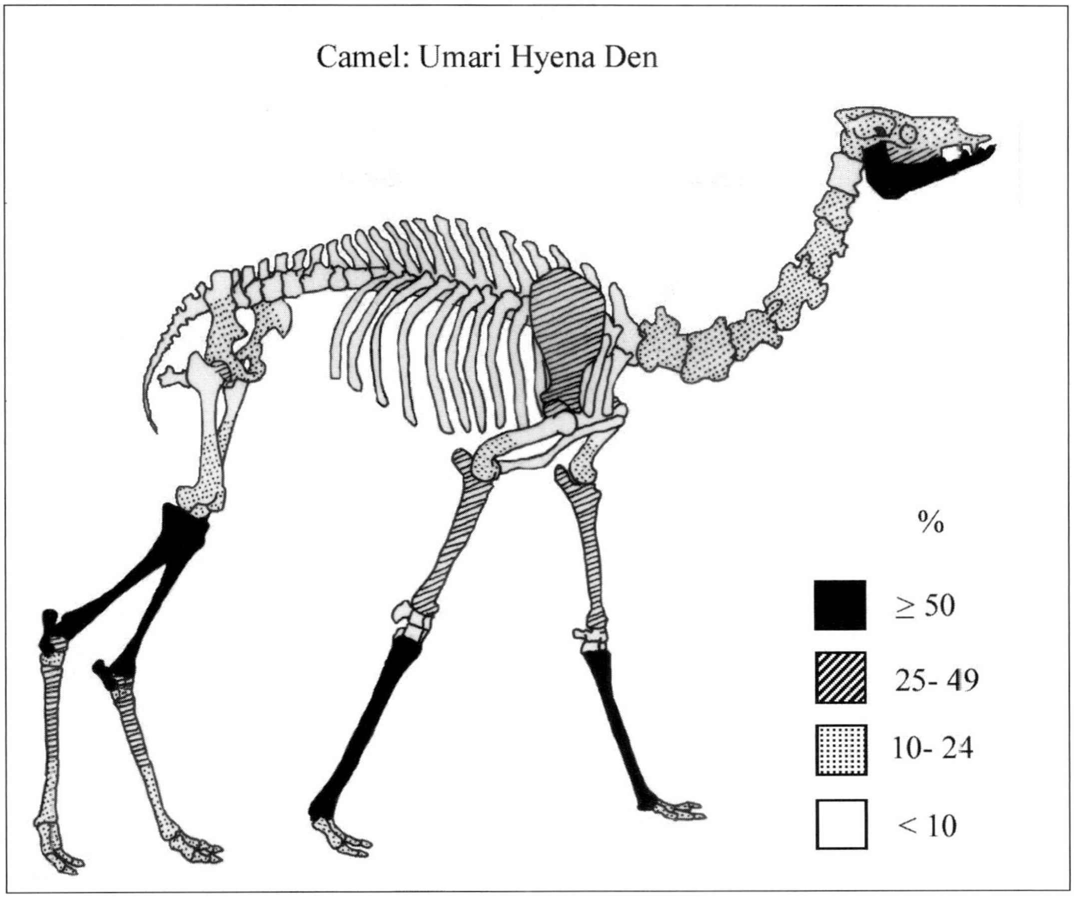


Figure 16. Camel element representation at the Umari den relative to expectations from MNI. Note that mandibles, metacarpals, tibiae, and calcanei are especially well represented.

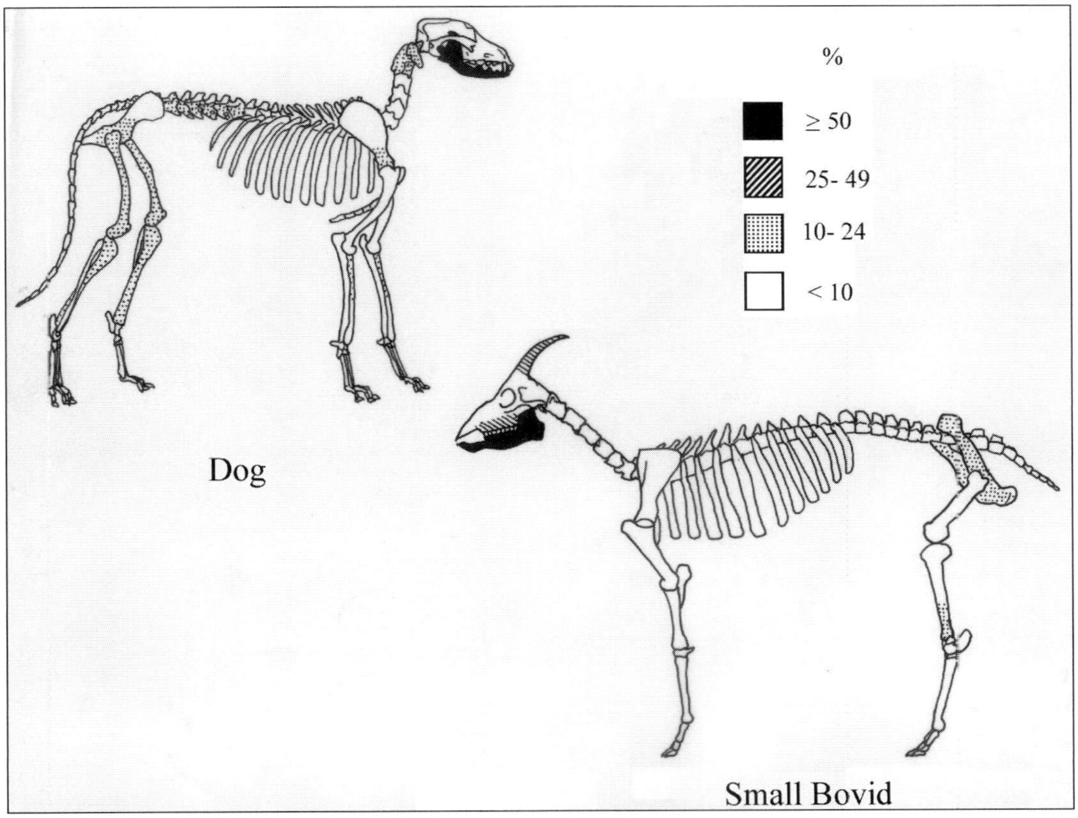


Figure 17. Element representation of dogs and of small bovids (sheep, goat and gazelle) at the Umari den. Note the high frequencies of mandibles as well as bovid maxillae and horn cores.

been excavated and sieved and thus having retrieved an extensive sample of broken limb bones from the 20 cm depth of buried accumulation.

Limb shaft fragments not identifiable to taxa or element were measured to see maximum thickness between the cortical surface and the marrow cavity wall (Figure 19). Measurements of identifiable limb shaft fragments showed that the majority of large mammal limbs tended to have a shaft thickness greater than 5 mm, while the majority of smaller mammal limbs had shaft thicknesses less than 5 mm, although there was some overlap between the two populations. The fact that over 70% of the non-identified shaft fragments were less than 5 mm strongly suggests that the majority of these fragments are from mammals smaller in body size than the hyenas. Conversely, the limbs of larger mammal were less likely to be comminuted into small unidentifiable fragments through hyena gnawing and remained as identifiable shaft fragments or sometimes even complete bones.

*Table 4. Ratio of cranial (excluding isolated teeth) to post-cranial elements for each taxonomic group from the Umari den. Note that overall the smaller animals (fox, gazelle, rabbit, goat/sheep, hedgehog, dog, and hyena) have much higher ratio of cranial to postcranial elements than do the larger animals (camel, horse, donkey, oryx, and human).*

| Animal | Cranial n | Postcranial n | Cranial/ Postcranial Ratio |
|---|---|---|---|
| Fox | 5 | 2 | 2.50 |
| Gazelle | 27 | 15 | 1.80 |
| Rabbit | 4 | 3 | 1.33 |
| Goat/sheep | 15 | 14 | 1.07 |
| Human | 2 | 2 | 1.00 |
| Hedgehog | 1 | 1 | 1.00 |
| Dog | 25 | 73 | 0.34 |
| Camel | 32 | 162 | 0.20 |
| Horse | 0 | 2 | 0.00 |
| Donkey | 0 | 15 | 0.00 |
| Oryx | 0 | 1 | 0.00 |
| Honey badger | 0 | 1 | 0.00 |
| Stork | 0 | 5 | 0.00 |
| Snake | 0 | 1 | 0.00 |
| Hyena | 1 | 0 | -- |
| Large animals | 34 | 182 | 0.19 |
| Small animals | 78 | 115 | 0.68 |

*Table 5. Axial/appendicular counts and ratios for the mammalian taxonomic groups at the Umari hyena den. Note that smaller animals tend to have a higher axial/appendicular ratio than larger animals.*

| Animal | Axial n | App. n | Ax/App. Ratio |
|---|---|---|---|
| Fox | 6 | 1 | 6.00 |
| Gazelle | 20 | 13 | 1.54 |
| Dog | 48 | 38 | 1.26 |
| Goat/sheep | 14 | 12 | 1.17 |
| Human | 2 | 2 | 1.00 |
| Camel | 24 | 131 | 0.18 |
| Honey badger | 0 | 1 | 0.00 |
| Oryx | 0 | 1 | 0.00 |
| Donkey | 0 | 13 | 0.00 |
| Horse | 0 | 2 | 0.00 |
| Hyena | 1 | 0 | -- |
| Larger animals | 26 | 149 | 0.17 |
| Smaller animals | 89 | 65 | 1.37 |

## Carnivore representation

The percentage of carnivores represented at the site based on MNI is 26%, and the percentage of carnivores represented based on NISP is 31%. The carnivore/ungulate ratio is 0.50. As has been suggested by other researchers, including Brain (1981), Cruz-Uribe (1991), and Pickering (2002), high frequencies of carnivores are characteristic of many hyena bone accumulations, but not all (Kuhn, 2005).

## Bone modification

### *Carnivore damage*

A total of 71% of the identifiable limbs specimens showed clear signs of carnivore damage: tooth scores, pits, punctures, and notching. This modification is consistent with damage from hyenas feeding behavior. Examples of carnivore tooth marks are shown in Figures 20–25. In addition, there are several bone flakes (Figure 26) as well as flake scars on limb shaft fragments (Figure 27) which are consistent with carnivore fracture.

We examined the effects of progressive bone weathering on the identifiability and abundance of toothmarks on cortical bone surfaces. Using a random sample of

*Table 6. Forelimb to hindlimb ratios for the mammalian taxonomic groups at the Umari hyena den. Small and large mammals have similar forelimb to hindlimb ratios.*

| Animal | Forelimb n | Hindlimb n | FL/HL ratio |
|---|---|---|---|
| Rabbit | 2 | 1 | 2.00 |
| Gazelle | 4 | 4 | 1.00 |
| Goat/sheep | 4 | 5 | 0.80 |
| Camel | 45 | 64 | 0.70 |
| Donkey | 3 | 5 | 0.60 |
| Dog | 6 | 13 | 0.46 |
| Human | 0 | 1 | 0.00 |
| Horse | 0 | 2 | 0.00 |
| Hyena | 0 | 0 | -- |
| Hedgehog | 0 | 1 | 0.00 |
| Oryx | 1 | 0 | -- |
| Honey badger | 1 | 0 | -- |
| Fox | 1 | 0 | -- |
| Large animals | 49 | 72 | 0.68 |
| Small animals | 18 | 24 | 0.75 |

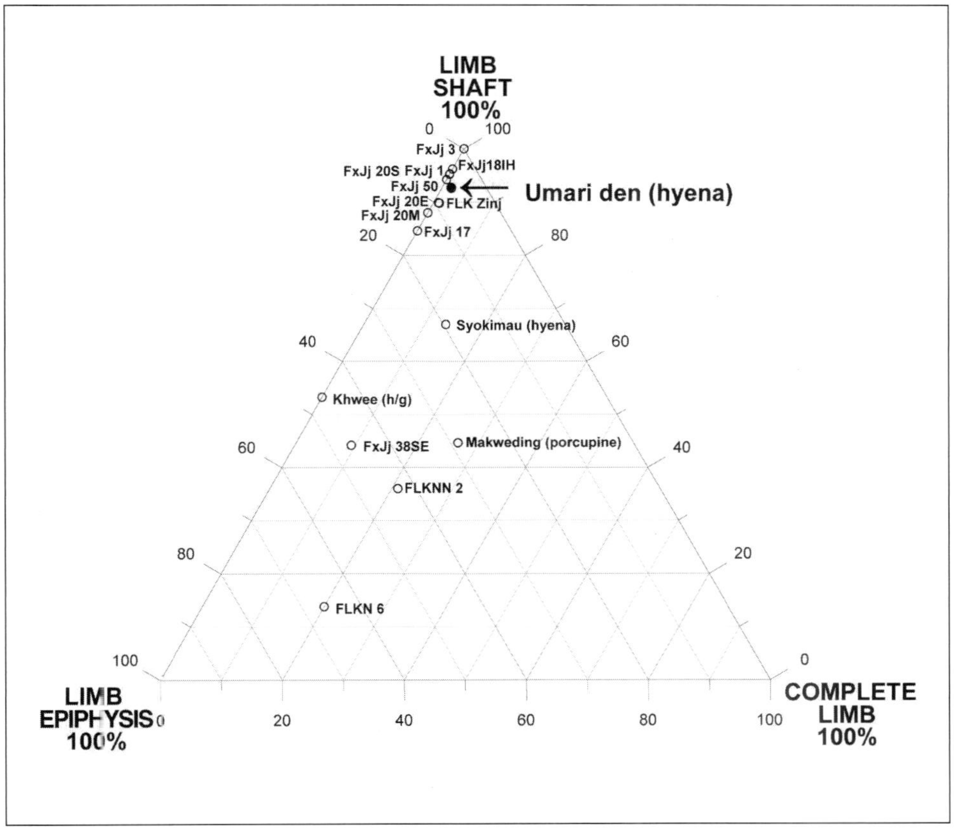


*Figure 18. Ternary graph of limb representation (complete limb v. limb shaft v. limb epiphysis) at the Umari den in comparison to ten Plio-Pleistocene archaeological sites at Koobi Fora and Olduvai Gorge, another hyena den (Syokimau), a porcupine den (Madweding), and a hunter-gatherer camp site (Khwee). Note that this excavated hyena den assemblage shows a very high limb shaft representation similar to many faunal assemblages at Plio-Pleistocene archaeological sites. (after Bunn, 1982).*

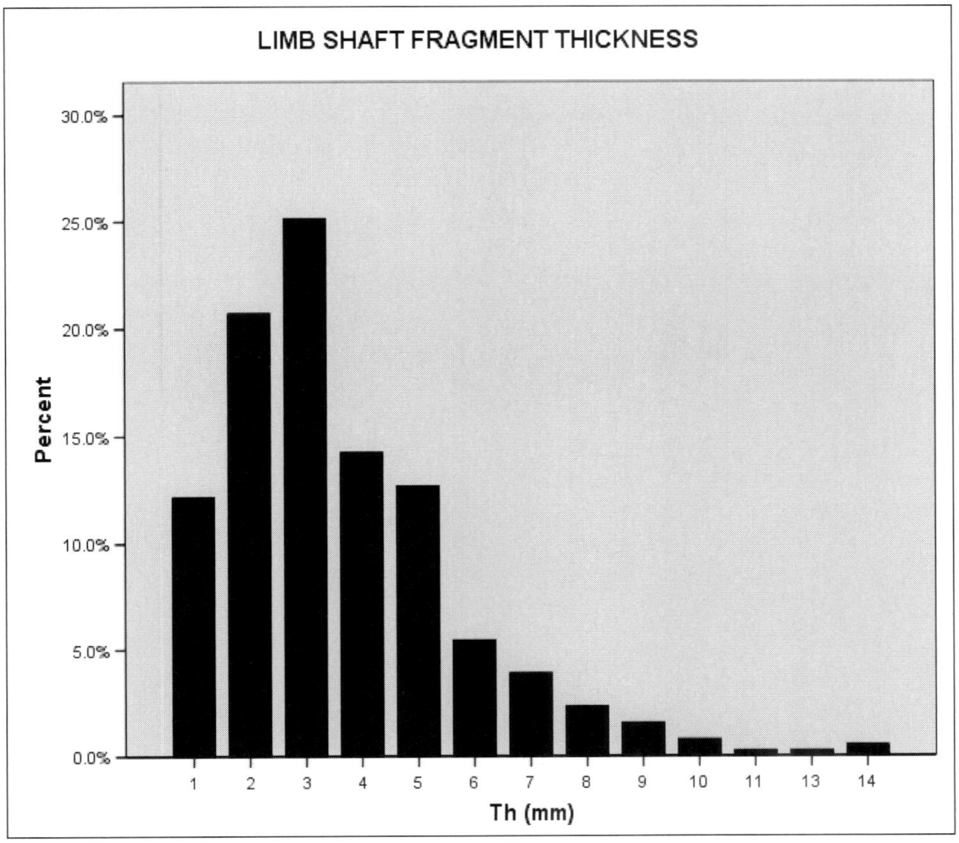


*Figure 19. Thickness of limb shaft fragments from a random sample of all size classes (n=388). Note that the great majority of the fragments are quite thin (mode of 3 mm) and would appear to represent primarily fragmented limbs of smaller animals. Only a small proportion of shaft fragments show a thickness of 6 mm or more. These thicker shaft fragments primarily represent bones of larger animals, whose shafts tended to be less heavily comminuted into nonidentifiable elements.*

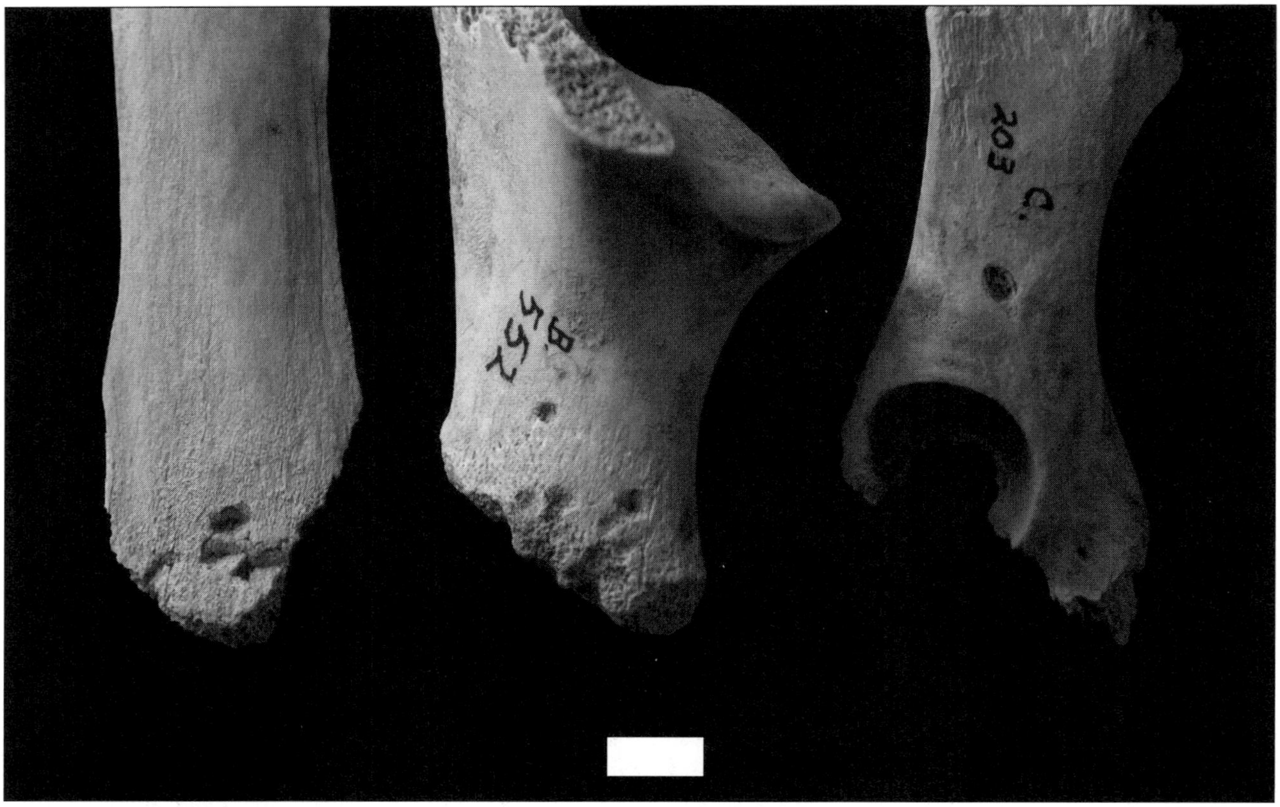

Figure 20. Examples of carnivore-induced tooth pits and punctures. (White bar = 1 cm).

non-identified shaft fragments, we estimated the cortical surface area of each specimen by multiplying shaft fragment length times mean fragment breadth. As can be seen in Table 7 and Figure 28, the number of identifiable toothmarks goes down markedly as bone surface weathering increases.

### Fracture patterning

A sample of limb shaft fragments was analyzed to identify whether fracture was green (i.e. broken when fresh and organic-rich), dry (broken when weathered), or a combination of both. As can be seen in Figure 29, bones with a weathering stage of 0–1 (Behrensmeyer, 1978) showed predominantly green fracture, while bones of weathering stage 3 showed predominantly dry fracture. This strongly suggests that much of the bone fracture at the hyena den was made during hyena feeding, but that a portion of the bones were subsequently broken (naturally or by trampling) as the bones dried out and weathered. For the broken limb ends of weathering stage 0–1 over half showed either spiral fracture or irregular spiral fracture.

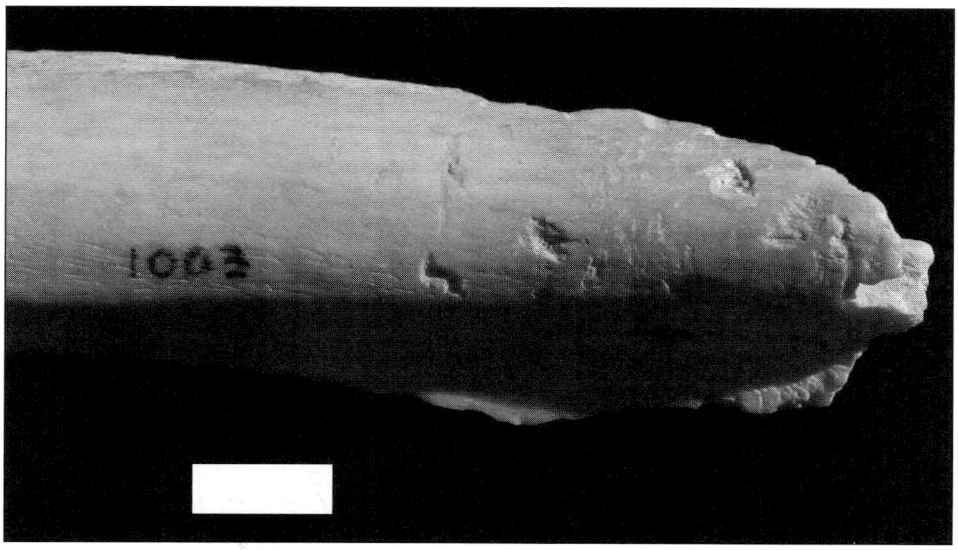

Figure 21. Examples of toothmarks and pits on a limb shaft fragment. (White bar = 1 cm).

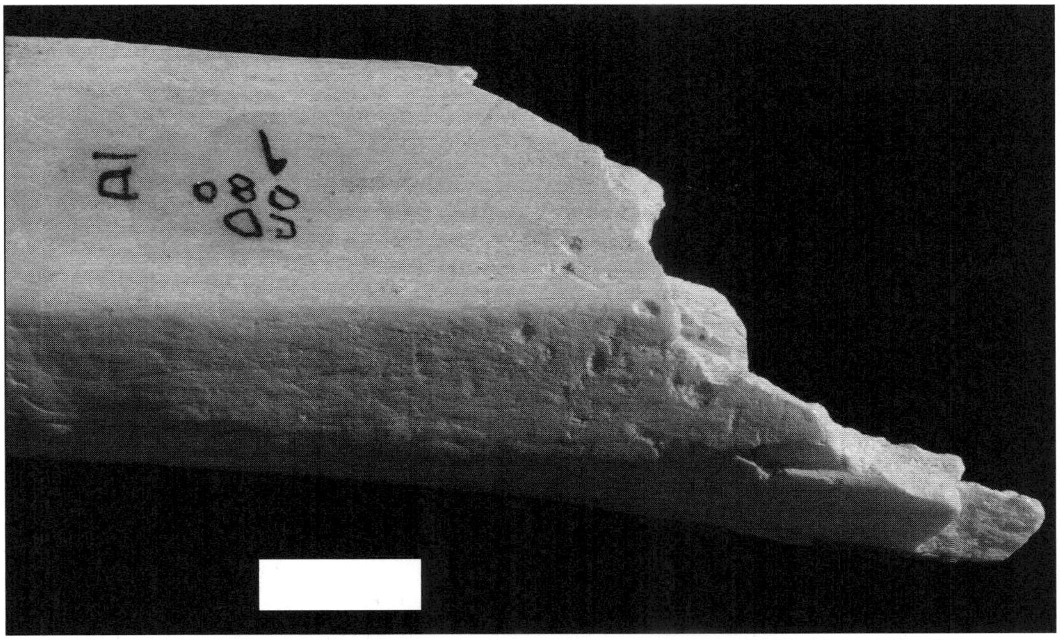


*Figure 22. Examples of toothmarks and limb shaft fracture. (White bar = 1 cm).*

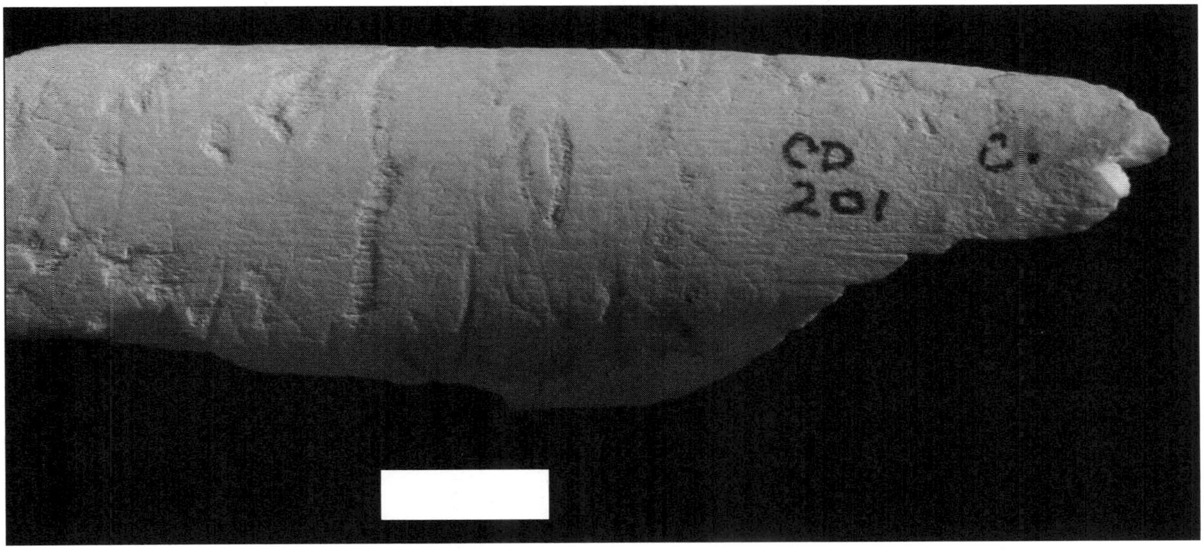


*Figure 23. Example of a shaft fragment with tooth scores. (White bar = 1 cm).*

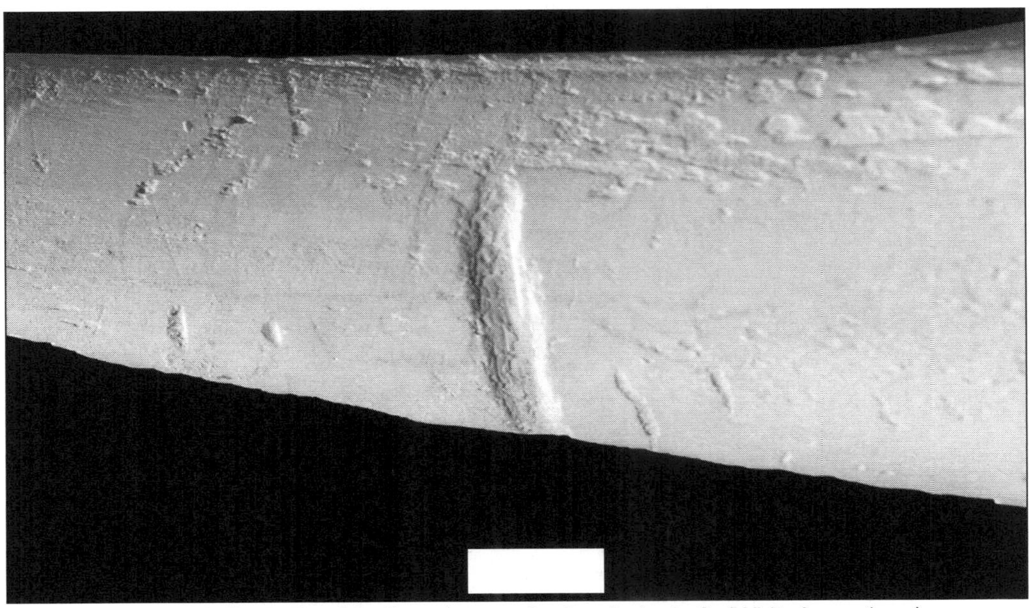

*Figure 24. Example of toothmarks on a broken limb shaft. (White bar = 1 cm).*

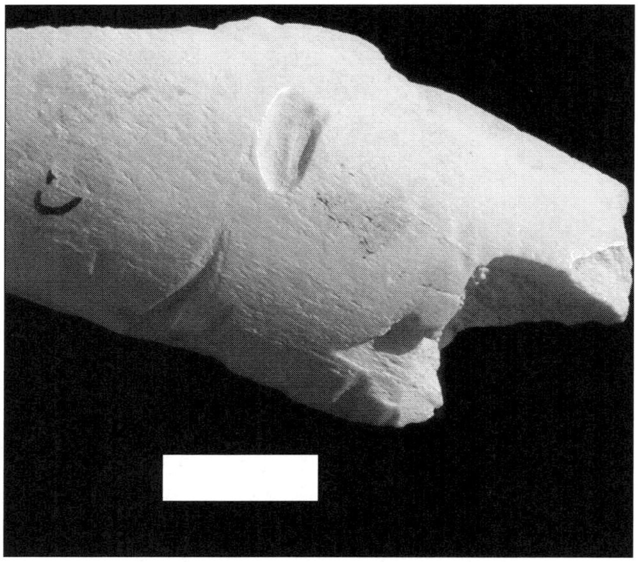

*Figure 25. Example of a large carnivore-induced tooth score. (White bar =1 cm).*

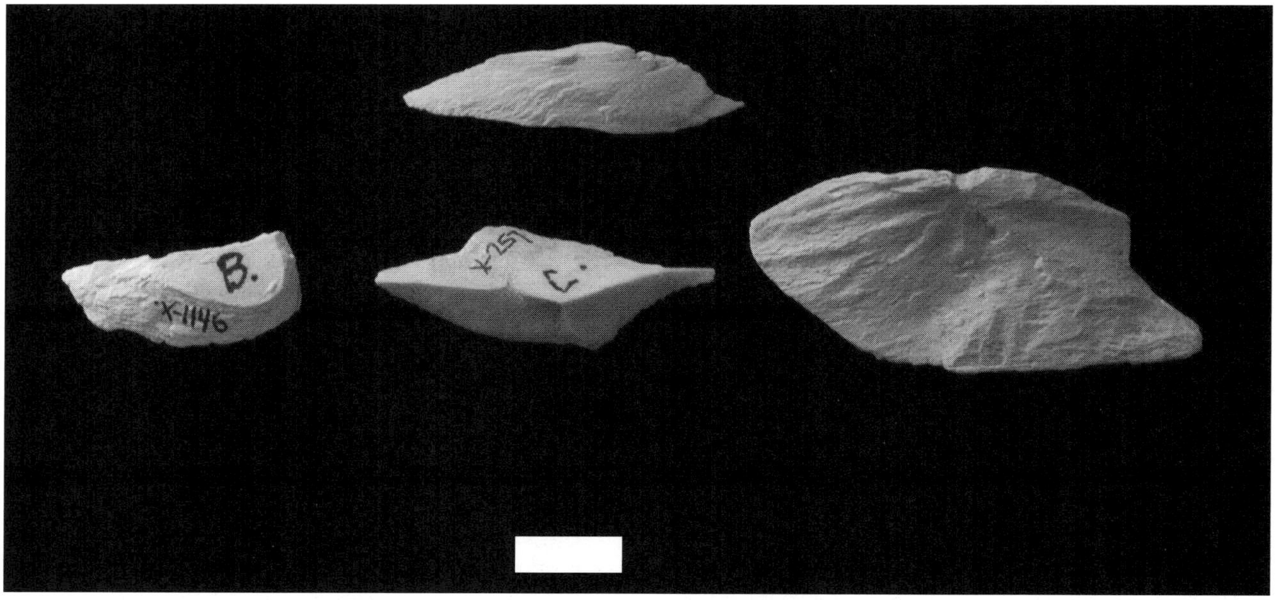


*Figure 26. Examples of bone flakes. (White bar = 1 cm).*

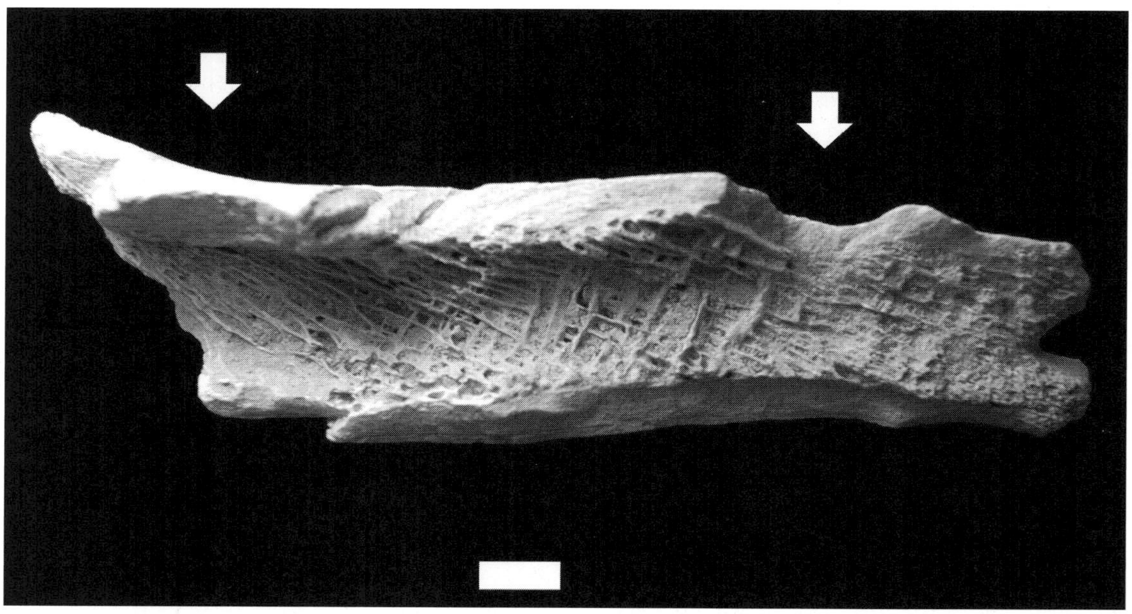

*Figure 27. Examples of negative flake scars or notches (arrows) on a shaft fragment. (White bar = 1 cm).*

*Table 7. Incidence of toothmarks on samples of bones from each of three weathering stages. Note that bones in weathering stage 0-1 had higher values in terms of mean number of toothmarks, percentage of bones with toothmarks, and extensiveness of toothmark modification (number per unit of bone area) than did the more heavily weathered bone. Less weathered bones (state 0–1) exhibit more than 15 times the number of toothmarks per bone, more than 13 times the proportion of bones with toothmarks, and more than 14 times the number of toothmarks per unit area of cortical bone than do heavily weathered bone (stage 3).*

| | Sample of Non-ID Shaft Fragments Weathering Stages | | | |
|---|---|---|---|---|
| | **Stage 0–1** | **Stage 2** | **Stage 3** | **All Stages Total** |
| **Sample size (# shaft fragments)** | 132 | 120 | 136 | 388 |
| **Number of toothmarks** | 329 | 162 | 19 | 510 |
| **Number of shaft fragments with toothmarks** | 65 | 41 | 5 | 111 |
| **Mean number of toothmarks/shaft fragments** | 2.49 | 1.35 | 0.14 | 1.31 |
| **% of shaft fragments with toothmarks** | 49.24 | 34.17 | 3.68 | 28.61 |
| **Approx. surface area ($cm^2$) of cortical bone** | .1583 | .1222 | .1344 | .4149 |
| **No. toothmarks per 10 cm x 10 cm area (100 sq. cm) of cortical bone** | 20.78 | 13.26 | 1.41 | 12.29 |

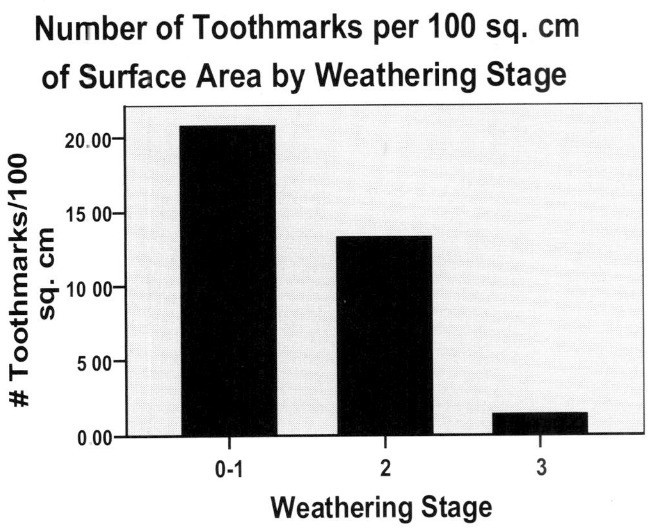


*Figure 28. Number of tooth-marks per 100 sq. cm. of bone surface, by weathering stage (Stage 0-1, Stage 2, and Stage 3) from a random sample of bone fragments. Note that the number of toothmarks evident decreases markedly with increased bone weathering.*

### *Other modification*

There were very few examples of human-induced cut-marks (n=3), burning (n=1), or rodent gnawing (n=2).This suggests that the hyenas were not scavenging human food refuse to any great degree, and that the overwhelming majority of the bones represent hyena collection and transport of animals and animal parts from locations of natural deaths, road kills, and hyena predation. Examples of these types of modification are shown in Figures 30, 31 and 32.

### *An atlas of element representation and modification*

Figures 33–50 show the elements, element portions, and bone and tooth fragments recovered from the hyena den. As noted above, the great majority (96.1%) of the 4,847 bone and tooth specimens were buried within 20 cm of the surface and were recovered through excavation and sieving of the soft sediment. These photographs show the degree of completeness of different skeletal parts and the fragmentation of cranial, dental, and limb shaft remains.

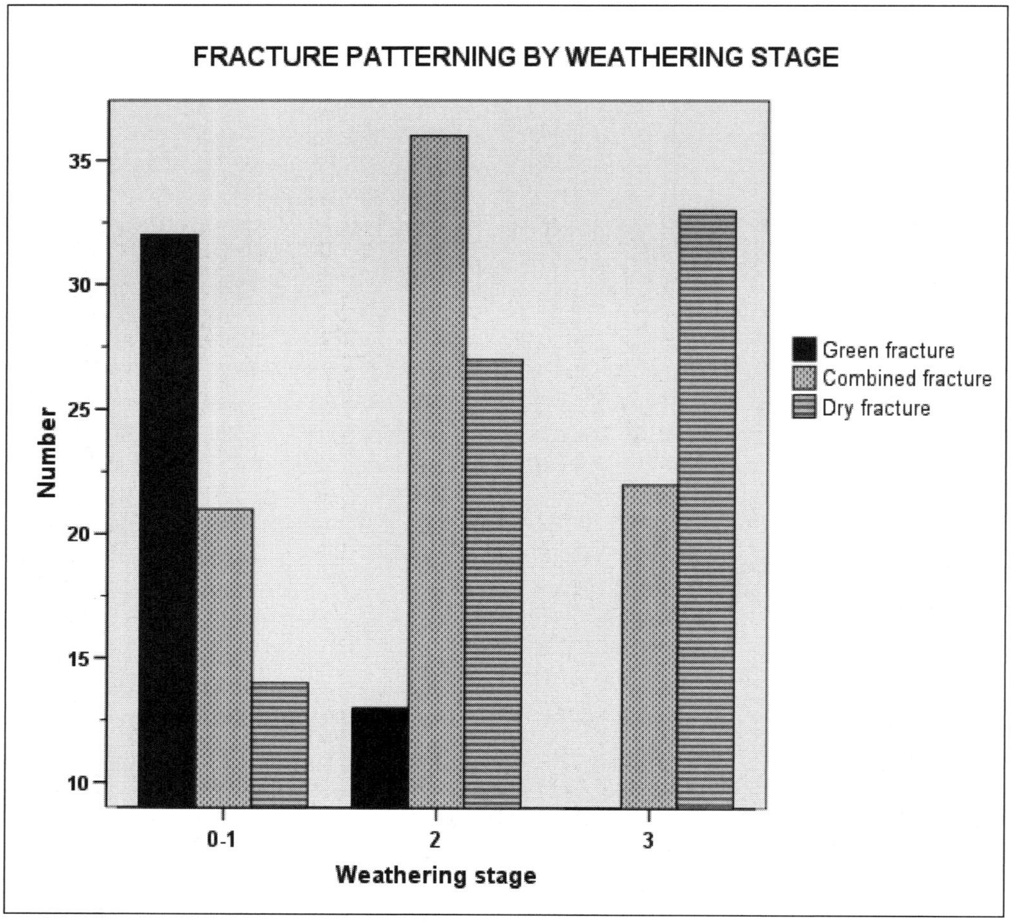


*Figure 29. Fracture patterning (green fracture, dry fracture, or a combination) observed on samples of shaft fragments at three different weathering stages. Note that fresher bone (Stage 0-1) is dominated by green fracture as well as green and dry combined, while bone in weathering stage 3 is dominated by dry fracture.*

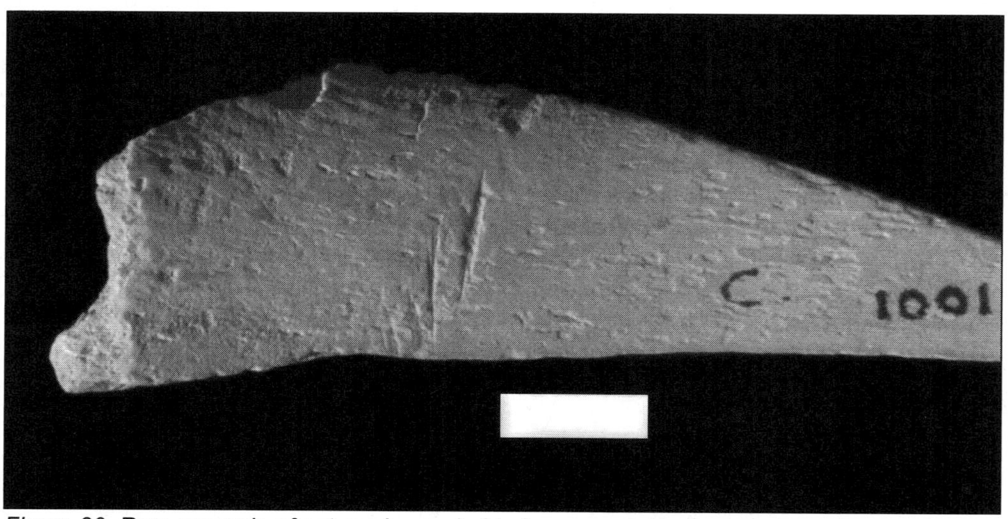


*Figure 30. Rare example of cut-marks, probably from a metal knife, on a bone fragment. (White bar = 1 cm).*

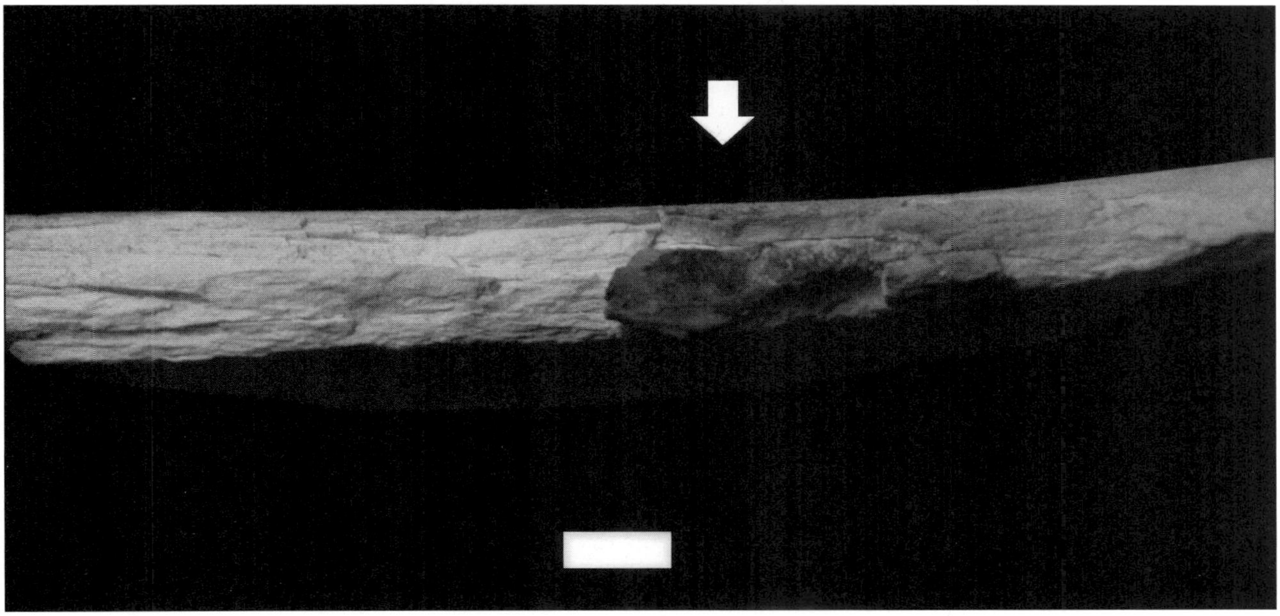

Figure 31. Rare example of dark discoloration, probably from burning, on the surface of a bone fragment. (White bar = 1 cm).

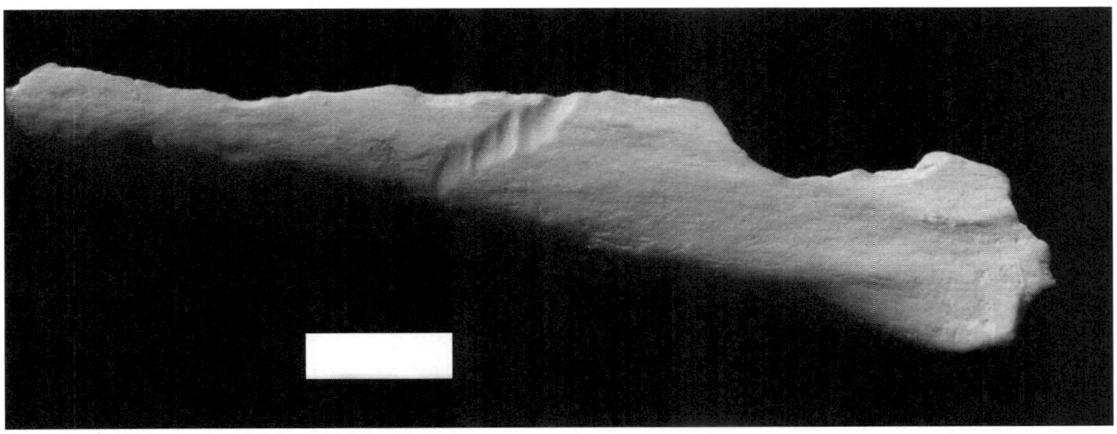

Figure 32. Rare example of probable rodent toothmarks on a shaft fragment. (White bar = 1 cm).

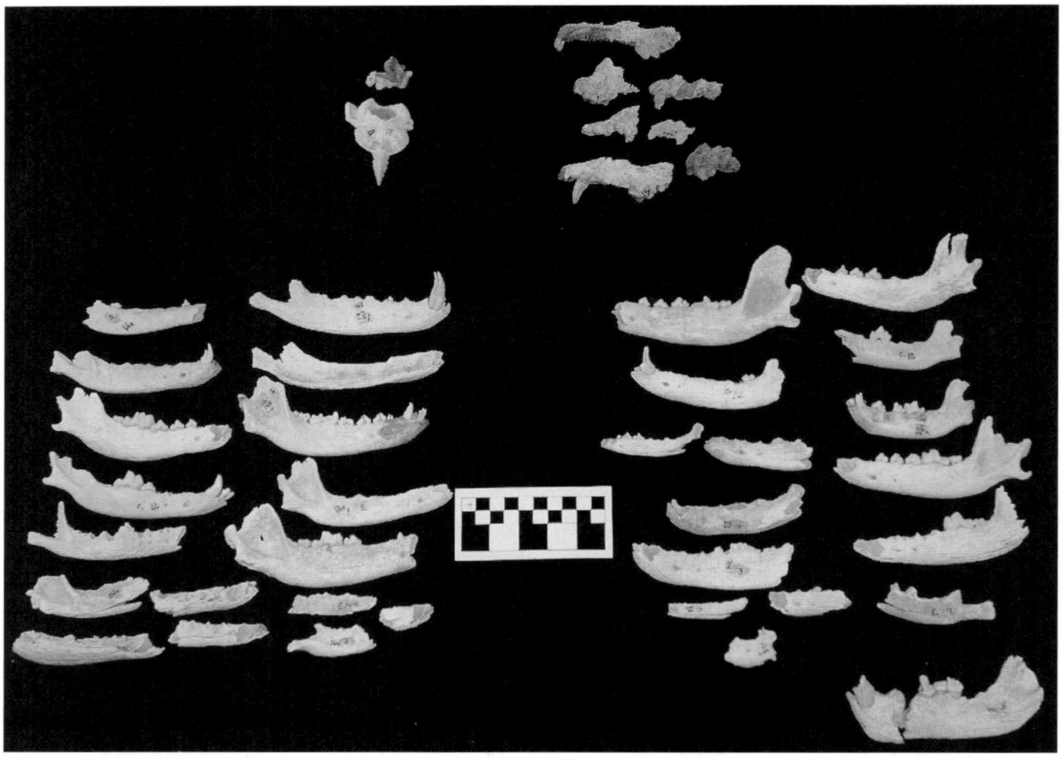

*Figure 33. Examples of carnivore crania and mandibles, including hyena mandible (lower right), from the excavated area. (Smaller squares on scale = 1 cm.)*

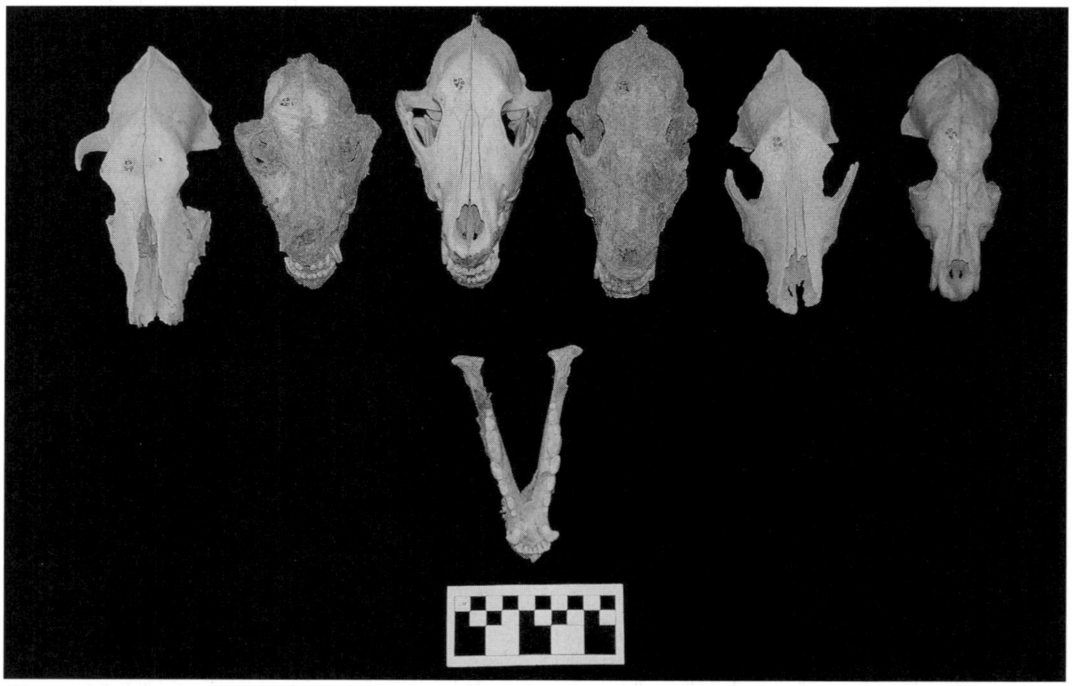

*Figure 34. More complete dog crania and a whole mandible collected from the surface outside of the excavation area. Note the two mummified skulls (second and fourth from left).*

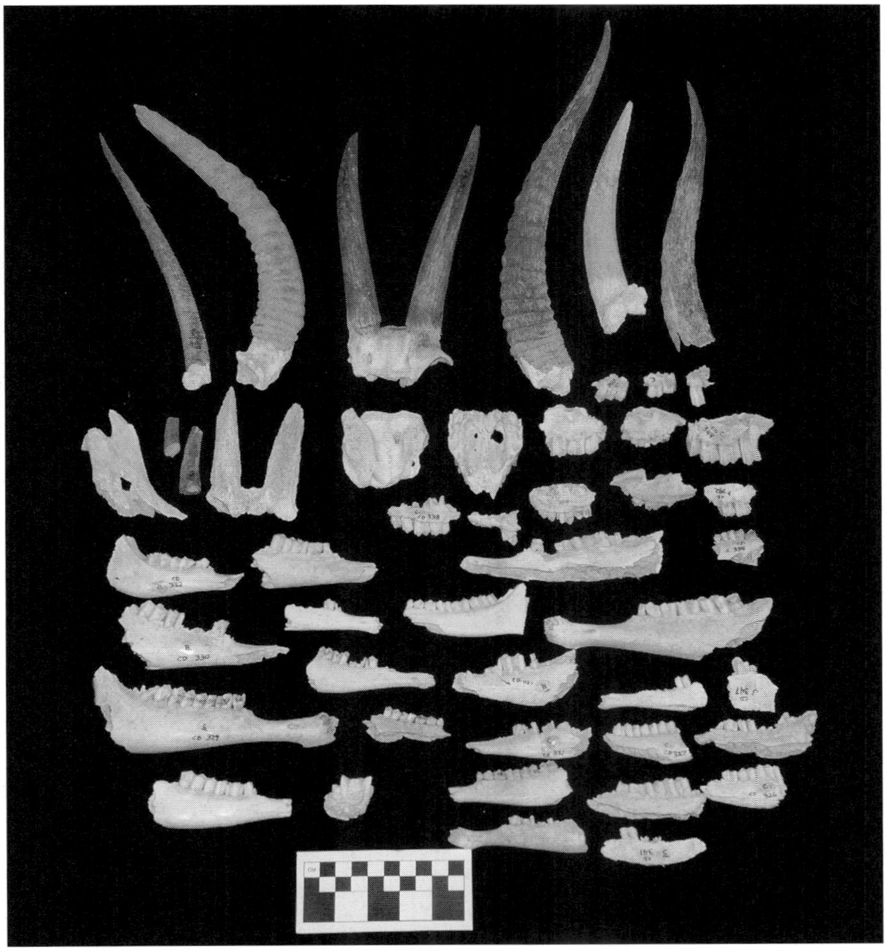

*Figure 35. Examples of cranial fragments, horn cores, and mandibular pieces from smaller bovids (goat/sheep, gazelle).*

*Figure 36. Examples of cranial and mandibular pieces from larger mammals (camels, donkeys and horses).*

Figure 37. Examples of smaller cranial fragments (all taxa).

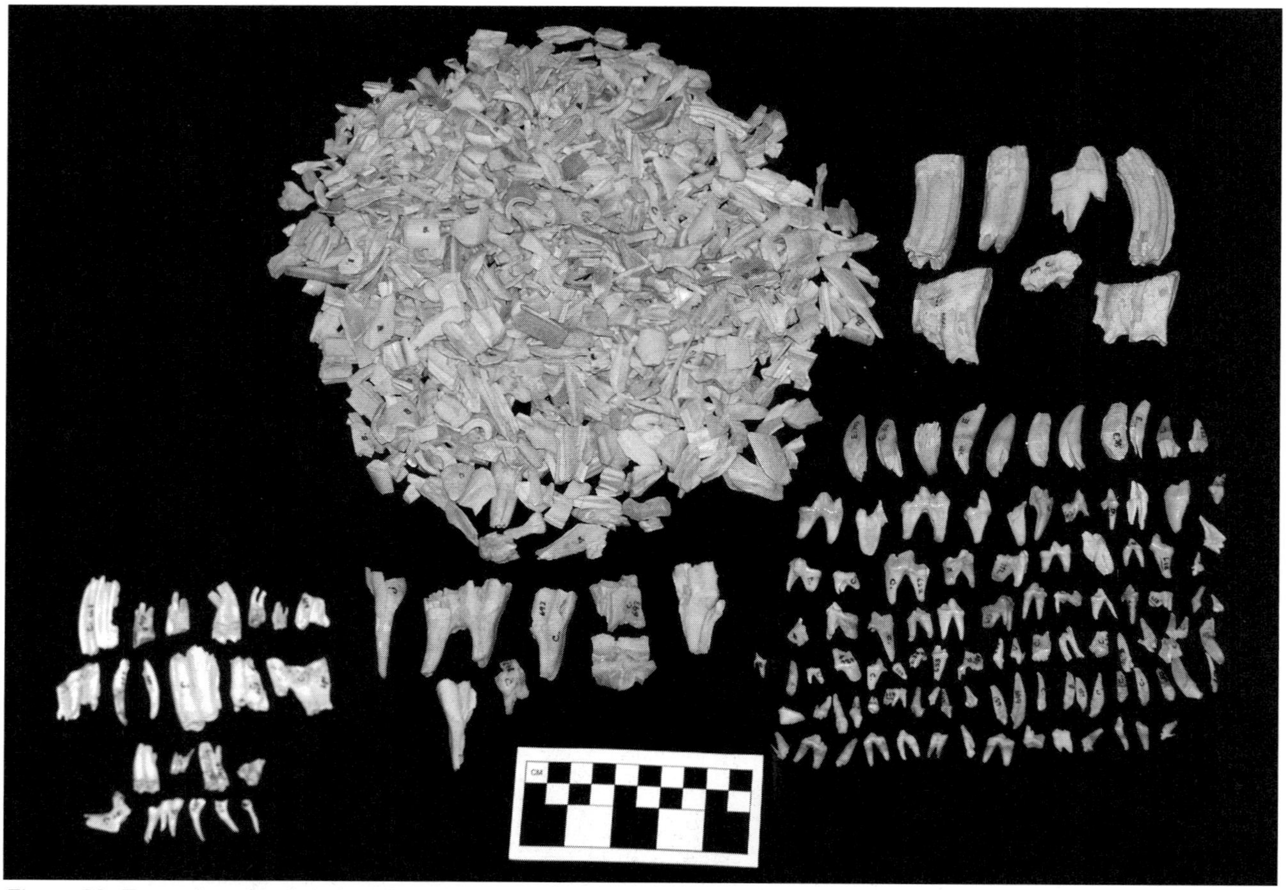

Figure 38. Examples of isolated teeth and tooth fragments (all taxa).

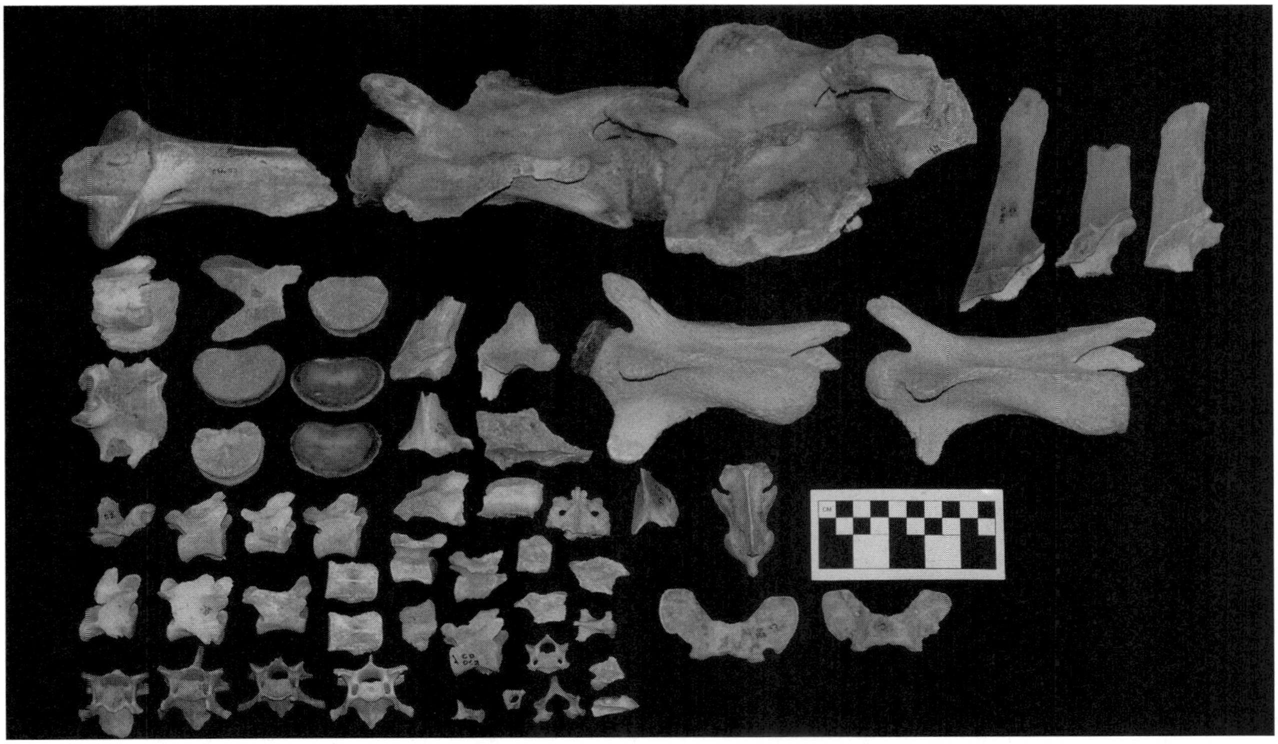

Figure 39. Examples of vertebrae (all taxa). Note the articulated vertebrae top center.

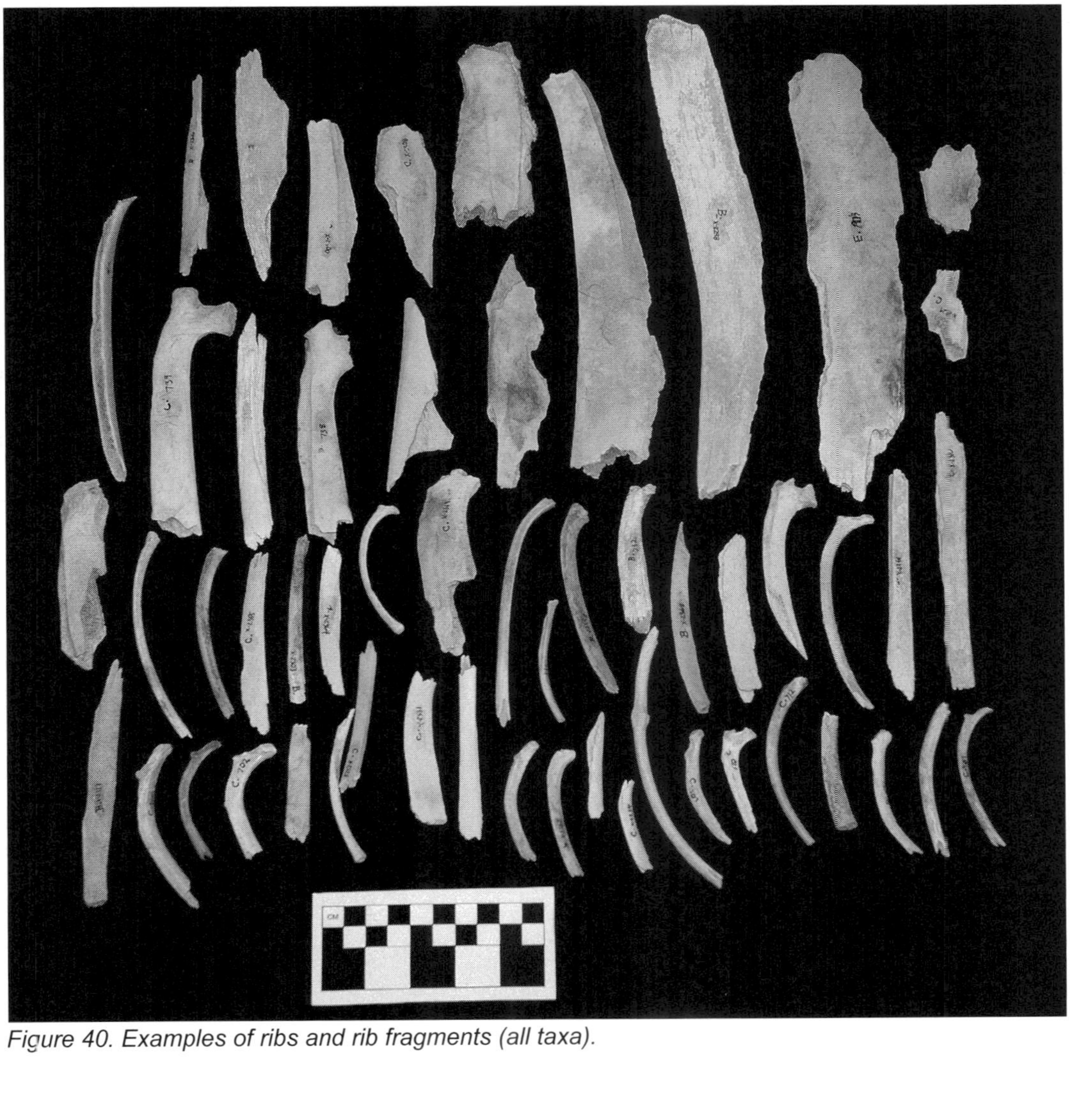

Figure 40. Examples of ribs and rib fragments (all taxa).

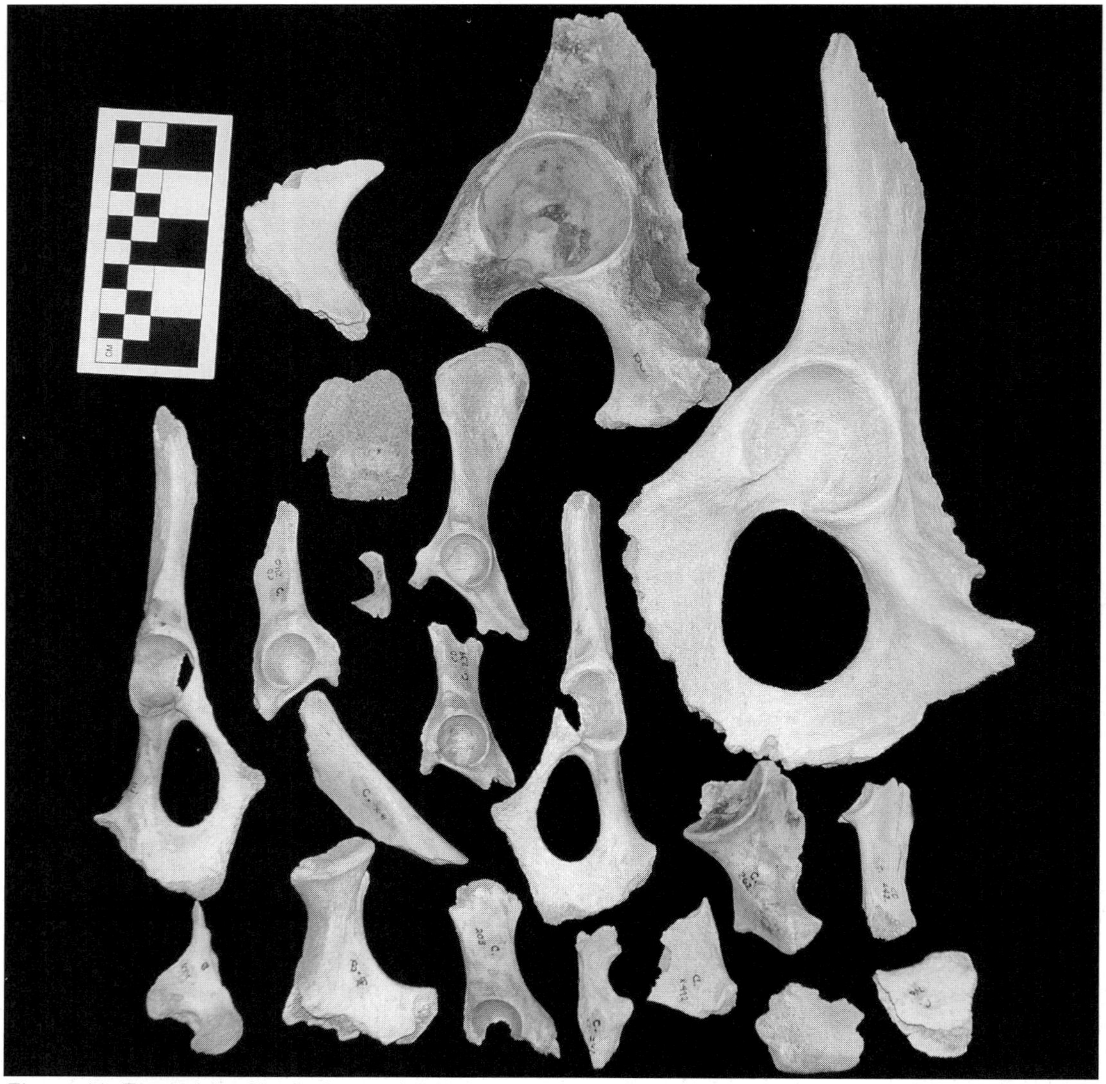

*Figure 41. Examples of innominates (all taxa).*

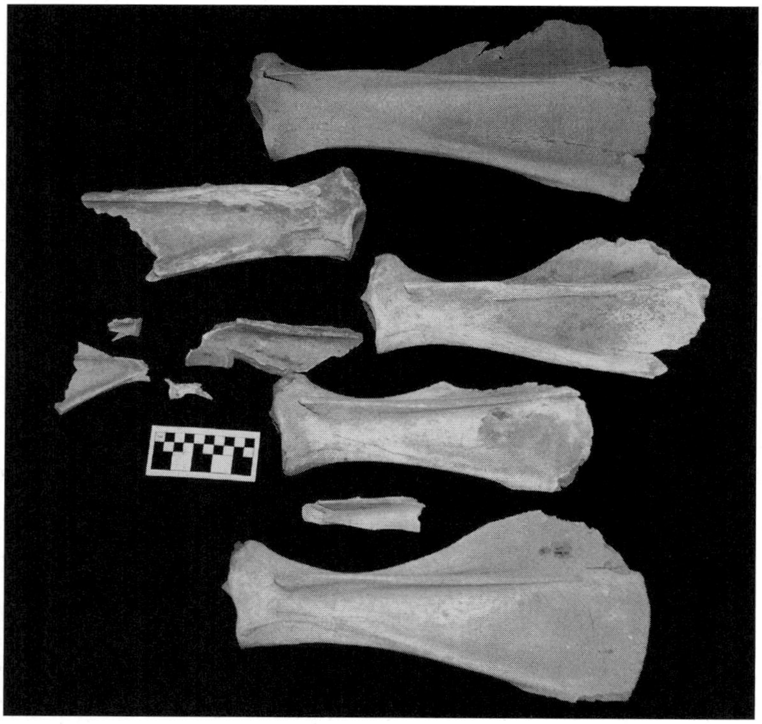

*Figure 42. Examples of scapulae (all taxa).*

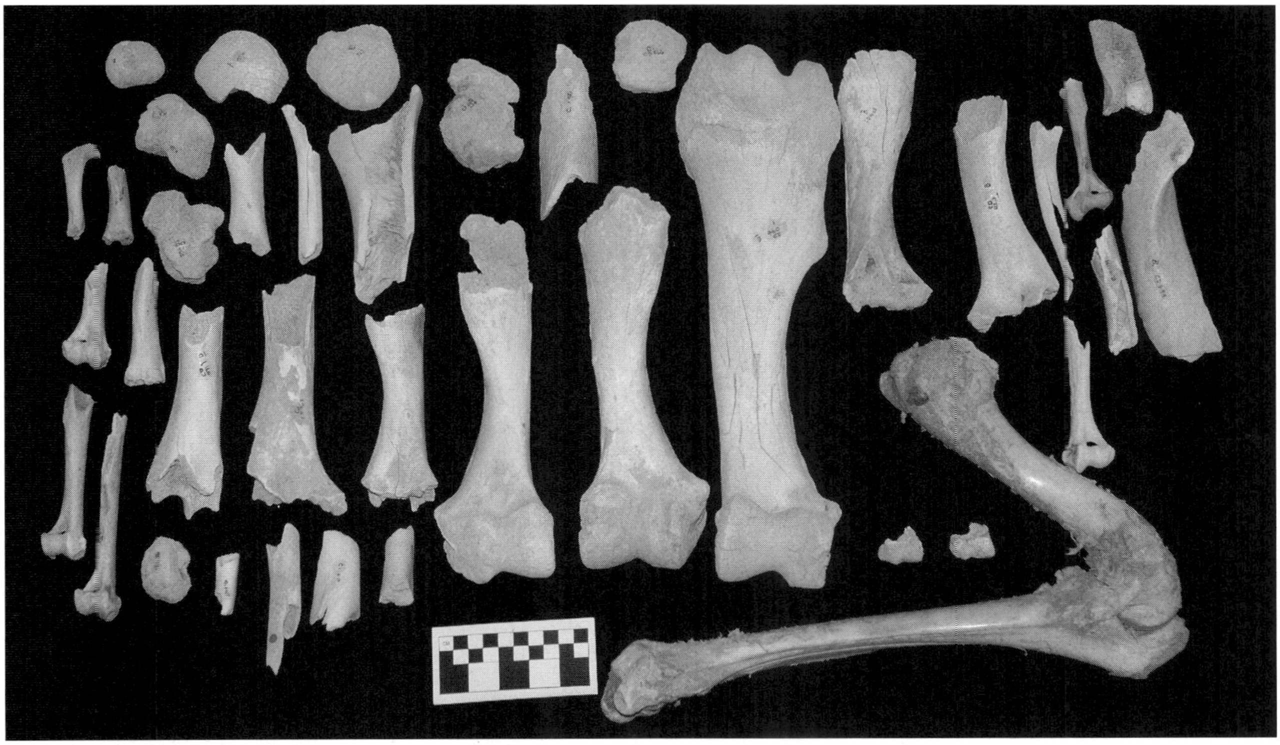

*Figure 43. Examples of humeri (all taxa); at lower right, still articulated with the radius-ulna.*

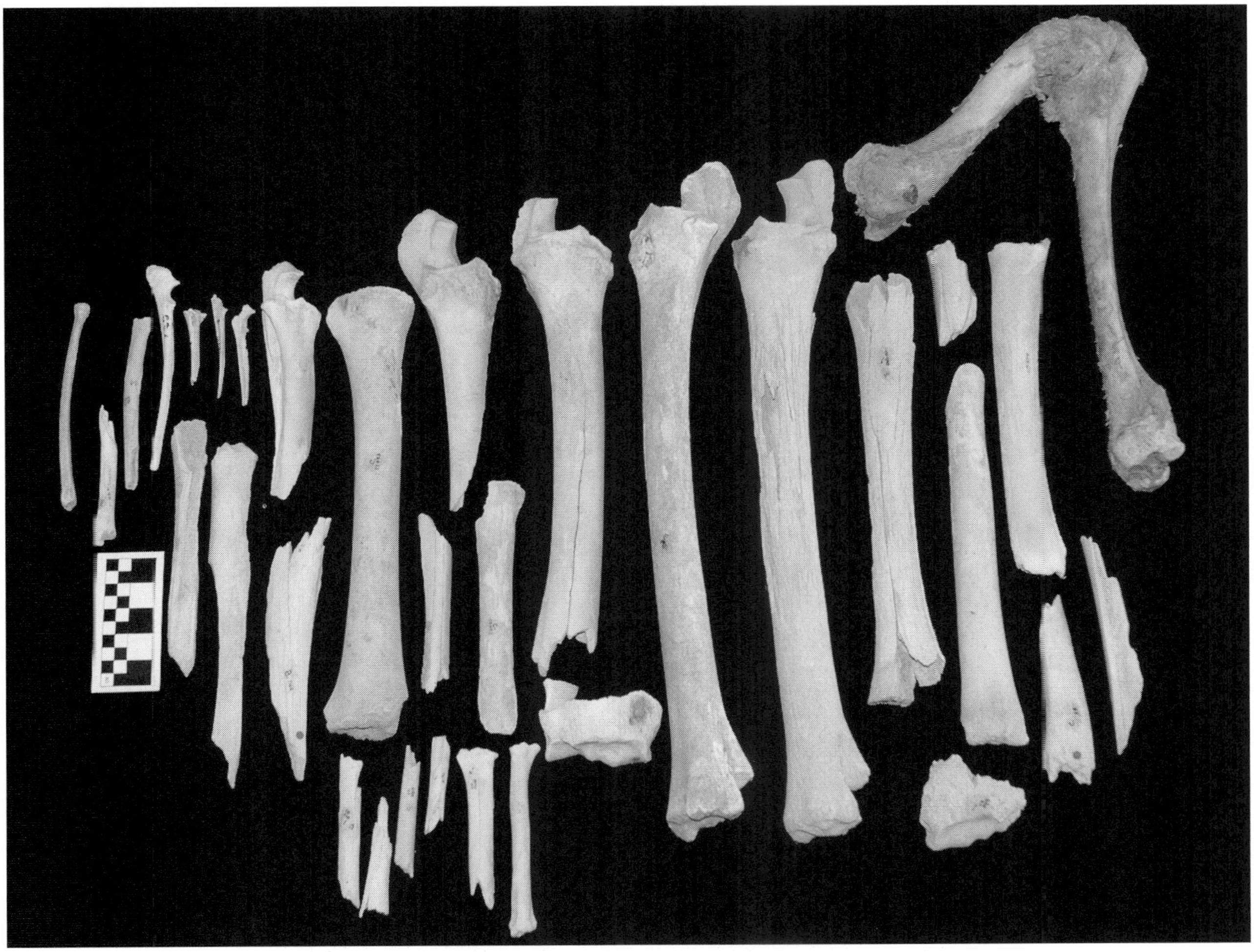

*Figure 44. Examples of radii and ulnae (all taxa); at upper right, still articulated with the humerus.*

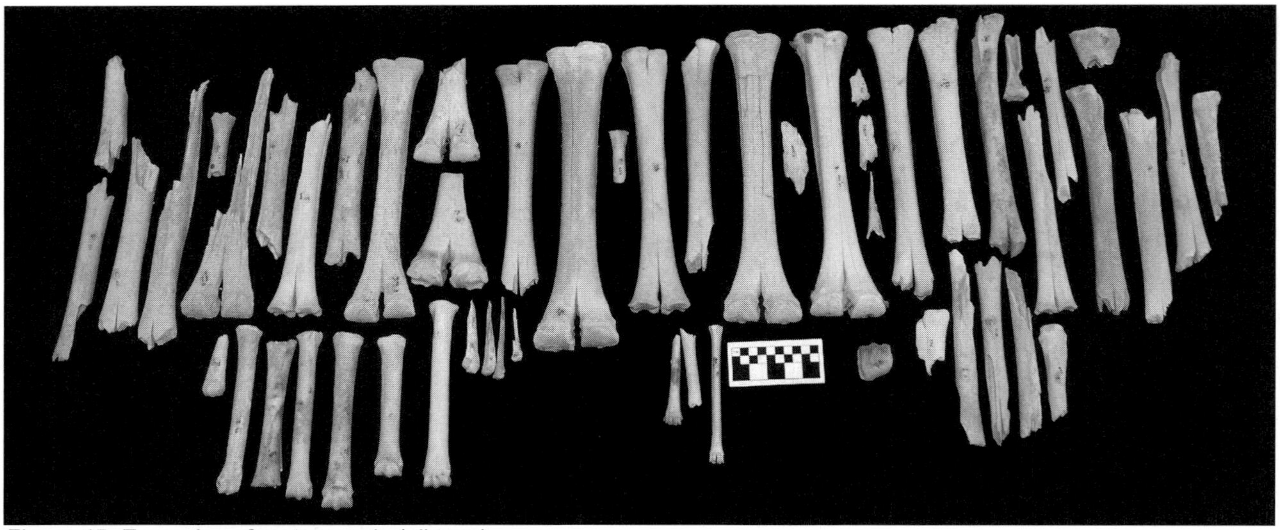

Figure 45. Examples of metacarpals (all taxa).

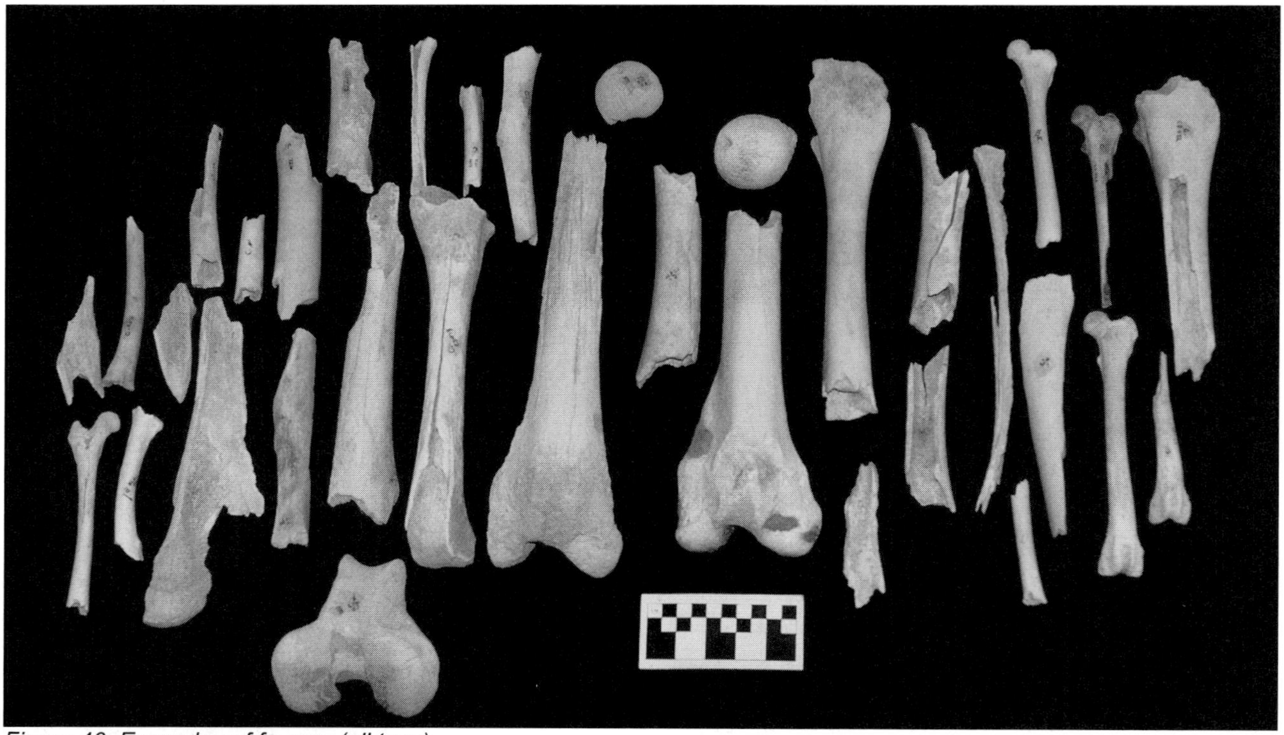

Figure 46. Examples of femora (all taxa).

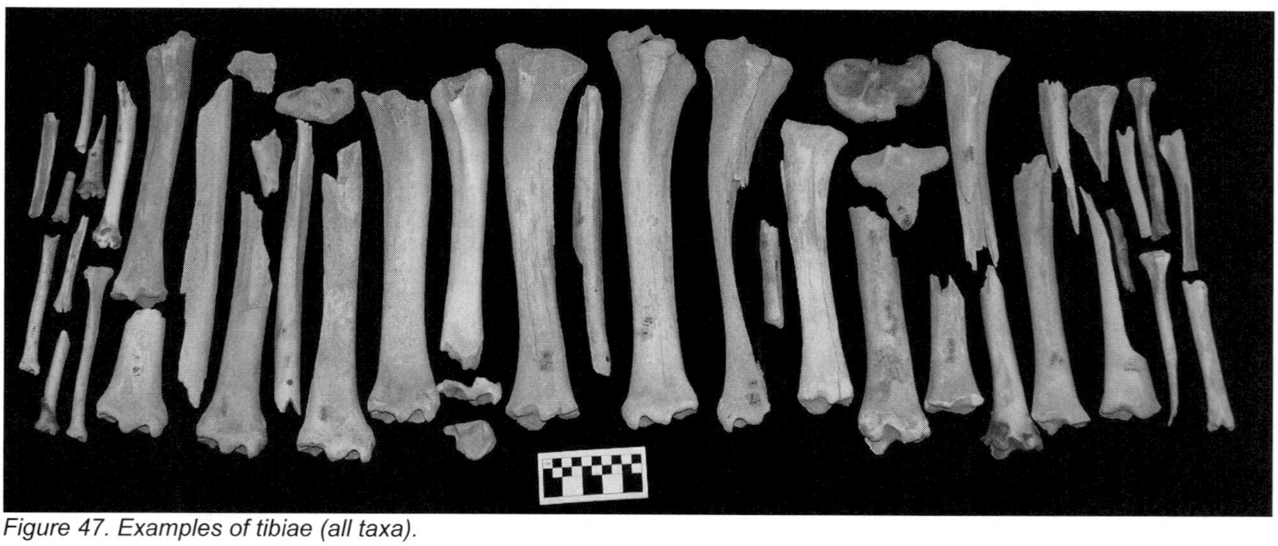

Figure 47. Examples of tibiae (all taxa).

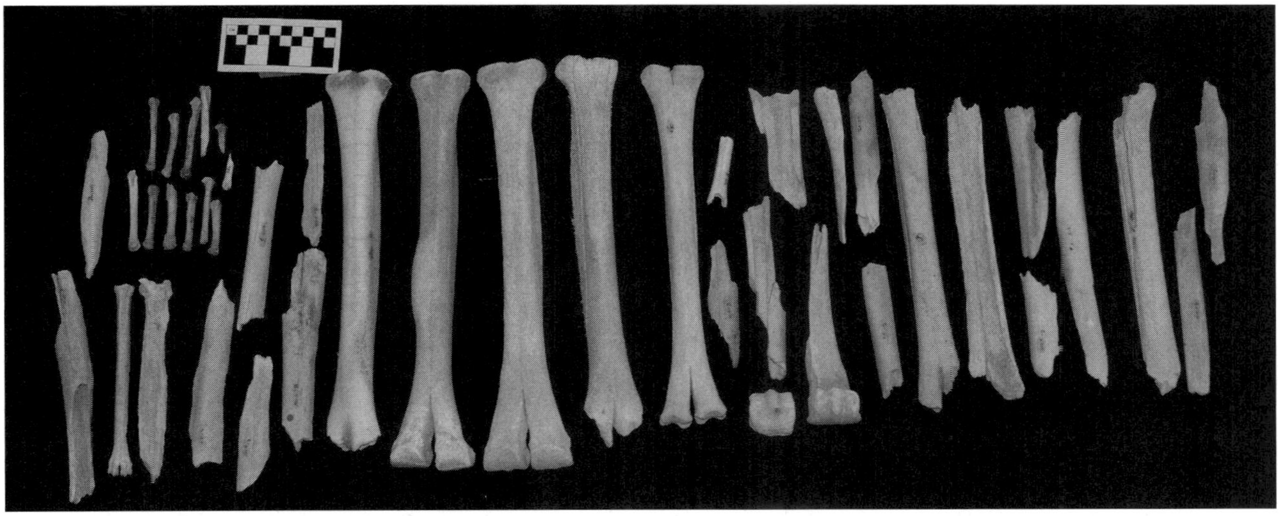

Figure 48. Examples of metatarsals (all taxa).

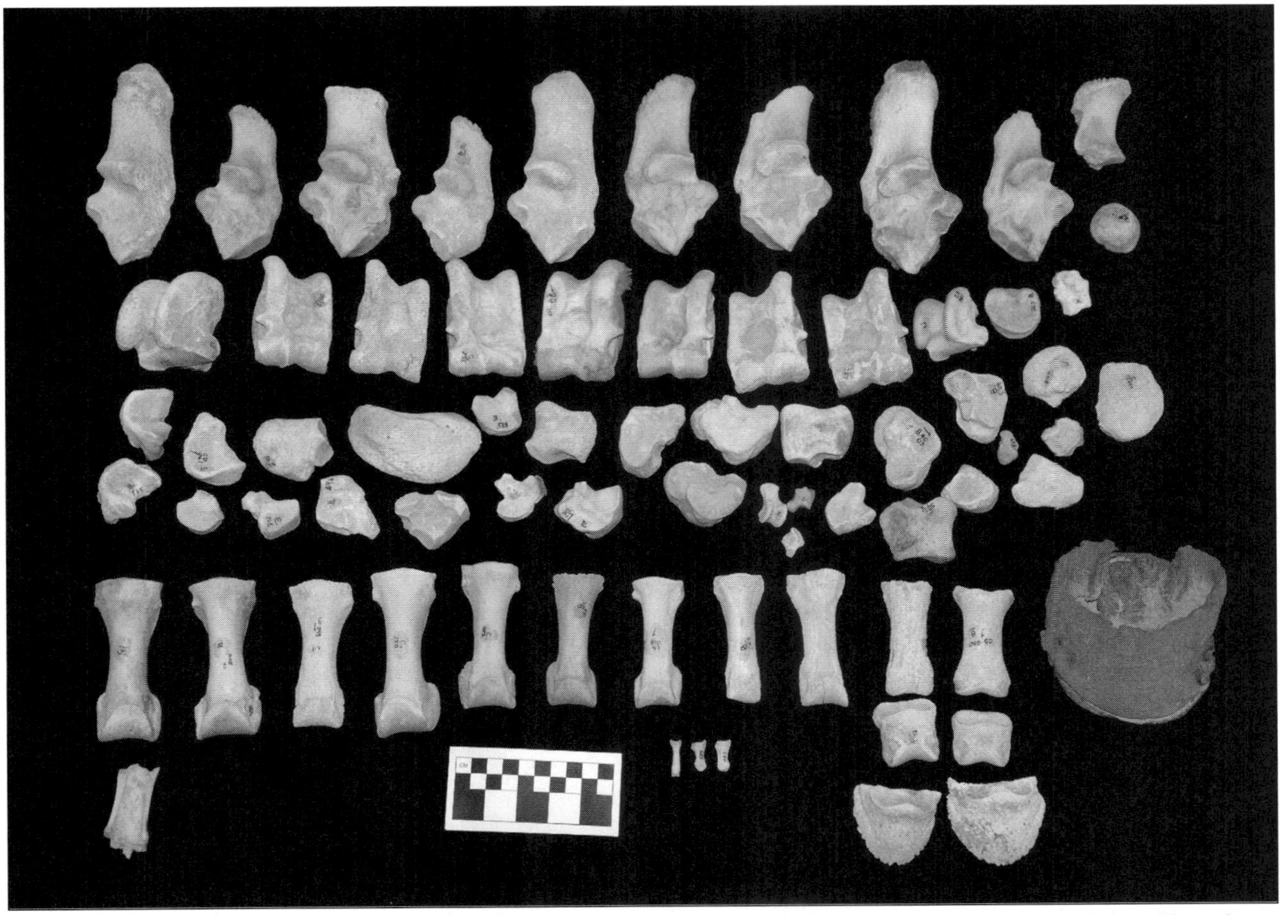

Figure 49. Examples of podials and phalanges (all taxa). Note hoof with attached horseshoe at lower right, without bone present.

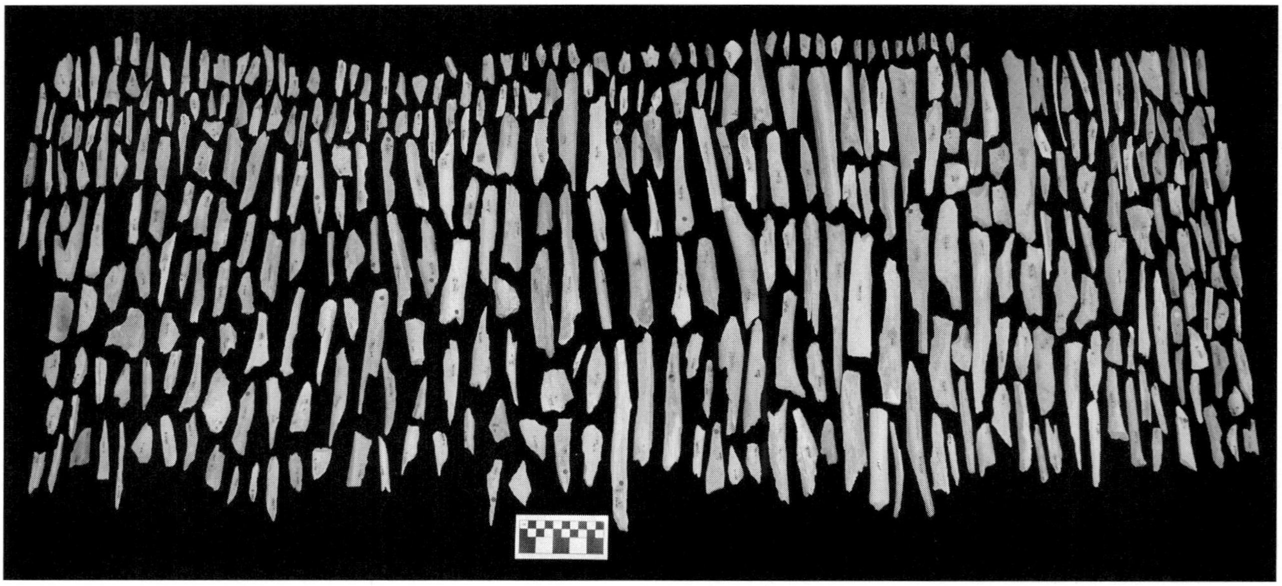

*Figure 50. Examples of limb shaft fragments (all taxa). The smallest fraction is only represented by a small sample in the top row.*

## Conclusion

This study is an analysis of the faunal assemblage of one of the few excavated hyena dens with an appreciable depth of deposit. Almost 5,000 bone specimens were recovered from an area of 16m$^2$. The major results of this study were:

- The primary taxonomic groups represented at the den are camel, dog, sheep/goat, and gazelle; other taxa include donkey, human, horse, fox, stork, hare, hedgehog, indeterminate birds, oryx, hyena, honey badger, and snake. The assemblage thus consists of a mix of domestic and wild forms, and both larger and smaller animals. It is likely that the larger animals were primarily acquired through scavenging, while the smaller animals could have been acquired through either hunting or scavenging.
- The proportion of carnivores in this assemblage is quite high (25.9% of the MNI and 31.4% of the NISP).
- A minimum number of 54 individuals representing at least 16 taxa are represented from the excavated area.
- The majority (96.1%) of the bone assemblage was buried, with only 3.9% exposed on the surface of the 16m$^2$ excavation area. The great majority of the buried faunal materials consist of fragmented bones and teeth, and the bones tend to be less weathered. The surface materials consist largely of larger, sometimes complete, bones, and tend to be much more heavily weathered than the buried portion of the assemblage. The buried sample would have had much greater likelihood of mineralization and fossilization over time.
- The average number of bone and tooth specimens (including both surface and sieved, buried materials) per m$^2$ is 303; the average number of specimens identifiable to both element and taxon was 32 per m$^2$.
- Smaller animals have a much higher cranial to postcranial ratio than large animals; this suggests that smaller animals may have been transported to the den as more complete carcasses than were the larger animals.
- Smaller mammals have a much larger axial to appendicular ratio than larger mammals; again, this appears to be due smaller animals being transported as more complete carcasses, while larger animals may have often had portions of the carcass, particularly limbs, transported to the den.
- The faunal assemblage shows a high degree of limb shaft fragmentation, comparable to that found at many Plio-Pleistocene archaeological sites. Bones of weathering stages 0–1 (primarily from the buried portion of the sample) predominantly exhibit green, often spiral, fracture.
- More weathered bones exhibit much higher frequencies of dry fracture and much lower frequencies of carnivore toothmarks than do relatively unweathered bones, suggesting that such surface modification had been obliterated by the weathering process.
- The vast majority of the modification observed on the bones of this assemblage consist of carnivore toothmarks and notches, strongly suggesting that hyenas were the principal agent of accumulation, consumption, and modification of these bones. Very few specimens (roughly one out of a thousand) show traces of human modification (cutmarks or burning) or rodent gnawing. Bone flakes are present but very rare in the assemblage.
- These patterns conform to the criteria emphasized by Pickering (2002) for differentiating faunal as-

semblages accumulated by hyenas as opposed to hominids.

This type of actualistic study, as emphasized by Brain (1981), provides the kinds of comparative evidence and patterning that can be used to evaluate prehistoric bone assemblages and assess the principal agents of bone accumulation and modification. This study adds to a growing corpus of hyena den studies and adds information as to the range of variation in hyena bone collecting and processing.

*Figure 51. Highway road sign in the eastern desert of Jordan.*

## ACKNOWLEDGMENTS

We would like to express our sincere gratitude to the individuals and institutions that have made this research possible. These include Tim White, Donald Johanson, Mujahed Muheisen, Bob Jones, Richard Meadow, the Institute of Human Origins, the University of California at Berkeley, Yarmouk University, the American Center for Oriental Research (ACOR) in Jordan, the Stone Age Institute, and Friends of CRAFT, Inc. We thank Blaire Hensley-Marschand and Andrew Bradley for research assistance. Support for fieldwork for this research was provided by a grant from the L.S.B. Leakey Foundation.

## REFERENCES

Beherensmeyer, A.K. 1978. Taphonomic and ecologic information from bone weathering. Paleobiology 4, 150-162.

Binford, L.R. 1981. Bones: Ancient Men and Modern Myths. Academic Press, New York.

Brain, C.K. 1981. The Hunters of the Hunted? An Introduction to African Cave Taphonomy. University of Chicago, Chicago.

Bunn, H.T. (1982). Meat-eating and human evolution: studies on the diet and subsistence patterns of Plio-Pleistocene hominids in East Africa. Ph.D. Dissertation, University of California at Berkeley.

Bunn, H.T. 1983. Comparative analysis of modern bone assemblages from a San hunter-gatherer camp in the Kalahari Desert, Botswana, and from a spotted hyena den near Nairobi, Kenya. In: Clutton-Brock, J., Grigson, C. (Eds.), Animals and Archaeology, Vol. 1: Hunters and Their Prey. British Archaeological Reports International Series 163, pp. 143-148.

Cruz-Uribe, K. 1991. Distinguishing hyena from hominid bone accumulations. Journal of Field Archaeology 18, 467-486.

Henschel, J., Tilson, R., von Blottnitz, F. 1979. Implications of a spotted hyaena bone assemblage in the Namib Desert. South African Archaeological Bulletin 34, 127-131.

Hill, A. 1989. Bone modification by modern spotted hyenas. In: Bonnichsen, R., Sorg, M.H. (Eds.), Bone Modification. Center for the Study of the First Americans, Orono (ME), pp. 169-178.

Horwitz, L.K., Smith P. 1988. The effects of striped hyaena activity on human remains. Journal of Archaeological Science 15, 471-481.

Hughes, A.R. 1954. Hyaenas versus australopithecines as agents of bone accumulations. American Journal of Physical Anthropology 12, 467-486.

Kerbis-Peterhans, J.C., Horwitz, L.K. 1992. A bone assemblage from a striped hyaena (*Hyaena hyaena*) den in the Negev Desert, Israel. Israel Journal of Zoology 37, 225-245.

Klein, R.G. 1975. Paleoanthropological implications of the nonarchaeological bone assemblage from Swartklip 1, south-western Cape Province, South Africa. Quaternary Research 5, 275-288.

Kuhn, B. 2005. The faunal assemblages and taphonomic signatures of five striped hyaena (*Hyaena hyaena syriaca*) dens in the desert of eastern Jordan. Levant 37, 221-234.

Lacruz, R., Maude, G. 2005. Bone accumulations of brown hyaena (*Parahyaena brunnea*) den sites in the Makgagikgadi Pans, northern Botwswana: taphonomic, behavioural and palaeoecological implications for interpreting palaeontological and archaeological assemblages. Journal of Taphonomy 3, 43-54.

Leakey, L.N., Milledge, S.A.H., Leakey, S.M., Edung, J., Haynes, P, Kiptoo, D.K., McGeorge, A. 1999. Diet of striped hyaena in northern Kenya. African Journal of Ecology 37, 314-326.

Maguire, J.M., Pemberton, D., Collett, M.H. 1980. The Makapansgat limeworks grey breccia: hominids, hyaenas, hystricids or hillwash? Palaeontologica Africana 23, 75-98.

Mills, M.G.L., Mills, M.E.J. 1977. An analysis of bones collected at hyaena breeding dens in the Gemsbok National Parks. Annals of the Transvaal Museum 30, 145-155.

Owens M., Owens, D. 1979. Communal denning and clan associations in brown hyaenas (*Hyaena brunnea*) of the Central Kalahari Desert. African Journal of Ecology 17, 35-44.

Pickering, T.R. 2002. Reconsideration of criteria for differentiating faunal assemblages accumulated by hyenas and hominids. International Journal of Osteoarchaeology 12, 127-141.

Scott, L., Klein, R.G. 1981. A hyena-accumulated bone assemblage from late Holocene deposits at Deelpan, Orange Free State. Annals of the South African Museum 86, 217-227.

Skinner, J.D., Davis, S., Ilani, G. 1980. Bone collecting by striped hyaenas, *Hyaena hyaena*, in Israel. Palaeontologia Africana 23, 99-104.

Skinner, J.D., Haupt, M.A., Hoffman, M., Dott, H.M. 1998. Bone collecting by brown hyaena *Hyaena brunnea* in the Namib Desert: rate of accumulation. Journal of Archaeological Science 25, 69-71.

Skinner, J.D., Henschel, J.R., van Jaarsveld, A.S. 1986. Bone-collecting habits of spotted hyaenas *Crocuta crocuta* in the Kruger National Park. South African Journal of Zoology 21, 303-308.

Stiner, M.C. 1991. Food procurement and transport by human and non-human predators. Journal of Archaeological Science 18, 455-482.

Sutcliffe, A.J. 1970. Spotted hyaena: crusher, gnawer, digester, and collector of bones. Nature 227, 1110-1113.

Toerien, M.J. 1952. The fossil hyaenas of the Makapansgat valley. South African Journal of Science 48, 293-300.

Wood, W. 1807. Zoography, or, The Beauties of Nature Displayed. In Select Descriptions from the Animal, and Vegetable, with Additions from the Mineral Kingdom. Cadell and Davies, London.

# CHAPTER 6

# Taphonomy and Site Formation of Two Early Miocene Sites on Rusinga Island, Kenya

Alan Walker

## Abstract

Many Early Miocene sites on Rusinga Island, Kenya, have produced fossils of the stem ape *Proconsul*. Two of them have unusual taphonomic histories. One, R114, that contained the type specimen of *Proconsul heseloni*, is the infilling with matrix and bones of a large hollow tree trunk. The other, the Kaswanga Primate Site, is either a small channel fill or the remains of an infilled carnivore burrow that was dug into soft sediment.

## Introduction

Thomas Whitworth (1953) found the R114 site while mapping the geology of the Gumba peninsula on the Southwest of Rusinga Island in Lake Victoria, Kenya. Louis Leakey and his associates excavated there in 1950 and removed large blocks of matrix. These contained parts of a *Proconsul* skull and associated bones of a forelimb and foot that were the subject of analysis by Napier and Davis (1959). Subsequently, parts of the same individual were found unidentified in the Kenya National Museum (Walker and Pickford, 1983). That discovery led to Johns Hopkins University/National Museums of Kenya expeditions from 1984-1988 to understand the taphonomic and sedimentary context and to search for more *Proconsul* bones. More were indeed found and the partial skeleton became the type of a new species, *P. heseloni* (Walker et al., 1993).

A new site was also discovered while prospecting in the first year of these expeditions. It contained several partial *Proconsul* individuals and was named the Kaswanga Primate Site. Only brief accounts of the sites have been published so far (Walker et al., 1986; Walker and Teaford, 1988), but some of the history behind them has been given (Walker, 1992; Walker and Shipman, 2005). This account expands on those. Figure 1 shows the locations of the sites on Rusinga Island.

## Site R114

Whitworth was asked by Louis Leakey to make a geological map of the Gumba Penninsula, the Western part of Rusinga Island that lies in the Winam Gulf of Lake Victoria. This is centered about 34° 6' East and 0° 25' South. In August 1950, he found site R114 which he described in his notes as "a tiny and isolated outcrop," and in his publication as "a small circular pipe," penetrating the flaggy Series (Whitworth, 1953). In the article he went on to say, "The profusion of articulated skeletons found in this limited deposit suggest [sic] that it may represent the infilling of a pothole in which animals were trapped." (Whitworth, 1953: 91). Napier and Davis (1959) in their monographic account of the *Proconsul* bones from the site, elaborated on this by suggesting that "the pot-hole may have acted as a trap for unwary animals that came there to drink." When Pickford and Walker found more parts of the same skeleton in the National Museum of Kenya that had been unidentified or misidentified, a search was made to find the site again (Walker and Pickford, 1983).

A preliminary exploration of the depositional environment by Pickford was published in that paper, but his results have been superseded by subsequent expeditions, so hardly anything in that account is correct. Although he recognized that the fossil deposit was not well exposed, he thought that the grit of which it was composed

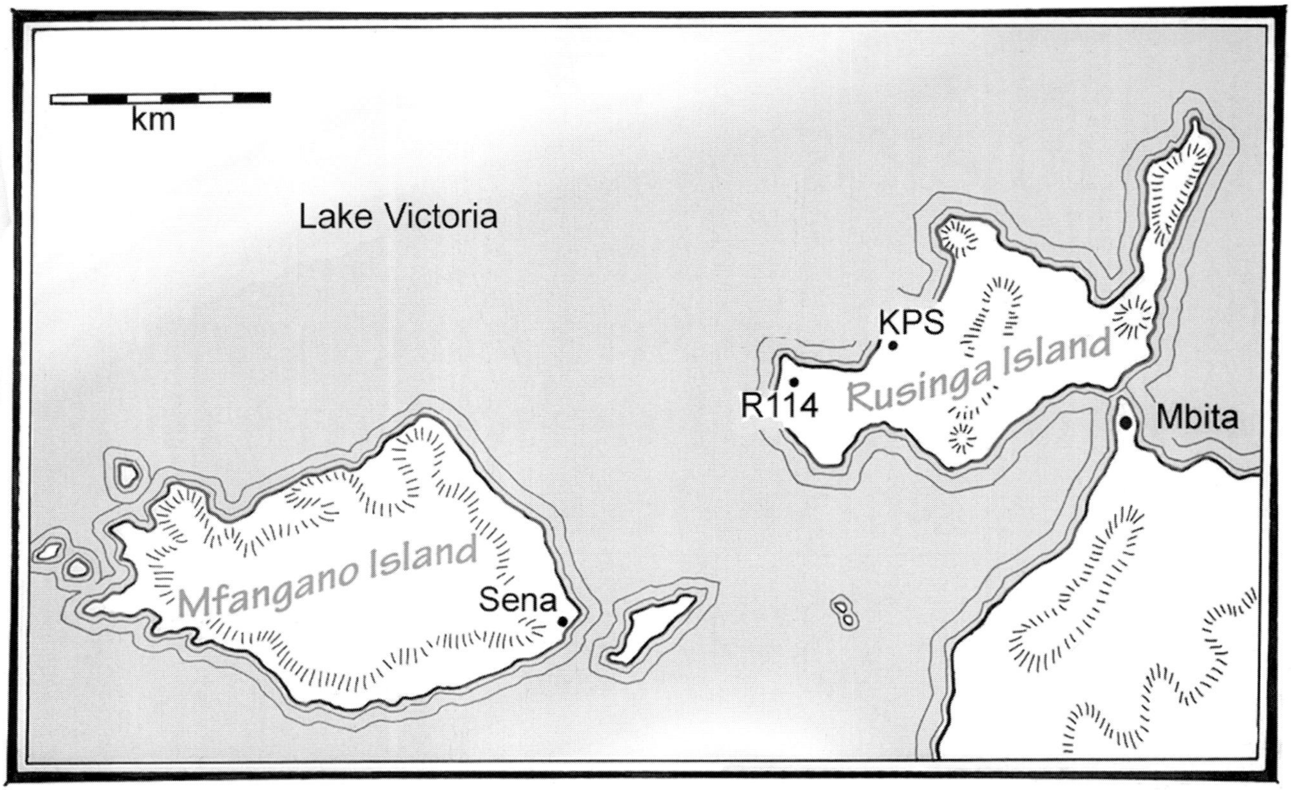


*Figure 1. Map of Rusinga and Mfangano Islands in the eastern part of Lake Victoria, Kenya, to show the locations of sites R114 and the Kaswanga Primate Site (KPS).*

"overlays a clay drape which thickens laterally where it contains a tragulid skeleton." The deposit proved to cut through the country rocks, rather than resting on them and the tragulid, which had no association with the deposit, turned out to be a springhare, *Megapedetes pentadactylus*. He further wrote, "the deposit was about 10 × 5 m in extent and about 0.3–0.5 m thick, with a gently sloping basal profile." The account of the taphonomy given by Walker and Pickford is likewise incomplete, but the observations are now seen with hindsight, to be mostly correct. Among the peculiarities noted then were: most fossils occurred as articulated or semi-articulated specimens with little or no mixing of bones from different individuals. The arms bones of *Proconsul*, though removed from the matrix in Leakey's time, were clearly articulated when buried and, like other parts of the skeleton including the lower incisors, warped *in situ*. It was also noted that the larger animals were juveniles while the small mammals were adult. And last, there was carnivore damage to the *Proconsul* skeleton. This can be seen by an obvious tooth marks on the proximal metaphysis of the right femur and the tibial articular surface of the right talus, and gnawing of the calcaneal tuber on the right calcaneum. Other possible tooth marks are present on other bones, but the crude preparation by hammers and nails in the early 1950s made most of them ambiguous.

Beginning in 1984, the site was cleaned up and many more blocks of rock with bones were found scattered around the area. These included more of the same subadult *Proconsul* skeleton. All blocks around the site and downslope of it were collected, washed and inspected. Bones showing on the surface of the blocks were extracted using Airscribes (Chicago Pneumatic Corp.,) powered by a gas powered air compressor on site. Those blocks without bones showing on the surface were broken down into smaller ones until they were either too small to contain mammal bones, or had bones within them. In this way more bones of the *Proconsul heseloni* type specimen were found (Walker et al., 1986).

The site was cleaned to reveal the situation that Louis Leakey and his associates left in 1950. The circular "pipe" was an obvious 1 m diameter feature in the center of a shallow depression excavated in the flaggy series. The excavators had made us of planes of slickensides to remove the country rock around the "pipe" and these were plotted for aximuth and dip. All the planes were dipping between 39° to 50° towards the center of the "pipe". When allowance is made for the regional dip, these all convert to almost exactly 45°. The circular "pipe" of Whitworth was still to be seen as a greenish, fenitized rock surrounded by grey volcanoclastic flagstones. Fenite is a metasomatically altered quartzo-feldspathic rock comprising mainly alkali feldspar and aegirine augite, and is common around carbonatite volcanos such as Kisingiri. Its typical green color made separation of the "pipe" rock from others very easy. Excavation of a wedge of flagstones down the outside of the "pipe" was undertaken and, when work was stopped in 1984, showed the "pipe" extending downwards into the country rock for at least 3 m. The country rock was asymmetrically deposited around the "pipe," with fine strata on one side and coarser rock with occasional pebbles on the south side. Figure 2 is a photograph of the "pipe" infilling, and shows that the infilling itself is coarsely stratified and that the flaggy series sedimentation was affected by the structure that the "pipe" represents.

These observations refuted the "pothole" theory and led to the conclusion that the structure that created the "pipe" was standing when the flaggy series rocks were deposited, and there are no obvious geological structures that also could contain bones that could be in that configuration. Further, several other, smaller vertical features were found in the Hiwegi Formation, some nearby site R114, and some also containing bones. Several of these had basal features that were clearly calcite-filled buttress roots of trees, and although the excavation at R114 did not reach the base of that infilling, that too was obviously the infilling of a large, hollow tree trunk. The pattern of slickensides also explained some of the distortion of bones, including the peculiar state of the original forelimb skeleton, whereby the arm was folded in its flexed death position so that both radius and humerus were bent in the same way. The sediment filling the hollow tree was initially loosely packed, whereas the surrounding flagstones were water deposited and were more rigid. Subsequent compression of the cylindrical infilling produced cone-in-cone faults with slickensides in the country rock immediately next to the tree at 45°. In another infilled tree site nearby, these cone-in-cone faults are seen to be both downwards and upwards, but at the R114 site the excavators had, naturally, only made us of the downward planes.

With the site established as the infilling of a hollow tree, several points are cleared up. First, the concentration of partial or nearly complete skeletons in a small area is clearly the result of animals being carried in or using the tree as a roost. Walker and Pickford (1983, Table 1) published a faunal list from the R114 site, and this has been added to with a large molossid bat, and a virtually complete tragulid skeleton. Roosting is almost certainly the case with the newly described bat species *Tadarida rusingae* (Arroya-Cabrales, et al., 2002), the python and monitor lizards (Walker and Pickford, 1983), and at least one of the three carnivore species that left tooth marks on the *Proconsul* skeleton. The 15 *Paraphiomys* rodent skeletons might also represent animals that used the tree when they were alive. A mammalian carnivore, probably a hyaenodontid creodont, is the most likely collection agent for the *Proconsul*, the seven small artiodactyls, five rabbit and one mole rat skeletons. Most of the larger skeletons were immature but the smaller mammals were adult, another point that supports the carnivore collection possibility. Although a detailed examination of all of the bones from this site has not yet been made, it seems as though the taphonomic history was a complex one, with a large tree becoming hollowed, probably after its death following the burial of the lower part by volcaniclastic sediments. Animals roosted in the hollow tree and were preserved in sediments filtering down from the upper surface of the newly deposited sediments. Similarly, animals brought in whole or in parts by carnivores would accumulate with the sediments trickling in from above. Second, the time elapsed for the hollow tree to be filled with sediment cannot have been very long. This means that the animals were part of the same community that lived in the immediate area and were not brought together in a "pothole" by stream action that might have been capturing runoff from large distances upstream. Third, this also means that the fauna is not any different in geological age from that of the rest of the Hiwegi Formation, as it might have been had a pothole formed much later than the hardened sediments into which it was cut. Record keeping by Louis Leakey during the collection of the original blocks from R114 was minimal. We know that the infilling stood out from the flaggy series as a low pillar and that Leakey and his colleagues broke this up into blocks (Walker, 1992). That several blocks that contained Proconsul bones were left on site, probably means that Leakey could have collected all of the skeleton had he taken more care. He, or others, also failed to recognize parts of the skeleton during preparation and we know, as was the case of the left first metacarpal that still had an old glue join on it, that parts have also been lost over the years.

*Figure 2. Photograph showing the infilled tree at R114 as viewed from a wedge-shaped trench excavated through the flaggy series surrounding it. Note that the infilling widens downwards and is stratified. Also note that the flaggy series beds are asymmetrical with respect to the infilling, showing that the tree acted to influence their sedimentation.*

Attempts have been made to recover more blocks of the infilling for preparation, and this has been successful in the case of a tragulid skeleton that has yet to be described. Much more remains to be done at this site. Some recovered blocks were unintentionally used to fill in potholes in the roads of the Museum grounds. More bones would certainly be found without undue time expenditure by staff if an acid preparation system such as that in use at the Transvaal Museum in Pretoria, South Africa, were started in Nairobi.

## The Kaswanga Primate Site

The site was discovered by Bwana Peter Nzube in 1984. The site is located on the northern set of exposures at Luanga near Kaswanga and is about 110 meters ESE of the Kenya Government meteorological station. It lies approximately 34°09' East, 0°24' South. Figure 3 shows its location relative to the regional geology. This area is part of site R5 of Andrews and Van Couvering, (1975) and Pickford (1986). The site number R5 was originally given by Le Gros Clark and Leakey (1951) to a "Red Band in upper part of Kathwanga Series" at Luanga. Additional numbers (30-40 and 80-89) referred to sites in the "upper and lower parts of the Kathwanga series," but none of them was ever used afterwards (J. A. Van Couvering, personal communication, 1988). Fossils collected on our expeditions are recorded on enlargements of aerial photographs that are kept in the Palaeontology Department of the National Museum of Kenya, so numbers for individual sites are not needed. Van Couvering and Miller (1969) give an account of the geology and dating of Rusinga Island with further details of the Kaswanga stratigraphy presented by Van Couvering (1972). Observations on the general geology and comments on the Kaswanga sites are given by Pickford (1986). The most recent geochronological studies of Rusinga (Drake et al., 1988), suggest that these deposits are of later Early Miocene age, about 17.8 million years old, significantly younger than those of Songhor and Koru.

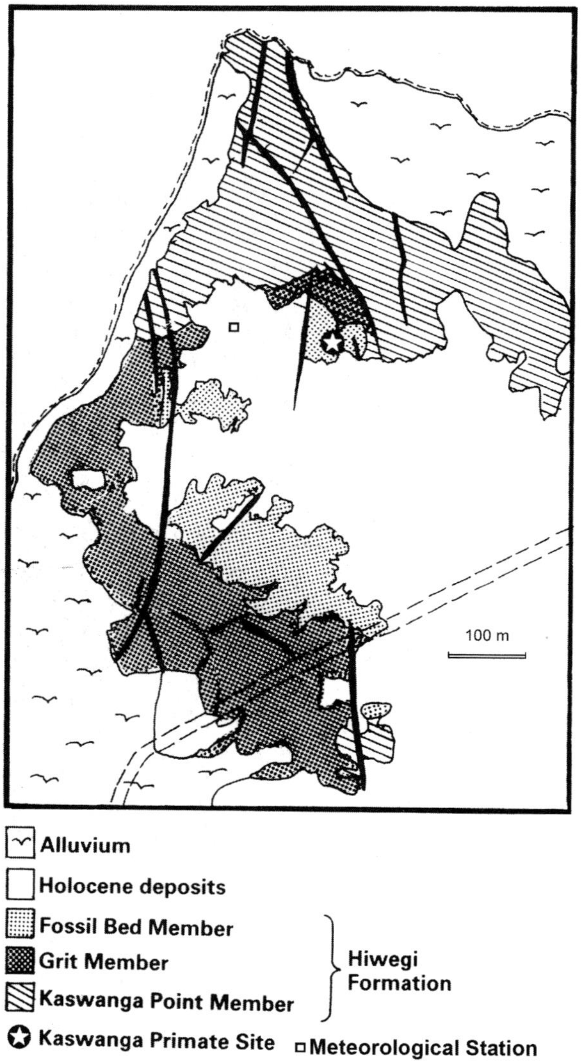


*Figure 3. Geological map of the Kaswaga area, Rusinga Island.*

The lower Hiwegi Formation forms the primary exposure in the area around the site. The excavation and detailed stratigraphic observations carried out in 1985 show that the fossils come from an infilling of Fossil Bed Member silts into a steep but shallow (1 m deep) channel or burrow cut into the underlying Grit Member of the Hiwegi Formation (Van Couvering and Miller, 1969). A topographic map of the site was made to 5 cm contours. This fine level of height resolution showed the disconformity between the two sedimentary units. The map and the local geology are shown in Figure 4. The sediments of the Grit Member either had a primary dip of about 4° to the northeast or were tilted in that direction before the channel or burrow was formed. It is difficult to prove that the feature was a burrow as the infilling matrix with bones is the identical to the surrounding rock, except for being less compacted, and because it seems that the roof of the burrow, if it was one, has long been eroded away. The features were then filled with fine-grained tuffaceous silts and clays that contained *Proconsul* remains, a very few other small mammal bones, leaves and fruits. The leaves and fruits are only found in the tops of the channel feature, but *Proconsul* bones are found throughout. The channels were probably filled in one, very brief, possibly slumping, episode, but the silts and clays point to it being a low energy sedimentary environment. As a result a few hominoid body parts were still in articulation when the original discovery was made. Although many bones were excavated from the depths of the feature, these were all disarticulated and their epiphyses were also found separated. There is no good reason at present to imagine anything other than post-mortem association between the individuals.

The first work at the site in 1984 consisted of collecting the surface bones and teeth that had weathered out. These included the two adult foot skeletons and the hand bones of Individual III and the infant leg and foot bones of Individual IV. Some of the loose surface soil and sediment was carried to Lake Victoria and water washed through screens of mosquito netting. Many bones and teeth were recovered in this way. The site was secured with a fence to prevent further disturbance until excava-

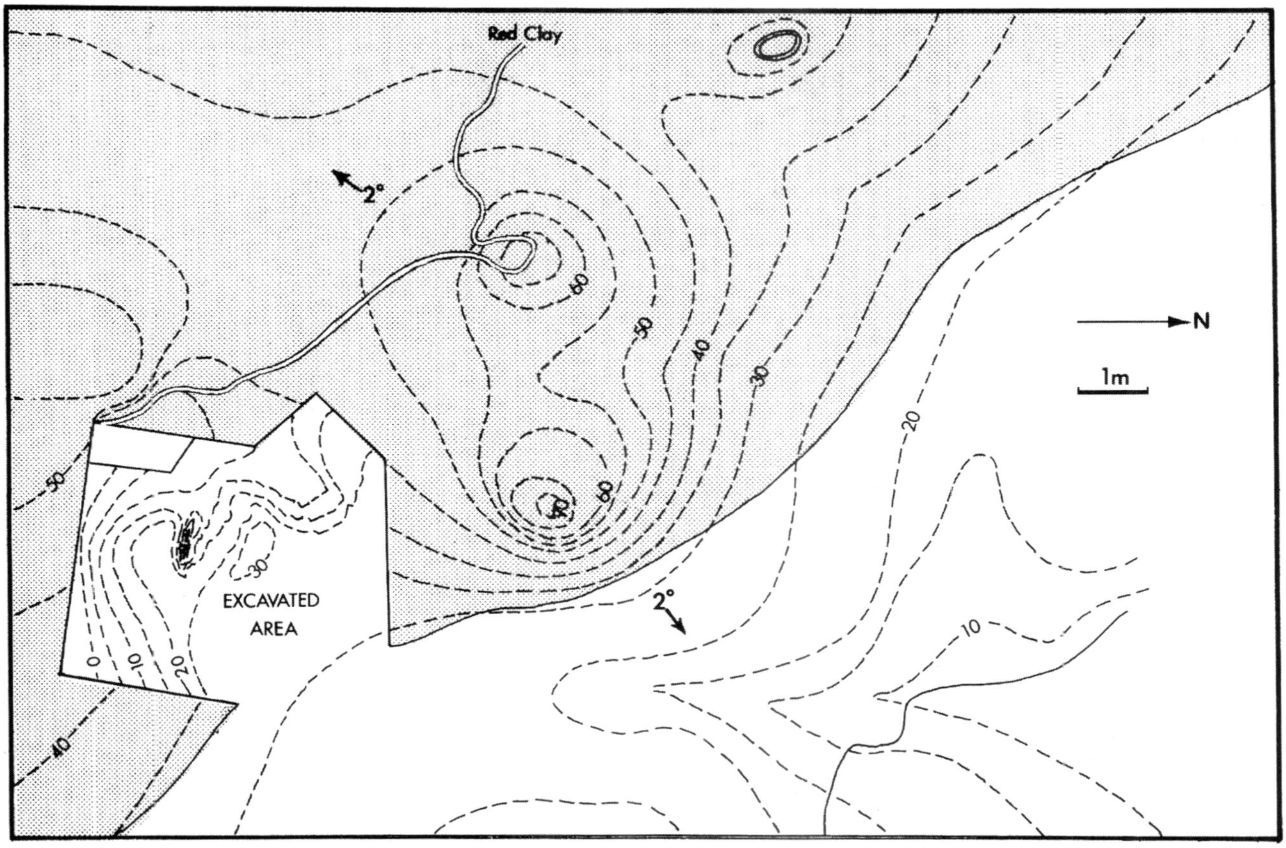


*Figure 4. Geological map of the Kaswanga Primate Site.*

tion could be started in 1985. Once again surface soil and sediment was water washed and excavations begun. Bones were found *in situ*, mainly to the west of the position of the adult foot skeletons, and their 3-D positions plotted. It is important to emphasize that a soil profile had developed across the site. Also that the similarity of filling material with bones to the surrounding sediment meant that the limits of the channel or burrow were better felt than seen. Bones were not found in hard, very consolidated tuffaceous sediment. They always occurred in softer, less consolidated rock. Subsequent sorting of the disarticulated bones that were excavated revealed the presence of only two individuals, one subadult, presumed male, mostly complete and the other, larger subadult male only represented by a few parts. This immediately suggested that the bones and sediment had mixed together while the feature was filling. Figure 5 is a site plan showing the excavated bones and those that were still in situ at the start of cleaning.

Orientation of the excavated bones at the site was limited to those that had a suitable shape and length. As the two individuals excavated were disarticulated subadults, all of their epiphyses were isolated, leaving shortened diaphyses that in some cases were distorted or broken. For the 27 bones for which good orientation measurements were collected, there is a consistent non-random trend that can be seen in the mirror-image rose diagram (Figure 6). A quarter (n = 7) of the bones were oriented between North and 20°. Nearly half the bones (n = 12) were oriented between North and 40°. These bones were sampled from all depths in the deposit and so this marked overall trend probably results from movement of the unconsolidated sediment body that already had disarticulated bones dispersed within it. This is in contrast to the articulated skeletons foot and leg skeletons of Individuals III and IV, for not only are they articulated, but the tibias of them were oriented at 75° and 145° respectively. It is very probable that the rest of these two skeletons were completely articulated before the site was eroded, and further, that if they has been moved in a sediment body they had been moved intact. Indeed, the position of the articulated hand skeleton of Individual III was found where it might have been expected had the articulated skeleton been complete. It is possible that careful examination of bones of those individuals that were not found *in situ* would show, through adjacent similar distortion or manganese dioxide dendritic staining, that they had lain together in the deposit.

The individual *Proconsuls* from the Kaswanga Primate Site were sorted by several means. Lower legs and feet of two of them were articulated *in situ*. This meant that these individuals (III and IV) could have other pieces glued to them that were recovered by screening or washing. Others had bones assigned to them by size, color, manganese dioxide staining patterns, age state, congruency of articulations or interstitial facets, and mirror-imaging. There still are hundreds of small pieces of bone that have not been assigned to individuals and these include 42 phalanges (Begun et al., 1994), and over 20 metapodials or parts of them. The difficulty is some-

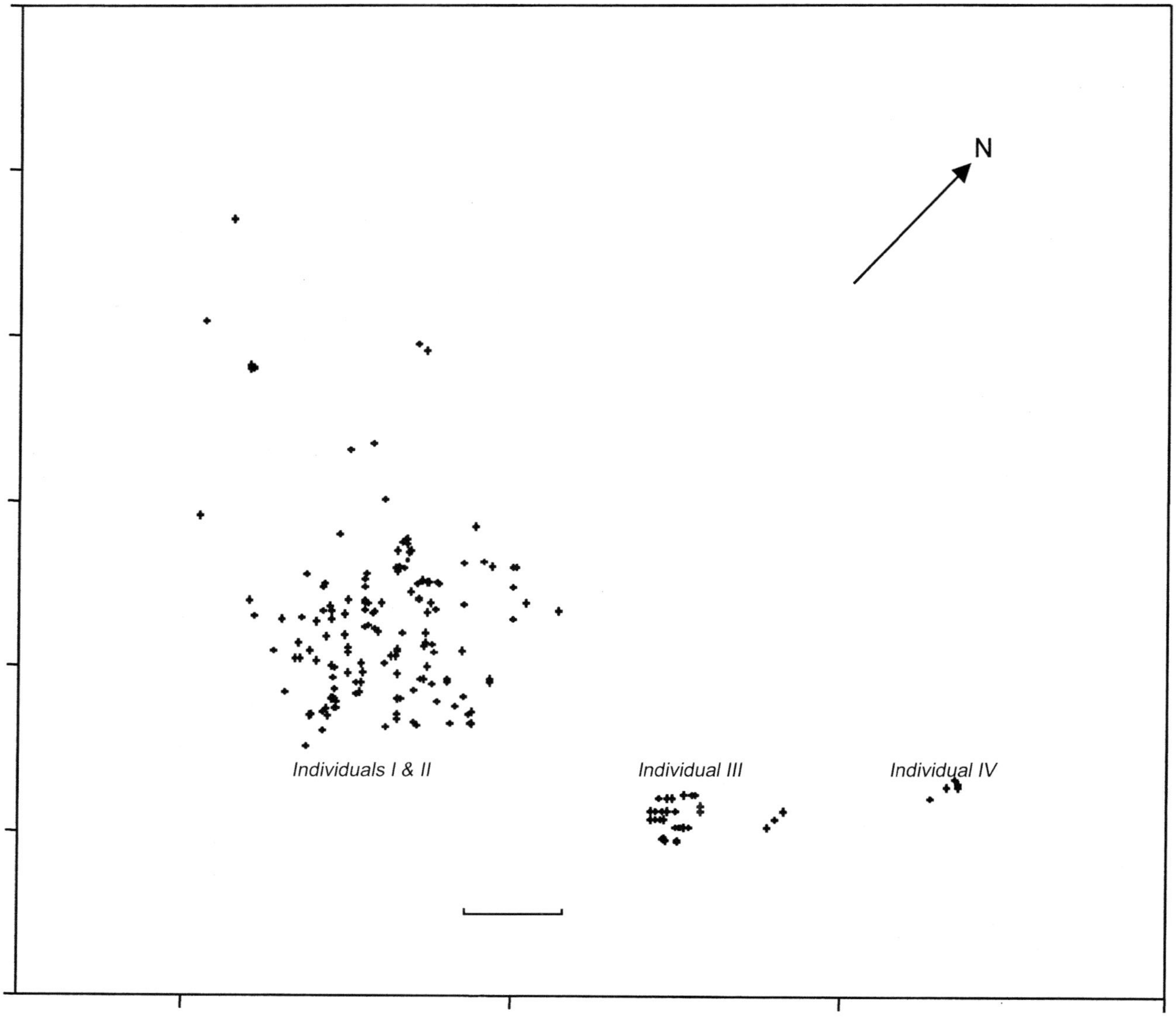


*Figure 5. Plot of the excavated part of the Kaswanga Primate Site from 3-D coordinated data. The positions of the excavated bones of Individuals I and II and the in situ lower leg and foot skeletons of Individuals III and IV. The scale bar represents 25 cm. Note that many points are plotted over each other.*

times made more difficult because we cannot tell which body part pieces are (e.g., a small cylindrical juvenile fibular fragment may be confused with an adult cylindrical metapodial one). Also, many of the bones are bent or otherwise distorted, sometimes misleading even experienced paleontologists. This was the case when Harrison (1998) decided that two specimens identified by me as lumbar vertebrae were in fact caudal ones. This misidentification presumably came about because although one of them (specimen V9) is half of a lumbar body, it is also distorted, and because although specimen V10 is a whole lumbar body, it too is distorted. Detailed study of these two bones (Nakatsukasa et al., 2004) shows quite clearly that the original identification was correct. It is still a concern that we made mistakes in attributing parts to individuals. For instance, it seems certain that two subadult males were mixed together in the part of the site that could be excavated. One of these has most of the skeleton preserved while the other has only a few scraps. Note particularly that we assigned a set of isolated lower teeth to one and a set of isolated uppers to another. This deserves a more thorough assessment as it could be that most of the bones belong to Individual I, and only a very few parts such as the distal end of a fibula, to Individual II.

The numbering of individual parts and individuals was a particular concern that has yet to be satisfactorily resolved. Standard practice at the National Museum of Kenya has been to write a unique identifying number on the specimen. This, for historical reasons, is preceded by a museum and site identifier, e.g., KNM-SO for Songhor, KNM-RU for Rusinga, although such crude site definitions have long given way to detailed site information, such as the field numbers written on the back of the relevant air photographs. The unique numbering system has for many years now been solidly in 5 figures for Department of Paleontology specimens. For those specimens with many isolated body parts the practice has been to

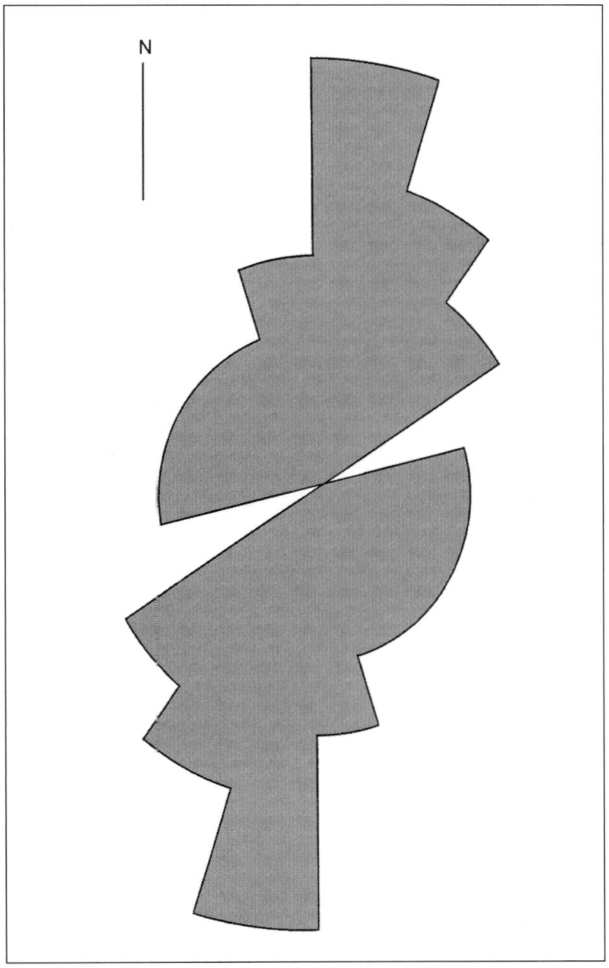


*Figure 6 Mirror-image rose diagram of 27 oriented bones excavated at the Kaswanga Primate Site.*

add a letter suffix, e.g., the type of *Proconsul heseloni*, KNM-RU 2036 has so many isolated bones that the suffixes have long since passed one alphabet, and so another letter has been added—e.g., the left tibia is KNM-RU 2036BA. To write a Museum Accession Number on Kaswanga Primate Site bones we would have to write a 13 digit number such as KNM-RUXXXXXAG. To put this on some large long bones might be possible, if unaesthetic, but to write such a number, as well as our field identifiers on tiny infant hand or foot bones would be virtually impossible. A scheme to use microdot numbers that could be glued on a specimen and read under a binocular microscope was not attractive to grant reviewers. So to date, only our field identifiers remain written on the bones. These are numbers that follow a letter code for body part (R for radius, P for pelvis, etc.) For ease of sorting and keeping track of material, we have added small water-soluble colored dots to each bone that we have given to a particular individual. These are as follows: Individual I–bright pink, Individual II–gold, Individual III–purple, Individual IV–blue, Individual V–dark green, Individual VI–yellow, Individual VII–red, Individual VIII–light green, Individual IX–maroon, Individual X–white. Figure 7 shows the individuals are constituted now. Hundreds more parts remain to be sorted.

## KPS individual I – subadult male

Teeth: Left $I^1$, fragment of incisor root, incisor root and bit of crown left $I^2$, left $P^3$, left $P^4$, right $P^4$ , left $M^{1-3}$, right $M^1$, right $M^3$.

Skull: Sk 2 – petrous temporal

Axial Skeleton: Tl – sternebra, RB3 – right first rib, RB1 – rib fragment, VI – vertebral lamina, V2 – vertebral body, V3 – vertebral body, V44 – vertebral body, V45 – lamina + part of spine and left lower articular process, V90-V96, V98 – misc. vertebral fragments, V97 – vertebral body.

Forelimb: HI – right distal half of humerus shaft, El – right capitular epiphysis, R2 – distal shaft of radius, proximal end (?left) radius, E3 – radial epiphysis, fragment of ulnar styloid, C10 – right pisiform (damaged), C3 – right capitate (damaged), C9 – Triquetrum, E4 – scaphoid tubercle, C48 – left trapezium (identified originally by Beard et al. (1986) as right trapezoid), C4 – left capitate (damaged), Cl – left centrale, C8 – left scaphoid, T14 – right trapezium, C6 – left hamate (damaged), C5 – left lunate (damaged), MT15 – right metacarpal 1, MT13 – right metacarpal II, MT9 – right metacarpal III, MT16 – right metacarpal IV, MT11 – right metacarpal V, MH1-MH12 – metapodial epiphyses, PH9 proximal phalanx ray 4, PH10 proximal phalanx ray 4, PH11 proximal phalanx ray 5, PH19 left proximal phalanx ray 1, PH21 proximal phalanx ray 3, PH24 proximal phalanx ray 2, PH37 proximal phalanx ray 3, PH26 middle phalanx ray 3, PH36 middle phalanx ray 4, PH96 terminal phalanx ray 1, 25 Phalangeal epiphyses.

Pelvis/Hindlimb: PI – right pubis, P17 – right ischium, P2 – left pubic ramus, P18 – left ischium, Fl – left femur, F2 – right femur, E45 – right distal epiphysis of femur, TB1 – left tibia, TB2 – anterior crest of (?left) tibia fragment, E44 – proximal epiphysis of left tibia, E9 - distal epiphysis of left tibia, E7 – distal epiphysis of right tibia, Rl – left fibula shaft fragment, FB3 – distal fibula (shaft), E8 - distal end of left fibula, T13 – right calcaneum, T11 – right talus, T3 – right cuboid, T16 – left lateral cuneiform, T5 – right navicular, T12 – left calcaneum, T10 – left talus, T6 – left navicular, T4 – left cuboid, T9 – left medial cuneiform, T7 – right medial cuneiform, MT14 – right metatarsal I, MT12 – right metatarsal II, MT4 – right metatarsal III, MT6 – right metatarsal IV, MT3 – right metatarsal V, MT2 – left metatarsal I, MT8 – left metatarsal II, MT5 – left metatarsal III, MTI – left metatarsal IV, PH1 left proximal phalanx ray 5, PH4 right proximal phalanx ray 4, PH6 left proximal phalanx ray 4, PH7 right proximal phalanx ray 4, PH8 left proximal phalanx ray 2, PH13 right proximal phalanx ray 1, PH25 left proximal phalanx ray 1, PH2 middle phalanx ray 2, PH15 middle

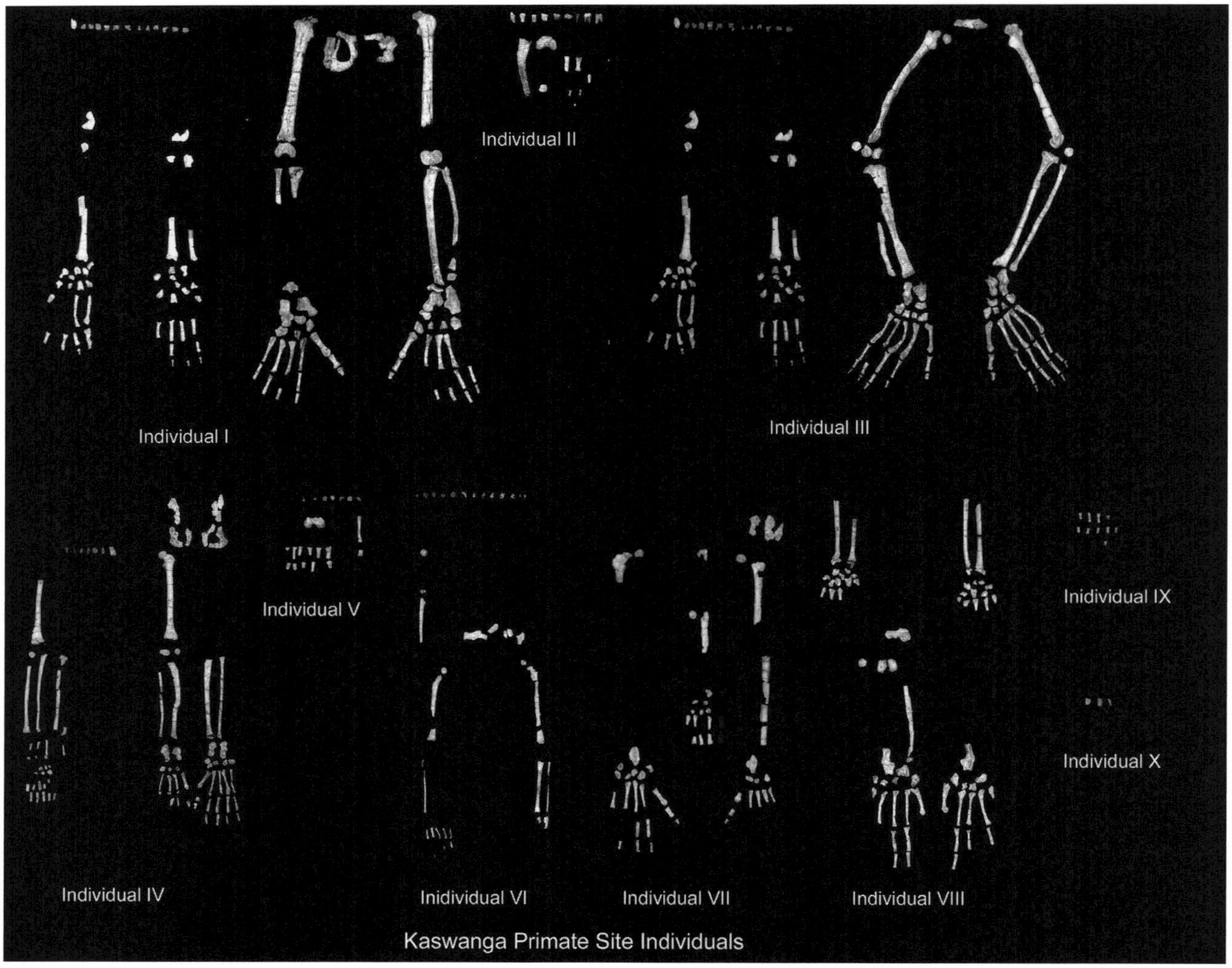


*Figure 7. The Kaswanga Primate Site individuals. All to the same scale.*

phalanx ray 3-4, PH17 middle phalanx ray 2, PH18 middle phalanx ray 3-4, PH29 middle phalanx ray 3-4, PH 33 middle phalanx ray 5, PH30 right terminal phalanx ray 1, PH16 left terminal phalanx ray 1.

## KPS individual II – subadult male, larger than individual I

Teeth: left $P_3$, left $M_{1-2}$ right $M_1$-$_3$, ?$I^1$, left $M^2$, right $M^2$.

Postcranial: R3 – distal radius, C2 – right scaphoid, PH219 – proximal hand phalanx ray 3-4, PH20 middle hand phalanx ray 3-4, PH32 terminal hand phalanx ray 1, E2- distal epiphysis of right femur, TB2 – left tibia shaft fragment, E6 - distal end of right fibula, PH 30 left proximal foot phalanx ray 1, PH22 middle hand foot phalanx ray 5.

## KPS individual III – adult female

Teeth: left lower C., left $P_3$, left $M_{1-3}$, right $M_2$, left $P^3$, left $M^{1-3}$, right $M^{1-3}$

Forelimb: H5 – left distal humeral epiphysis, H6 – right distal humeral epiphysis, U6 – right proximal ulna, U5 – left distal ulna shaft, R13 – radial head, R14 – radial head, R12 – right distal radius shaft, Rll – left distal radius shaft, C15 – right scaphoid, C28 – right capitate, C31 – right centrale, C39 – right triquetrum, C44 – right trapezoid, C35 – right pisiform, C19 – right trapezium, C23 – right lunate, C13 – right hamate, C38 – left triquetrum, C22 – left lunate, C26 – left capitate, C30 – left centrale, C42 – left trapezoid, C34 – left pisiform, C18 – left trapezium, C12 – left hamate, C14 – left scaphoid, PH188 – proximal end of MC1 (mistakenly labeled in the field as phalanx), PH186 – contralateral MC1, MT42 + MT66 associated metacarpals mistakenly labeled in the field as metatarsals, PH99 – left proximal phalanx ray 4, PH100 – left proximal phalanx ray 1, PH101 – left proximal phalanx ray 5, PH221-left proximal phalanx ray 3, PH224 – left proximal phalanx ray 2. PH220 – left middle phalanx ray 4, PH222 – left middle phalanx ray 3, PH 103 – left middle phalanx ray 2, PH104 – left middle phalanx ray 5, PH105 – right terminal phalanx ray 3, PH106 – left terminal phalanx ray 5, PH107 – left terminal phalanx ray 2, PH108 – left terminal phalanx ray 4, PH 223 left terminal phalanx ray 3.

Pelvis/Hindlimb: P9 + P5 – right ischium fragment, P3 – left patella, P4 – right patella, F13 – femoral condyles, F15 – femoral condyles, F12 – right femoral head, Fll – right femur, F10 – left femur, TB7 – right tibia, TB6 – left tibia, FB6 – right fibula, FB5 – left fibula, complete articulated right and left feet.

## KPS individual IV – infant

Teeth: left upper dc, left $M^1$, right $di^1$, right $di^2$, right upper dc, right lower dc, right $dm_1$.

Forelimb: H2 – right humerus (distal 2/3), H3 – distal epiphysis of right humerus, R8 - ?right radius, (distal 2/3), R10 – radial head epiphysis, R9 - ?left radius shaft, U3 – right ulna – sigmoid notch to distal end, U7 – left ulna, sigmoid notch region, U9 – left ulna shaft fragment. U* - left ulnar styloid process, C50 – left scaphoid, C51 – right scaphoid, C52 – left lunate, C53 – right centrale, C54 – left hamate, C55 – left centrale, C56 – left capitate, MT * - right metacarpals 2-5, MH22-28 – metacarpal heads, PH151 – left proximal phalanx ray 1, PH154 – proximal phalanx ray 3-4, PH155 proximal phalanx 2-5, PH161 – proximal phalanx ray 2-5, PH162 proximal phalanx 2-5, PH76 – middle phalanx ray 4, PH77 – middle phalanx ray 3, PH126 – middle phalanx ray 3- 4, PH149 – middle phalanx ray 4,PH168 – middle phalanx ray 2-5,PH170 – middle phalanx ray 2-5,PH173 – middle phalanx ray 3,PH218 middle phalanx ray 2-5.

Pelvis/Hindlimb: P8 – left ilium fragment, P5 – right ilium fragment, P10 – left ischium fragment, P12 – left pubis fragment, Pll – right ischium fragment, P13 – right pubis fragment, F7 – left femur with part of distal epiphysis, PT2 – Patella, TB4 – right tibia (no epiphyses), TB3 – left tibia TB3 with distal epiphysis, FB7 – right fibula distal end with eparate proximal piece, FB4 – left fibula and distal epiphysis, T43 – right talus, T36 – left talus, T42 – right calcaneum, T35 – left calcaneum, T37 – left navicular, T62 – right navicular, T38 – left cuboid, T39 – left medial cuneiform, T41 – left lateral cuneiform, T40 – left intermediate cuneiform, C4 – right intermediate cuneiform (field identification of right trapezoid), MT20 – right metatarsal V, MT63 – right metatarsal IV, MT64 – (?), MT61 – right metatarsal I, MT56 – left metatarsal I with epiphysis, MT57 – left metatarsal II, MT58 – left metarsal III, MT59 – left metatarsal IV, MT60 – left metatarsal V, PH109 – Left proximal phalanx ray 1, PH111 – left proximal phalanx ray 2, PH112 – left proximal phalanx ray 3, PH113 – left proximal phalanx ray 4, PH114 - left proximal phalanx ray 5, PH152 – right proximal phalanx ray 2, PH157 – right proximal phalanx ray 4, PH 115 – left middle phalanx ray 2-5, PH116 – left middle phalanx ray 4-3, PH117 – left middle phalanx ray 2-5, PH118 – left middle phalanx ray 3-4, PH 127 – right middle phalanx ray 2-5, PH216 – right middle phalanx ray 3-4, PH110 – right terminal phalanx ray 1, PH119 – left terminal phalanx ray 3-4, PH120 – left terminal phalanx ray 3-4, , PH121 – right terminal phalanx ray 3-4, PH122 – right terminal phalanx ray 3-4, PH123 – right terminal phalanx ray 5, PH124 – left terminal phalanx ray 1, PH125 – right terminal phalanx ray 2.

## KPS individual V – old adult female

Teeth: left lower C., left $P_3$, lower molar fragment, left $M^1$, right $M^1$, ?right $M^2$ fragment, right $M^3$ .

Postcranial: E17 – distal end of right femur, MT5 – left metatarsal V, PH62, PH192, PH196 – middle hand phalanges, PH97 – right proximal foot phalanx ray 1, PH98 – left proximal foot phalanx ray 1, PH70 – proximal foot phalanx ray 2-5, PH 179 proximal foot phalanx ray 3-4, PH180 proximal foot phalanx ray 3-4, PH184 proximal foot phalanx ray 2-5, PH191 middle foot phalanx, PH39 right terminal foot phalanx, PH40 terminal foot phalanx.

## KPS individual VI – infant

Teeth: right $dm^1$, right $dm^2$, left $dm^2$, right $dm_1$, left$M^1$, - also tooth germs extracted from maxilla.

Note record casts and photographs were taken of maxilla and mandible pieces from which germs were extracted.

Forelimb: U4 – right ulnar shaft, proximal part, H4 – humeral head epiphysis, H8 – humerus – capitulum. PH 75 - proximal hand phalanx ray 1.

Pelvis/Hindlimb: PI 4 – right ilium fragment, P19 – left ilium fragment, P19 – right ischium fragment, P16 – left ischium fragment, P20 – pubis fragment, F8 – femoral head, F9 – left femoral neck and area around lesser trochanter, F7 – right femoral shaft with neck and distal end of femur, - assorted bits of left femoral shaft and distal end of femur, TB5 – right tibia shaft, pieces of fibula from both sides, T64 – right calcaneum, T60 – right talus (originally MH15), PH73, PH74, PH147, PH148, PH150/211, PH174 – proximal foot phalanges, PH23, PH128 terminal foot phalanges, phalangeal epiphyses.

## KPS individual VII – juvenile female

Forelimb: - scapula fragment, part of spine and glenoid, R7 – radial head epiphysis, distal end styloid process of ulna, C47 – right lunate, C49 – left triquetrum, C46 - left scaphoid tubercle, C56 – right hamate, MT20 – metacarpal, MT25 – metacarpal. Four damaged metacarpal epiphyses, PH166 proximal hand phalanx ray 1.

Hindlimb: P6 – right ischium fragment, P7 – left ischium fragment, ilium fragment, F3 – right proximal femur, including neck and head epiphysis, H9 + F4

– left proximal femur – (but no neck or head), F5 – femoral head epiphysis, F6 – femoral head epiphysis, tibia fragments – shaft bits, T18 – right calcaneum, T26 – talus fragment, T30 – left navicular, T44 – left lateral cuneiform, T21 – left cuboid, T34 – left medial cuneiform, T63 – head of talus, T17 – left calcaneum, T31 – right navicular, T33 – right lateral cuneiform, MT 52 – left metatarsal I – distal end, MT27 – right metatarsal I, MT29 – metatarsal, MT30 – metatarsal, MT41 – metatarsal, MT65 – metatarsal, MT24 – metatarsal, MT28 – metatarsal, 2 metatarsals that are unnumbered, PH 90 proximal foot phalanx ray 1, PH91 proximal foot phalanx ray1, PH 88 – terminal foot phalanx ray 1, PH89 – terminal foot phalanx ray 1. The following phalanges have not yet been assigned to hand or foot. PH92 – proximal ray 3-4, PH93 – proximal ray 3-4, PH95 – proximal ray 3-4, PH175 – proximal ray 3-4, PH94 – proximal ray 2-5, PH 153 – proximal ray 2-5, PH154b – proximal ray 2-5, PH156 – proximal ray 2-5, PH167 – proximal ray 2-5, PH160/203 – middle phalanx, PH 159/214 middle phalanx, PH171 – middle phalanx, 158 – middle phalanx, PH164 – middle phalanx, PH 165 – middle phalanx, PH178 – middle phalanx, PH204 – middle phalanx, PH207 – middle phalanx, PH208 – middle phalanx, PH210 – middle phalanx, PH212 – middle phalanx, PH215 – middle phalanx.

### KPS individual VIII – subadult female

Forelimb: R6 – left distal radius, R5 – right distal radius, Ul – left distal ulna with epiphysis, U2 – right distal ulna with epiphysis, C24 – right lunate (damaged), C27 – right capitate, C16 – right hamate (damaged), C40 – right triquetrum (damaged), C32 – right centrale, C43 – right trapezoid, C33 – right pisiform (damaged), C45 – right scaphoid tubercle, C26 – left capitate, C36 – left pisiform (damaged), C37 – left triquetrum (damaged), C21 – left lunate (damaged), C17 – left trapezium (damaged), C29 – left centrale, C25 – left capitate, C41 – left trapezoid , PH 69/131 – proximal hand phalanx ray 2-5, PH71 – proximal hand phalanx ray 3-4, PH181 proximal hand phalanx, PH183 proximal hand phalanx, PH185 proximal hand phalanx, PH187 proximal hand phalanx, PH 64 – middle hand phalanx, PH 65 – middle hand phalanx, PH 67 – middle hand phalanx.

Hindlimb: F14 – femoral condyles, F16 – femoral condyles, PT1 – patella, FB4 – right fibula, distal end with epiphysis, T20 – right calcaneum, T19 – left calcaneum, T22 – right talus, T27 – right medial cuneiform, T28 – right intermediate cuneiform, T29 – right lateral cuneiform, T23 – left navicular, T24 – left medial cuneiform, T25 – left intermediate cuneiform, T23 – left lateral cuneiform (note duplicated field number). MT36 – right metatarsal I, MT37 – right metatarsal II, MT38 – right metatarsal III, MT39 – right metatarsal IV, MT40 – left metatarsal V, MT31 – left metatarsal I, MT32 – left metatarsal II, MT33 – left metatarsal III, MT34 – left metatarsal IV, MT35 – left metatarsal V, PH68 proximal foot phalanx ray 3-4, PH72 – proximal foot phalanx ray 2-5, PH182 proximal foot phalanx, PH189 proximal foot phalanx, PH 190 proximal foot phalanx, PH197 – middle foot phalanx.

### KPS individual IX – adult female

Hindlimb: PH66 – foot middle phalanx ray 2, PH102 – foot middle phalanx ray 5. The following middle phalanges have not been assigned to hand or foot. PH193, PH194, PH198, PH199, PH200, PH201.

### KPS individual X – adult male

Teeth: right $I^2$, right $M^3$, root frag.

## Site Formation at Kaswanga

It is worth emphasizing that most of this *Proconsul* material came from screening of downslope weathered sediment and soil. The site must have been deflating for many years, and it is possible that *Proconsul* fossils may have been collected here as long ago as the early 1930s. MacInnes (1943) described a poorly preserved mandible from site R5 and it is worth considering that it belongs to one of the individuals listed here. This specimen is now KNM-RU 1710, and close examination of its teeth might match it to a Kaswanga Primate Site specimen. Two of them were subadult small individuals to which we assigned no teeth.

Because most of the individuals had washed out before we found them, we do not know how complete they were at the time of burial. However, bearing in mind that we might only have excavated one subadult male, rather than two, and that more work needs to be done on identifying fragments of the collection, and taking the rest of the collection at face value, it appears that four points are worth noting.

1. Hardly any skull or mandible parts are preserved, although many isolated teeth are.
2. The proximal parts of the hind limbs are better preserved than those of the forelimbs. Foot and hand bones were often left articulated on the ends of the limb skeleton.
3. There are hardly any ribs, very few sternebrae, and not many vertebrae in this total assemblage.
4. There are hardly any other mammal fossils occurring with this one primate species here, other than the usually background scatter of fossils. Only two small lagomorph partial skeletons were found in the same general area.

These points can be taken in order. We did find small parts of skull bones and mandibles, but all were fragmentary. The mammalian fossils of the Hiwegi Formation are usually badly cracked and often distorted. Much care is needed to excavate them from the sediment, especially if penetrated by plant roots. In the case of this site, the sediments shrink when dry and expand when wet, causing even short bones such as phalanges to be broken into pieces. The large number of isolated teeth attest to maxillae and mandibles being present before deposition, in the case of Individuals I and II, and at least before erosion in the case of the others. So the lack of complete mandibles and skulls is almost certainly due to destruction either before burial or during erosion. The next two points can be taken together as they seem to be answered by Brain's (1981) carnivore feeding experiments. Baboon skeletons fed to cheetahs look like a perfect model for the Kaswanga primate remains (see Brain 1981, Figure 22a, 23e). Vertebrae (except for the tail), ribs and sternum were all preferentially consumed. Scapulae, being thinner and not as firmly attached to the torso as the pelvis, were destroyed or badly damaged. Hands and feet were sometimes eaten and sometimes not. Skulls were left intact.

Of course, cheetahs were not around in the early Miocene, but the striking similarity between what Brain's cheetahs left and what remains of *Proconsuls* here seems to point to a carnivorous mammal as the agent of accumulation. The Miocene species had no tail, so no tail vertebrae were found (Nakatsukasa et al., 2004), but otherwise the anatomical resemblance between the two anthropoids *Papio* and *Proconsul* is strong. The question of which of the several species of carnivorous mammals known from the Hiwegi Formation was the culprit in this case, is difficult. Most of the genera and species listed (e.g. in Pickford, 1986) are very poorly known—mostly from jaw and tooth fragments—(Savage, 1978) and none from associated postcranial bones. However, some of them like *Hyainailuros* are extremely large, larger than modern living felids, and big enough to swallow a *Proconsul* whole, while others, like the mongoose *Kechechia* are obviously too small. Yet another, *Teratodon*, appears to have been a specialist feeder on land snails. It is more likely that the predator was a creodont, possibly the wolf-sized *Anasinopa leakeyi* or *Isohyaenodon andrewsi,* like the one hypothesized to have carried the R114 site *Proconsul* into its hollow tree lair.

The last point concerns the dominance of *Proconsul* in the assemblage. Primates make up a high proportion of the small to mid sized mammals in the Hiwegi Formation, but there are many other taxa that could be taken by a predator. This concentration on one species that has a mean body mass estimate of about 11 kg (Rafferty et al., 1995) suggests strong prey selection on the part of a predator, rather than any other cause of death.

## Summary

Both the R114 and Kaswanga Primate Site *Proconsuls* seem to have been prey of a selective carnivorous mammal, probably a medium-sized hyaenodontid creodont. In the case of the former, the predator carried a carcass into a dead or dying hollow tree. In the case of the latter site the predator probably concentrated the skeletons in a burrow or narrow gully.

## Acknowledgements

I am very pleased and honored to be able to dedicate this note to Bob Brain, whom I have known since I was a graduate student. I have long been an admirer of his eclectic studies of site formation and have always enjoyed his quiet sense of humor. I once sat next to a man on flight from London to Nairobi. He asked me one of those silly "Do you know so-and-so who lives in South Africa?" questions. We both knew Bob Brain. But he only knew him as a world expert on rotifers and I only knew him as a taphonomist! I thank the Government of Kenya for permission to carry out the field research. I am also indebted to the Governors and staff of the National Museums of Kenya, and especially Richard Leakey, for help during all phases of the work. Bwana Kamoya Kimeu and his team were instrumental in the discovery and collection of the fossils listed here. Chief John Omolo Nditi gave us invaluable support and help on Rusinga Island. I thank the following people who worked sieving and excavating the site: the late Mary Leakey, Mark Teaford, Blaire Van Valkenburgh, Kamoya Kimeu, Peter Nzube, Wambua Mangao, Musa Kyeva, Benson Kyongo, Mwongela Muoka and Blake Edgar. Dr Richard Hay kindly helped us with the interpretation of the geology and Lars Werdelin taught me about Rusinga carnivores. Dr. Emma Mbua gave invaluable curatorial help. The work was supported by The National Museums of Kenya and National Science Foundation Grant BNS 8418567.

## References

Andrews, P., Van Couvering, J. A. H. 1975. Palaeoenvironments in the East African Miocene. Contributions to Primatology 5, 62-103

Arroya-Cabrales, J., Gregorin, R., Schlitter, D.A., Walker, A. 2002. A new molossid bat from the Miocene of Kenya. Journal of Vertebrate Paleontology 22, 380-387.

Beard, K.C., Teaford, M.F., Walker, A. 1986. New Wrist Bones of *Proconsul africanus* and *P. nyanzae* from Rusinga Island, Kenya. Folia primatologica 47, 97-118.

Begun, D.R., Teaford, M.F., Walker, A. 1994. Comparative and functional anatomy of *Proconsul* phalanges from the Kaswanga Primate Site, Rusinga Island, Kenya. Journal of Human Evolution 26, 89-165.

Brain, C.K. 1981. The Hunters or the Hunted? An Introduction to African Cave Taphonomy. University of Chicago Press, Chicago.

Drake, R. L., Van Couvering, J. A., Pickford, M., Curtis, G. H., Harris, J. A. 1988. New chronology for the Early Miocene mammalian faunas of Kisingiri, Western Kenya. Journal of the Geological Society of London 145, 479-491.

Harrison, T. 1998. Evidence for a tail in *Proconsul heseloni.* American Journal of Physical Anthropology 26 (Supplement), 93-94.

Le Gros Clark, W. E., Leakey, L. S. B. 1951. The Miocene Hominoidea of East Africa. British Museum (Natural History) Fossil Mammals of Africa 1, 1-117.

Nakatsukasa, M., Ward, C.V., Walker, A., Teaford, M.F., Kunimatsu, Y., Ogihara, N. 2004 Journal of Human Evolution 46, 777-784.

Napier, J.R., Davis, P.R. 1959. The forelimb skeleton and associated remains of *Proconsul africanus*. British Museum (Natural History) Fossil Mammals of Africa 16, 1-69.

Pickford, M. 1986. Cainozoic palacontological sites of western Kenya. Münchner Geowissenschafliche Abhandlungen (A) 8, 1-151.

Rafferty, K., Walker, A., Ruff, C.B., Rose, M.D., Andrews, A. 1995. Postcranial estimates of body weight in *Proconsul,*with a note on a distal tibia of *P. major* from Napak, Uganda. American Journal of Physical Anthropology 97, 391-402.

Savage, R.J. G. 1978. Carnivora. In: Maglio, V.J., Cooke, H.B.S. (Eds.), Evolution of African Mammals, Harvard University Press, Cambridge, pp. 249-267.

Van Couvering, J. A. 1972. Geology of Rusinga Island and correlation of the Kenya Mid-Tertiary fauna. Ph.D. Dissertation, Cambridge University.

Van Couvering, J. A., Miller, J. A. 1969. Miocene stratigraphy and age determinations, Rusinga Island, Kenya. Nature 221, 628-632.

Walker, A. 1992. Louis Leakey, John Napier and the history of *Proconsul*. Journal of Human Evolution, 22, 245-254.

Walker, A.C., Pickford, M. 1983. New post-cranial fossils of *Proconsul africanus* and *Proconsul nyanzae*. In: Ciochon, R.L., Corruccini, R.S. (Eds.), New Interpretations of Ape and Human Ancestry, Plenum Press, New York, pp. 325-351.

Walker, A., Shipman, P. 2005. The Ape in the Tree. Harvard University Press, Cambridge.

Walker, A., Teaford, M. 1988. The Kaswanga Primate Site: An Early Miocene hominoid site on Rusinga Island, Kenya. Journal of Human Evolution, 17, 539-544.

Walker, A., Teaford, M.F., Martin, L., Andrews, P. 1993. A new species of *Proconsul* from the early Miocene of Rusinga/Mfangano Islands, Kenya. Journal of Human Evolution 25, 43-56.

Walker, A., Teaford, M.F., Leakey, R.E. 1986. New information concerning the R114 *Proconsul* site, Rusinga Island, Kenya. In: Else, J. Lee, P. (Eds.), Primate Evolution, Cambridge University Press, Cambridge, pp. 143-149.

Whitworth, T. 1953. A contribution to the geology of Rusinga Island. Quarterly Journal of the Geological Society of London, 109, 75-96.

# CHAPTER 7

# Are You In or Out (of Africa)? Site Formation at Dmanisi and Actualistic Studies in Africa

Martha Tappen, David Lordkipanidze, Maia Bukshianidze, Reid Ferring and Abesalom Vekua

## Abstract

Researchers from all over the world use the actualistic studies conducted in Africa to interpret the formation of faunal localities. Aspects of applying African models to the early Pleistocene faunas from Dmanisi in Georgia are discussed. Currently, there is no compelling evidence that the mammalian taxa from Dmanisi migrated from Africa at the same time as *Homo*. Preliminary analyses of taphonomic and stratigraphic evidence indicate that Dmanisi has a complex, but not a long taphonomic history. Several taphonomic agents were involved in modifying the fauna and hominin fossil assemblage, rather than one agent predominating, yet the bones accumulated relatively rapidly. Hominins left cut marks, carnivores left tooth marks, and porcupines gnawed some of the bones. Many specimens seem to have little alteration at all, including no weathering, no evidence for geologic transport (rounding or microstriations), and remain as articulated subunits, but not as whole animals. In many ways Dmanisi does not fit classic models of human habitation sites, hyena dens, or mass death sites. Specifically, the areas excavated at Dmanisi so far do not have enough stone tool damage to be primarily accumulated by hominins, nor do they have as much carnivore tooth scoring as modern dens. Detailed spatial analysis of taphonomic modifications is still underway, and may allow spatial parsing of the site into areas by taphonomic agent.

## Introduction

The interpretations of fossil bone assemblages derive from observations of the taphonomic journeys of present day animal remains. While actualistic studies of taphonomy are conducted all over the world, most have been conducted in Africa, where parks conserve relatively healthy ecosystems retaining large carnivores that are key preburial taphonomic agents, and where the early archaeological record begins. But how universally can the observations from actualistic studies conducted by Africanists be applied to localities outside of Africa? Which features of bone accumulation and modification are robust enough to be generalized, with constrained amounts of variation possible or probable? This question is important to untangling the ecology of hominins and the taphonomic histories of early Palaeolithic sites found outside of Africa, and gets to the heart of factors that allowed the original range expansion into higher latitudes, because it requires precise paleoecological and behavioral reconstruction. Was the initial spread of *Homo* beyond the australopithecine geographic range due to the spreading of African-like biomes, or was it due to novel, intrinsic behaviors of *Homo*? In this paper we discuss some of these issues, and make some comparisons between the paleoecologic and taphonomic record of the early Pleistocene site of Dmanisi, in Georgia (Trans-Caucasus), and actualistic taphonomic studies in Africa. We outline some of the major taphonomic characteristics of Dmanisi, and argue that the signature from the site as a whole is not a good match for our archetypal models of bone formation processes of hominin sites, carnivore dens, or porcupine dens. We compare some key taphonomic features to time-averaged attritional death bone deposition studies as models for predation arenas from Africa to Dmanisi. While there are still more analyses to conduct, the predation arena model also does not fit Dmanisi's signature in some key ways.

## Dmanisi: General Background

Dmanisi is located around a thousand miles from of the NE tip of Africa at 41° N latitude, south of the Caucasus Mountains in Georgia. Numerous hominin fossils have been found in direct association with a large assemblage of mammalian fauna and simply flaked stone tools. David Lordkipanidze leads the excavations with an international team of principal investigators including the authors of this paper, with Philip Rightmire of SUNY Binghamton, Marcia Ponce de Leon and Christophe Zollikofer of the University of Zurich, and others. The hominin occupations date to the earliest Pleistocene, shortly after the Olduvai Normal Subchron which ended at 1.78 mya (Van Couvering, 1997). Geological and paleobiological evidence suggests that the these levels with Mode 1 tools date to before 1.7 mya, and are closer to 1.77 mya (Gabunia and Vekua, 1995; Gabunia et al., 2000; Vekua et al., 2002).

Today rainfall, seasonality and habitat characteristics in Georgia vary greatly with elevation and from west to east. Eastern Georgia tends to be drier and have a more continental pattern, while in Western Georgia the weather has a more Mediterranean pattern. The site is in the South Central region (Kvemo Kartli Province) at an elevation of 915 meters above sea level. The region around Dmanisi has a good deal of topographic relief with hills and valleys that would have enhanced the potential for a mosaic of habitats, and the fauna indicate there were both wooded and open areas (Gabunia et al., 2000). Dmanisi is an open air site on a promontory overlooking the confluence of the Pinazauri and Mashavera rivers. These rivers have eroded down, in place, through 80–100 meters of basalt since the early Pleistocene, leaving the site high above them today (Figure 1).

Determining what processes accumulated the fossils is the core taphonomic question at Dmanisi. The site has eight species of large predator in direct association with *Homo erectus sensu lato*, thus Bob Brain's celebrated question of "The hunter or the hunted?" is immediately *apropos*. Interestingly, there are several other analogies to the South African karstic cave sites that Brain so carefully deciphered: some of the bones accumulated in underground hydraulically formed pipes that eventually breached the surface, which can be thought of as mini-analogs to karstic cave formation, and many of the bones were introduced into the pipes by predators.

*Figure 1. Dmanisi Promontory, above the confluence of the Mashavera and Pineazauri Rivers. The Medieval fortress and church can be seen in the back of the isthmus, the Paleolithic excavations are located in the middle of the plateau among the trees. With thanks to Ken Garret.*

The stratigraphy of the site is being worked out by Reid Ferring of the University of North Texas, with preliminary dating by Carl Swisher, and detailed work continues. On top of basalt that dates to within the Olduvai subchron, there are two main strata, A and B. The A layers are normal and from within the Olduvai subchron, and the B layers were deposited immediately after this; they have reversed polarity and so post-date 1.78 myr. A series of hydraulic pipes formed within the A sediments, creating tunnels that then were filled with bones as well as reversed B1 sediment. Microstratigraphic analysis by Ferring indicates that pipes filled and some collapsed, forming low spots and then small drainage runnels. As of now, his evidence indicates that most of the hominins as well as artifacts and the majority of the fauna come from within these pipes and from on top of pipes that had collapsed and thus would have been low-lying land surfaces and small sediment traps when bones were deposited. The B2 stratum sediment that was deposited above also contains fauna and Mode 1 tools. Later a post depositional carbonate zone (K) was formed that sealed in the lower deposits of the site, especially the lower pipes and pipe collapses, protecting the bones from compaction and preserving them well (Vekua et al., 2002, Figure 2). It seems that the complex configuration of the microstratigraphy at the site was caused by the very complicated configuration of the surface of the underlying basalt, causing the small drainages and pipes to form. Bones accumulated in especially dense concentrations where the basalt is low lying.

## Biogeography of the Dmanisi Large Mammals

The species composition of the fauna is the first evidence we address about both site formation and environments at Dmanisi. Did the Dmanisi hominins migrate out of Africa because of an expansion of African biomes, or from the agency of cf. *Homo erectus* itself? In the recent deliberations regarding the migration of hominins out of Africa, several mammalian paleontologists have discussed the importance of associated African faunas. For example, Tchernov suggested that "any hominid dispersal was a natural part of any emigration until the late Upper Pleistocene period." (Tchernov, 1998:80). Turner likewise stated that "The dispersion of *Homo* can be seen as part of larger pattern of dispersion by members of the terrestrial mammalian fauna" (Turner, 1999). Rook (2004) has also pointed out a movement of African species, including *Megantereon, Theropithicus*, and *Hippotamus* and suggested an association with *Homo*.

*Figure 2. Excavations in 2003 in Block 2 at Dmanisi. At the top right in the lighter color sediments students can been seen excavating in the B2 sediments. Below this the carbonate layer can be seen. In the foreground, Georgian excavator David Takakishvili (in striped shirt) can be seen excavating in the darker B1x and y sediments. Just to the right of Takakishvili's shoulder is a round excavation pit where the skull 3444 was taken out in a block of sediments. He is sitting on the basal basalt, and the very rough, irregular and unweathered surface of the basalt is readily visible.*

The association of African faunas with *Homo* out of Africa has been considered significant for several reasons, but especially because 'Ubediya in the Levant was long been considered one of the earliest sites out of Africa, and it does indeed have African elements in its fauna, such as *Kolpochoerus olduvaiensis, Hippopotamus gorgops, Pelorovis olduwayensis, Crocuta crocuta*, (Tchernov, 1992a, 1992b, 1999) and *Theropithecus* (Belmaker, 2002) although by far most of the fauna is Eurasian. These forms indicate a significant flow out of Africa into this corner of Eurasia, but this site is probably at least 400,000 years later in time than Dmanisi, and evidently represents a later dispersal event. These African species are also in Italy and Spain (at Pirro Nord, Fuente Nueva-3 and Barranco Leon-5), but these sites likewise may be later in time than the Dmanisi. There is evidence for several influxes of hominins with fauna from Africa with later hominins also, for example the dispersal of early modern humans at Qafzeh are associated with an influx of African forms (Tchernov, 1998). But as discussed below, these associations are not in evidence for the expansion represented at Dmanisi.

Since the docking of Africa with Eurasia in the Early Miocene, there have been various moments when transfer of faunas was encouraged or discouraged by marine transgressions, the relative aridification of the Arabian Peninsula, and the Taurus-Zagros orogeny (Tchernov, 1998). There are several dispersals of African mammals that could be associated with the dispersal of hominins. In particular between 3.1-2.6 million years ago several bovid species spread out of Africa into Asia (Vrba, 1995). During this time there was a major retreat of the seas, called the "Acquatraversan erosional phase" correlating with the "Elephant-*Equus*" event in Eurasia (Azzaroli, 1995). Turner likewise stressed the late Pliocene as a time when hominin dispersion was most feasible (Turner, 1999). In this paper we take on the question with the perspective that if *Homo* spread due to a general emigration or spread of African savanna species and biomes, there would be several other large mammal species found at the site that had evidently had made it out of Africa about 2-1.7 million years ago. We therefore are addressing this question here from a strict and literal point of view: that the Plio-Pleistocene border was in fact the time when *Homo* made it out of Africa, and not earlier. There is a reasonable possibility that we will find earlier hominins, perhaps at the origins of tool use around 2.5 mya in Eurasia in the future, but here we proceed with the evidence that exists as it does today using the fauna directly associated with *Homo* at Dmanisi.

So are the Dmanisi large mammals directly from Africa (did these species recently emigrate to Eurasia)? Table 1 presents the current list of fauna identified from the Lower Pleistocene strata at Dmanisi. The list of taxa present is not yet static as each year of excavation we have found new species. For example, there are several new bovid species identified by Maia Bukhsianidze in the last year. Furthermore, some taxonomic assignments may be revised and more precisely defined as more and more detailed analyses take place within each taxon. Even with these caveats, important observations can be made about the Dmanisi taxa.

Starting with the herbivores, one can see immediately that for many of the lineages, such as the Equidae, Rhinocerotidae and Cervidae, an African origin is out of question. The stenoid representatives of the genus *Equus* Linnaeus, 1758 first appeared during the early Pliocene in North America. In the early Villafranchian they entered Eurasia (about 2.6 myr), where they passed through an adaptive radiation that led to the gradual displacement of the hipparions, that had been common until then, and then through Eurasia entered into Africa (Agu-

*Table 1. Large mammal taxa from Dmanisi*

Perrisodactyla Owen, 1848
- Equidae Gray, 1821
  - *Equus stenonis* Cocchi, 1867
  - *Equus* sp. aff. *altidens* Reichenae, 1915
- Rhinocerotidae Owen, 1845
  - *Stephanorhinus etruscus etruscus* Gloger, 1841

Artiodactyla Owen, 1848
- Cervidae Gray, 1821
  - *Cervus perrieri* Croizet and Jobert, 1828
  - *Cervus* sp. (ex. gr. *Arvernoceros ardei* Croizet and Jobert, 1828)
  - *Eucladoceros* aff. *senezensis* Depéret
  - *Cervus (Dama)* cf. *nestii* Major
- Giraffidae, Gray, 1821
  - *Palaeotragus* sp.
- Bovidae Gray, 1821
  - *Bison (Eobison) georgicus* Burchak-Abramovich and Vekua, 1994
  - *Gallogoral menighinii sickenbergii* Kostopoulos, 1996
  - *Capra* sp. nov. sp.
  - *Sorgelia* cf. *minor* (Moyà-Solà, 1987)
  - Ovibovini gen.et sp. indet
  - Antilopini gen.et sp. indet. (A)
  - Antilopini gen et sp. indet. (B)

Proboscidea Illiger, 1811
- Elephantidae Gray, 1821
- *Mammuthus meridionalis* Nesti, 1825

Carnivora Bowdich, 1821
- Canidae Gray, 1821
  - *Canis etuscus* Major, 1877
- Ursidae Gray, 1825
  - *Ursus etruscus* Cuvier, 1812
  - *Ursus* sp.
- Mustelidae Swainson, 1835
  - *Martes* sp. Frisch, 1775
- Hyaenidae Gray, 1869
  - *Pachycrocuta* Croizet, 1828
- Felidae Gray, 1821
  - *Lynx issiodorensis* Croizet and Jobert, 1828
  - *Panthera gombaszoegensis* Kretzoi, 1938
  - *Megantereon cultridens* Cuvier, 1824
  - *Homotherium crenatidens* Fabrini, 1890

irre et al., 1997; Lindsay, 1997; Eisenmann, 2004). For the Rhinoceratidae, *Stephanorhinus etruscus* Falconer 1868 represents a typical species characteristic the entire Villafranchian (MN16-MN19) of Europe and Middle Asia. (Fortelius et al., 1993). The Cervidae is of course an autochtonous Eurasian group. Cervids have sometimes spread into North Africa, but are fundamentally Eurasian.

All the bovid taxa from Dmanisi are palearctic: *Bison (Eobison) georgicus (=Dmanisibos georgicus* Burchak-Abramovich and Vekua, 1994*)* is one of the earliest representatives of bison (the oldest is *Bison sivalensis* Lydekker ex. Falconer M.S. 1868, 1878 from Pinjor zone of Upper Siwaliks, late Pliocene), Caprinae: *Gallogoral meneghinii sickenbergii* (Kostopoulos, 1996)—the genus *Gallogoral* Guérin 1965 is known only from European Plio-Pleistocene sites (MN17-MN18) and supposedly has Asian origins (Guérin, 1965), *Capra* sp. nov sp.—perhaps the oldest representative of the genus *Capra* Linnaeus, 1758 known so far in the world (The first undoubted remains of *Capra* are known from Petralona, Greece, Rissian, [Sickenberg, 1971, Tsoukala, 1991]), the genus *Capra* is considered to have originated in Eurasia, during the Plio-Pleistocene (Pilgrim, 1947). The Ovibovini (*Soergelia* cf. *minor* and Ovibovini indet.) is likewise an Eurasian group. Furthermore, the two representatives of Antilopini reveal affinities to the Eurasian spiral horn antelopes (Bukshianidze, 2005).

Due to the present African distribution of Giraffids it is important to stress that the small giraffid from Dmanisi belongs to the genus *Palaeotragus* Gaudry 1861. The origin and early stages of the evolution of the genus are not clear but it seems to go extinct in Africa at the end of Miocene, while from the end of middle Miocene to the end of Pliocene this genus has a wide distribution in the Old World (Godina, 1979). The Dmanisi *Palaeotragus* is one of the last representatives of this genus and its affinities should be looked for among the Eurasian Palaeotraginae.

The earliest (Ethiopian) representatives of *Mammuthus* (sensu Maglio, 1973) migrated into Eurasia in the Early Pliocene (Palmqvist et al., 1999; Lister and van Essen, 2002, 2004; Kahlke, 2003). The Dmanisi *Mammuthus meridionalis* is a typical representative of the European Villafranchian, and one of the primitive forms of the archidiscodonts' Eurasian evolutionary lineage (Vekua, 1995; Gabunia et al., 2000) leading through the steppe mammoth—*Mammuthus trogontherii* to the wooly mammoth—*M. primigenius*.

Carnivores as possible "fellow travelers" are of special interest. Many of the Villafranchian carnivores, especially the canids, ursids, and mustelids, do not have African roots and are essentially Eurasian. Others are so widespread that it may be premature to trace their speciation events and dispersals without more fossil sites.

Among the carnivores found in Dmanisi the small wolf-like Canid–*Canis etruscus* Major 1877 is the most abundant. The present level of knowledge of the history of the genus *Canis* does not allow unambiguous localization of its place of origin. The taxon represented at Dmanisi—*Canis etruscus*—appears in Eurasia at the limit of middle and late Villafranchian. This species is strictly Eurasian and is in the evolutionary line leading to extant *C. lupus*. The spread of *Canis etruscus* and the extinction of the racoon dog *Nyctereuntes megamastoides* Pomel 1842, marks the "wolf event" and represents a main faunal turnover in the course of the Villafranchian (Azzaroli, 1983). This event occurred at the time interval between the Reunion and the end of the Olduvai magnetic subchrons (Azzaroli et al., 1988). In Europe the first arrival of this species is registered in Olivola faunal unit (MN18, Azzaroli et al., 1988, Azzaroli, 1983). Among Georgian localities, *Canis etruscus* is first registered in Diliska (2.2 ma) followed by Kotsakhuri (1.8-1.9 ma), Dmanisi (1.75 ma) and Tsalka (early lower Pleistocene, Vekua, 1991; Vekua et al., 1985). *Canis etruscus* is thus not among the possible fellow travelers of early *Homo* in its migration out of Africa.

Furthermore, the Eurasian origin of mustelids (*Martes* sp.) and ursids (*Ursus etruscus, Ursus* sp.) is not doubted. The genus *Martes* Pinel, 1792 is exclusively palearctic, known from numerous Eurasian sites including several species starting from Miocene (MN3) up to present (Fortelius, 2003). *Ursus etruscus* Cuvier 1823 represented at the site is a typical Villafranchian European form (Rustioni and Mazza, 1992). It first appears in MN16–Lower Villafranchian and is present up to the end of MN19.

The Dmanisi hyena belongs to the genus *Pachycrocuta* Cretzoi. The roots of the genus *Pachycrocuta* are not clear, either a Eurasian or African origin is possible (Howell and Petter, 1980). In the case of Dmanisi it is important to stress that the first representatives of the genus are present in Eurasia from the Ruscinian (MN15 localities such as Sera en Voke, Lyana, Odessa Catacombs summarized in Sotnikova, 1989). During the earlier Villafranchian, *P. perrieri* Croizet and Jobert 1828, a species derived from this stock became the dominant hyena in western European and circummediterranean faunas (Howell and Petter, 1980) and was then replaced by larger hyena *P. brevirostris* Aymard, 1864, a species closely allied to *P. perrieri* from which it differs by its larger size (Howell and Petter, 1980). The place of origin of *P. brevirostris* is also unclear: it is "difficult to determine whether *P. brevirostris* originated in Asia or Africa, though the age determinations presently known suggest an African origin" (Torre et al., 1992). *P. brevirostris* is present in Africa before 3 Ma (Turner, 1992). Later on it appears in Asia - Pinjor zone, India indicating an age not earlier then 2.5 Ma (Torre et al., 1992). To summarize, the origins and migration of *Pachycrocuta* species need to be more precisely defined.

There are at least four species of felid at Dmanisi. *Panthera gombaszoegensis* Kretzoi, 1938 identified in the Dmanisi fossil assemblage represents one of the earliest appearances of this species. The earliest appearance of this *Panthera* is registered from Late Villafranchian European localities such as: Olivola, Upper Valdarno, Tegelen, and Erpfingen. Supposedly this species originated in the western palearctic (Hemmer, 1981) and although it is very close to the basal forms of the genus *Panthera*, the place of the origin of the genus *Panthera* is not clear: Africa after Howell and Petter (1976), Sotnikova (1989), and Asia after Hemmer (1981). The Dmanisi species is unknown in Africa (Turner, 1992) and is strictly Eurasian.

*Lynx issiodorensis* Croizet and Jobert 1828 is a typical late Pliocene early Pleistocene form and has a vast geographical distribution (Africa, Eurasia, N. America). An African origin of the genus is possible, as the oldest remains of the genus (*L. issiodorensis*) are known from the Pliocene site of Langebaanweg, South Africa, > 4 myr, (Hendey, 1974) but they are also present in the Pliocene of Europe (at Etouaires, more than 3.3 Ma, and at the Odessa catacombs, the age of which is considered to be older then of Etouaires [Sotnikova, 1989]). However, since the finds in Europe are much more numerous than in Africa, the opposite point of view—an Eurasian center of origin, can be also supported (Sotnikova, 1989). There is an important time interval between the first appearance of the species and Dmanisi, and in the Caucasus *Lynx issiodorensis* is identified in Kvabebi, an Early Villafranchian fauna, from the middle-late Pliocene of Georgia, Vekua (1972), so the Dmanisi species of *Lynx* was already present at this earlier time.

From the late Pliocene (the limit of Rouscinian and Villafranchian faunal zones) the genus *Homotherium* Fabrini, 1890 appears almost simultaneously all over the Old World (Turner, 1997; Sotnikova, 1989), seeming to originate in Eurasia. The Chinese early Pliocene *Machairodus* Kaup 1883 species are hypothetical ancestral forms of *Homotherium* (Sotnikova, 1989), and consequently migration from Eurasia into Africa is supported. However, *Homotherium* appears in the African Pliocene at Langebaanweg as well (Hendey, 1974), and thus an African origin is certainly possible, more so as genus *Machairodus*, its most likely ancestor, is also present there (Turner, 1990). All previously described species now are united into one species by some authors (*H. crenatidens* Fabrini, 1890–after Ficcarelli, 1979 and *H. latidens* Owen, 1846–after Turner, 1997) and this idea is widely accepted among paleontologists. Although *Homotherium* could have an African origin it is clear that as *Homotherium* is present in Eurasia since the late Pliocene times there is no reason to see the Dmanisi form as newly appeared form. In addition, it should also be mentioned that the genus is present in Georgia since the earlier Kvabebi fauna (Vekua, 1972; Sotnikova, 1989).

*Megantereon cultridens* Cuvier 1824 is the second Machairodontinae species present in Dmanisi, is smaller than *Homotherium*, and as a rule coexisted with the latter. The genus *Megantereon* Croizet and Jobert 1828 first appears at 4.5 Ma in the North America in the Bone Valley Formation, Florida (Berta and Galiano, 1983). *Megantereon* disperses from North America before 3.5 Ma and spreads all over the Old World. In Europe the oldest remains of the genus are known from the very end of Ruscinien (Catacombs of Odessa, Sotnikova, 1989). The first record from China is from Yushe basin Shansi province (Teilhard and Leroy, 1945) that corresponds the early Villafranchian of Europe; in India from Pinjor zone of Upper Siwaliks (Pilgrim, 1932) corresponding the late Pliocene, and this genus is also known from Java (Koenigswald, 1974). In Africa the oldest remains of the genus is dated ~ 3.5 Ma deposits of the Nachukui Formation south of the Turkwel River in West Turkana, Northern Kenya (Werdelin and Lewis, 2000). The genus lasts up to the end of the Villafranchian in Europe, in central Asia to the end of early Pleistocene, and in America to the Holocene, where most probably it gives rise to the extant genus *Smilodon* Lund 1842. Initially a number of *Megantereon* species were described, but in a comprehensive review of the systematics of *Megantereon* in the New and Old World, Turner (1987) considered *M. cultridens* Cuvier 1824 to be the only valid species of this genus. While some authors consider that the genus *Megantereon* comprises three species: *M. cultridens* Cuvier 1824 (North America, Asia, Europe) and its descendents on the one hand in the Indian subcontinent–*M. falconeri* Pomel 1853 and on the other hand in Africa–*M. whitei* Broom 1937 (Martinez-Navarro and Palmqvist, 1995, 1996). According these authors the latter species (*M. whitei*) colonizes Europe (it is present at European sites: Dmanisi, Venta-Micena and Apollonia) at the time of the Plio-Pleistocene boundary, and the spread of this species is probably related to the first arrival of *Homo* in Eurasia. The question whether the Dmanisi *Megantereon* has an African origin or not greatly depends on the sufficiency of the taxonomic importance of the observed morphological characters for the identification of new species. The main considerations for the separation and identification of the species *M. whitei* according to the above mentioned authors is on the tendency of reduction of p3 and p4 and the longer diastema between p3 and p4. It should be stressed that the reduction of p3 and p4 is the general character of *Megantereon* (Sotnikova, 1989). The judgment based on such a small set of rather variable characters is not yet convincing.

The giant ostrich–*Struthio dmanisensis (*Burchak-Abramovich and Vekua), is of special interest, it is very close to the Olduvai *Struthio olduvaiensis* in size, but giant ostrich populations are widely distributed in Eurasia in Pliocene, and they are known from the Caucasian region as well: Kvabebi–*Struthio transcaucasicus* (Vekua, 1972), Taribana (Gabunia and Vekua, 1963), Late Pliocene, and Palan-Tukan in West Azerbaijan late Pliocene (Burchak-Abramovich, 1953). Therefore it is more reasonable to consider Dmanisi ostrich within the context of late Pliocene populations of giant ostriches in the Caucasian region.

Thus, the large animals that coexisted with *Homo* in Dmanisi are either purely Eurasian taxa or taxa that have migrated out of Africa a long time before the *Homo* has reached Eurasia. Most lineages can immediately be recognized as deeply Eurasian (Cervidae, Ursidae, Mustelidae). Some confusion has arisen in perception because those who have taken a cursory look at the faunal list have noted lineages that have extant relatives in Africa (*Struthio*, Giraffidae, Hyaenidae). Other taxanomic assignations, such as within the Bovidae, have recently been revised. Finally, the fossil record of the felid taxa present at the site is so broad in space and time, that answers to questions of origins and of dispersals are premature.

The implications of the fact that *Homo* is the only large mammalian African species among Eurasian forms are very important and need to be emphasized. It shows that something vital and intrinsic to *Homo* lead to this early phase of dispersal. Of course, *Homo* had to make it out of Africa when it was *possible* to make it out of Africa. Nonetheless, *Homo* at Dmanisi without other African taxa suggests *Homo* was able to extend on their own. Even if in the future some of the species are demonstrated to have simultaneously dispersed from Africa, the Dmanisi fauna can not be considered to be a real extension of "Africa-ness."

In addition, there are some key taxa that are present in nearly every East African hominin site that are entirely absent from the Dmanisi fauna to date. These are the suids, monkeys, hippopotami, and crocodiles. These species would have interacted in significant ways with hominins. Crocodile and hippopotamus could have been a significant danger to hominins when near water, and pigs and monkeys would have been food competitors and perhaps prey. The lack of pigs and monkeys may indicate habitats where tree cover was less than continuous. Also notable is the scarcity of aquatic species at Dmanisi, which are so common in East African hominin sites. Dmanisi indicates a significant broadening of habitats occupied by *Homo* and supports hypotheses that base the biogeographic spread on intrinsic characteristics of *Homo*.

Would this Eurasian fauna require different strategies in the behavior and ecological adaptations of the hominins? We do not know surely how cold it became during winter in Georgia in the Plio-Pleistocene, but there may have been significantly more seasonality in temperatures than nearer to the equator. Like Africa, there were both carcass producing felids and bone destroying carnivores.

Would hominins have to change meat acquisition strategies in a habitat with a greater number of cervids than bovids? To some extent we know that predators adapt mainly to prey size, but still there were probably many behavioral differences in the faunas that would have changed hominin food procurement strategies, and research should focus on these issues. Studies on the ecomorphology of escape behaviors of extant and extinct cervids is currently being conducted by Sabrina Curran of the University of Minnesota will help elucidate aspects of these questions.

## General Taphonomic Observations of Dmanisi

Over the past decade of excavations in the Paleolithic levels at Dmanisi, thousands of fossil bones have been recovered, and excavations continue each summer. Taphonomic observations and analysis are preliminary. For example, many bones from the last seasons are still encased in hard matrix, and therefore some have yet to be fully identified, so characterizations and conclusions regarding the fauna are preliminary. All specimens are being examined for surface modifications under magnifications between 5-50x (usually 10-20x). Enough of the assemblage has been systematically examined (N > 2000) to make some important taphonomic characterizations of the fauna, although final frequencies of bone modifications are not yet known. This paper discusses preliminary observations on surface modifications and bone damage, while mortality profiles and skeletal element frequencies will be reported in subsequent papers.

The Dmanisi fauna in the units below the carbonate horizon (K) is very well preserved, and many bones have survived unbroken. The area below the carbonate horizon typically includes the A strata (normal) and the lower portions of the B1 stratum (reversed), although K formed in various layers depending on the location in the site. The bone surfaces are well preserved with little subareal weathering: For the site overall, Seventy-two percent of the specimens are unweathered, i.e. they are in weathering stage 0; 21% are in stage 1, and 5% are in stage 2 (stages after Behrensmeyer, 1978). The lack of weathering indicates the bones were removed from subaerial exposure rapidly after death of the animals, perhaps immediately, or within a year or two.

Detailed spatial analyses are being conducted, and all specimens are being recorded in three dimensions using a total mapping station that allows computerized three-dimensional presentation. These images will be forthcoming in future papers by the team, along with detailed parsing of taphonomic features by microstratigraphic layers. Some general patterns are evident, for example, above and within the carbonate horizon, the bone assemblage is less well preserved than below it, largely because of longer subaerial exposure (more weathering), and the precipitation of carbonates within weathering cracks in bones causing splitting. In general bones and artifacts are more evenly scattered above the carbonate layer especially in the B2 layer, while very dense clumps of bone occur in the pipe fills and gullies of B1 sediment within A2. It is in these latter areas that the hominins are mostly found, and Mode 1 artifacts are directly associated with these dense concentrations (as well as above them). Figure 3 shows one view of a dense concentration of bone including the new edentulous hominin mandible uncovered in 2003 (Lordkipanidze et al., 2005).

*Figure 3. Dense pocket of bone, including vertebrae and long bones of herbivores as well as the edentulous hominin mandible D3900.*

Most bones are not articulated, but articulated body segments are not uncommon, and in addition bones of the same body segment are very often found near one another. Examples of articulated units include long (e.g., 10 or more) and short (e.g., two) units of vertebrae (including two hominin cervical vertebrae found articulated), an entire hindlimb of a lagomorph, and partial to whole limbs of large herbivores and carnivores. These specimens were deposited while ligaments still survived, and the partially articulated carcass segments indicate that during the preburial phase they were separated from other body portions (Figure 4).

Fully one third of the bones plotted from Dmanisi are whole (unbroken), and many that are broken retain a substantial portion of the original element. There are also several hundred long bone shaft fragments, but the assemblage is not nearly as comminuted as at most archaeological sites. This pattern of many whole or minimally fractured bones contrasts with most other early hominin archaeological sites at Olduvai and East Turkana, where typically 70-95% of the fauna are non-identifiable fragments (Potts, 1988; Bunn, 1997, [although perhaps with the new emphasis on including more detailed analysis of shaft fragments would render larger portions of these assemblages identifiable, (e.g., Marean et al., 2001)]). The fragments at Dmanisi often retain identifiable landmarks, and were not further crushed by sediment compaction below the K horizon. However, recent comparisons by Villa et al. of late Pleistocene European hyena dens and Middle and Upper Paleolithic cave sites suggests increased breakage and decreased identifiability should not be considered a characteristic of archaeological sites when contrasted with hyena dens (Villa et al., 2004). The numbers reported here from Dmanisi are of plotted specimens and do not yet include the fragments of bone less than 2 cm in length recovered from screening. (The screened bones have been examined but the tallies are not yet complete.) Most often, breaks on bones at Dmanisi occurred while the bone was still fresh: 50% of the breaks have curved outlines and typically have oblique fracture edges. A further 21% are "intermediate" in character (categorization after Villa and Mahieu, 1991). Post-fossilization breaks are uncommon, except for those breaks that inevitably occurred during the excavation of the often delicate fossils within hard matrix. Dmanisi's pattern of whole bones and large fragments is partially due to the protection from sediment compaction by the carbonate horizon, but also reflects relatively little breakage during the primary accumulation.

Together the evidence of little weathering, articulated units, and relatively little breakage indicates that bones were rapidly deposited after the deaths of the animals and that there was little subsequent movement. The lack of much post-fossilization breakage and the relatively brittle nature of the fossils further confirms that there was no substantial geological displacement. The larger scale issues of fluvial transport of bones (Behrensmeyer, 1988; Behrensmeyer, 1991; Behrensmeyer and Quade, 1993; Behrensmeyer, 2002) are absent from Dmanisi, where only short distance water transport in pipes or small gullies was possible. There are only a handful of specimens from the site that have evidence of abrasion indicative of trampling or geological transport, and the bones are not in fluvial sediments. If the bones were washed together, it was from trivial distances on the scale of meters, and therefore there must have been another bone concentrating mechanism for their initial accumulation.

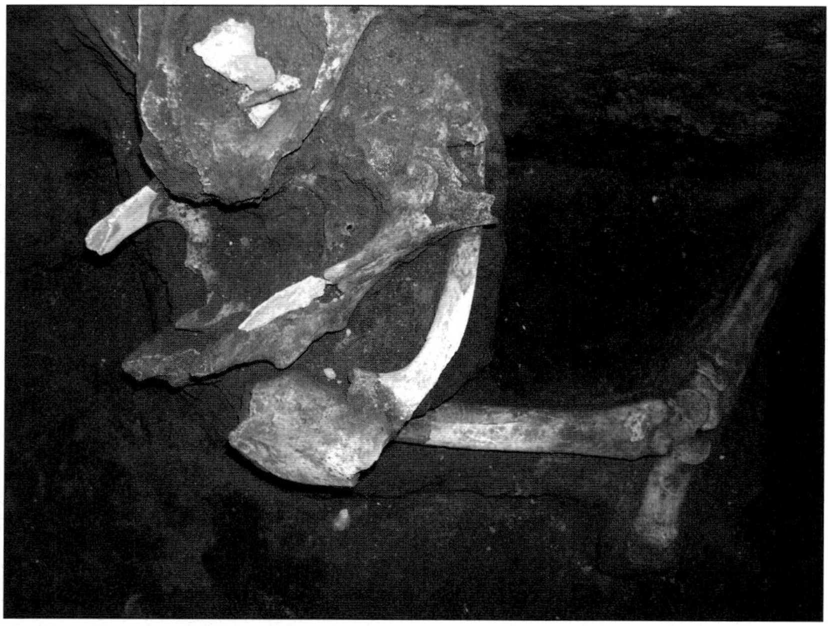

*Figure 4. Articulated hind limb of a size 2 bovid,* Gallogoral *with unfused epiphyses, an articulated pelvis can also be seen.*

Some other commonly acknowledged bone accumulating mechanisms are unlikely to have been significant at Dmanisi. Porcupines were present at the site, but less than 2% of the specimens examined so far are rodent gnawed. Rare specimens of *Hystrix* are present, but, based on the actualistic studies by Brain (1981), the levels of gnawing are relatively low, and the bones are too unweathered to suspect that a large portion of the Dmanisi fauna was accumulated by porcupines. So far there is no geological evidence for a deep fall that would trap large animals. Although the top of the A sediments into which the hydraulic pipes formed has been eroded, it seems that where most of the pipes breached the surface the pipes would have been on the order of one meter and possibly two meters deep, and so did not have the depth required to trap large animals. In this area of repeated volcanic activity, there were repeated ashfalls devastating the land suggesting that mass death could have also contributed to the site. If so, one expects many entire carcasses, and catastrophic age profiles of the fauna. A consideration of the mortality profiles will be presented in another paper by the team, but the separation of carcasses in units suggests consumption by predators, which will be the main consideration here. Thus, in the absence of evidence for transport by geological processes, or the possibility of a drowning event, or even much trampling, the separation of carcass parts into units and their aggregation was likely caused by carnivores and/or hominins.

## Hominins

With plenty of hominin fossils, stone tools, and manuports present, we know hominins were living at the site or in the immediate vicinity; but how much of the bone assemblage was accumulated by hominins? The most secure method for identifying hominin involvement is via surface modifications by stone tools such as cut marks, striae fields, and percussion marks (Potts and Shipman, 1981; Bunn, 1982; Isaac, 1983; Blumenschine and Selvaggio, 1988; Potts, 1988, 1998). For the analysis at Dmanisi, surface modifications such as scores and pits are described in terms of their cross sectional shape, presence of microscopic striations, and other morphological features; size; location on the bone (using a 20 unit system for long bones, similar to that used by (Marean and Spencer, 1991); and orientation in respect to the long axis of the bone. Interpretation of the cause of the marks (e.g., tool, tooth, trampling, etc.) is then made and recorded in a separate category from the descriptions, along with a confidence rating of the interpretation of 1 (certain) 2 (most likely) or 3 (possibly). Tappen has been very conservative with cut mark identifications because the surfaces of the bones are not hard and are easily scratched during excavation and preparation. Often preparation or excavation marks are easily distinguished from ancient marks, but if marked when the bone is still damp from being in the ground, the coloration difference is often masked. When the bones are washed in water, sediment can be moved into preparation scratches and give them the appearance of being old. For these reasons great caution is taken when identifying cutmarks or assessing the antiquity of scores, and the cleaning methods of each bone are tracked to the degree possible. Most bones are cleaned by light brushing or an airscribe, and then sometimes rinsed in water. Data are collected on whether tenacious matrix, calcrete, manganese, or root marks that coat or pass through surface marks. If they do, the marks are confidently considered to be ancient. This conservative method is necessary at Dmanisi because of the texture and hardness of the fossils.

Using this procedure, there are less than ten scores and pits from Dmanisi that are interpreted as ancient tool marks, and approximately 20 that are classified as certainty level 2 tool marks. Some of these tool marks can be seen in Figure 5. Several conclusions can be drawn from this small amount of tool mark evidence from the site. First, because of the low frequency of tool marks, hominins were unlikely to have been the main bone accumulators at the site. As more of the assemblage is viewed and spatial analysis is conducted, there may nonetheless be locations within the site that can be attributed to hominins, and our team is looking at this issue because of its potential for giving us more behavioral information. Second, some of the marks represent filleting marks for removal of meat from prime meat bearing bones, such as the femur and humerus (e.g., Figure 5 a,b). These marks show that hominins had early access to the carcasses, and that these carcasses were not first consumed by large cats or hyenas, who leave little edible meat (Domínguez-Rodrigo, 1999; Domínguez-Rodrigo, 2002). Third, other marks represent pits from hammerstone blows for marrow removal.

Both carnivores and humans create notches when breaking long bone shafts to obtain marrow, and these resulting notches on shaft fragments overlap in size and shape substantially (Capaldo and Blumenschine, 1994). There is a tendency for dynamic hammerstone impact to create broader and more arcuate notches than those created by carnivore teeth, although overlap in size is substantial; and the platform angle of the negative flake scar may be more acute. At Dmanisi most notches can be explained as carnivore notches. The mean breadth to width ratio of notches on shaft fragments is 6.27 mm (N = 39), which falls between the ratio of notches studied by Capaldo and Blumenschine created by carnivores and those by hammerstone. Dmanisi has more variation in shape than their experimental carnivore and hammerstone sample (sd

A

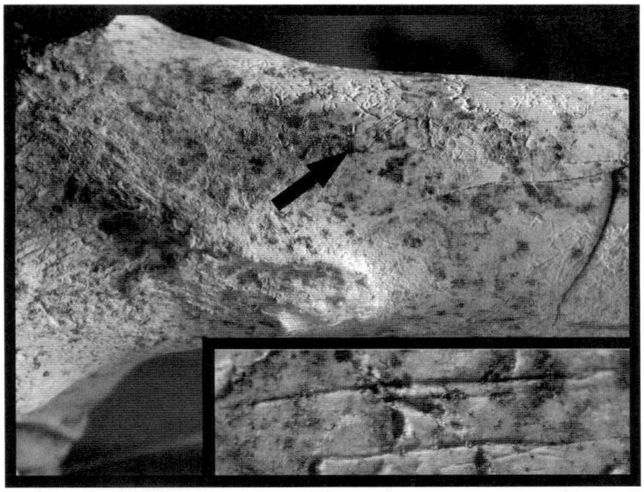

B

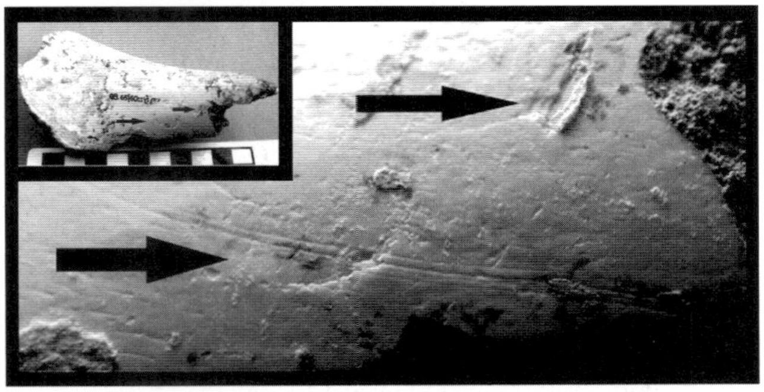

C

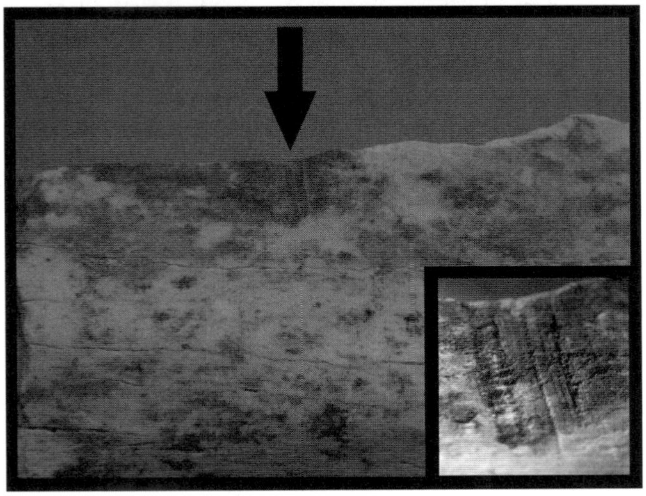

*Figure 5.*

*A. Cervidae size class 2 (classes after Brain, 1981) femur with stone tool cut marks on the distal lateral shaft. The location of the marks signifies meat removal rather than marrow processing. Inset: Close up of marks. Root marks and manganese cross the marks, indicating the tool marks are ancient. A barb, characteristic of tool marks, can be seen at the end of the lower mark.*

*B. Long cut mark with internal striations on a proximal humerus midshaft fragment (includes distal portion of deltoid tuberosity). This mark was found under a calcrete layer that coated the bone. Note adjacent chop mark or tooth mark. Inset: View of entire fragment.*

*C. Percussion striations on the edge of a long bone shaft fragment, for removal of marrow. Inset: close-up of striations, occupying an area of 7 by 3.5 mm.*

= 9.77), as expected if more than one process broke the bones. Fracture Angles are still to be measured. There are a few striae fields and pits with associated striations that are indicative of hammerstone damage at the site (Figure 5c). A medium sized humerus shaft fragment is also illustrated with cut marks indicative of filleting meat, and also has a deep impact mark near the break that may indicate marrow removal by hominins, although striae are not present, and so it could also represent a pit caused by carnivore teeth (Figure 5b). This latter interpretation would suggest hominins filleted the meat before carnivores broke the bone.

## Carnivores

Carnivores could have used and expanded the piping features for denning (Tappen et al., 2002) and the presence of large carnivore coprolites indicates that superpredators were living directly at the site. Forty-five coprolites have been preserved; most are round and/or round with a pit—the morphology of hyena coprolites. Others are more elongated and may be coprolites of felids or canids. Evidently, conditions were right for coprolite preservation, but there are few compared to many den sites, such as San Teodoro Cave in Sicily where 68% of the nearly 6,000 plotted specimens are coprolites! (Marra, Villa et al., 2004), or Bois Roche, France (Villa and Bartram, 1996; Bartram and Villa, 1998; Villa, Castel et al., 2004); apparently they have documented latrine areas at these sites. Bones with clear signs of digestion by hyenas or other larger carnivores, which would include bones with thinned edges, pinholes, polish, and severe irregular erosion are rare at Dmanisi; only three positive identifications have yet been made of digested bone. This number could increase when the screened bone has been tallied, but preliminary examination does not suggest that a high percentage of screened bone was digested. (This contrasts sharply with percentages of digested bones in screen fractions at Bois Roche 87% and at San Teodoro 27% (Marra et al., 2004). Digested bones were also abundant at a modern den studied at Masai Mara and Arad, Israel (Peterhans, 1990).

Linear drag marks and pits caused by teeth on bone surfaces are the best evidence for carnivore activities. Indeed such marks are preserved at Dmanisi, but less frequently than in modern dens. Tooth scores and pits with a certainty level of 1 are on 6.1% of the specimens analyzed, and adding those classified at the second level of certainty, brings the total number of carnivore marked specimens to 7.56%. A number of authors have remarked that ancient fossil dens have lower frequencies of gnawing than modern ones, often gnawing is found only on one or two percent of the bones (Cruz-Uribe, 1991; Pickering, 2002). The lack of congruence between actualistic and ancient assemblages suggests that multiple agents were involved at many of the ancient sites. Another explanation for this phenomenon noted by several authors is that less well preserved surfaces and very broken bones at archaeological sites relative to actualistic assemblages decreases the percentage of bones with observable gnawing (Milo, 1998). Dmanisi's bone surfaces are mostly well enough preserved to exhibit surface marks such as gnaw and cut marks. Still, there are portions of bone surfaces *not* visible to the analyst, caused by: 1) tenacious matrix, usually cemented on the surface by carbonates, 2) some exfoliation (flaking of the bone surface), or 3) post-depositional dendritic "root marks" (which are in fact likely saprophytic fungal rhizomes, N.C. Tappen, personal communication). The percentage of the bone surface that was visible and preserved well enough to exhibit cut marks and gnaw marks for each specimen was recorded using a quartile system (these estimations were made for periosteal and articular surfaces, not endosteal surfaces, which are not expected to have as many marks). On bones with less than 50% of their surfaces intact (36.2% of the specimens) 4.55% had carnivore gnaw marks, and 0.72% have tool marks; while on bones with 50% or more of their surfaces exposed and well preserved (63.8% of

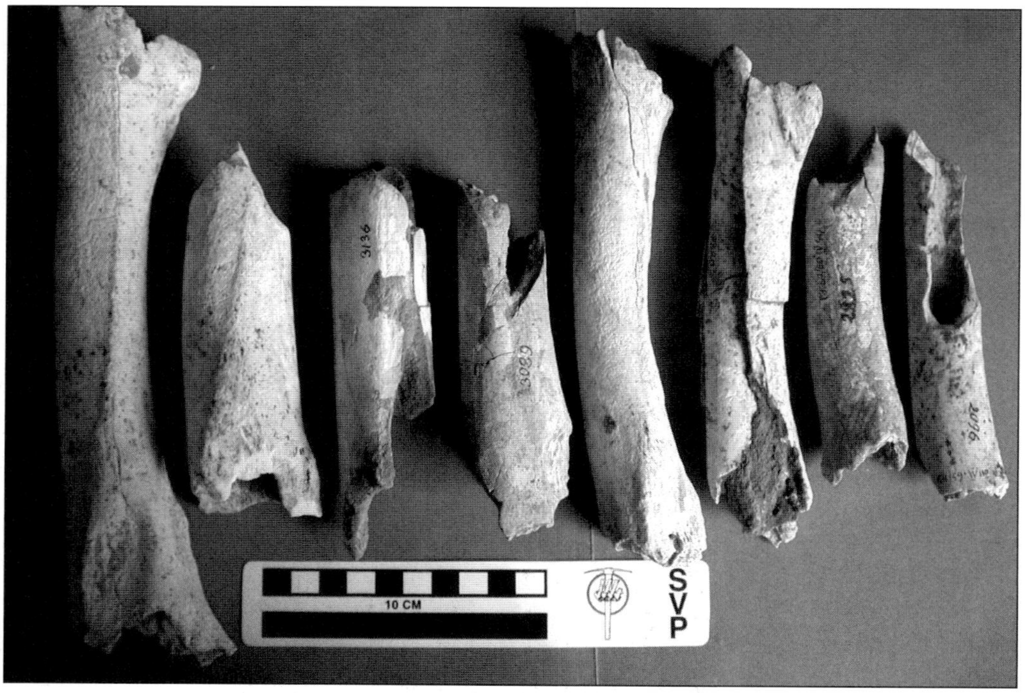


*Figure 6. Femur long bone cylinders from Dmanisi.*

specimens) 9.26% had carnivore tooth marks and 3.5% have tool marks. Therefore, it is likely that some marks made by carnivores and humans are unobserved, and so reported frequencies should be considered minima.

There are also bones with crenulated break edges and characteristic carnivore breakage, but without confirming surface scores or pits. When these are added to the tooth marked specimens, the number of carnivore-modified specimens increases by another to 2.7%. Hyenas and other carnivores tend to gnaw soft ends of bones, and if they do not break through the shafts, the result is a long bone cylinder (e.g., Bunn, 1983; Potts, 1988), whereas humans tend to break long bones more often midshaft. Hyenas can break bones midshaft too, especially size class 2 and even 3 mammals. It is probably more rare for hominids to break off the ends without also breaking the bone in middle, and so bone cylinders are considered indicators of carnivore damage. At Dmanisi, femurs occur as cylinders more than of the other long bone elements examined so far. Of 56 femurs, 13 (23%) are long bone cylinders broken by carnivores, most likely hyenas (Figure 6).

In sum, carnivore damage is present on about 6–10% of non-tooth NISP, depending on how conservative one prefers to be. Dens, on the other hand, usually have at least 30% carnivore damage and often twice this amount or more (Bunn, 1983; Pickering, 2002; Marra et al., 2004). Therefore, although carnivore gnaw marks are more common than tool marks, they are not present in the high frequencies of modern dens (Figure 7).

Our knowledge of carnivore gnawing frequencies in dens is mostly from modern hyena dens. At Dmanisi we have more remains of felids than of hyenas. Although felids sometimes den, they do so less often than hyenas. It has been found that while felids break bones less than hyenas, there still tends to be abundant tooth marks (Martin and Borrero, 1997). At a *Homotherium* den in

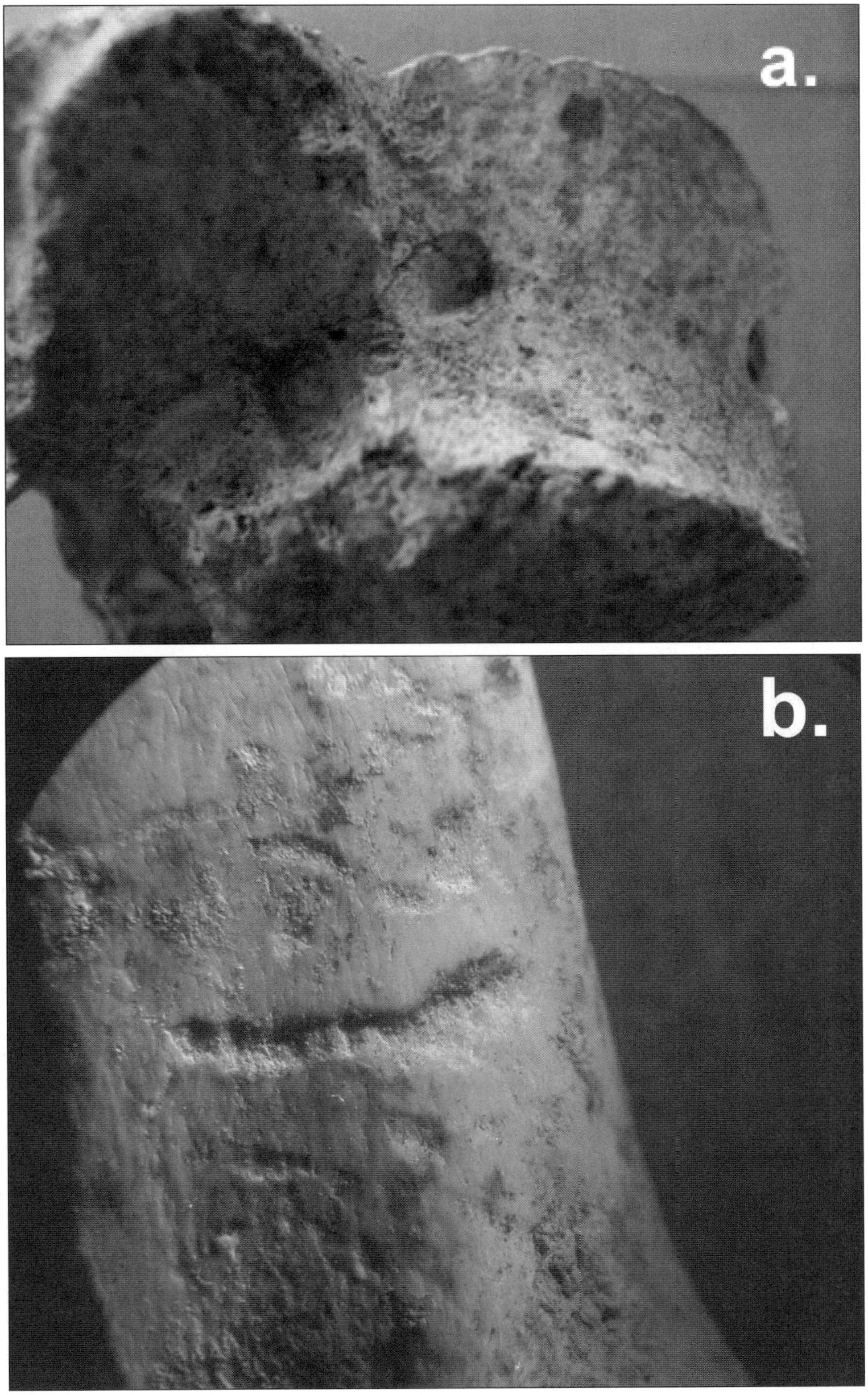


*Figure 7. Carnivore tooth marks. a. Tooth punctures on a size class 2 thoracic vertebra b. Drag mark and tooth pits on the anterior portion of a cervid mandible.*

Texas 21% of bones had tooth marks (Marean and Ehrhardt, 1995). *Homotherium's* very large size suggests that it had prey preferences for larger animals (e.g., for baby pachyderms, as seen in the Texas cave). It may have been involved in some of the larger herbivore deaths and bone deposition at Dmanisi, (Tappen et al., 2002), but the majority of the fauna are from smaller size class 2 or 3 herbivores, and so *Homotherium* is less likely to have been the major agent at Dmanisi than hyenas and the other cats, such as *Megantereon* and cf. *Panthera gombazsogensis*. In a puma lair in South America 47% of isolated bones have carnivore damage, but articulated segments do not have as much marking, (Martin and Borrero, 1997) which would lower this percentage considerably. Thus, although carnivore denning is one of the probable causes of accumulation at Dmanisi, carnivore damage occurs at lower frequencies than expected at dens, so denning is likely only part of the explanation for the bone accumulation.

## Predation Hot Spots

In addition to denning, carnivores can create bone concentrations at what have been variously called "predator arenas," "predation patches," or "serial predation spots" (Behrensmeyer, 1983; Haynes, 1988; Tappen, 1995). These predation "hot spots" occur when there is an area on the landscape where predation is facilitated and therefore repeated frequently enough to accumulate bones in high densities. Predation hot spots are thought to require a special landscape feature that improves hunting success, because all studies conducted so far show that modern attritional deaths of medium and large mammals do not normally result in spatial concentrations of bone that mimic the high densities at large archaeological and paleontological sites. These lower spatial densities were found in Amboseli Park, Kenya (Behrensmeyer, 1975; Behrensmeyer et al., 1979; Behrensmeyer and Dechant-Boaz, 1980; Behrensmeyer, 1981, 1983, 1991), Ngorongoro Crater and Serengeti National Park, Tanzania (Blumenschine, 1989), Parc National des Virunga, in the Congo (Tappen, 1995; Tappen, 2001). Predation hot spot bone accumulations therefore require a change in elevation or some other geomorphological feature; a change in vegetation physiogamy such as a windy hilltop, a treeline, the bottom of a cliff, a drinking spot, or perhaps a natural cul de sac. Over a period of some years, repeated carnivore kills result in the bones of many animals concentrating in a restricted area. How common such predation hot spots are, how concentrated bones can become in such sites, and how often they have resulted in paleontological sites is still not well understood. For an accumulation to occur, predation rates, removal from sunshine and burial rates must be higher than subaerial weathering and the alternate wetting and drying that are so damaging to bone preservation (Behrensmeyer, 1983).

The Dmanisi site is located on a promontory on top of thick basalt that has been cut on two sides by the Mashavera and Pinezauri Rivers. The lava of the basal basalt flowed down the Mashavera River during the Olduvai normal event and when it reached the confluence with the Pinezauri River, Ferring has documented that it blocked the river and created a lake. Thus, the promontory has always been an isthmus projecting out between either two rivers or between a river and a lake. Large hills further surround the locale. The area of the promontory today is about 13,000 sq m in area, and because the rivers have incised downward (rather than moving much laterally), in prehistoric times the size of the isthmus would have been similar. So a special landscape feature, as required by a serial predation patch, did indeed exist at Dmanisi. In one scenario, prey species could have wandered into this cul-de-sac, perhaps attracted to the smell of water, and found themselves ambushed with nowhere to escape on three sides. Another possibility is predators could have actively driven animals out on the promontory where they were relatively easy to catch. Hominins conceivably could have also used such a strategy, been the victims of such a strategy, or both.

The taphonomic signatures of predation hot spots are not well documented. As carnivore kill and consumption sites, but not dens, we can make predictions of their characteristics. At both dens and hot spots bones are accumulated by carnivores, and so many of their characteristics would be similar, and many of their differences in character would be related to being open air versus covered, analogous to comparisons of human open air versus cave sites. Table 2 considers some of these factors.

Landscape attritional death assemblages such as in the studies cited above include the bones of vertebrates that died from many processes, including disease and predation. However, such assemblages are comprised largely of the remains of mammals consumed by predators, and in many ways the taphonomic features of bones at predation hot spots will be simply concentrations of bones with the signature of attritional death bone deposition across landscapes (Behrensmeyer, 1983). Furthermore, since landscape bone deposition studies have yielded large samples of bones, they more likely incorporate the variation possible for predation hot spots than do the very small samples we have from such hot spots. Carnivore gnawing frequencies, bone cylinder frequencies, and weathering rates of an attritional death bone deposition study by Tappen conducted in Central Africa are compared here to these taphonomic features at Dmanisi. The actualistic landscape study was in Park National des Virunga (PNV) along the Semliki River at the base of the Western Rift Valley of the Congo to Dmanisi (Tappen, 1995, 2001). PNV is a semi-arid savanna, with non- migratory ungulate populations: principally reduncines especially the kob (*Kobus kob*), waterbuck (*Kobus ellipsiprymus*) and reedbuck (*Redunca redunca*). Buffalo (*Synceros caffer*), bushbuck (*Tragelaphus scriptus*), sitatunga (*Tragelaphus spekii* ) warthog (*Phacochoerus*

*Table 2. Some predicted characteristics of carnivore dens compared with predation hot spots*

| Characteristic | Den | Predation Hotspot |
|---|---|---|
| Landscape feature [1] | Hole or cave | Various possibilities |
| Spatial Distribution | Concentrated by edges of feature | More diffuse |
| Bone transport [2] | Bones transported further | Bones less transported |
| Carnivore surface modification[3] | More gnawing?<br>More juvenile gnawing | Less gnawing ?<br>Less juvenile gnawing |
| Numbers of Juvenile carnivores[4] | More juvenile carnivores | Fewer juvenile carnivores |
| Weathering Stages[5] | Little weathering | Variation in weathering |
| Coprolites[6] | Coprolites better preserved | Coprolites present but fewer |
| Trampling | By carnivores only | Also by large ungulates |

1. *Discussed in the text.*
2. *Since by definition predation hot spots represent a place of repeated carnivore kills, there would be less transport from the kill than in dens, and hot spots would have the signature of kill sites multiplied several times. Behrensmeyer predicted that relative to dens at hotspots there would be more vertebrae and fewer skulls, which tend to get transported (1983). This would vary with the size of the prey and the number of carnivores competing (Tappen, 1995).*
3. *While both dens and predation hot spots are carnivore consumption areas, there may be lower frequencies of carnivore gnawing in predation hot spots than in dens. This would be predicted based on less time spent at the site and therefore less chewing on bones picked up again and again. Continuing investigation by Behrensmeyer (this volume) of bone deposition and modification at Amboseli Park over 25 years indicates that the amount of carnivore competition, and the relative number of hyenas, largely dictates the degree of bone destruction by carnivores. Gnawing by juveniles would be higher in maternal dens. This would be evident in more narrow tooth scores and smaller, less deep tooth punctures, evidence of mouthing, and perhaps more gnawing on soft cancellous bones with fewer breaks in shafts, and possibly more bone cylinders relative to long bone shaft fragments if the juvenile means they were unable to break bones.*
4. *Maternal dens would have more juvenile carnivores present (e.g., Stiner, 1994).*
5. *A time-averaged open air situation, accumulated over several years, predicts significant weathering, and a large variation in weathering stages at hot spots. Dens, on the other hand, would contain bones immediately removed from ultraviolet light, and thus would have fewer signs of subaerial weathering.*
6. *Coprolites in the open would be destroyed more quickly than in the closed situation of a den.*

*aethiopicus*) and hippopotamus (*Hippopotamus amphibius*) are also abundant. Lions (*Panthera leo*), spotted hyena (*Crocuta crocuta*) and leopard (*Panthera pardus*) are the large predators common in the park. As an extant African habitat, there are no mammalian species in common with Dmanisi, and the bone assemblages are only comparable to the extent to which the taphonomic characteristics compared are shaped simply by a landscape with super predators and ungulates of a variety of body sizes.

Fifteen percent of NISP display clear traces of carnivore gnawing in the PNV attritional death landscape assemblage. This percentage of NISP displaying tooth marks should be considered a minimum at PNV,[1] whereas actualistic studies of hyena dens suggest usually between 38–100% of specimens have tooth marking (summarized in Pickering, 2002). Using these attritional death bone assemblage data as a proxy for predation hot spots, shows that they have fewer tooth marks than do dens, yet still more than that of Dmanisi.

At PNV, 10% (N=13) of the MNE of 133 femurs were "long bone cylinders"—which is a lower frequency than observed at Dmanisi. This could be because the relative size of the long bones to the power of the carnivores' jaws are greater in PNV, or perhaps trampling reduced long bone cylinders to shaft fragments very easily in the PNV open air situation. We should be able to resolve this issue as the investigation of Dmanisi proceeds.

Another factor that could discriminate between hot spots and dens is variation in weathering stages, because in dens most bones are removed from the most severe weathering agent, the sun, while open-air landscape attritional death assemblages typically would have longer exposure. The frequencies of weathering stages of bones at PNV were: Stage 0–19%, Stage 1–25%, Stage 2–18%, Stage 3–22%, Stage 4–11%, Stage 5–5 %. The bones are fairly evenly distributed throughout the stages, until Stage 4 and 5 where there is a decrease. This pattern is congruent with the concept of bones being continuously added to the landscape, with increased chance of burial or obliteration by trampling the longer they are exposed. It contrasts sharply with the Dmanisi pattern presented above, where it was shown that weathering is uncommon and varies little at the site, and so time averaging of attritional deaths across an open air landscape seems unlikely at Dmanisi. However, attritional deaths by carnivores are a possibility if the carcasses fell into the pipes immediately, perhaps in a "Transvaal-Cave" type of scenario

(Brain, 1981; Brain, 1993). In this scenerio Dmanisi was an habitual eating site of large carnivores, and preserved bones fell into the hydraulic pipes and/or were quickly buried in the low spots.

## Discussion

Like with so many Stone Age sites, reconstructing the taphonomic puzzle of Dmanisi site formation has been inspired and informed by Bob Brain's taphonomic and actualistic research in South Africa. While there are still many analyses to be done and questions to be answered about Dmanisi site formation, initial observations reported here are indicative of some of the processes involved and suggest further lines of inquiry.

The large species present in the Dmanisi fauna are Eurasian, and most if not all did not leave Africa 1.8 myr at the time the evidence for the expansion of *Homo* at Dmanisi indicates. There is the possibility that some species, such as *Megantereon,* may have dispersed from Africa at this time, but the evidence now is ambiguous. There is, of course, also the possibility that some African species will be identified at Dmanisi, and the early Pleistocene is a well-known period of drying in Europe and Africa, so climate change is likely to have been important. But clearly the Dmanisi hominins are in a very different kind faunal community than they are at the Early Stone Age sites from Africa. From the perspective of the immediate region around Dmanisi, *Homo* significantly expanded its range of ecozones, and the spread of African-like fauna is not the main correlate or determinant. Adaptations of *Homo* must have significantly changed, and we should consider that there may be too much stress on the "fellow travelers" concept (Turner, 1984) and not enough on the agency of *Homo* in deliberations on this topic.

Geological evidence shows the Dmanisi fauna was buried within a series of open air ashfalls as well as portions deposited in hydraulic pipes and into gullies that formed over collapsed pipes. Bone modifications indicate that hominins contributed to but were not the main accumulators of bones at the site, and there is considerable evidence that carnivores were important taphonomic agents at the site. Nonetheless, the presence of some cut marks, abundant Mode 1 tools and manuports, speaks to hominins' presence directly at the site. Furthermore, the location of defleshing marks on the shafts of a humerus and femur add to the growing body of data that hominins at this time period had early access to meat, and were not only passive scavengers of abandoned carnivore kills.

Comparisons of weathering to attritional landscape bone assemblages indicate that the site was probably not formed merely as an attritional death hot spot. Still, the geomorphological position of the site on an isthmus between bodies of water likely contributed to the attraction of the area for hominins, predators, and prey. The frequency of carnivore marks and toolmarks co-occurring on bone may be the best indicator of the level of interdependence on the same carcasses by hyenas and hominins (Egeland et al., 2004), and may through our further analyses, indicate the amount of direct interaction, resource competition, close encounters, etc. that occurred at the site between carnivores and hominins. Our continued work on the taphonomy of the site, of the hominins themselves, and parsing of microstratigraphic units, spatial analyses, and skeletal element frequency and mortality profiles will surely inform us of these important issues as the excavations and analyses progress.

## Notes

[1] The PNV gnawing counts include only bones with "conspicuous tooth marks" (Blumenschine et al., 1996), as identifications were made in the field without the aid of a microscope. The percent gnawed does not include rib fragments, the vast majority of which had ragged breaks which were attributed largely to carnivore chewing. Nor does it include long bone shaft fragments, or "bone flakes." These shaft fragments were often found as small piles with no epiphyseal ends, and were found distributed across the landscape, sometimes by themselves and sometimes within larger bone patches. At the time of the survey they were simply tallied, and noted if some of the group had evidence for gnaw marks. There were 298 of these bunches of shaft fragments.

## Acknowledgements

Tappen would like to thank Travis Pickering, Nick Toth and Kathy Schick for the invitation to this remarkable meeting of taphonomists, in honor of Bob Brain who has inspired all of us. Thanks also to Sabrina Curran and Claire Kirchoff for helpful comments on the text as well as assistance in the field research. Taphonomic research at Dmanisi has been funded by the Leakey Foundation, National Science Foundation, the Fulbright Foundation, and the University of Minnesota. Research at Parc National des Virunga, the Congo, was also funded by National Science Foundation, Leakey Foundation, Harvard University, Sigma Xi and The Holt Family Charitable Trust.

## References

Aguirre, E., Vangengeim, E.A., Morales, J., Sotnikova, M.V., Zazhigin, V.S. 1997. Plio-Pleistocene mammal faunas: An overview. In: Van Couvering, J.A. (Ed.), The Pleistocene Boundary and the Beginning of the Quaternary. Cambridge University Press, Cambridge, pp. 114-128.

Azzaroli, A. 1983. Quaternary mammals and the "End-Villafranchian" dispersal event: A turning point in the history of Eurasia. Paleogeography, Palaeoclimatology, Palaeoecology 44, 117-139.

Azzaroli, A. 1995. The "Elephant-Equus" and the "End-Villafranchian" Events in Eurasia. In: Vrba, E., Denton, G., Partridge. T., Burckle, L. (Eds.), Paleoclimate and Evolution with Emphasis on Human Origin. Yale University Press, New Haven, pp. 311-318.

Bartram, L., Villa, P. 1998. The archaeological excavation of prehistoric hyena dens: Why bother? Economie Prehistorique: Les comportements de subsistance au Paleolithique, XIIIe Recontres Internationales d'Archeologie det d'Histoire d'Antibes, Editions APDCA.

Behrensmeyer, A. K. 1975. The bones of Amboseli Park as a key to East African paleoecology, National Geographic Society Research Reports.

Behrensmeyer, A. K. 1978. Taphonomic and ecologic information from bone weathering. Paleobiology 2, 150-162.

Behrensmeyer, A. K. 1981. Vertebrate paleoecology in a recent East African ecosystem. In: Gray, J., Boucot, A., Berry, J. (Eds.), Communities of the Past. Hutchinson Pross, Stroudsburg (PA), pp. 591 - 615.

Behrensmeyer, A. K. 1983. Patterns of natural bone distribution on recent land surfaces: implications for archaeological site formation. In: Clutton-Brock, J., Grigson, C. (Eds.), Animals and Archaeology, Vol. 1: Hunters and Their Prey. British Archaeological Reports International Series 163, 93-106.

Behrensmeyer, A. K. 1988. Vertebrate preservation in fluvial channels. Paleogeography, Paleoclimatology, and Paleoecology. 63, 183-199.

Behrensmeyer, A. K. 1991. Terrestrial vertebrate accumulations. In: Allison, P.A., Briggs, D.E. (Eds.) Taphonomy: Releasing the Data Locked in the Fossil Record. Plenum Press, New York, 291-335.

Behrensmeyer, A. K. 2002. Bones through time: The importance of biotic versus abiotic taphonomic processes in the vertebrate fossil record. Current Topics on Taphonomy and Fossilization. Ajuntament de Valencia, Valencia, pp. 297-304.

Behrensmeyer, A. K., Dechant-Boaz, D. 1980. The recent bones of Amboseli National Park, Kenya, in relation to East African paleoecology. In: Behrensmeyer, A.K., Hill, A. (Eds.), Fossils in the Making. University of Chicago Press, Chicago, pp. 72-93.

Behrensmeyer, A. K., Quade, J. 1993. Floodplains and paleosols of Pakistan Neogene and Wyoming Paleogene deposits: Implications for the taphonomy and paleoecology of faunas.

Behrensmeyer, A.K., Western, D., Dechant-Boaz, D. 1979. New perspectives in vertebrate paleoecology from a recent bone assemblage. Paleobiology 5, 12-21.

Belmaker, M. 2002. First evidence of the presence of *Theropithecus* sp. in the southern Levant. Israel Journal Zoology 48, 165.

Berta, A., Galiano, H. 1983. Megantereon hesperus from the late Hemphillian of Florida with remarks on the phylogenetic relationships of machairodonts (Mammalia, Felidae, Machairodontinae). Journal of Paleontology 7, 892-899.

Blumenschine, R. J. 1989. A landscape taphonomic model of the scale of prehistoric scavenging opportunities. Journal of Human Evolution 18, 345-371.

Blumenschine, R. J., Marean, C.W., Capaldo, S.D. 1996. Blind tests of inter-analyst correspondence and accuracy in the identification of cut marks, percussion marks, and carnivore tooth marks on bone surfaces. Journal of Archaeological Science 23, 493-507.

Blumenschine, R. J., Selvaggio, M.M. 1988. Percussion marks on bone surfaces as a new diagnostic of hominid behavior. Nature 333, 763 - 765.

Brain, C. K. 1981. The Hunters or the Hunted? An Introduction to African Cave Taphonomy. University of Chicago, Chicago.

Brain, C. K. (Ed.) 1993. Swartkrans: A Cave's Chronicle of Early Man. Transvaal Museum, Pretoria.

Bukshianidze, M. 2005. The fossil Bovidae of Dmanisi. Ph.D Dissertation, Ferrara University.

Bunn, H. T. 1982. Archaeological evidence for meat-eating by Plio-Pleistocene hominids from Koobi Fora and Olduvai Gorge. Nature 291, 574-577.

Bunn, H. T. 1983. Comparative analysis of modern bone assemblages from a San hunter-gatherer camp in the Kalahari Desert, Botswana, and from a spotted hyena den near Nairobi, Kenya, In: Clutton-Brock, J., Grigson, C. (Eds.), Animals and Archaeology, Vol. 1: Hunters and their Prey. British Archaeological Reports International Series 163, 143-148.

Bunn, H.T. 1997. The bone assemblages from the excavated sites. In Isaac, B., Isaac, G.Ll. (Eds.), Koobi Fora Research Project, Vol. 5: Plio-Pleistocene Archaeology. Oxford University Press, Oxford, pp. 402-444.

Burchak-Abramovich, N.I. 1953. Iskopayemye strausi Kavkaza I yuga Ukrainy. Trudy Yestestvenno-Istorichychyeskogo Museyya im. Zardaby 7, 1-124. (In Russian).

Burchak-Abramovich, N.I., Vekua, A.K. 1994. Novyy pryedstavityel' plyeystotsyenobykh bykov iz vostochnoy Gruzii. Voprosy teriologii, Rossiyskaya Akademiya Nauk, Tyeriologicyeskoye obshchyestvo, "Nauka" pp. 253-261. Moskva 1994. (In Russian)

Capaldo, S. D., Blumenschine, R.J. 1994. Quantitative diagnosis of notches made by hammerstone percussion and carnivore gnawing on bovid long bones. American Antiquity 59, 724-748.

Cruz-Uribe, K. 1991. Distinguishing hyena from hominid bone accumulations. Journal of Field Archaeology 18, 467-486.

Dominguez-Rodrigo, M. 1999. Flesh availability and bone modification in carcasses consumed by lions: Palaeoecological relevance in hominid foraging patterns. Palaeogeography, Palaeoclimatology, Palaeoecology 149, 373-388.

Dominguez-Rodrigo, M. 2002. Hunting and scavenging in early humans: The state of the debate. Journal of World Archaeology 13, 1-54.

Egeland, C.P., Pickering, T.R., Dominguez-Rodrigo, M., Brain, C.K. 2004. Disentangling Early Stone Age palimpsests: Determining the functional independence of hominid and carnivore derived portions of archaeofaunas. Journal of Human Evolution 47, 343-358.

Eisenmann, V. 2004. *Equus*: An evolution without lineages? 18th International Senckenberg Conference: Late Neogene and quaternary biodiversity and evolution: Regional developments and interregional correlations, Weimar, Germany, Forschungsinstitut und Naturnuseum Seneckenberg.

Fortelius, M., Mazza P., Sala B., 1993. *Stephanorhinus* (Mammalia, Rhinocerotidae) of the western European Pleistocene, with a revision of *S. etruscus* (Falconer, 1868). Palaeontographia Italica 80, 63-155.

Gabunia, L.K., Vekua, A.K. 1963. Iskopayemyy slon iz Taribany. Tbilisi, 1963

Gabunia, L.K., Vekua, A.K. 1995. A Plio-Pleistocene hominid from Dmanisi, East Georgia, Caucasus. Nature 373, 509-512.

Gabunia, L. K., Vekua, A., Lordkipanidze, D., Swisher, C.C., Ferring, C.R., Justus, A., Nioradze, M., Tvalchrelidze, M., Anton, S.C., Bosinski, G., Jöris, O., de Lumley, M.A., Majsuradze, G., Mouskhelishvili, A. 2000. Earliest Pleistocene cranial remains from Dmanisi, Republic of Georgia: Taxonomy, geological setting, and age. Science 288, 1019-1025.

Godina, A.Y. 1979. Istorichyeskoye razvitiye zhiraf. Trudy Palyeontologichyeskogo Instituta, AN SSSR 177, 1-114.

Guerin, C. 1965. *Gallogoral* (nov. gen.) *meneghinii* (Rütimeyer, 1878), un rupicapriné du Villafranchien d'Europe occidentale. Lyon, Doc. Lab. Géol. Fac. Sci. 11.

Haynes, G. 1988. Mass deaths and serial predation: Comparative taphonomic studies of modern large mammal death sites. Journal of Archaeological Science 15, 219-235.

Hemmer, H. 1981. Die Evolution der Pantherkatzen. Modell sur überprüfung der Brauchbarkeit der Hennigschen Prinzipien der pgylogenetischen Systematik für wirbeltierpalaontologische Studien. Paläont. Z., 55, 109-116.

Hendey, Q.B. 1974. The late Cenozoic carnivora of the southwestern Cape Province. Annals of the South African Museum 63, 1-369.

Howell, F.C., Petter, G. 1976. Carnivora from Omo Group Formations, southern Ethiopia. In: Leakey, R.E.F. (Ed.), Earliest Man and Environments in the Lake Rudolf Basin. University of Chicago, Chicago, pp. 314-331.

Isaac, G. Ll. 1983. Bones in contention: Competing explanations for the juxtaposition of Early Pleistocene artifacts and faunal remains. In: Clutton-Brock, J., Grigson, C. (Eds.), Animals and Archaeology, Vol. 1: Hunters and their Prey. British Archaeological Reports International Series 163, 3-19.

Kahlke, R. 2003. Contemporary mammals of *Mammuthus meridionalis* during the Late Lower Pleistocene in Europe: The diverse Epivillafranchian Intermassfeld Fauna. Third International Conference on Mammoths, Abstracts.

Kostopoulos, D.S. 1996. The Plio-Pleistocene artiodactyls of Macedonia (Greece): Systematics, palaeoecology, biochronology, biostratigraphy. Ph.D. Dissertation, University of Thessaloniki.

Lindsay, E. 1997. The Plio-Pleistocene Boundary in continental sequences of North America. In: Van Couvering, J.A. (Ed.), The Pleistocene Boundary and the Beginning of the Quaternary. Cambridge University Press, Cambridge, pp. 278-289.

Lister, A., van Essen, H. 2002. The earliest mammoths in Europe. 18th International Senckenberg Conference, Late Neogene and Quaternary Biodiversity and Evolution: Regional Developments and Interregional Correlations. Weimar, Germany.

Lordkipanidze, D., Vekua, A., Ferring, C.R., Rightmire, P., Augusti, J., Kiladze, G., Mouskhelishvili, A., Nioradze, M., Ponce de Leon, M., Tappen, M., Zollikofer, C. 2005. The earliest toothless hominin skull. Nature 434, 717-718.

Maglio, V.J. 1973. Origin and evolution of the Elephantidae. Transactions of the American Philosophical Society 63, 1-149.

Marean, C. W., Y. Abe, Nilssen, P. J., Stone, E.C. 2001. Estimating the minimum number of skeletal elements (MNE) in zooarchaeology: A review and a new image-analysis GIS approach. American Antiquity 66, 333-348.

Marean, C. W., Ehrhardt, C.L. 1995. Paleoanthropological and paleoecological implications of the taphonomy of a sabertooth's den. Journal of Human Evolution 29, 515-547.

Marean, C. W., Spencer, L.M. 1991. Impact of carnivore ravaging on zooarchaeological measures of element abundance. American Antiquity 56, 645-658.

Marra, A., Villa, P., Beauval, C., Bonfiglio, L., Goldberg, P. 2004. Same predator, variable prey: Taphonomy of two Upper Pleistocene hyena dens in Sicily and SW France. Revue de Paleobiologie, Geneve 23, 787-801.

Martin, F., Borrero, L. 1997. A puma lair in Southern Patagonia: Implications for the archaeological record. Current Anthropology 38, 453-461.

Milo, R. 1998. Evidence for hominid predation at Klasies River Mouth, South Africa, and its implications for the behaviour of early modern humans. Journal of Archaeological Science 25, 99-133.

Palmqvist, P., Arribas, A., Martìnez-Navarro, B. 1999. Ecomorphological study of large canids from the Lower Pleistocene of southeastern Spain. Lethaia 32, 75-87.

Peterhans, J. K. 1990. Roles of porcupines, leopards and hyenas in ungulate carcass dispersal: Implications for paleoanthropology. Ph.D. Dissertation, University of Chicago.

Pickering, T. 2002. Reconsideration of criteria for differentiating faunal assemblages accumulated by hyenas and hominids. International Journal of Osteoarchaeology 12, 127-141.

Pilgrim, G.E., 1932: The fossil Carnivora of India. Palaeontologia India 18, 3-232.

Potts, R. 1988. Early Hominid Activities at Olduvai. Aldine de Gruyter, New York.

Potts, R., Shipman, P. 1981. Cutmarks made by stone tools on bones from Olduvai Gorge, Tanzania. Nature 291, 577-580.

Rook, L. 2004. Occurrence of Theropithecus sp. in the Late Villafranchian of southern Italy and implication for Early Pleistocene "out of Africa" dispersals. Journal of Human Evolution 47, 267-277.

Rustioni, M., Mazza, P. 1992. The genus *Ursus* in Eurasia; Dispersal events and stratigraphical significance. Rivista Italiana di Paleontologia e Stratigrafia 98, 487-494.

Sickenberg, O. 1971. Revision der Wirbeltier Fauna der Hohle Petralona (Griech. Mazedonien). Annales Géologiques des Pays Helleniques 23, 230-264.

Sotnikova, M.V. 1989: Hishchnye mlekopitayushchie pliocena – rannego pleistocena. Stratigraficheskoe znachenie, Nauka, Moskva (In Russian).

Stiner, M.C. 1994. Honor Among Theives. Princeton University, Princeton.

Tappen, M. 1995. Savanna ecology and natural bone deposition. Current Anthropology 36, 223-260.

Tappen, M. 2001. Deconstructing the Serengeti. In: Stanford, C.B., Bunn, H.T. (Eds.), Meat-Eating and Human Evolution. Oxford University Press, Oxford, pp. 13-32.

Tappen, M., Adler, D.S., Ferring, C.R., Gabunia, M., Vekua, A., Swisher, C. 2002. Akhalkalaki: An Early Pleistocene locality in the Republic of Georgia. Journal of Archaeological Science 29, 1367-1391.

Tappen, M., Ferring, R., Lordkipanidze, D., Vekua, A., Kiladze, G. 2002. Preliminary observations on the vertebrate taphonomy of the Dmanisi locality in the Republic of Georgia. In: De Renzi, M., Alonso, M.V.P., Belinchónet, M. (Eds.), Current Topics on Taphonomy and Fossilization. Ajuntament de Valencia, Valencia, pp. 161-170.

Tchernov, E. 1992a. The Afro-Arabian component in the Levantine mammalian fauna- a short biogeographical review. Israel Journal of Zoology 38, 155-192.

Tchernov, E. 1992b. Eurasian-African biotic exchanges through the Levantine Corridor during the Neogene and Quaternary. Courier Forsch-Inst. Senckenberg 153, 103-123.

Tchernov, E. 1998. The faunal sequence of the Southwest Asian Middle Paleolithic in relation to hominid dispersal events. In: Akazawa, T., Aoki, K., Bar-Yosef, O. (Eds.), Neanderthals and Modern Humans in Western Asia. Plenum Press, New York, pp. 77-90.

Tchernov, E. 1999. The earliest hominids in the southern Levant. Proceedings of the International Conference of Human Palaeontology, Orce, Spain, 1995, Orce Spain, Museo de Prehistoria y Paleontologia.

Teilhard de Chardin, P., Leroy, P. 1945. Les Félidés de Chine. Publ. Inst. Géobiol. Peking, 11, 58.

Torre, D., Ficcarelli, G., Rook, L., Sala, B. 1992. Mammal dispersal events in the Early Pleistocene of western Europe. Courier Forsch. Inst. Senckenberg 153, 51-58.

Tsoukala, E. 1991. Contribution to the study of the Pleistocene fauna of large mammals (Carnivora, Perissodactyla, Artiodactyla) from Petralona (Chalkidiki, N. Greece). Comptes Rendus de l'Academie des Sciences Paris 312, 331-336.

Turner, A. 1984. Hominids and fellow travellers: Human migration into high latitudes as part of a large mammal community. In: Foley, R. (Ed.), Hominid Evolution and Community Ecology. Academic Press, London, pp. 193-217.

Turner, A., 1987. *Megantereon cultridens* (Cuvier) (Mammalia, Felidae, Machairodontinae) from Plio-Pleistocene deposits in Africa and Eurasia, with comments on dispersal and the possibility of a New World origin. Journal of Paleontology 61,1256-1268.

Turner, A. 1990. The evolution of the guild of larger terrestrial carnivores during the Plio-Pleistocene in Africa. Geobios 23, 349-368.

Turner, A. 1992. Large carnivores and earliest European hominids: Changing determinants of resource availability during the Lower and Middle Pleistocene. Journal of Human Evolution 22, 109-126.

Turner, A., 1997. The Big Cats and Their Fossil Relatives. Columbia University Press, New York.

Turner, A. 1999. Assessing earliest human settlement of Eurasia: Late Pliocene dispersions from Africa. Antiquity 73, 563-570.

Van Couvering, J.A. (Ed.). 1997. The Pleistocene Boundary and the Beginning of the Quaternary.Cambridge University, Cambridge.

Vekua, A.K. 1991. Istoriya jivotnogo mira – pozvonoshnyye. gl. 5 pp. 308-426.In: Gruziya v antropogyenye (ed.: Ninua, Ts.) pp.608. Edition: "Sakartvelo" Tbilisi, 1991 (in Russian).

Vekua, A.K. 1995. Die Wirbeltierfauna des Villafranchium von Dmanisi und ihre biostratigraphische Bedeutung. Mainz, Jahrbuch Des Romisch-Germannischen Zentralmuseums.

Vekua, A.K., Djigauri D.G., Torozov, P.I. 1985. Novye palyeontologicheskiye nakhodki v okryestnostyakh Tsalki. Soobshch. AN GSSR. 118 (2).

Vekua, A., Lordkipanidze, D., Rightmire, P., Agusti, J., Ferring, C.R., Maisuradze, G., Mouskhelishvili, A., Nioradze, M., Ponce de Leon, M., Tappen, M., Tvalchrelidze, M., Zollikofer, C. 2002. A new skull from early *Homo* from Dmanisi, Georgia. Science 297, 85-89.

Villa, P., Bartram, L.E. 1996. Flaked bone from a hyena den. Paleo 8, 143-159.

Villa, P., Castel, J.C., Beauval, C., Bourdillat, V., Goldberg, P. 2004. Human and carnivore sites in the European Middle and Upper Paleolithic: Similarities and differences in bone modification and fragmentation. Revue de Paleobiologie, Geneve 23, 705-730.

Villa, P., Mahieu, E. 1991. Breakage patterns of human long bones. Journal of Human Evolution 21, 27-48.

Vrba, E. 1995. The fossil record of African antelopes (Mammalia, Bovidae) in relation to human evolution and paleoclimate. In: Vrba, E., Denton, G., Partridge, T., Burckle, L (Eds.), Paleoclimate and Evolution with Emphasis on Human Origins. Yale Univeristy, New Haven, pp. 385-424.

Werdelin, L., Lewis, M.E. 2000. Carnivora from the South Turkwell hominid site, northern Kenya. Journal of Paleontology 74, 1173-1180.

# CHAPTER 8

# Changes Through Time in Carcass Survival in the Amboseli Ecosystem, Southern Kenya

A.K. Behrensmeyer

## Abstract

Bob Brain pioneered using modern bone assemblages to build a body of comparative information that could be used to interpret taphonomic processes affecting faunal remains in the paleoanthropological record. His original research inspired other neo-taphonomic studies that have been used to formulate scavenging vs. hunting models for early hominin subsistence strategies. Study of the Amboseli ecosystem in southern Kenya (1975–2004) demonstrates how decade-scale changes in predator diversity and population dynamics affect the taphonomic features of bone assemblages and carcass survival. Twenty transects in 3 different habitats were surveyed in 1975 and 2002-2003, resulting in a sample of over 800 individuals (MNI) and nearly 10,000 skeletal elements (MNE). Data for each MNI include species, skeletal elements present, bone completeness and other modification features. Relatively high rates of carcass survival occurred during the 1970s and 1980s when lions, spotted hyenas and other carnivores competed for prey and the spotted hyena population was low. Herbivore die-offs during droughts also provided occasional surfeits of carcasses for scavengers. Under such conditions, hominins would have had access to carcasses either through accidental discovery or power-scavenging, as well as ample opportunities to hunt and retain control of their prey. Relative to the 1975 sample, a 75% decrease in bones per individual and higher levels of damage to surviving elements have been the norm in Amboseli from 1990-2004, especially in the prey size range from 25–250 kg. This change correlates with a population increase of *Crocuta* and low numbers of other predators. Skeletal part survival and bone modification patterns provide direct evidence for these different ecological conditions. The taphonomic changes in Amboseli show that over several decades, the survival of carcasses and bones can vary markedly within the same ecosystem and within the same habitat. Thus, the availability of scavengeable remains could have changed markedly over the lifetime of individual hominins. The results also indicate that differences in habitat–carcass associations can virtually disappear when a bone-consumer such as *Crocuta* becomes the dominant predator. In the fossil and archeological record, unbiased samples of skeletal part abundance and completeness, adult/juvenile ratios, and damage patterns in relation to prey body size could be used to indicate different levels of bone-consuming predator pressure.

## Introduction

A rich history is recorded in the patterns of preservation of animal skeletons. Vertebrate taphonomists work throughout the Phanerozoic fossil record, but the growth in understanding of the meaning of taphonomic patterns has been greatly stimulated by interest in human evolution and the study of bones and stones that our ancestors left scattered on the ancient landscapes. Bob Brain exemplifies this approach and has contributed more than anyone of his generation to the ideas and knowledge that turn evidence from bones into understanding of taphonomic processes, both human and non-human. A major strength of his research has been the actualistic observations and experiments that he conducted to understand taphonomic cause and effect.

Taphonomic studies of bones in modern ecosystems have been championed by Brain (1967, 1969, 1981) as well as a number of other researchers, starting with Weigelt in the 1920s (Weigelt, 1927) and continuing with the work of Hill (1975; 1980), Gifford (Gifford and Behrensmeyer, 1977), Yellen (1977), Behrensmeyer (Behrensmeyer et al., 1979; Behrensmeyer and Boaz 1980; Behrensmeyer, 1993), Haynes (1985, 1988), Bunn (Bunn, 1982, Bunn et al., 1988, Bunn and Ezzo, 1993), Blumenschine (1989), Dominguez-Rodrigo (2001) and Tappen (1995, 2001). Controlled experiments on taphonomic processes have also contributed to information that can be brought to bear on the past (e.g., Shipman, 1981; Marean et al., 1992, Marean, 1997). Through such actualistic research, paleontologists and paleoanthropologists have built a large body of information on processes that modify bones and leave identifiable traces, allowing us to decode some of the patterns in the fossil record and to distinguish non-human from human damage features. This research has also resulted in a huge leap in understanding of what happens to bones in the post-mortem environment and the realization that different processes can result in similar end-products. The bones themselves, their size, shape and strength, exert definitive though not exclusive control on which body parts, and which bone portions, are most likely to survive to become fossils. Taphonomic "reality checks" provided by experiments and observations in the modern world have shaped many scientific careers, including Bob Brain's, and will continue to challenge and intrigue archeologists, paleoanthropologists, and paleontologists into the future.

Although much has been learned about specific bone-modifying agents and other taphonomic processes over the past century, questions about the relationships of bone assemblages to the ecology of living animals remain unanswered. How faithfully do surface bone assemblages represent the vertebrate species richness, population abundances, and habitat structure of an ecosystem? How stable are the taphonomic features of a bone assemblage, and how is ecological change reflected in these assemblages? Such questions relate to problems of interest to paleoanthropologists, such as whether early hominins might have found some habitats more favorable than others for hunting or scavenging meat from carnivore kills (Blumenschine, 1989; Potts, 2003), and how taphonomic features of fossil bone assemblages correlate with varying levels of predator and scavenger pressure on prey populations.

This paper focuses on the specific problem of carcass completeness and survival in modern East African ecosystems. It builds upon pioneering research by Blumenschine (1989), Tappen (1995; 2001) and Dominguez-Rodrigo (2001), who studied different analogue communities to understand variables and processes that might have controlled carcass availability for early humans. The long-term taphonomic research in Amboseli documents change over several decades in carcass survival and indicates that predator impact as well as overall ecological change likely were important in controlling resources for early hominins living in similar ecosystems.

## Background: Taphonomic Research in Amboseli

Amboseli is a national park in southern Kenya that has been under continuous ecological study since the 1960's (Western, 1973, Western and von Praet, 1973; Western and Maitumo, 2004) (Figures 1 and 2). It has a rich vertebrate fauna, supported largely by springs emerging along the base of Mt. Kilimanjaro, which lies to the south on the Tanzanian side of the Kenya-Tanza-

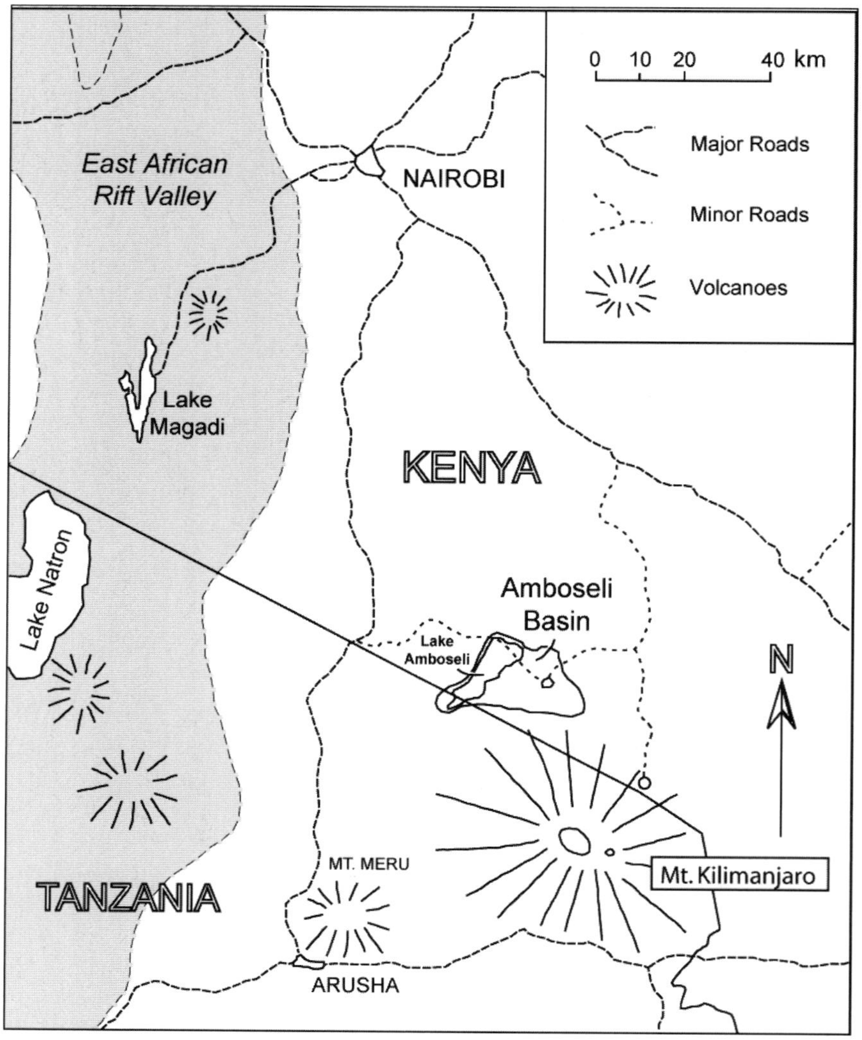


*Figure 1. Map of Kenya showing location of the Amboseli Basin in southern Kenya.*

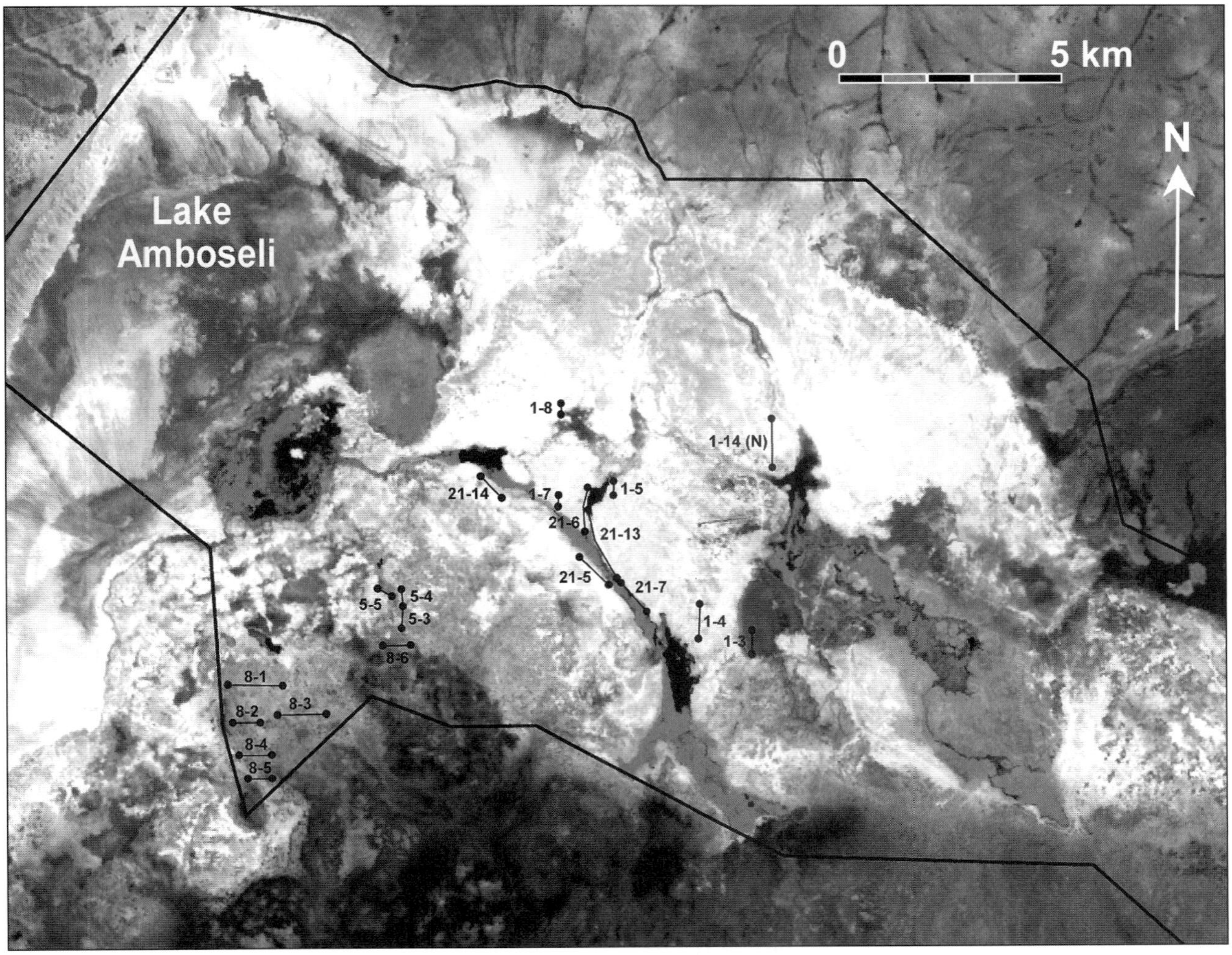


*Figure 2. Landsat (1997) image of the Amboseli Basin showing the positions of the 20 bone sampling transects analyzed in this paper. Polygon shows the outline of the Amboseli National Park boundary.*

nia border. During the past 30–40 years, Amboseli has experienced major ecological change due to the loss of *Acacia* woodlands and the expansion of grassland habitats. There also has been a substantial increase in the mean annual temperature (MAT) over the past 30 years (Altmann et al., 2002). Human impact has played a role in ecosystem change, with the exclusion of Maasai herds from the central basin, the expansion of tourism and road proliferation, followed by the return of domestic stock to portions of the central basin and increased levels of conflict between animals and wildlife in the areas surrounding Amboseli (Western, 2004, Western and Maitumo, 2004). All of these processes and circumstances have affected the vegetation and the vertebrate community in the park. The Amboseli ecosystem thus is an appropriate natural laboratory for examining how bone assemblages track ecological change and, conversely, how ecological change affects the taphonomic features of these assemblages.

The Amboseli basin covers approximately 600 km$^2$, with Amboseli National Park restricted to the central 388 km$^2$, which includes a number of spring-fed wetlands and part of seasonal Lake Amboseli (Figure 2). The climate is semi-arid, with average monthly temperatures between 26 and 34°C. There are two rainy seasons, November-December and March-May, and an average yearly total of 350–400 mm of rainfall. Humidity is generally low, and the dry seasons are both dusty and windy. The permanent springs and accompanying high water table sustain primary productivity in the central basin and support large populations of plant and animal species. Some of the ungulates are resident, but most migrate in during the dry season and disperse during the wet season when there is water and forage elsewhere. Amboseli was known for its beautiful mosaic of woodland, grassland, and bush habitats when the bone study began in 1975. Over 20 different habitat types were originally designated in the 1960s by D. Western (1973), based on characteristic flora. The woodlands are now much reduced in area, and the park is dominated by open plains and salt bush habitats.

The goals of the long-term Amboseli bone study include monitoring bone weathering and destruction rates and patterns, documenting bone frequencies and identities for comparison with live census data, and establishing transects in different habitats for repeated re-sampling over periods of years to decades to record changes in the vertebrate remains (Figure 3). These goals were designed to address the following major questions: (1) What is the relationship of the faunal composition of the

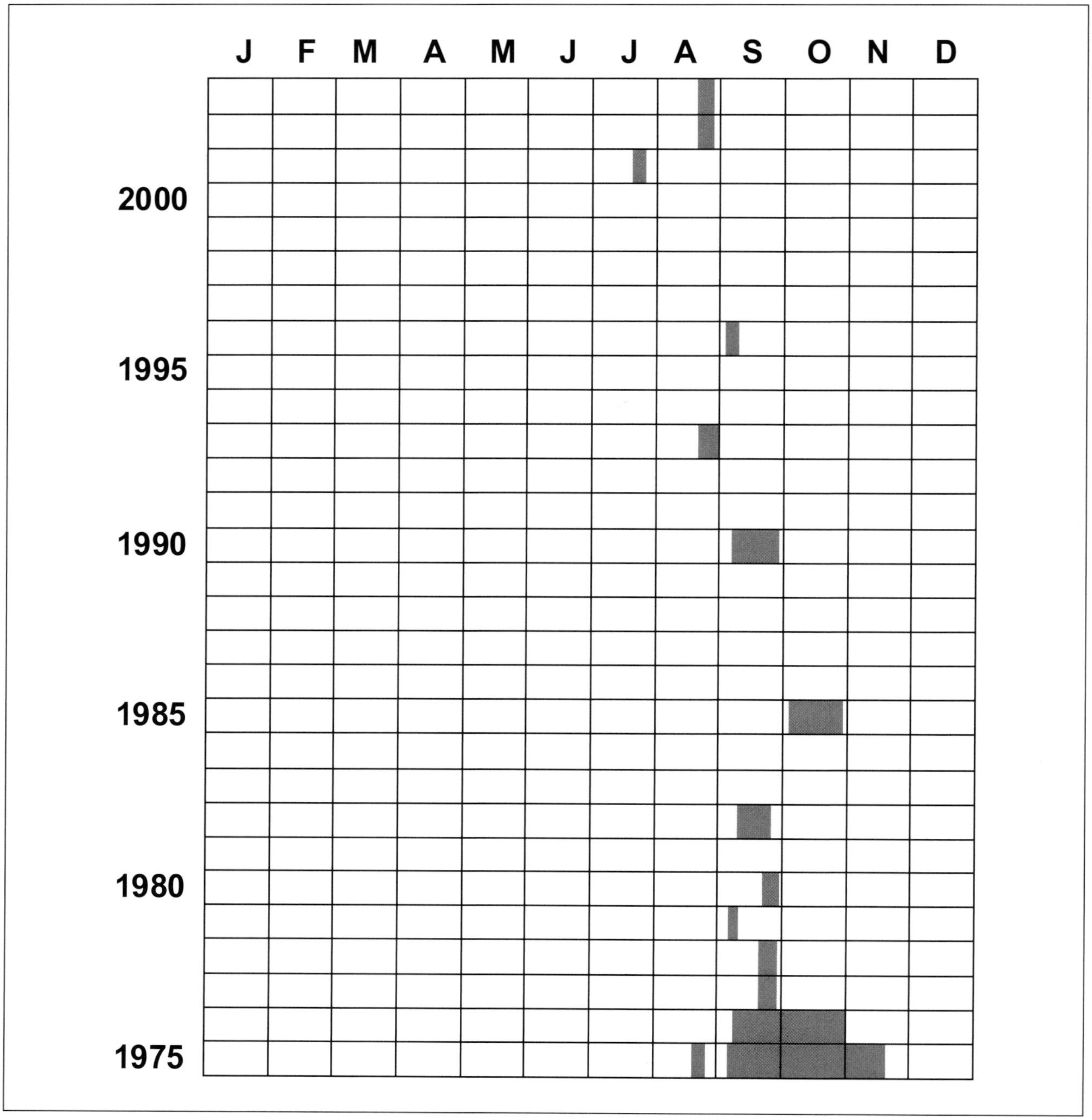


*Figure 3. Overview of timing and duration of taphonomic field research in Amboseli through 2003 (darkened bars). Month abbreviations across the top of the chart.*

surface bone assemblage to the living community from which it is derived? (2) How rapidly do bones weather under natural circumstances? (3) How do biological, physical, and chemical processes affecting vertebrate remains bias the fossil record, and do bone assemblages have a taphonomic signature that can indicate specific biases? (4) How are bones concentrated and buried in an ecosystem lacking fluvial processes? Because it has been possible to continue this study for nearly 30 years, the results can also be used to examine, (5) How does a bone assemblage track ecological change in the animal community from which it is derived? Results addressing the first four questions have been published, though the research is on-going (See Behrensmeyer, 1978; Behrensmeyer et al., 1979; Behrensmeyer and Boaz, 1980, Behrensmeyer, 1993; Tuross et al., 1987, Koch et al., 2001; Cutler et al., 1999). The results reported below particularly address the fifth question and have not been previously published.

Six major habitat types were sampled for surface bones, based on Western's vegetation map, and three of these are used in the analyses reported here: plains, woodland, and swamp. The latter two categories include several different vegetation types. "Swamp" refers to the accessible swamp margin, within 30 m of the actual water or muddy areas that could not be searched effectively. This swamp margin area still retained considerable bush and tree cover in 1975, resembling a riparian woodland in terms of the structure of the vegetation and the density of cover for the animals. This cover also limited bone visibility and increased our level of caution during the surface surveys. In 2002–03, all of this cover had disappeared on the transects we sampled, transforming the swamp margin into open, grassy habitat more similar to the plains, except for the nearby access to water. Likewise, the *Acacia xanthophloea* (yellow fever tree) woodland still had many living or partially moribund large trees in 1975, with shrubs and bushes plus a thick ground cover of grass growing in the shade of these trees. This woodland was a favored place for the Maasai and their domestic stock, which limited the number of wild animals. In 2002–03, all of the trees had disappeared, and the terrain can be described more accurately as a saltbush plain, with some low areas moist enough to sustain a healthy grass cover. Although the original habitat names have been retained for the purposes of this paper, readers should be aware that "Woodland-*A.x.*" was no longer woodland in 2002–03. The plains habitat has changed little in terms of vegetation, remaining very open with a variable cover of grasses. The *Acacia tortilis* woodland was still recognizable in 2002–03 although there were more dead or dying trees than in 1975, the understory of bushes had thinned, and dusty areas had expanded due to the loss of grass cover. Overall, however, the changes were not nearly as pronounced as in the swamp margin or fever tree woodland.

When the Amboseli bone study was initiated in 1975, the land surface of the basin (then a wildlife reserve) was littered with bones and mummified carcasses, and it seemed at the time to be an ideal place to study vertebrate taphonomy. Understanding of the decomposing and recycling components of this ecosystem based on several decades of research have led to the realization that considerable change can occur within the complex of ecological and taphonomic processes and material results that characterize Amboseli. These make up the Amboseli "taphosystem"—a set of biological, chemical and physical processes that shape the living ecosystem but are particularly important in its decomposing/recycling functions, thereby controlling what organic remains and information are available for preservation in the fossil record. Just as every ecosystem has unique features that distinguish it from other similar ecosystems, taphosystems may also have distinctive signatures that are recorded in the assemblages of organic remains.

## Methods

Field methods were originally designed to obtain statistically large samples of the surface (modern) bone assemblage in order to characterize species and skeletal parts and their relative abundances in different habitats (Behrensmeyer and Dechant Boaz, 1980). Transects that were distributed within each of six major habitats, defined originally by Western (1973) based on vegetation types and species. Air photographs and a vegetation map were used to determine areas to be sampled, and starting points were chosen based on landmarks such as trees, roads, or other features that could aid in later re-location (in 1975, of course, we had no GPS). Transects usually were oriented north-south or east-west to make it easier to keep on a straight line using a compass bearing (Figure 2). In 2002–03, GIS coordinates allow calibration of the area searched and also precisely document the positions of transects for future surveys. However, in 1975, we paced off widths and lengths of the transects. The area to be searched was pre-determined by visibility; in dense vegetation, transect width was 30 meters either side of the midline (i.e., where a vehicle was driven); in open vegetation 50 m either side. During transect sampling, two to four individuals walked the transects, covering as much of the ground as possible. One person (AKB) was responsible for recording all bone occurrences on standardized data sheets; bones that could not be identified without comparative materials or those of special taphonomic interest were collected for later checking. An occurrence was defined as one to many bones belonging to one individual animal in close spatial proximity. Body parts likely belonging to the same individual but dispersed more than 15–20 m away from each other were given separate occurrence numbers. Also, when two different animals occurred at the same place, they were given separate occurrence numbers.

Successful bone surveying requires one or more team members able to identify fragmentary skeletal remains to taxon and skeletal element. Data recorded include: taxon, age (adult, juvenile, state of tooth eruption), skeletal parts present, habitat, weathering stage, breakage and other damage features such as tooth marks and degree of burial. In Amboseli, we continued the transect until we had ~20 individuals ("MNI" = mininum number of individuals). MNI is based on the number of different individual animals that can account for the documented bones; decisions were made in the field, based on body size, species ID, growth stage (juvenile vs. adult), weathering stage, etc. The general approach in Amboseli is to assume that an unknown bone is not a separate individual unless it can be demonstrated to be—a conservative stance that worked against inflation of the MNI count. For more information on the prototype sampling methods in Amboseli Park, Kenya, see Behrensmeyer and Dechant, 1980; Behrensmeyer, 1993.

Generally, a sample of at least 100 MNI is necessary to characterize the presence and relative abundances of common species in a particular habitat (i.e., 5–6 transects), though more may be required to capture the rare species. In Amboseli, we were able to do 2–3 transects in a day, thus it was possible to obtain an adequate sample of several different habitats in a week of bone surveying. This depends, of course, on the density of bones on the ground and the time required by the team to locate and identify these bones.

For the purposes of this study, bones in all weathering stages (WS) (Behrensmeyer, 1978) were used to provide the largest possible samples of skeletal elements and species for each habitat. Most of the identifiable bone occurrences were in WS 0–3, representing the 10 years prior to the time of transect sampling, i.e., 1965 to1975 and 1992-3 to 2002–03 (Behrensmeyer, 1978). The average WS for the four habitats is between 1.7 and 2.3 for 1975 and between 1.9 and 2.3 for 2002–03. Some of the bone occurrences were WS 4-5 in the 2002–2003 surveys, but it is unlikely that a signifcant number of the same bones recorded in 1975 were identified and recorded again in 2002–03. Thus, these samples represent essentially independent records of skeletal elements and taxa at two successive time intervals, representing a maximum of ~20 years each but dominated by bones that accumulated over the 10 years prior to sampling.

The bone transect data have been entered into electronic databases using a variety of formats since 1976. Analysis in this paper focuses on 20 transects in 4 different habitats (plains, swamp, *Acacia xanthophloea* (yellow fever tree) woodland, *Acacia tortilis* woodland) that were sampled in both 1975 and 2002–03.

*Table 1. Summary of data from 1975 and 2002-03 surveys of surface bones on 20 transects in 4 different Amboseli habitats. The counts are for all bones and individuals that were identifiable to mammals, birds, reptiles, or fish, and exclude remains that could not be certainly assigned to these groups. Adult/juvenile ratio and percentage > 20 bones per occurrence were calculated for wild mammals only. MNI = Minimum Number of Individuals, MNE = Minimum Number of Elements.*

| | 1975 | 2002-03 | Decrease | |
|---|---|---|---|---|
| Occurences | 641 | 575 | 10% | |
| Individuals (MNI) | 458 | 365 | 20% | |
| Bones (MNE) | 8160 | 1700 | 79% | |
| Bones/MNI | 17.82 | 4.66 | 74% | |
| Adult/Juvenile | 1.73 | 3.04 | | |
| >20 bones/Occurrence (Wild only) | 21% | 2% | | |
| Occurrences with 1 bone | 22% | 57% | | |
| **By habitat (all mammals)** | **1975** | | | |
| | **MNE** | **MNI** | **MNE/MNI** | |
| Plains | 2307 | 116 | 19.89 | |
| Swamp | 2658 | 105 | 25.31 | |
| Woodland - A x. | 848 | 69 | 12.29 | |
| Woodland - A t. | 2347 | 168 | 13.97 | |
| | **2002-03** | | | |
| | **MNE** | **MNI** | **MNE/MNI** | **Decrease in ratio** |
| Plains | 630 | 106 | 5.94 | 70% |
| Swamp | 702 | 139 | 5.05 | 80% |
| Woodland - A x. | 75 | 31 | 2.42 | 80% |
| Woodland - A t. | 293 | 89 | 3.29 | 76% |

## Results

### Characteristics of the surface bone assemblage

There are striking changes in the bones recorded in 2002-03 compared with 1975 (Table 1). The number of occurrences and individuals has decreased, but the most notable difference is in the total number of bones (MNE = Minimum Number of Elements), which has declined by 79%. This translates into a 74% decrease in the average number of bones per individual. The decrease is MNE/MNI has occurred in all transects except T1-8 in the central plains habitat (Table 2, Figures 4 and 5), with the greatest average difference in the swamp habitat. The higher variability of the MNE/MNI ratio in 1975 also contrasts with lower variability in 2002-03; not only has the number of bones per individual decreased, but the surface bone assemblages in the sampled habitats have become much more homogeneous in this respect.

The major shift in carcass and bone survival between 1975 and 2002–03 is further underscored by the lack of complete or partial skeletons in the later sample and the dominance of isolated bone occurrences. In 1975, 21% of the recorded occurrences of wild species had more than 20 associated bones, whereas in 2002–03, this dropped to only 2% (Table 1). Moreover, in 1975, only 21% of the wild species occurrences were singleton bones, but this increased to 56% in 2002–03. These figures indicate wider dispersal as well as increased destruction of skeletal elements in the later sample. The numbers of bones that could be assigned to adult versus juvenile changed as well, with increases in the number of adults and decreases in juveniles, both for the total wild mammal sample and for the two most common species, wildebeest (*Connochaetes taurinus*) and zebra (*Equus burchelli*) (Table 1, Figure 6).

The shift in bone survival between 1975 and 2002–03 is most pronounced in the middle body sizes between 25 kg (Thompson's Gazelle) and 500 kg (Cape buffalo), with category 3 (Brain, 1981) showing the biggest change (Figure 7). There is also a marked decrease in MNE/MNI for elephant. Exceptions to the overall trend occur in body size category 5 (giraffe, rhino, hippo), which is the only one that increases in bones per individual, and category 1 (< 25 kg), which shows little change.

*Table 2. Bone survey data for 1975 and 2002-03. A. Summary of 1975 MNE (Minimum Number of Elements) and MNI (Minimum Number of Individuals) data by transect and break-down for each major vertebrate group. B. Summary of 2002-03 MNE and MNI and break-down for vertebrate groups. T1=Plains, T5=Woodland (Acacia xanthophloea), T8=Woodland (Acacia tortilis), T21=Swamp; habitat number designations based on Western (1973) and some no longer apply to the 2002-03 sample areas (see text)*

**A. 1975 summary** | **Major vertebrate groups**

| | MNE 1975 | MNI 1975 | MNE/MNI 1975 | MNE 1975 | MNE 1975 | MNE 1975 | MNE 1975 | MNE 1975 | MNI 1975 | MNI 1975 | MNI 1975 | MNI 1975 | MNI 1975 |
|---|---|---|---|---|---|---|---|---|---|---|---|---|---|
| TRANSECT | Total | Total | | Fish | Reptile | Bird | Mammal | Total | Fish | Reptile | Bird | Mammal | Total |
| T1-3 | 430 | 24 | 17.92 | · | · | · | 430 | 430 | · | · | · | 24 | 24 |
| T1-4 | 504 | 24 | 21.00 | · | · | · | 504 | 504 | · | · | · | 24 | 24 |
| T1-5 | 447 | 21 | 21.29 | · | · | · | 447 | 447 | · | · | · | 21 | 21 |
| T1-7 | 550 | 24 | 22.92 | · | · | 26 | 524 | 550 | · | · | 1 | 23 | 24 |
| T1-8 | 89 | 10 | 8.90 | · | · | · | 89 | 89 | · | · | · | 10 | 10 |
| T1-14 | 287 | 13 | 22.08 | · | 3 | · | 284 | 287 | · | 1 | · | 12 | 13 |
| T5-3 | 337 | 33 | 10.21 | · | · | 3 | 334 | 337 | · | 1 | 1 | 31 | 33 |
| T5-4 | 291 | 20 | 14.55 | · | · | 3 | 288 | 291 | · | · | 1 | 19 | 20 |
| T5-5 | 220 | 16 | 13.75 | · | · | 3 | 217 | 220 | · | 1 | 1 | 14 | 16 |
| T8-1 | 310 | 34 | 9.12 | · | · | 2 | 308 | 310 | · | · | 2 | 32 | 34 |
| T8-2 | 647 | 31 | 20.87 | · | · | · | 647 | 647 | · | · | · | 31 | 31 |
| T8-3 | 157 | 25 | 6.28 | · | · | · | 157 | 157 | · | · | · | 25 | 25 |
| T8-4 | 266 | 23 | 11.57 | · | · | · | 266 | 266 | · | · | · | 23 | 23 |
| T8-5 | 120 | 25 | 4.80 | · | · | 3 | 117 | 120 | · | · | 1 | 24 | 25 |
| T8-6 | 847 | 30 | 28.23 | · | · | · | 847 | 847 | · | · | · | 30 | 30 |
| T21-5 | 747 | 29 | 25.76 | 14 | · | 7 | 726 | 747 | 4 | · | 1 | 24 | 29 |
| T21-6 | 377 | 18 | 20.94 | · | · | · | 377 | 377 | · | · | · | 18 | 18 |
| T21-7 | 814 | 32 | 25.44 | · | · | 60 | 754 | 814 | · | · | 2 | 30 | 32 |
| T21-13 | 202 | 12 | 16.83 | · | · | · | 202 | 202 | · | · | · | 12 | 12 |
| T21-14 | 518 | 14 | 37.00 | · | · | · | 518 | 518 | · | · | · | 14 | 14 |
| | 8160 | 458 | 17.82 | 14 | 3 | 107 | 8036 | 8160 | 4 | 3 | 10 | 441 | 458 |

**B. 2002-03 Summary** | **Major vertebrate groups**

| | MNE 2002-03 | MNI 2002-03 | MNE/MNI 2002-03 | MNE 2002-03 | MNE 2002-03 | MNE 2002-03 | MNE 2002-03 | MNE 2002-03 | MNI 2002-03 | MNI 2002-03 | MNI 2002-03 | MNI 2002-03 | MNI 2002-03 |
|---|---|---|---|---|---|---|---|---|---|---|---|---|---|
| TRANSECT | Total | Total | | Fish | Reptile | Bird | Mammal | Total | Fish | Reptile | Bird | Mammal | Total |
| T1-3 | 38 | 12 | 3.17 | · | · | · | 38 | 38 | · | · | · | 12 | 12 |
| T1-4 | 147 | 26 | 5.65 | · | · | · | 147 | 147 | · | · | · | 26 | 26 |
| T1-5 | 120 | 16 | 7.50 | · | · | · | 120 | 120 | · | · | · | 16 | 16 |
| T1-7 | 34 | 8 | 4.25 | · | · | · | 34 | 34 | · | · | · | 8 | 8 |
| T1-8 | 196 | 21 | 9.33 | · | · | · | 196 | 196 | · | · | · | 21 | 21 |
| T1-14 | 95 | 23 | 4.13 | · | · | · | 95 | 95 | · | · | · | 23 | 23 |
| T5-3 | 29 | 9 | 3.22 | · | · | · | 29 | 29 | · | · | · | 9 | 9 |
| T5-4 | 13 | 7 | 1.86 | · | · | 1 | 12 | 13 | · | · | 1 | 6 | 7 |
| T5-5 | 33 | 15 | 2.20 | · | · | · | 33 | 33 | · | 2 | · | 13 | 15 |
| T8-1 | 120 | 28 | 4.29 | · | · | · | 120 | 120 | · | 2 | 0 | 26 | 28 |
| T8-2 | 3 | 2 | 1.50 | · | · | · | 3 | 3 | · | · | · | 2 | 2 |
| T8-3 | 35 | 15 | 2.33 | · | · | 5 | 30 | 35 | · | · | 3 | 12 | 15 |
| T8-4 | 57 | 19 | 3.00 | · | · | · | 57 | 57 | · | · | · | 19 | 19 |
| T8-5 | 40 | 14 | 2.86 | · | · | 3 | 37 | 40 | · | · | 2 | 12 | 14 |
| T8-6 | 38 | 11 | 3.45 | · | · | · | 38 | 38 | · | · | · | 11 | 11 |
| T21-5 | 243 | 36 | 6.75 | · | · | · | 243 | 243 | · | · | · | 36 | 36 |
| T21-6 | 126 | 25 | 5.04 | 9 | · | · | 116 | 126 | 5 | · | · | 20 | 25 |
| T21-7 | 58 | 15 | 3.87 | · | · | · | 58 | 58 | · | · | · | 15 | 15 |
| T21-13 | 138 | 38 | 3.63 | · | · | · | 138 | 138 | · | · | · | 38 | 38 |
| T21-14 | 137 | 25 | 5.48 | · | · | · | 137 | 137 | · | · | · | 25 | 25 |
| | 1700 | 365 | 4.66 | 9 | 0 | 9 | 1681 | 1700 | 5 | 4 | 6 | 350 | 365 |

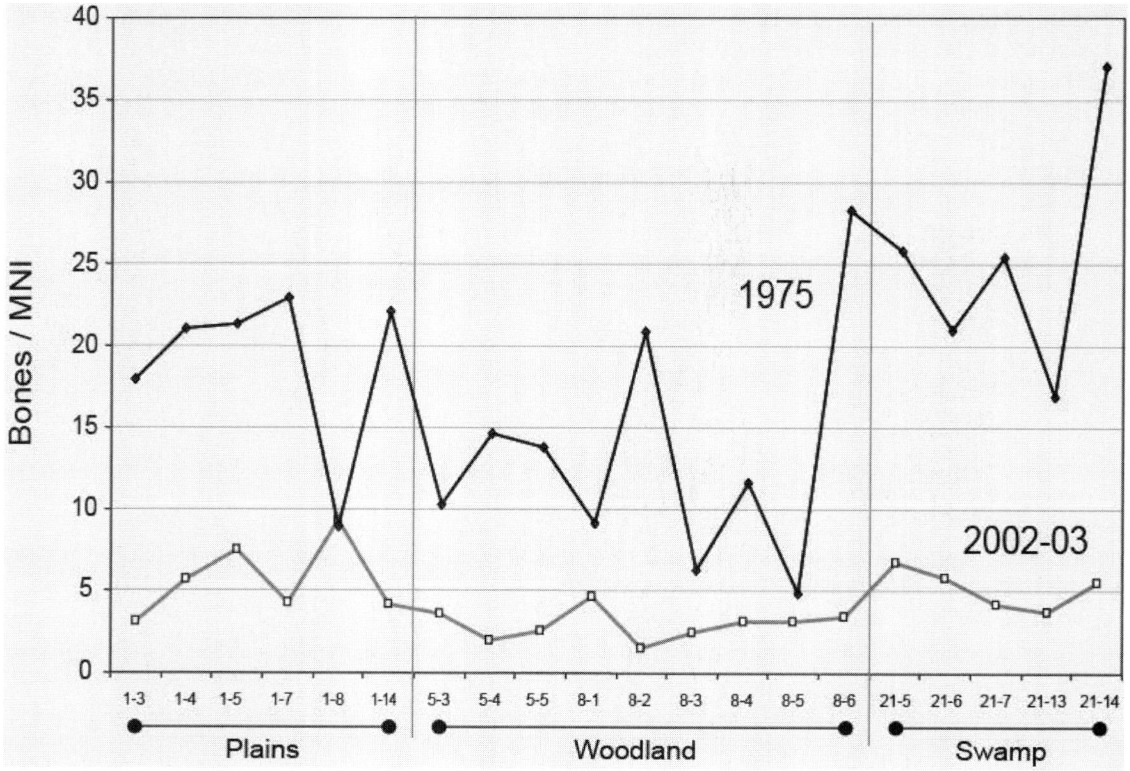


*Figure 4. Change in bone survival for 20 transects sampled both in 1975 and 2002-03, for all bone occurrences identifiable to major vertebrate group (fish, reptile, bird, mammal).*

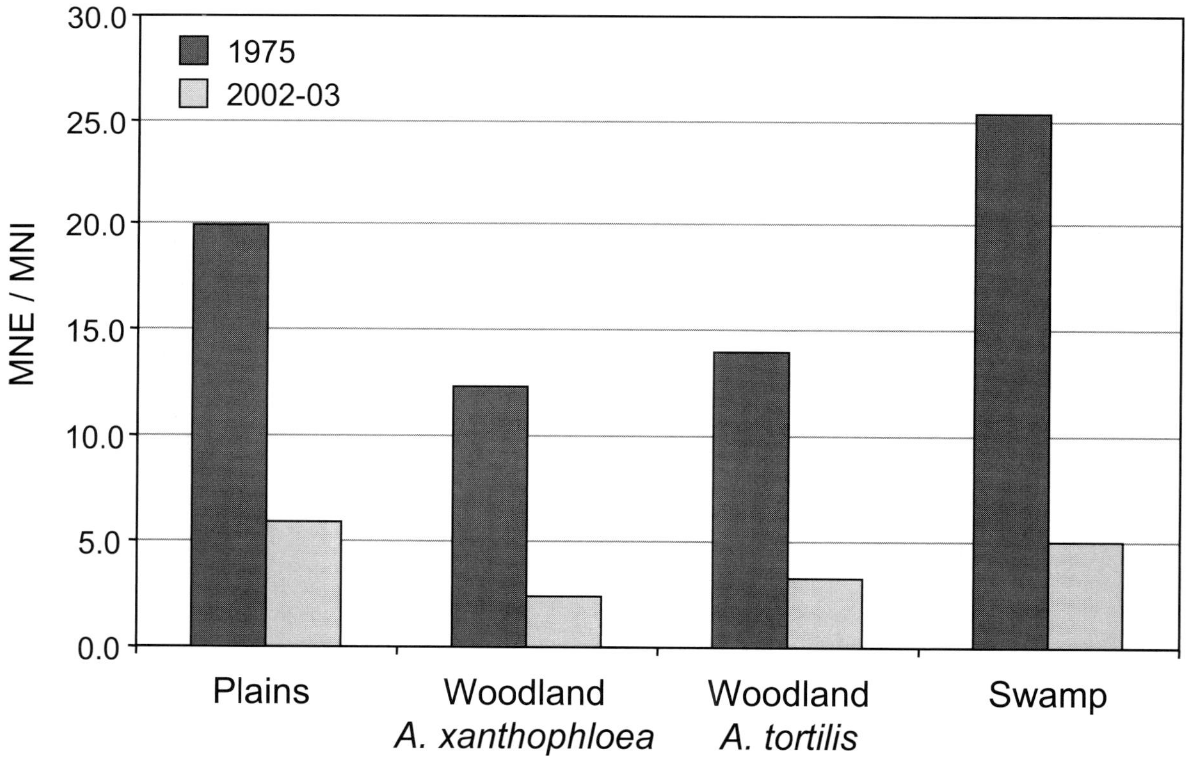


*Figure 5. Summary of change in bone survival in different habitats (data from Table 1). There was a 74% decrease overall in the number of bones per individual between 1975 and 2002-03.*

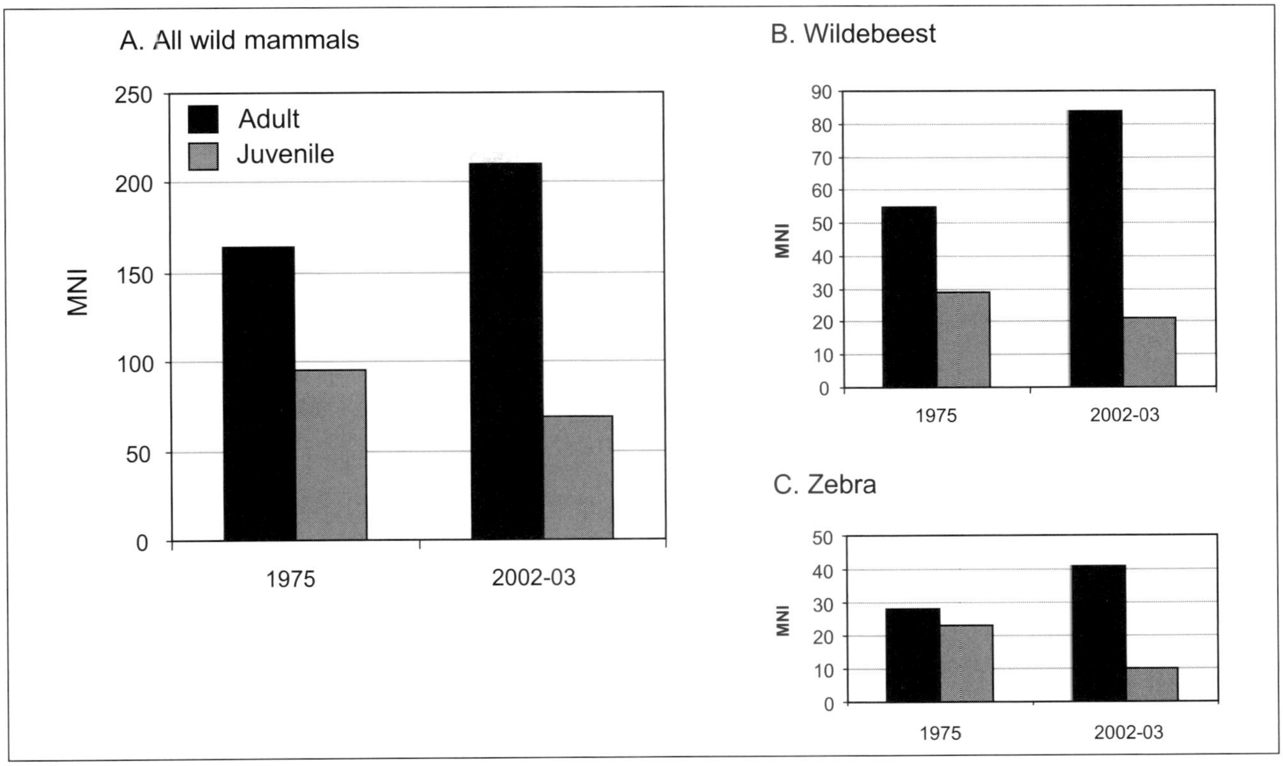


*Figure 6. Comparison of age group representation for 1975 vs. 2002-03 in the wild mammal samples. A. Results for the total sample that could be categorized to relative age (1975: MNI=320, 2002-03: MNI=339) showing decrease in numbers of juveniles and increase in adults in 2002-03. Individuals were counted as adults if they had fused epiphyses and/or adult dentition, and juveniles were counted if they had unfused epiphyses and/or sub-adult dentition. B. Results for wildebeest only. C. Results for zebra only. Note that scales for B and C are the same to show the absolute lower numbers of zebra MNI.*

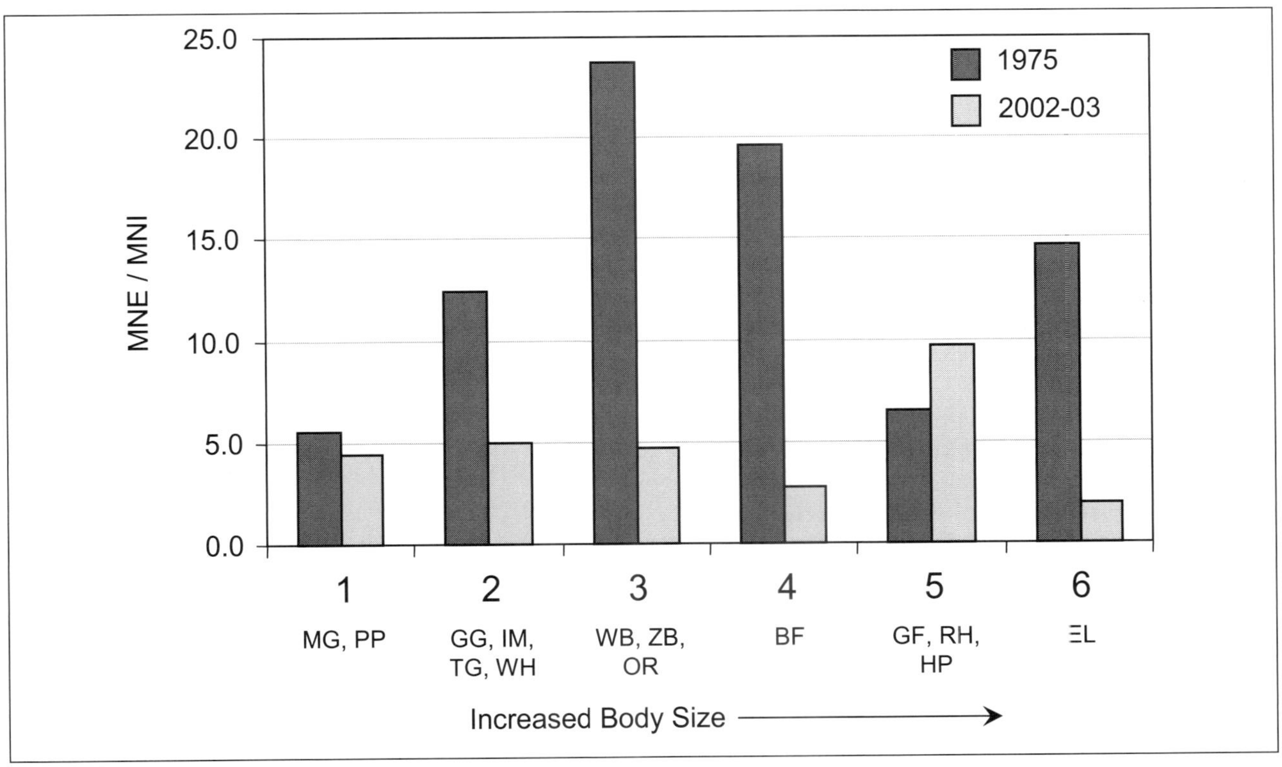


*Figure 7 Bones per individual (MNEs / MNIs) by body size category for wild mammals (domestic species omitted), showing the greatest change for sizes 3, 4 and 6 between 1975 and 2002-03. Size categories based on Brain (1981). Key to abbreviations on X axis: WB=wildebeeste, ZB=zebra, BF=Cape buffalo, CW=cow, GG=Grant's Gazelle, TG=Thompson's Gazelle, HP=hippo, WH=warthog, EL=elephant, IM=impala, GF=giraffe, RH=black rhino, OR=oryx, PP=porcupine, MG=mongoose.*

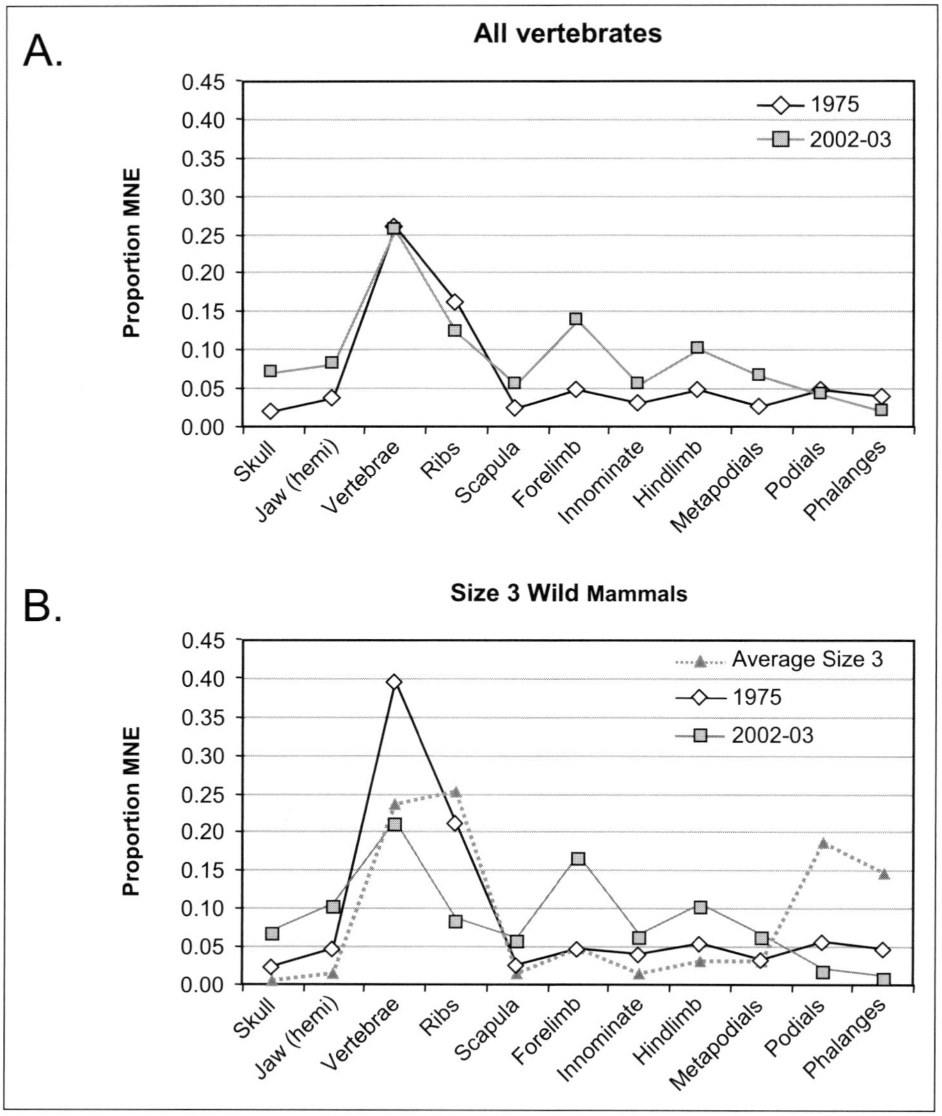


*Figure 8. Skeletal part representation in the Amboseli surface bone assemblage, for all 20 transects. A. Comparison of 1975 and 2002-03 for all vertebrates (Table 3). B. Comparison of 1975 and 2002-03 for size 3 wild mammals only (mainly wildebeest and zebra; domestic cows and donkeys not included) and the proportions of skeletal parts in a single average ungulate skeleton (zebra + wildebeest).*

Differences in the relative proportions of skeletal parts are not pronounced for the vertebrate sample as a whole between 1975 and 2002–03 (Table 3; Figure 8A). There is a relative increase in cranial and limb elements, with the latter biased toward resistant limb-ends such as distal humeri, proximal radius-ulnae, distal tibiae and metapodials in the 2002–03 sample. Considering only the most common wild ungulate species in Brain's body size category 3 (Brain, 1981) (Tables 3 and 4, Figure 8B), it is notable that there is a disproportionate number of vertebrae relative to other elements in 1975, making a much stronger peak for this size group than for the vertebrate assemblage as a whole. The 1975 skeletal part sample retains more overall similarity to the proportions of a single average skeleton (Figure 8B), than the 2002–03 sample, indicating that the taphonomic processes operating in Amboseli have shifted to stronger overprinting of the body-part signal from the original whole skeletons.

Skeletal part representation for size 3 wild mammals varies across the 4 different habitat samples (Table 3, Figure 9). In 1975, the patterns are similar among the habitats, except that proportionally fewer vertebrae occurred in the *Acacia xanthophloea* woodland. In 2002–03, however, the two woodland samples retain a higher proportion of vertebrae and also show increased numbers of durable limb elements, which is characteristic of the more open habitats in 1975 and the 2002–03 sample as a whole. Of all the habitats, the *Acacia tortilis* woodland has changed the least in terms of the relative abundance of the different body segments.

The degree of bone modification by predator/scavengers has yet to be quantified in the two samples (1975 vs. 2002–03), and a thorough treatment is beyond the scope of this paper. Data collected on tooth marks and other damage to individual bones was more systematic in the 2002–03 sample than in 1975, making a detailed level of comparison problematic. However, skeletal part representation, limb element completeness and the survival of different bone portions (proximal, distal, shaft only, etc.) were recorded in the same way in the two samples (Table 4). Analysis of skeletal element survival and damage in one species (zebra) reduces cross-taxon variability and serves to demonstrate the major taphonomic changes between 1975 and 2002–03. Skeletal parts were tallied from a subset of 11 transects (swamp and plains) and analyzed in terms of observed versus expected numbers of different elements (Figure 10). In 1975, there is clear evidence of post-mortem deletion of bones in certain body segments, such as distal limbs, vertebrae and ribs, caused by destruction and/or burial of bones. Overall, however, the pattern reflects relatively predictable survival based on bone strength and size, with cranial elements dominating and relatively lower survival of vertebrae and ribs, forelimb relative to

*Table 3. A. Counts of bones from all vertebrate remains on the 20 transects, grouped by skeletal region. B. Totals for Amboseli Size 3 wild mammals only, including both adults and juveniles. Numbers may differ from Table 1 because totals here include only bones that could be identified to the specified skeletal element. Note: podials include fibulae, innominate = left or right half of the pelvis, and forelimb includes radii, ulnae, and fused radius-ulnae.*

**A. All vertebrates**

| | All Habitats (MNE) | | Proportions MNE | | By Habitat | | | | | | | |
|---|---|---|---|---|---|---|---|---|---|---|---|---|
| | 1975 | 2002-03 | 1975 | 2002-03 | PLAINS 1975 | PLAINS 2002-03 | SWAMP 1975 | SWAMP 2002-03 | WD-A.x. 1975 | WD-A.x. 2002-03 | WD-A.t. 1975 | WD-A.t. 2002-03 |
| Skull | 155 | 68 | 0.020 | 0.070 | 37 | 26 | 49 | 21 | 15 | 4 | 54 | 17 |
| Jaw (hemi) | 287 | 78 | 0.037 | 0.081 | 60 | 26 | 89 | 37 | 44 | 1 | 94 | 14 |
| Vertebrae | 1993 | 248 | 0.259 | 0.256 | 592 | 78 | 756 | 105 | 179 | 17 | 493 | 48 |
| Ribs | 1250 | 118 | 0.163 | 0.122 | 327 | 56 | 458 | 46 | 127 | 3 | 338 | 13 |
| Scapula | 193 | 52 | 0.025 | 0.054 | 58 | 20 | 55 | 26 | 25 | 5 | 55 | 3 |
| Forelimb | 365 | 134 | 0.047 | 0.138 | 98 | 41 | 97 | 62 | 63 | 14 | 107 | 17 |
| Innominate | 242 | 52 | 0.031 | 0.054 | 72 | 21 | 68 | 26 | 35 | 5 | 67 | 3 |
| Hindlimb | 362 | 97 | 0.047 | 0.100 | 118 | 27 | 106 | 47 | 34 | 7 | 104 | 16 |
| Metapodials | 201 | 63 | 0.026 | 0.065 | 72 | 28 | 458 | 23 | 36 | 2 | 61 | 10 |
| Podials | 366 | 40 | 0.048 | 0.041 | 140 | 20 | 112 | 13 | 51 | 0 | 110 | 7 |
| Phalanges | 297 | 18 | 0.039 | 0.019 | 122 | 8 | 80 | 4 | 48 | 0 | 47 | 6 |
| MNE | 7686 | 968 | | | 1696 | 351 | 2328 | 410 | 657 | 58 | 1530 | 154 |
| MNI | 458 | 365 | | | 116 | 106 | 105 | 139 | 69 | 31 | 168 | 89 |
| MNE/MNI | 16.78 | 2.65 | | | 14.62 | 3.31 | 22.17 | 2.95 | 9.52 | 1.87 | 9.11 | 1.73 |

**B. Size 3 wild mammals**

| | All Habitats (MNE) | | Proportion MNE | | By Habitat | | | | | | | |
|---|---|---|---|---|---|---|---|---|---|---|---|---|
| | 1975 | 2002-03 | 1975 | 2002-03 | PLAINS 1975 | PLAINS 2002-03 | SWAMP 1975 | SWAMP 2002-03 | WD-A.x. 1975 | WD-A.x. 2002-03 | WD-A.t. 1975 | WD-A.t. 2002-03 |
| Skull | 67 | 39 | 0.02 | 0.07 | 27 | 23 | 32 | 11 | 2 | 0 | 6 | 5 |
| Jaw (hemi) | 134 | 57 | 0.05 | 0.11 | 50 | 22 | 69 | 24 | 3 | 0 | 12 | 11 |
| Vertebrae | 1073 | 114 | 0.40 | 0.21 | 492 | 49 | 524 | 30 | 17 | 8 | 40 | 27 |
| Ribs | 576 | 47 | 0.21 | 0.09 | 279 | 28 | 286 | 8 | 6 | 0 | 5 | 11 |
| Scapulae | 77 | 33 | 0.03 | 0.06 | 41 | 15 | 28 | 15 | 3 | 2 | 5 | 1 |
| Forelimb | 135 | 91 | 0.05 | 0.17 | 67 | 29 | 52 | 37 | 8 | 11 | 8 | 14 |
| Innominate | 113 | 36 | 0.04 | 0.07 | 59 | 16 | 46 | 15 | 5 | 2 | 3 | 3 |
| Hindlimb | 151 | 57 | 0.06 | 0.11 | 81 | 18 | 61 | 29 | 2 | 4 | 7 | 6 |
| Metapodials | 98 | 36 | 0.04 | 0.07 | 58 | 17 | 32 | 14 | 4 | 2 | 4 | 3 |
| Podials | 156 | 13 | 0.06 | 0.02 | 102 | 9 | 47 | 4 | 7 | 0 | 0 | 0 |
| Phalanges | 135 | 8 | 0.05 | 0.02 | 95 | 1 | 36 | 3 | 4 | 0 | 0 | 4 |
| MNE | 2715 | 531 | | | 1351 | 227 | 1213 | 190 | 61 | 29 | 90 | 85 |
| MNI | 226 | 342 | | | 110 | 133 | 78 | 129 | 10 | 24 | 28 | 56 |
| MNE/MNI | 12.01 | 1.55 | | | 12.28 | 1.71 | 15.55 | 1.47 | 6.10 | 1.21 | 3.21 | 1.52 |

*Table 3. (continued)*

**C. Proportions from B; size 3 wild mammals**

| | PLAINS 1975 | PLAINS 2002-03 | SWAMP 1975 | SWAMP 2002-03 | WD-A.x. 1975 | WD-A.x. 2002-03 | WD-A.t. 1975 | WD-A.t. 2002-03 |
|---|---|---|---|---|---|---|---|---|
| Skull | 0.020 | 0.101 | 0.026 | 0.058 | 0.033 | 0.000 | 0.067 | 0.059 |
| Jaw (hemi) | 0.037 | 0.097 | 0.057 | 0.126 | 0.049 | 0.000 | 0.133 | 0.129 |
| Vertebrae | 0.364 | 0.216 | 0.432 | 0.158 | 0.279 | 0.276 | 0.444 | 0.318 |
| Ribs | 0.207 | 0.123 | 0.236 | 0.042 | 0.098 | 0.000 | 0.056 | 0.129 |
| Scapulae | 0.030 | 0.066 | 0.023 | 0.079 | 0.049 | 0.069 | 0.056 | 0.012 |
| Forelimb | 0.050 | 0.128 | 0.043 | 0.195 | 0.131 | 0.379 | 0.089 | 0.165 |
| Innominate | 0.044 | 0.070 | 0.038 | 0.079 | 0.082 | 0.069 | 0.033 | 0.035 |
| Hindlimb | 0.060 | 0.079 | 0.050 | 0.153 | 0.033 | 0.138 | 0.078 | 0.071 |
| Metapodials | 0.043 | 0.075 | 0.026 | 0.074 | 0.066 | 0.069 | 0.044 | 0.035 |
| Podials | 0.076 | 0.040 | 0.039 | 0.021 | 0.115 | 0.000 | 0.000 | 0.000 |
| Phalanges | 0.070 | 0.004 | 0.030 | 0.016 | 0.066 | 0.000 | 0.000 | 0.047 |
| MNE | 1351 | 227 | 1213 | 190 | 61 | 29 | 90 | 85 |

**D. Single whole skeleton**

| | Single ZB | Single WB | Average Size 3 | Proportion MNE |
|---|---|---|---|---|
| Skull | 1 | 1 | 1 | 0.01 |
| Jaw (hemi) | 2 | 2 | 2 | 0.02 |
| Vertebrae | 32 | 27 | 29 | 0.24 |
| Ribs | 36 | 26 | 31 | 0.25 |
| Scapula | 2 | 2 | 2 | 0.02 |
| Forelimb | 6 | 6 | 6 | 0.05 |
| Innominate | 2 | 2 | 2 | 0.02 |
| Hindlimb | 4 | 4 | 4 | 0.03 |
| Metapodials | 4 | 4 | 4 | 0.03 |
| Podials | 26 | 20 | 23 | 0.19 |
| Phalanges | 12 | 24 | 18 | 0.15 |
| MNE | 127 | 118 | 122 | |
| MNI | 1 | 1 | 1 | |

hind limb, and distal versus proximal limb bones. In the 2002–03 sample, cranial and mandibular elements are similar in terms of observed vs. expected ratio (~20%), axial elements are nearly absent, and fore- and hind-limb bones are similar in terms of survival rates. Overall, the survival of zebra remains is reduced by about 80% although the number of individuals on the sampled transects is only 20% less than in 1975. Damage to individual bones, based on humeri and femora, also changed markedly from 1975 to 2002–03 (Figure 11), with a decrease in the number of whole elements accompanied by an increase in durable portions such as distal humeri. Based on more detailed examination of bone modification features, there is an accompanying shift in the amount of moderate to heavy chewing and fragmentation in the 2002-03 sample (Figure 11B).

### *Species diversity in the bone assemblage*

The diversity of species represented in the surface bone assemblages likely reflects the diversity of the living populations in the different habitats, which was probably important to early hominin scavengers or hunters. Based on field and laboratory identifications of the fragmentary remains on the 20 transects, there are a total of 19 mammal, one fish, one reptile, and several bird species in the sampled bone assemblage (Table 5; Figures 12 and 13). A total mammal species richness of 14 in 1975 increased to 17 in 2002-03. Species abundance is based on field assessment of MNI rather than on the most common skeletal element (Behrensmeyer and Dechant Boaz, 1980). Comparison of the MNI abundances for mammals highlights the impact of the removal of domestic animals from the park in the 1980s and also indicates a shift to a somewhat different abundance distribution in the 2002–03 sample, with a stronger dominance of wildebeeste in the bone assemblage. Fisher's alpha, which characterizes diversity in terms of abundance distribution for a given number of species, is very similar for the top 12 species –2.412 (1975) and 2.581 (2002–03). This indicates surprising stability in this ecological parameter, in spite of major changes in the ecosystem.

The four habitats show different patterns of change

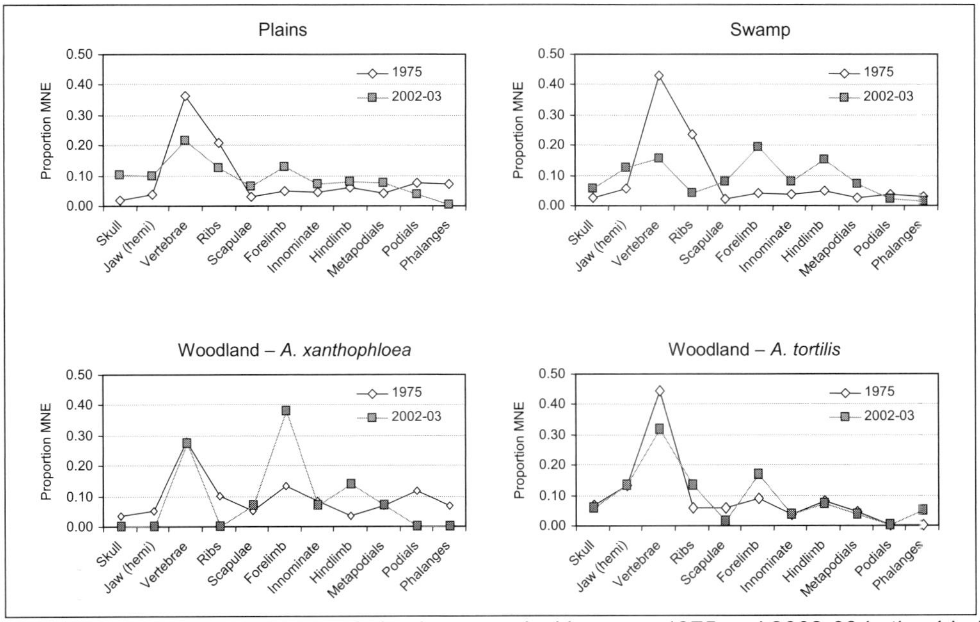


*Figure 9. Line graphs showing differences in skeletal part survival between 1975 and 2002-03 in the 4 habitats sampled in Amboseli, for size 3 wild mammals only (Table 4).*

*Table 4. Data for Amboseli adult and juvenile zebra remains recorded on transects in the plains and swamp habitats, for 1975 and 2002-03 (Figures 10-11). A. Counts of MNEs for different elements, and proportions of observed over expected based on the number of bones in a single zebra, B. Counts of humeri and femora in different stages of completeness, C. Counts of MNEs in carnivore damage categories of B. Pobiner (Pobiner and Blumenschine, 2003; pers. comm. 2004): A: Minimal = toothmarks, both ends still present, B: Moderate = one end missing, C: Heavy = both ends missing, shaft only, D: Fragments only.*

**A. Skeletal representation**

| | 1975 | Single Zebra | Expected | Proportion O/E | 2002-03 | Single Zebra | Expected | Proportion O/E |
|---|---|---|---|---|---|---|---|---|
| Skull | 18 | 1 | 45 | 0.40 | 7 | 1 | 36 | 0.19 |
| Jaw (hemi) | 28 | 2 | 90 | 0.31 | 15 | 2 | 72 | 0.21 |
| Vertebrae | 236 | 32 | 1440 | 0.16 | 9 | 32 | 1152 | 0.01 |
| Ribs | 95 | 36 | 1620 | 0.06 | 3 | 36 | 1296 | 0.00 |
| Scapula | 19 | 2 | 90 | 0.21 | 7 | 2 | 72 | 0.10 |
| Humerus | 20 | 2 | 90 | 0.22 | 11 | 2 | 72 | 0.15 |
| Radius/ulna | 29 | 4 | 180 | 0.16 | 14 | 4 | 144 | 0.10 |
| Metacarpal | 8 | 2 | 90 | 0.09 | 8 | 2 | 72 | 0.11 |
| Innominate | 32 | 2 | 90 | 0.36 | 10 | 2 | 72 | 0.14 |
| Femur | 28 | 2 | 90 | 0.31 | 7 | 2 | 72 | 0.10 |
| Tibia | 26 | 2 | 90 | 0.29 | 9 | 2 | 72 | 0.13 |
| Metatarsal | 16 | 2 | 90 | 0.18 | 5 | 2 | 72 | 0.07 |
| Patella | 2 | 2 | 90 | 0.02 | 0 | 2 | 72 | 0.00 |
| Podials | 41 | 26 | 1170 | 0.04 | 8 | 26 | 936 | 0.01 |
| Phalanges | 18 | 12 | 540 | 0.03 | 3 | 12 | 432 | 0.01 |
| MNE | 616 | 129 | 5805 | | 116 | 129 | 4644 | |
| MNI | 45 | 1 | 45 | | 36 | 1 | 36 | |

**B. Completeness**

| | 1975 | 2002-03 |
|---|---|---|
| Whole | 26 | 3 |
| Prox-Dist Pair | 2 | 1 |
| Shaft only | 5 | 3 |
| Prox only | 1 | 0 |
| Prox+Shaft | 1 | 0 |
| Dist only | 9 | 7 |
| Dist+Shaft | 4 | 3 |
| Total | 48 | 17 |

**C. Damage**

| | 1975 | 2002-03 |
|---|---|---|
| No Damage | 20 | 1 |
| A | 17 | 2 |
| B | 11 | 11 |
| C | 0 | 2 |
| D | 0 | 1 |
| Total | 48 | 17 |

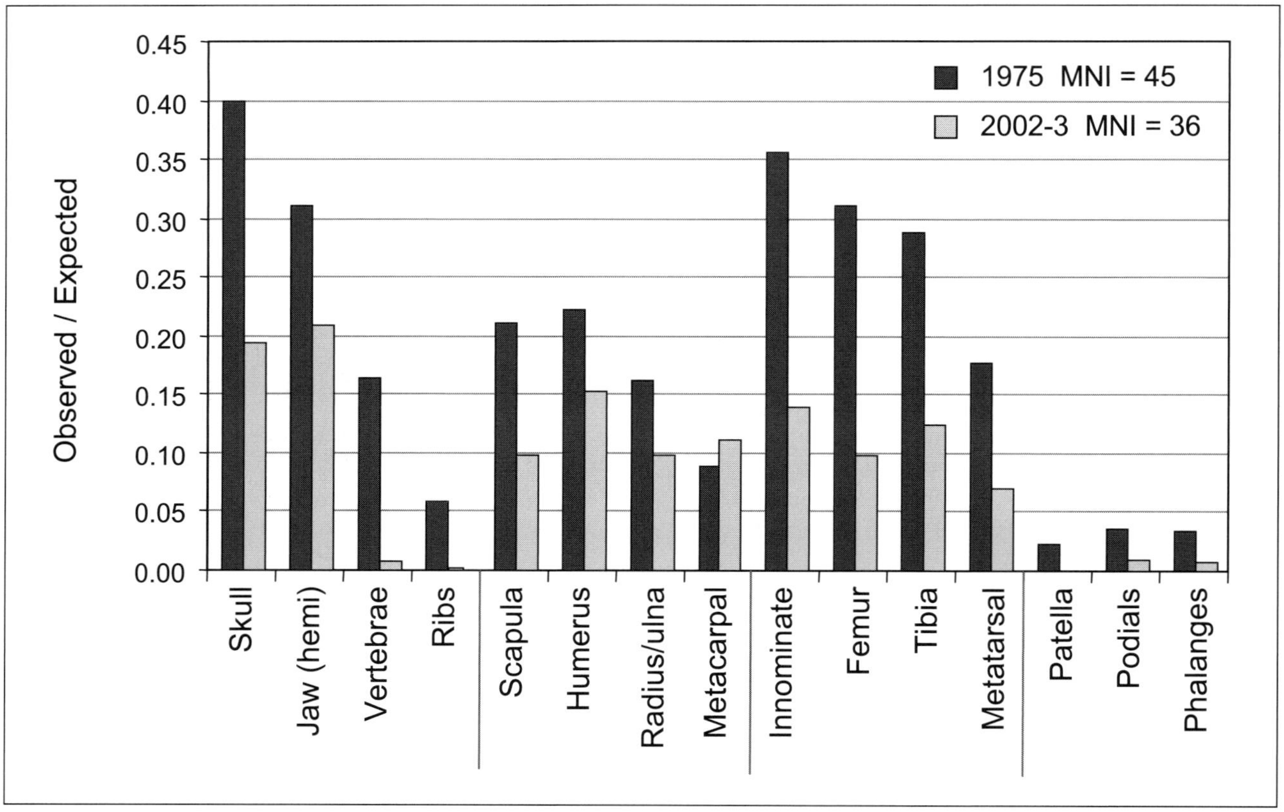


*Figure 10. Skeletal part representation for Amboseli zebra remains from 20 bone survey transects in 1975 and 2002-03, as a proportion of the expected number of bones for the observed MNI. Teeth are not counted.*

in the common mammal species (Figure 14, Table 6). The removal of cows from the central area of the park in 1981 has a clear signal in the bone assemblages, especially in the two woodland habitats preferred by the Maasai herdsmen, where the MNIs for cow drops by 85->90%. Interestingly, the species richness in each of the 4 habitats changed little between 1975 and 2002–03, in spite of the removal of domestic stock. There are shifts in abundance of some species, such as decreased numbers of impala, Grant's gazelle and giraffe in the woodland habitats in 2002–03 accompanied by slight increases in wildebeest and zebra. Rhinoceros have been extinct in the Amboseli ecosystem since the mid-1980s, but their bones continue to be found on the transects in low numbers. MNIs for zebra decreased in the plains habitat, while buffalo and warthog increased. Slightly larger MNIs occur for all wild species in the swamp habitat 2002-03 transects, except elephant and rhino. Overall diversity, as measured by Fisher's Alpha, increases for the two woodland habitats and slightly for plains but remains approximately the same for swamp. However, if cows are removed from the calculations, diversity remains stable in the bone assemblages of the four habitats between 1975 and 2002–03, with only plains showing a slight increase in the evenness of the abundance distribution (Table 6).

The most common species in Amboseli are plotted on Figure 14 in order of body size, providing an overview of the dominance of size 3 herbivores in the bone assemblage (zebra, wildebeest, and cow). The relationship of this distribution to abundances in the living populations has yet to be determined, but the MNI counts are known to be affected by taphonomic as well as sampling biases in the low numbers of size 2 species (Behrensmeyer et al., 1979). The ecology of all the species—their habitat preferences, seasonal movements, turnover rates, and areas of high vs. low mortality—also affect the distributions and numbers of MNIs in the different habitat samples (Western, 1980; Behrensmeyer and Dechant Boaz, 1980), resulting in a complex relationship between the living animal community and the information recorded in the surface bone assemblage.

## Comparison of skeletal element survival in Amboseli and the Serengeti

Data from the bone transect surveys in Amboseli and the Serengeti ecosystem (Blumenschine, 1989) can be used to investigate similarities among the different habitat samples based on the proportions of different body parts (MNEs) (Table 5, Figures 15 and 16). The survey methods were comparable, although bone visibility may have been less in the Serengeti surveys due to denser ground cover. The size and shape of Blumenshine's transects also differed from those in Amboseli. Sample sizes (MNEs and MNIs) are generally similar (Table 5a), especially for the woodland habitats.

The cluster diagram (Figure 15) groups Serengeti riparian woodland most closely with the Amboseli's *Acacia tortilis* woodland (1975 sample). The other Serengeti

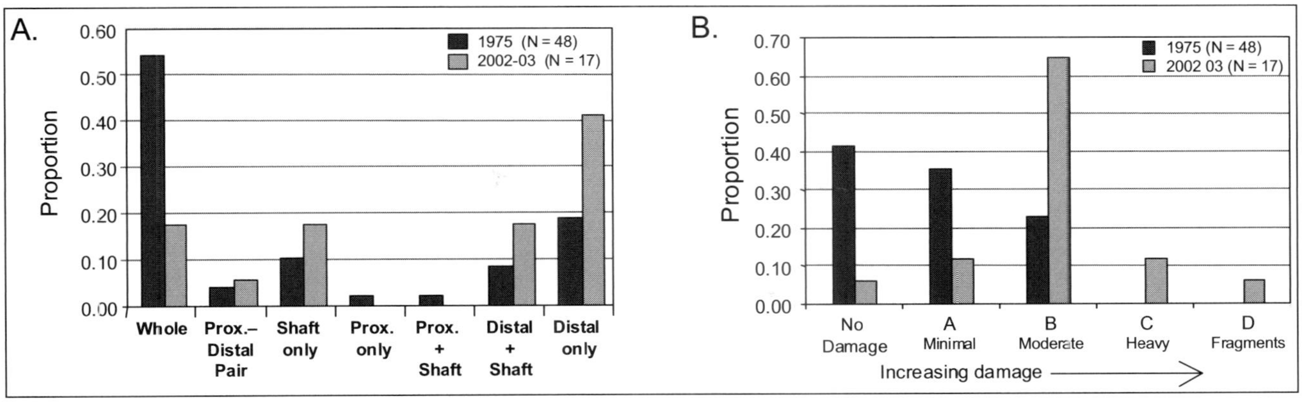


*Figure 11. Change in bone modification patterns for zebra humeri and femora from 1975 to 2002-03, based on MNE from the 20 Amboseli transects. A. Proportions of more and less complete individual elements. B. Damage levels for same bone sample as in A., based on categories of Briana Pobiner (Pobiner and Blumenschine, 2003; pers. comm. 2004): Minimal = toothmarks, both ends still present, Moderate = one end missing, Heavy = both ends missing, shaft only.*

*Table 5. Comparison of skeletal part representation in the 1980s Serengeti bone surveys of Blumenschine (1989; 16 transects) and Amboseli, based on data in Table 3. A. Raw MNE counts. B. Proportions based on total MNEs for each habitat. C. Contingency Table of similarity (Pearson correlation coefficient) for data in 5B (MNE Proportions) organized by habitat; all cells in bold have significant values at the level of significance alpha=0.050 (two-tailed test). D. Contingency Table (Pearson correlation coefficient) for data in 5B (MNE Proportions) organized by year; all cells in bold have significant values at the level of significance alpha=0.050 (two-tailed test).*

**A. MNE Counts**

| | Serengeti | | | Amboseli | | | | | | | |
|---|---|---|---|---|---|---|---|---|---|---|---|
| **Skeletal Parts (MNE)** | **NGO** | **AW& GP** | **RW** | **PLAINS 1975** | **PLAINS 2002-03** | **SWAMP 1975** | **SWAMP 2002-03** | **WD-A.x. 1975** | **WD-A.x. 2002-03** | **WD-A.t. 1975** | **WD-A.t. 2002-03** |
| Skull | 40 | 44 | 11 | 27 | 23 | 32 | 11 | 2 | 0 | 6 | 5 |
| Hemi-Mandible | 32 | 27 | 9 | 50 | 22 | 69 | 24 | 3 | 0 | 12 | 11 |
| Vertebrae | 64 | 111 | 76 | 492 | 49 | 524 | 30 | 17 | 8 | 40 | 27 |
| Ribs | 5 | 14 | 13 | 279 | 28 | 286 | 8 | 6 | 0 | 5 | 11 |
| Scapula | 12 | 25 | 9 | 41 | 15 | 28 | 15 | 3 | 2 | 5 | 1 |
| Forelimb | 18 | 23 | 3 | 67 | 29 | 52 | 37 | 8 | 11 | 8 | 14 |
| Innominate | 4 | 32 | 9 | 59 | 16 | 46 | 15 | 5 | 2 | 3 | 3 |
| Hindlimb | 13 | 30 | 8 | 81 | 18 | 61 | 29 | 2 | 4 | 7 | 6 |
| Metapodials | 7 | 21 | 4 | 58 | 17 | 32 | 14 | 4 | 2 | 4 | 3 |
| Podials | 2 | 4 | 3 | 102 | 9 | 47 | 4 | 7 | 0 | 0 | 0 |
| Phalanges | 0 | 1 | 0 | 95 | 1 | 36 | 3 | 4 | 0 | 0 | 4 |
| MNE | 197 | 332 | 145 | 1351 | 227 | 1213 | 190 | 61 | 29 | 90 | 85 |
| MNI | 46 | 51 | 12 | 110 | 133 | 78 | 129 | 10 | 24 | 28 | 56 |
| MNE/MNI | 4.28 | 6.51 | 12.08 | 12.28 | 1.71 | 15.55 | 1.47 | 6.10 | 1.21 | 3.21 | 1.52 |

**B. MNE Proportions**

| **Skeletal Parts (MNE)** | **NGO** | **AW& GP** | **RW** | **PL 75** | **PL 02-03** | **SW 75** | **SW 02-03** | **WD-Ax 75** | **WD-Ax 02-03** | **WD-At 75** | **WD-At 02-03** |
|---|---|---|---|---|---|---|---|---|---|---|---|
| Skull | 0.203 | 0.133 | 0.076 | 0.020 | 0.101 | 0.026 | 0.058 | 0.033 | 0.000 | 0.067 | 0.059 |
| Hemi-Mandible | 0.162 | 0.081 | 0.062 | 0.037 | 0.097 | 0.057 | 0.126 | 0.049 | 0.000 | 0.133 | 0.129 |
| Vertebrae | 0.325 | 0.334 | 0.524 | 0.364 | 0.216 | 0.432 | 0.158 | 0.279 | 0.276 | 0.444 | 0.318 |
| Ribs | 0.025 | 0.042 | 0.090 | 0.207 | 0.123 | 0.236 | 0.042 | 0.098 | 0.000 | 0.056 | 0.129 |
| Scapula | 0.061 | 0.075 | 0.062 | 0.030 | 0.066 | 0.023 | 0.079 | 0.049 | 0.069 | 0.056 | 0.012 |
| Forelimb | 0.091 | 0.069 | 0.021 | 0.050 | 0.128 | 0.043 | 0.195 | 0.131 | 0.379 | 0.089 | 0.165 |
| Innominate | 0.020 | 0.096 | 0.062 | 0.044 | 0.070 | 0.038 | 0.079 | 0.082 | 0.069 | 0.033 | 0.035 |
| Hindlimb | 0.066 | 0.090 | 0.055 | 0.060 | 0.079 | 0.050 | 0.153 | 0.033 | 0.138 | 0.078 | 0.071 |
| Metapodials | 0.036 | 0.063 | 0.028 | 0.043 | 0.075 | 0.026 | 0.074 | 0.066 | 0.069 | 0.044 | 0.035 |
| Podials | 0.010 | 0.012 | 0.021 | 0.076 | 0.040 | 0.039 | 0.021 | 0.115 | 0.000 | 0.000 | 0.000 |
| Phalanges | 0.000 | 0.003 | 0.000 | 0.070 | 0.004 | 0.030 | 0.016 | 0.066 | 0.000 | 0.000 | 0.047 |
| MNE | 197 | 332 | 145 | 1351 | 227 | 1213 | 190 | 61 | 29 | 90 | 85 |
| MNI | 46 | 51 | 12 | 110 | 133 | 78 | 129 | 10 | 24 | 28 | 56 |
| MNE/MNI | 4.28 | 6.51 | 12.08 | 12.28 | 1.71 | 15.55 | 1.47 | 6.10 | 1.21 | 3.21 | 1.52 |

*Table 5. (continued)*

**C. Contingency Table organized by habitat.**

| | | NGO | AW& GP | RW | PL 75 | PL 02-03 | SW 75 | SW 02-03 | WD-Ax 75 | WD-Ax 02-03 | WD-At 75 | WD-At 02-03 |
|---|---|---|---|---|---|---|---|---|---|---|---|---|
| Serengeti | NGO | 1.000 | **0.897** | **0.809** | 0.539 | **0.805** | **0.626** | 0.519 | 0.555 | 0.394 | **0.879** | **0.770** |
| | AW&GP | **0.897** | 1.000 | **0.941** | **0.699** | **0.842** | **0.761** | 0.522 | **0.709** | 0.497 | **0.941** | **0.781** |
| | RW | **0.809** | **0.941** | 1.000 | **0.884** | **0.820** | **0.912** | 0.377 | **0.843** | 0.427 | **0.961** | **0.828** |
| Amboseli | PL 75 | 0.539 | **0.699** | **0.884** | 1.000 | **0.734** | **0.988** | 0.190 | **0.867** | 0.338 | **0.804** | **0.806** |
| | PL 02-03 | **0.805** | **0.842** | **0.820** | **0.734** | 1.000 | **0.814** | **0.642** | **0.723** | **0.614** | **0.869** | **0.899** |
| | SW 75 | **0.626** | **0.761** | **0.912** | **0.988** | **0.814** | 1.000 | 0.264 | **0.845** | 0.352 | **0.853** | **0.853** |
| | SW 02-03 | 0.519 | 0.522 | 0.377 | 0.190 | **0.642** | 0.264 | 1.000 | 0.351 | **0.844** | 0.573 | **0.635** |
| | WD-Ax 75 | 0.555 | **0.709** | **0.843** | **0.867** | **0.723** | **0.845** | 0.351 | 1.000 | **0.603** | **0.807** | **0.801** |
| | WD-Ax 02-03 | 0.394 | 0.497 | 0.427 | 0.338 | **0.614** | 0.352 | **0.844** | **0.603** | 1.000 | 0.543 | **0.639** |
| | WD-At 75 | **0.879** | **0.941** | **0.961** | **0.804** | **0.869** | **0.853** | 0.573 | **0.807** | 0.543 | 1.000 | **0.905** |
| | WD-At 02-03 | **0.770** | **0.781** | **0.828** | **0.806** | **0.899** | **0.853** | **0.635** | **0.801** | **0.639** | **0.905** | 1.000 |

**D. Contingency Table organized by year.**

| | | NGO | AW& GP | RW | PL 75 | SW 75 | WD-Ax75 | WD-At75 | PL 02-03 | SW 02-03 | WD-Ax02-03 | WD-At02-03 |
|---|---|---|---|---|---|---|---|---|---|---|---|---|
| Serengeti | NGO | 1.000 | **0.897** | **0.809** | 0.539 | **0.626** | 0.555 | **0.879** | **0.805** | 0.519 | 0.394 | **0.770** |
| | AW&GP | **0.897** | 1.000 | **0.941** | **0.699** | **0.761** | **0.709** | **0.941** | **0.842** | 0.522 | 0.497 | **0.781** |
| | RW | **0.809** | **0.941** | 1.000 | **0.884** | **0.912** | **0.843** | **0.961** | **0.820** | 0.377 | 0.427 | **0.828** |
| Amboseli | PL 75 | 0.539 | **0.699** | **0.884** | 1.000 | **0.988** | **0.867** | **0.804** | **0.734** | 0.190 | 0.338 | **0.806** |
| | SW 75 | **0.626** | **0.761** | **0.912** | **0.988** | 1.000 | **0.845** | **0.853** | **0.814** | 0.264 | 0.352 | **0.853** |
| | WD-Ax75 | 0.555 | **0.709** | **0.843** | **0.867** | **0.845** | 1.000 | **0.807** | **0.723** | 0.351 | **0.603** | **0.801** |
| | WD-At75 | **0.879** | **0.941** | **0.961** | **0.804** | **0.853** | **0.807** | 1.000 | **0.869** | 0.573 | 0.543 | **0.905** |
| | PL 02-03 | **0.805** | **0.842** | **0.820** | **0.734** | **0.814** | **0.723** | **0.869** | 1.000 | **0.642** | **0.614** | **0.899** |
| | SW 02-03 | 0.519 | 0.522 | 0.377 | 0.190 | 0.264 | 0.351 | 0.573 | **0.642** | 1.000 | **0.844** | **0.635** |
| | WD-Ax02-03 | 0.394 | 0.497 | 0.427 | 0.338 | 0.352 | **0.603** | 0.543 | **0.614** | **0.844** | 1.000 | **0.639** |
| | WD-At02-03 | **0.770** | **0.781** | **0.828** | **0.806** | **0.853** | **0.801** | **0.905** | **0.899** | **0.635** | **0.639** | 1.000 |

*Table 6. Numbers of individuals for all vertebrate species found on the 20 transects. A. Data for the 12 most common mammals. Key: EL = elephant (*Loxodonta africana*), HP = hippo (*Hippopotamus amphibious*), RH = black rhino (*Diceros bicornis*), GF = giraffe (*Giraffa camelopardalis*), BF = Cape buffalo (*Syncerus caffer*), ZB = zebra (*Equus burchelli*), WB = wildebeest (*Connochaetes taurinus*), CW = cow (*Bos taurus*), GG = Grant's gazelle (*Gazella granti*), IM = impala (*Aepyceros melampus*), WH = warthog (*Phacochoerus aethiopicus*), TG = Thompson's gazelle (*Gazella thompsoni*). B. Data for additional 7 rare mammals. C. Data for non-mammals. D. Rank order (based on 2002-03).*

**A. Common Mammals (MNI)**

| Habitat, Year | EL | HP | RH | GF | BF | ZB | WB | CW | GG | IM | WH | TG | N (MNI) | Fisher's Alpha All | Fisher's Alpha Wild Only |
|---|---|---|---|---|---|---|---|---|---|---|---|---|---|---|---|
| Plains1975 | 0 | 1 | 0 | 3 | 2 | 30 | 45 | 4 | 6 | 1 | 0 | 3 | 95 | 2.441 | 2.113 |
| Plains02-03 | 0 | 1 | 0 | 2 | 7 | 16 | 41 | 1 | 1 | 1 | 3 | 2 | 75 | 3.099 | 2.685 |
| WoodA.x.1975 | 0 | 0 | 1 | 3 | 3 | 3 | 2 | 24 | 1 | 5 | 0 | 1 | 43 | 3.468 | 5.205 |
| WoodA.x.02-03 | 1 | 1 | 1 | 2 | 2 | 4 | 7 | 1 | 0 | 0 | 0 | 1 | 20 | 6.296 | 5.205 |
| WoodA.t.1975 | 1 | 0 | 3 | 6 | 3 | 10 | 9 | 62 | 9 | 18 | 1 | 2 | 124 | 2.915 | 3.374 |
| WoodA.t.02-03 | 2 | 0 | 1 | 1 | 3 | 13 | 14 | 9 | 3 | 6 | 4 | 1 | 57 | 4.057 | 3.843 |
| Swamp1975 | 9 | 1 | 3 | 0 | 9 | 15 | 37 | 8 | 2 | 0 | 0 | 1 | 85 | 2.543 | 2.245 |
| Swamp02-03 | 5 | 8 | 1 | 0 | 9 | 21 | 50 | 0 | 5 | 0 | 2 | 6 | 107 | 2.341 | 2.341 |
| 1975 | 10 | 2 | 7 | 12 | 17 | 58 | 93 | 98 | 18 | 24 | 1 | 7 | 347 | 2.412 | 2.355 |
| 2002-03 | 9 | 10 | 3 | 5 | 21 | 54 | 114 | 13 | 11 | 7 | 10 | 10 | 267 | 2.581 | 2.343 |

**B. Rare Mammals**

| | ORYX | DONKEY | BUSHBUCK | HYENA | SHEEP-GOAT | PORCUPINE | MONGOOSE | N (MNI) |
|---|---|---|---|---|---|---|---|---|
| Plains 1975 | | | | | | | | 0 |
| Plains 02-03 | | | | | | | | 0 |
| Wood A.x. 1975 | | | | | | | | 0 |
| Wood A.x. 02-03 | 1 | | 1 | 1 | | | | 3 |
| Wood A.t. 1975 | | 1 | | | 3 | | | 4 |
| Wood A.t. 02-03 | | | 1 | 1 | | 1 | 1 | 4 |
| Swamp 1975 | | | | | | | | 0 |
| Swamp 02-03 | | | | | | | | 0 |
| 1975 | 0 | 1 | 0 | 0 | 3 | 0 | 0 | 4 |
| 2002-03 | 1 | 0 | 2 | 2 | 0 | 1 | 1 | 7 |

**C. Non-Mammals**

| FISH (Catfish) | REPTILE (Tortoise) | BIRD (non-Ostrich) | OSTRICH | N (MNI) |
|---|---|---|---|---|
| | 1 | | | 0 |
| | 2 | 1 | | 1 |
| | | 3 | | 3 |
| | 2 | 1 | | 1 |
| | | 3 | | 3 |
| | 2 | 4 | 1 | 4 |
| 4 | | 3 | | 3 |
| 5 | | | | 0 |
| 4 | 2 | 10 | 0 | 10 |
| 5 | 4 | 5 | 1 | 5 |

**D. Rank order (based on 2002-03)**

| | 1975 | 2002-03 |
|---|---|---|
| WB | 93 | 114 |
| ZB | 58 | 54 |
| BF | 17 | 21 |
| CW | 98 | 13 |
| GG | 18 | 11 |
| TG | 7 | 10 |
| HP | 2 | 10 |
| WH | 1 | 10 |
| EL | 10 | 9 |
| IM | 24 | 7 |
| GF | 12 | 5 |
| RH | 7 | 3 |
| N (MNI) | 347 | 267 |

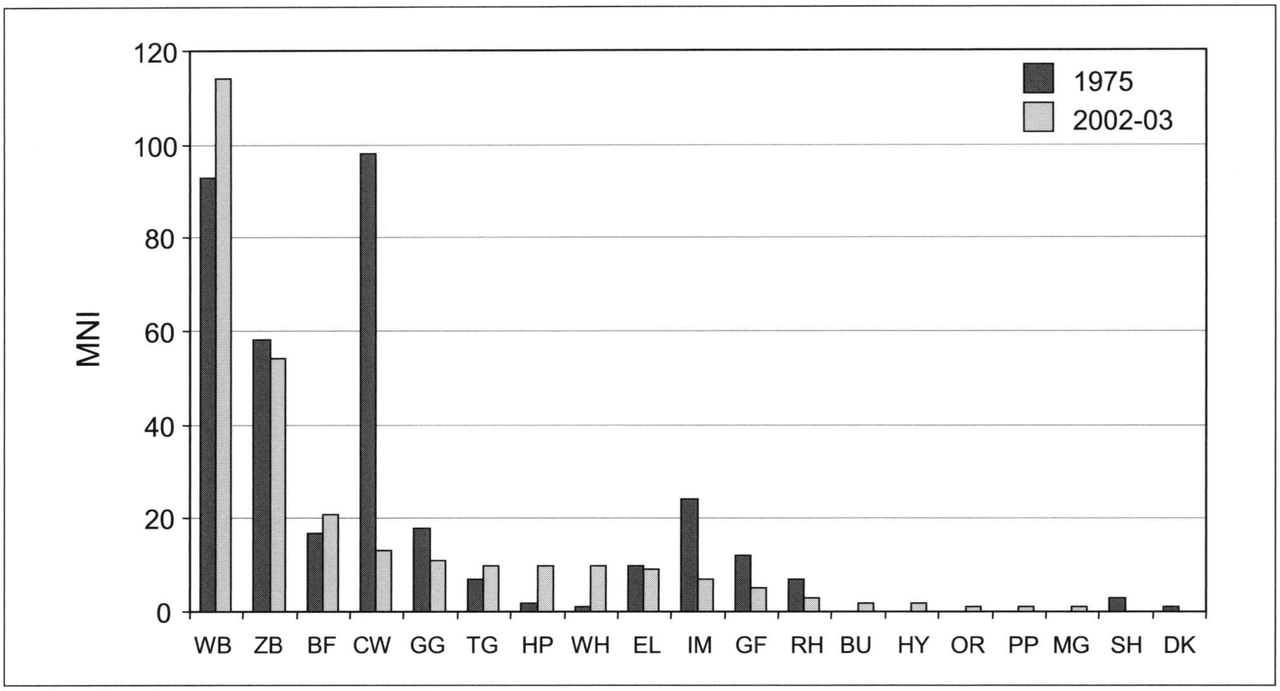


*Figure 12. Abundances of all 19 mammal species identified on 20 sampled transects in 1975 and 2002-03, rank order based on 2002-03 sample. The number of species increased from 14 in 1975 to 17 in 2002-03 on these transects, while the proportion of domestic stock (CW, SH, DK) markedly decreased. Key: WB = wildebeeste, ZB = zebra, BF = cape buffalo, CW = cow, CG = Grant's gazelle, TG = Thompson's gazelle, HP = hippo, WH = warthog, EL = elephant, IM = impala, GF = giraffe, RH = black rhino, BU = bushbuck, HY = spotted hyena, OR = oryx, PP = porcupine, MG = mongoose (white-tailed), SH = sheep/goat, DK = donkey*

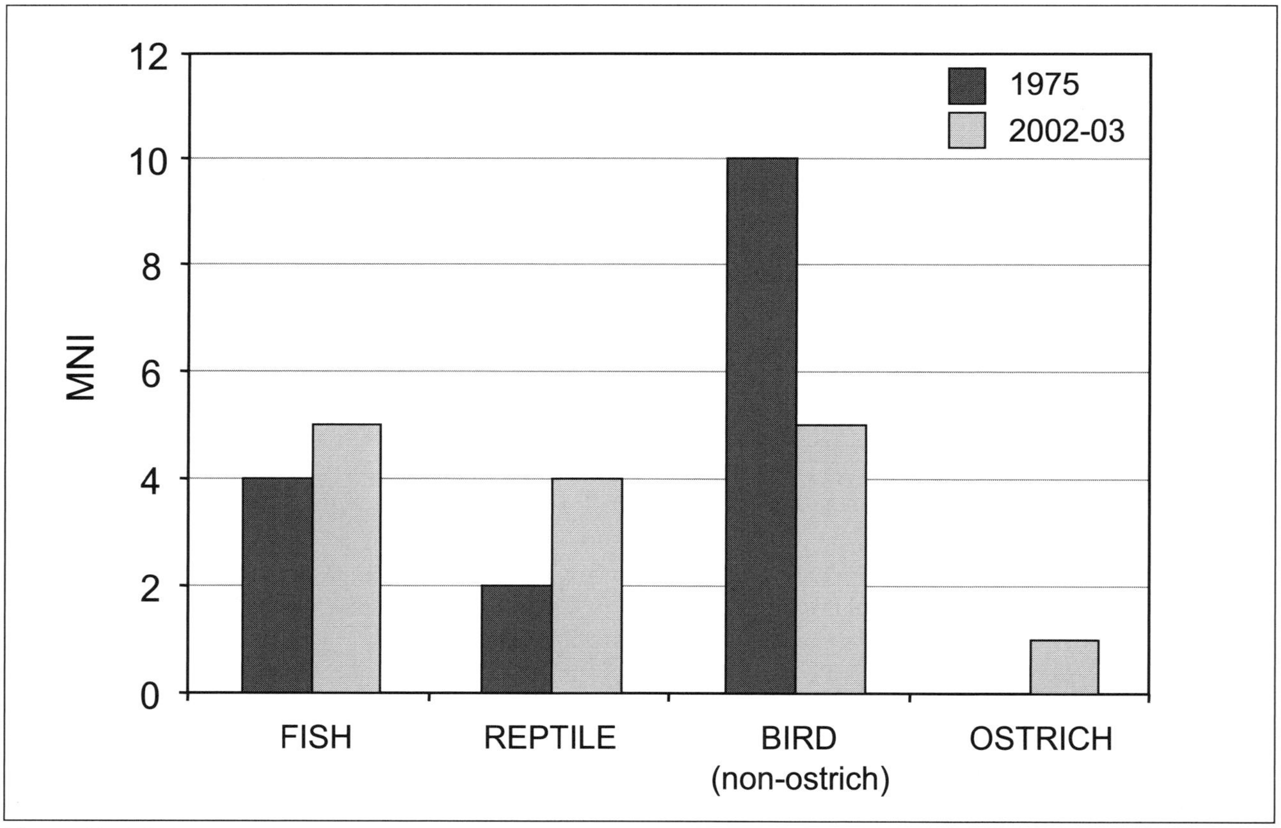


*Figure 13. MNI counts of non-mammals in the 20 Amboseli bone transects for fish (catfish), reptiles (tortoise), and birds, with ostrich plotted separately.*

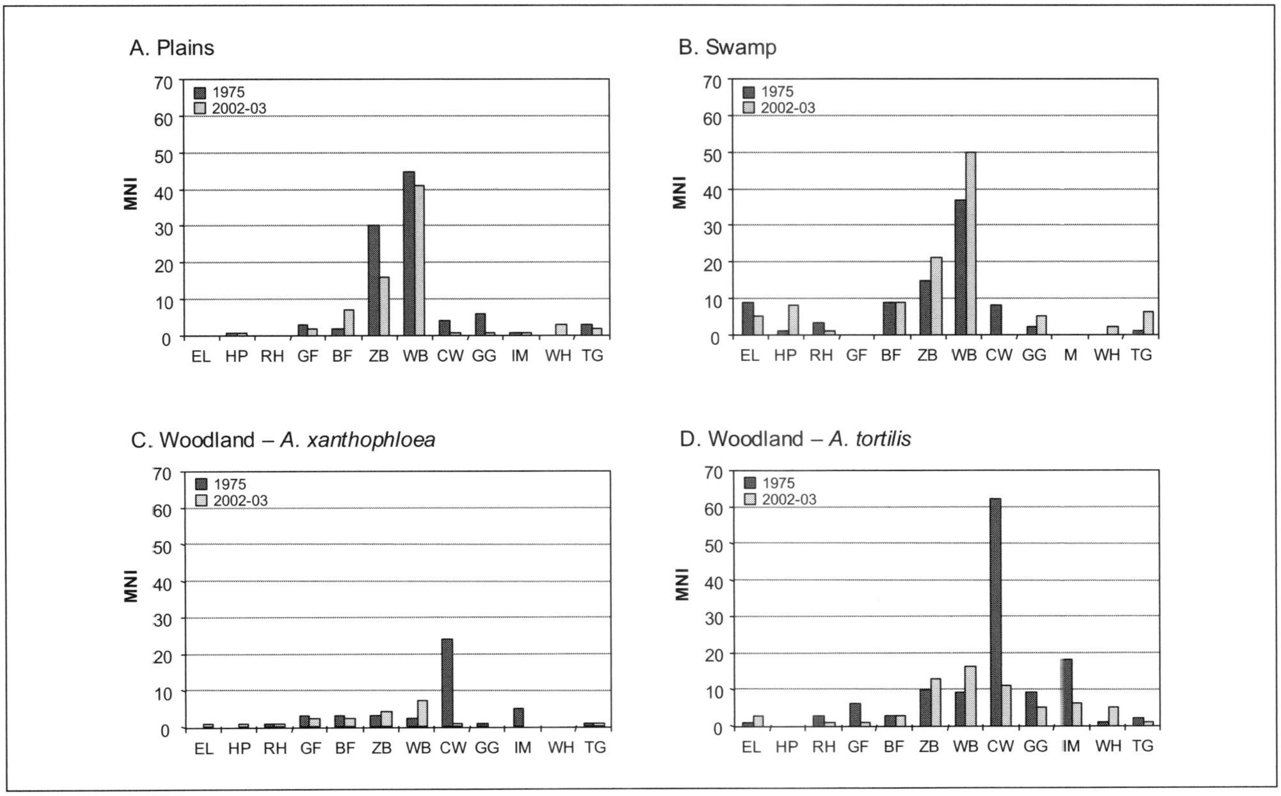


*Figure 14. Comparison of MNI abundances for the 12 most common mammal species in 4 different habitats, based on bone surveys in 20 transects. See Figure 13 for labels on X axis. All Y axes are the same scale to show the overall differences of total MNI among habitats as well as abundances of the different species.*

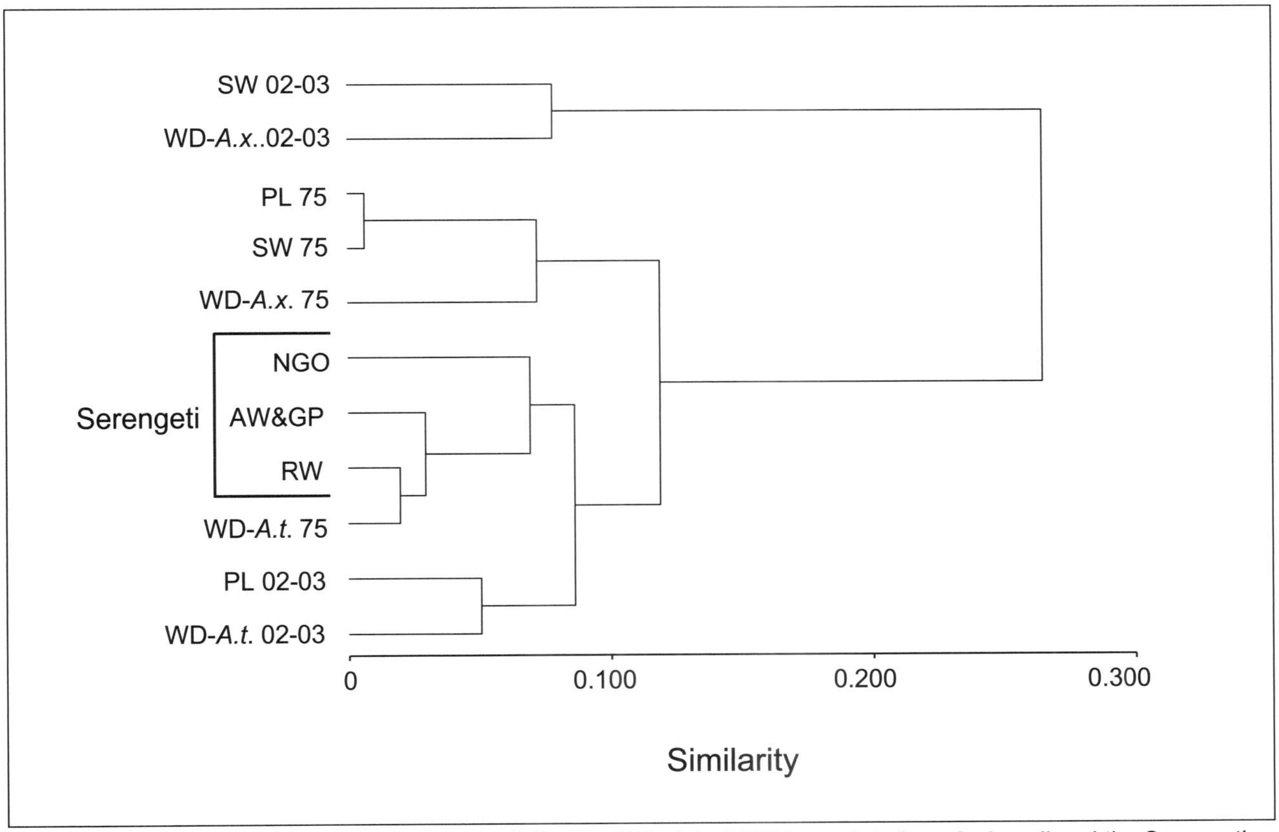


*Figure 15. Cluster analysis using Pearson correlation coefficient for MNE bone data from Amboseli and the Serengeti, based on skeletal parts of Size 3 mammals (excluding cow). See Table 6; Serengeti data from Blumenschine, 1989. Habitat abbreviations for Serengeti: NGO = Ngorongoro, AW&GP =* Acacia *woodland and grassy plains, RW = riparian woodland; all other branch ends are Amboseli data: WD-A.t. =* Acacia tortilis *woodland, WD-A.x. =* Acacia xanthophloea *woodland.*

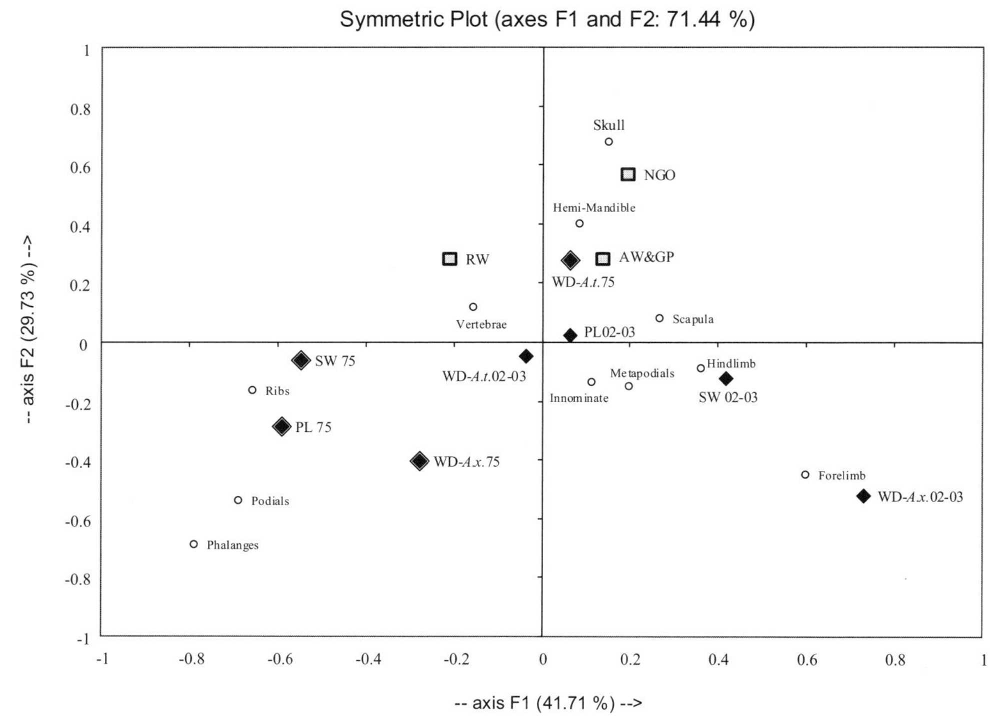


*Figure 16. Correspondence Analysis comparing habitats in the Serengeti (gray diamonds) and Amboseli (double line (1975) and black (2002-03) diamonds), based on proportional MNE skeletal part data of Size 3 mammals (excluding cow). Skeletal parts plotted on the diagram show which of these have the most impact in the distribution of the 7 habitat samples on the two major axes. Based on data from Table 5; Serengeti data from Blumenschine (1989). Habitat abbreviations: Serengeti: NGO = Ngorongoro, AW&GP =* Acacia *woodland and grassy plains, RW = riparian woodland. Amboseli: WD-A.t. =* Acacia tortilis *woodland, WD-A.x. =* Acacia xanthophloea *woodland, PL = Plains, SW = Swamp (1975 and 2002-03).*

samples form a cluster with these two woodlands, and together they are associated with Amboseli's plains and *A. tortilis* woodland in 2002–03. The Pearson coefficient of similarity values (Table 5C, 5D) shows that most of these samples are significantly alike, and slight differences in the numbers are only an indication of which pairs are more or less similar based on the skeletal part data. The correspondence analysis in Figure 16 provides more insight into which skeletal elements are responsible for the observed hierarchy in the cluster diagram. The high relative abundance of skulls and mandibles groups the Serengeti samples and 1975 Amboseli *A. tortilis* woodland, abundant distal limb elements (podials and phalanges) group the other three 1975 Amboseli samples, and a higher proportion of hind- and forelimb elements pull the 2002–03 Amboseli swamp and *Acacia xanthophloea* woodland away from the others.

The 2002–03 Amboseli samples are the outliers on the cluster diagram, while the 1975 samples group with each other and with the three Serengeti habitats. These patterns show that there is no dominant habitat signal in the Amboseli surface bone assemblages—instead, greater differences occur between the same habitats in the 1975 vs. 2002–03 samples than among the habitats themselves.

## Discussion

Changes in the Amboseli bone assemblages over nearly 30 years provide a basis for considering the potential impact of resource fluctuations that might have affected predator/scavengers in the Plio-Pleistocene. What would be the impact of predator-prey cycles or climate-driven habitat changes over decades, on the availability of species for scavenging or hunting? Blumenshine (1989) proposed that woodland habitats in the Serengeti ecosystem provide more opportunities for hominin scavenging than open plains habitats because there are lower numbers of other carcass consumers (e.g., hyenas) in woodlands. Recent work by Domingquez-Rodrigo (2001) indicates that competition with other predators is lower in riparian woodland vs. open plains habitats, further supporting Blumenschine's hypothesis. Tappen (1995, 2001), in contrast, has countered that in the savanna ecosystem of the western rift valley of Zaire, there are few scavenging opportunities in any of the habitats she sampled, indicating that deliberate foraging for carcasses would not have been a productive use of time and energy for early homnins.

The results of the Amboseli analysis show that, from 1975 to 2002–03, bones per individual have decreased by nearly 75%, skeletal part survival has shifted toward higher proportions of the more durable limb elements, and the degree of damage to individual bones has increased, with fewer whole elements and more partial or fragmentary remains. These changes are particularly marked for species in body sizes 3 and 4 (wildebeest, zebra, cow, buffalo). Isolated identifiable bone fragments have shifted from 21% to 57% of the recorded occurrences, and variability in bones per individual has evened out across the different habitats. There also is a higher proportion of adult vs. juvenile remains in the later sample. All of these results indicate a marked increase in the impact of taphonomic processes that disperse and destroy skeletal remains over the past several decades. In contrast, the numbers of bone occurrences and MNIs on the transects have changed relatively little (10% and 20% decreases, respectively) and wild species diversity recorded in the bone assemblages in the four habitats has remained remarkably stable. This implies that shifts in the ecology of the basin, including the removal of domestic stock and the loss of woodlands, have had less of an effect on the changes in the bone assemblages than destructive processes acting on these assemblages.

Changes in the predator populations in Amboseli over the past 15 years, combined with an unusual drought die-off in the early 1970s, provide an explanation for the marked shift in the taphonomy of the bone assemblages. In 1975, lions were the dominant predator in the basin, spotted hyena (*Crocuta crocuta*) was relatively uncommon, and leopard and cheetah also were present. A severe drought weakened the ungulate populations, resulting in large numbers of dead wild and domestic animals in 1972-73; a number of these mummified and were still partially intact when transect sampling began in 1975. In the late 1980s, lions were temporarily absent from the ecosystem and the *Crocuta* population began an increase that has continued into the 21st century, with estimates of 250–300 individuals, one of the highest densities of spotted hyena thus far recorded (Watts, Personal Communication, 2004). Intense intra-specific competition for carcasses means that hyenas disperse and consume much more of their own kills and those of other predators, leaving behind primarily the most durable portions of the skeleton. Species in body size categories 2–3 are the preferred prey for *Crocuta*, which is able to completely consume most parts of the skeleton. *Crocuta* is less capable of destroying the bones of larger species, resulting in a gradient of decreasing damage from size categories 3 through 6. Fresh kills can attract up to 40 hyenas at a time, and for body size 3 and smaller animals, there is virtually nothing left at the kill site after 10–20 minutes of feeding frenzy; much is consumed on the spot but any hyena that can tear off a portion of the carcass takes it away immediately for less stressful feeding elsewhere.

In summary, the shift from relatively complete carcasses in 1975 to dispersed, highly processed, fragmentary remains in 2002–03 represents a contrast between low and high scavenging pressure on the Amboseli bone assemblage. The most reasonable explanation is that this has been caused by a major shift in the predator structure, from diverse species dominated by a meat-consumer (lion) to super-dominance by a meat+bone consumer (spotted hyena), combined with a drought-generated carcass surfeit in 1975, when scavenging pressure on the bone assemblage was unusually low. There have been important changes in the vegetation of the Amboseli Basin during the same time interval, and the increase in open habitats probably contributes to the current dominance of spotted hyena, which is primarily a pursuit predator. There is less cover for ambush predators such as lion, although they again are present in moderate numbers; cheetah numbers also appear to be increasing in the central park area (Personal observation; H. Watts, Personal Communication, 2004).

If there were a strong habitat signal in the Amboseli skeletal part data, one would expect that the woodlands to be similar in both 1975 and 2002–03, and that the other habitats would retain characteristic taphonomic signatures distinct from the woodlands and consistent through time (1975 to 2002–03). Contrary to this expectation, however, results indicate that the dominant signal reflects major changes over three decades in biological taphonomic processes affecting the bone assemblages across all habitats. Even between contemporaneous woodland samples in 1975 (i.e., *Acacia xanthophloea* and *Acacia tortilis* woodlands), there is no indication of a distinctive taphonomic "imprint" of this habitat type. All habitat samples were fairly similar in 1975 (Table 5C, 5D), and this similarity actually decreased among some sample pairs in 2002–03, suggesting that the impact of hyena consumption may be somewhat variable among the different habitats, with the most intense carcass destruction in the 2002–03 swamp habitat and the saltbush plain that replaced the former *A. xanthophloea* woodland. The *A. tortilis* woodland assemblages remained the most alike between 1975 and 2002–03. Thus, it is clear that shifts in skeletal survival rate can be controlled by the changes in the dominant predator and are expressed across all the Amboseli habitats analyzed in this study.

Comparisons with the data of Blumenschine (1989) indicate that the proportions of recorded skeletal parts in all three Serengeti habitats are most similar to the 1975 Amboseli *A. tortilis* woodland, but beyond this, there is no particular pattern of clustering with other Amboseli habitat samples of either time interval (Figure 15). The relatively high proportions of cranial remains (skull and mandible) that are responsible for grouping all three Serengeti habitats with the Amboseli, 1975 *A. tortilis* woodland (Figure 16) could relate to visibility of remains in these habitats (i.e., a sampling bias signal), and/or to actual similarity in skeletal part survival patterns in these two ecosystems. The proportions of innominates, also a large and visible skeletal element, are somewhat higher the Serengeti AW&GP and RP habitats vs. the *A. tortilis* woodland samples, but low in the NGO (more open) habitat. This suggests that there may indeed be a sampling bias toward more visible elements in the more densely vegetated habitats. (It should be noted, however, that visibility may also be a factor in predator scavenging and could have affected the foraging strategies of early hominins as well.)

Returning to the question of scavenging opportunities for hominins in woodlands vs. other habitats, the results from Amboseli demonstrate that while there may be a tendency for less destruction and deletion of skeletal parts in relatively stable woodlands (e.g., the Amboseli *Acacia tortilis* woodland), a major shift to bone-consuming predator dominance could overwhelm this pattern, leaving little that would be worth scavenging for periods of years to decades. Therefore, woodlands were not necessarily the optimal habitat for hominin scavenging activities through the Plio-Pleistocene because other processes could intervene to upset the balance of foraging benefits vs. risks. If hominins were more involved in hunting than scavenging, they would also have been affected by the population dynamics and habitat preferences of other predators, and it is unlikely that early hominins would have been effective competitors as scavengers or hunters in habitats, or times, dominated by either felids or hyenids. However, the cycle of carcass abundance vs. scarcity over three decades in Amboseli also points to the periodic opportunities for meat-eating during the times when predator pressure was relatively low. An omnivore with the ability to shift dietary and foraging strategies seems ideally suited to take advantage of changes through time in available resources as well as recognizing and effectively utilizing differences among habitats and their resident predators at any given point in time.

## Conclusions

The overall goal of the Amboseli study was to determine how a bone assemblage tracks decade-scale ecological change in the animal community from which it is derived. Results were also used to test the hypothesis that more complete animal remains typically occur in woodland habitats in the African savanna ecosystems, which has been used in support of a model for early hominin foraging strategies in woodland versus grassland habitats (Blumenschine, 1989). Analysis of the Amboseli data involved counts of skeletal elements and other evidence for carcass modification and bone completeness to determine: 1) the degree of correlation between skeletal part survival and habitat, 2) the impact of ecological change through time on skeletal occurrence patterns and survival. The taphonomic changes in Amboseli show that over several decades, the survival of carcasses and bones can vary markedly within the same ecosystem and within the same habitat. This suggests that the availability of scavengeable remains for hominins not only may vary from habitat to habitat (Blumenschine, 1989), but also change over time-spans shorter than or comparable to the lifetime of individual hominins. The results also indicate that predator pressure exerts a major control on carcass completeness and bone survival patterns in Amboseli and that differences in habitat—carcass associations can disappear under increased abundance of a bone-consuming carnivore. Thus, it is not likely that any particular habitat, such as woodlands, would have been consistently the best place for early hominins to forage for carcasses or hunt for animals. Rather, hominins and other predators or scavengers would have had to adjust their foraging strategies to cycles of shifting abundance in both prey and competing predator/scavenger species, as well as environmentally induced changes in habitat structure.

It is possible that the changes observed in the bone assemblage of Amboseli are a unique result of the combination of human impact, environmental change, and chance, making the shift from carcass glut to scattered fragmentary bones limited as an analogue for taphonomic cycles of the past. However, some variability would be expected over ecological and geological time in biological and ecological processes affecting skeletal remains in any ecosystem. It is clear that changes in the predator populations *can* have major impact on bone assemblages in tropical East Africa, and the same would likely hold true for other vertebrate communities that have both meat and bone-consuming carnivore species. Predator diversity and dominance patterns are unlikely to be stable for long periods of time because they are linked to alternating periods of abundant vs. scarce prey, disease, changes in interspecific competition, or shifts in vegetation structure affecting hunting success.

An unexpected outcome of this study is the relative stability of species richness and evenness (as measured by Fisher's Alpha) through time in the Amboseli habitats, in spite of the loss of woodlands and other habitat changes and the increase in destructive taphonomic processes. This indicates that although the Amboseli surface bone assemblages were greatly reduced in quantity and quality between 1975 and 2002–03, the taxonomic and ecological information in the fragmentary but identifiable remains would show the same basic species abundance distribution for each habitat. The evidence for ecological change (i.e., shift to bone-consuming predator dominance) is in the skeletal element assemblages themselves, and in differences in rank order reflecting more subtle increases and decreases in the abundance of particular species (e.g., increase of wildebeeste and zebra and decline of giraffe and impala remains in the woodlands). If the past 30 years of Amboseli's history were preserved in two successive, stratified assemblages with minimal additional diagenetic bias, there likely would be sufficient evidence to reconstruct the major taphonomic and ecological processes that formed this record.

Skeletal part abundances and completeness, adult/juvenile ratios, and damage patterns and how these vary in relation to prey body size can provide evidence for different levels of predator pressure on bone assemblages. Similar information on skeletal elements and bone modification is preserved in the paleontological/archeological record, but the value of such data depends on collecting strategies that control for taxonomic, taphonomic, and sampling biases in both excavated and surface assemblages. If such controlled samples could be documented, then the actualistic evidence from Amboseli, the Serengeti, Zaire and other ecosystems could be used to develop and test hypotheses regarding predator pressure and scavenging and/or hunting behaviors that affected ancient bone assemblages. This approach would be particularly appropriate for fossil assemblages that can be sampled in the context of preserved paleolandscapes. Given the problems of time-averaging and variability in the ecological and taphonomic processes that may have affected skeletal survival rates in the past, it will be essential to develop and compare as many actualistic studies as possible to obtain a realistic sense of variation in predator/scavenger impacts on bone assemblages during the Plio-Pleistocene.

## Acknowledgments

I am deeply grateful for the early inspiration of Bob Brain, who showed that the best way to explore the mysteries and revelations of taphonomic processes was through careful experiments and observations in modern ecosystems. I also sincerely thank David Western for the initial opportunity to work on the taphonomy of Amboseli and for his continuing interest in and collaboration in this research. Dorothy Dechant Boaz, Bill, Kristina, and Sarah Keyser, Clive Trueman, Paul Koch, Alan Cutler, Kathleen Stewart, Sarah Lansing, Erustus Kanga, Fred Lala, Fiona Marshall, Jeanne Altmann, Andrew Hill, Henry Bunn, Briana Pobiner, Diane Gifford-Gonzalez, Ralph Chapman, and many others have collaborated on field and lab work over the three decades of the study. Heather Watts and Kay Holekamp, and Susan Alberts have provided interesting and important insights on the living hyena population. Institutions supporting the research include The National Museums of Kenya, The Kenya Wildlife Service, and the Smithsonian Institution, and particularly the National Geographic Society with long-term funding of this research (Grants #1508, #4339-90, and #7525-03). Josh Miller and Tyler Faith provided helpful comments on the manuscript. I thank Travis Pickering, Nick Toth and Kathy Schick for organizing the 2004 conference in honor of Bob Brain and for their support and patience during the process of manuscript preparation.

## References

Altmann, J., Alberts, S. C., Altmann, S. A., Roy, S. B. 2002. Dramatic change in local climate patterns in the Amboseli basin, Kenya. African Journal of Ecology 40, 248-251.

Behrensmeyer, A. K. 1978. Taphonomic and ecologic information from bone weathering. Paleobiology 4, 150-162.

Behrensmeyer, A. K. 1993. The bones of Amboseli: Bone assemblages and ecological change in a modern African ecosystem. National Geographic Research, 402-421.

Behrensmeyer, A. K., Western, D. Dechant Boaz, D. E. 1979. New perspectives in paleoecology from a recent bone assemblage, Amboseli Park, Kenya. Paleobiology 5, 12-21.

Behrensmeyer, A. K., Dechant, D. E. 1980. The recent bones of Amboseli Park, Kenya in relation to East African paleoecology. In: Behrensmeyer, A. K., Hill, A. (Eds.), Fossils in the Making. University of Chicago, Chicago, pp. 72-93.

Behrensmeyer, A. K., Stayton, C. T., Chapman, R. E. 2002. Taphonomy and ecology of modern avifaunal remains from Amboseli Park, Kenya. Paleobiology 29, 52-70.

Blumenschine, R. J. 1986. Early hominid scavenging opportunities: implications of carcass availability in the Serengeti and Ngorongoro ecosystems. British Archaeological Reports, International Series 283, 1-163.

Blumenschine, R. J. 1989. A landscape taphonomic model of the scale of prehistoric scavenging opportunities. Journal of Human Evolution 18, 345-371.

Blumenschine, R. J. 1991. Prey size and age models of prehistoric hominid scavenging: test cases from the Serengeti. In: Stiner, M. C. (Ed.), Human Predators and Prey Mortality, Westview Press, Boulder (CO), pp. 121-147.

Blumenschine, R. J., Cavallo, J. A. 1992. Scavenging and human evolution. Scientific American 267, 90-96.

Blumenschine, R. J., Marean, C. W. 1993. A carnivore's view of archaeological bone assemblages. In: Hudson, J., (Ed.), From Bones to Behavior: Ethnoarchaeological and Experimental Contributions to the Interpretation of Faunal Remains, pp. 273-300. Center for Archaeological Investigations, Carbondale (IL), pp 273-300.

Blumenschine, R. J., Peters, C. R., Masao, F. T., Clarke, R. J., Deino, A. L., Hay, R. L., Swisher, C. C., Stanistreet, I. G., Ashley, G. M., McHenry, L. J., Sikes, N. E., van der Merwe, N. J., Tactikos, J. C., Cushing, A. E., Deocampo, D. M., Njau, J. K., Ebert, J. I. 2003. Late Pliocene *Homo* and hominid land use from western Olduvai Gorge, Tanzania. Science 299,1217-1221.

Brain, C. K. 1967. Hottentot food remains and their bearing on the interpretations of fossil bone assemblages. Scientific Papers of the Namib Research Station 32, 1-7.

Brain, C. K. 1969. The contribution of the Namib Desert Hottentots to an understanding of australopithecine bone accumulations. Scientific Papers of the Namib Research Station 39, 13-22.

Brain, C. K. 1981. The Hunters or the Hunted? University of Chicago, Chicago.

Bunn, H. T. 1982. Meat-eating and human evolution: studies on the diet and subsistence patterns of Plio-Pleistocene hominids in East Africa. Ph.D. Dissertation. University of California, Berkeley.

Bunn, H. T., Bartram, L. E., Kroll, E. M. 1988. Variability in bone assemblage formation from Hadza hunting, scavenging and carcass processing. Journal Anthropological Archaeology 7, 412-457.

Bunn, H. T., Ezzo, J. A. 1993. Hunting and scavenging by Plio-Pleistocene hominids: Nutritional constraints, archaeological patterns, and behavioral implications. Journal of Archaeological Science 20,365-398.

Cutler, A. H., Behrensmeyer, A. K., Chapman, R. E. 1999. Environmental information in a recent bone assemblage: Roles of taphonomic processes and ecological change. In: Martin, R., Goldstein, S., Patterson, R. T. (Eds.), Fossil Taphonomy: Paleoenvironmental Reconstruction and Environmental Assessment. Palaeogeography, Palaeoclimatology, Palaeoecology 149, 359-372.

Dominguez-Rodrigo, M. 2001. A study of carnivore competition in riparian and open habitats of modern savannas and its implications for hominid behavioral modeling. Journal of Human Evolution 40, 77-98.

Gifford, D. P. 1980. Ethnoarcheological contributions to the taphonomy of human sites. In: Behrensmeyer, A. K., Hill, A. (Eds.), Fossils in the Making. University of Chicago, Chicago, pp. 94-107.

Gifford, D. P., Behrensmeyer, A. K. 1977. Observed formation and burial of a recent human occupation site in Kenya. Quaternary Research 8, 245-266.

Haynes, G. 1985. On watering holes, mineral licks, death, and predation. In: Meltzer, D., Mead, J. I. (Eds), Environments and Extinctions in Late Glacial North America. Center for the Study of Early Man, Orono (ME), pp. 53-71.

Haynes, G. 1988. Mass deaths and serial predation: comparative taphonomic studies of modern large-mammal death-sites. Journal of Archaeological Science 15, 219-235.

Hill, A. 1975. Taphonomy of contemporary and late Cenozoic East African vertebrates. Ph.D. Disssertation. University of London.

Hill, A. 1980. Early post-mortem damage to the remains of some contemporary East African mammals. In: Behrensmeyer, A. K., Hill, A. (Eds.), Fossils in the Making. University of Chicago, Chicago, pp. 131-155.

Hill, A. 1990. Bone modification by modern spotted hyenas. In: Bonnichsen, R., Sorg, E. M. (Eds.), Bone Modification. Center for the Study of the First Americans, Orono (ME), pp. 169-178.

Koch, P. L., Behrensmeyer, A. K., Tuross, N., Stott, A. W., Evershed, R. P., Fogel, M. L. 2001. The effects of surficial weathering on the stable isotope composition of bones. Ancient Biomolecules 3,117-134.

Marean, C. W. 1997. Hunter-gatherer foraging strategies in tropical grasslands: Model building and testing in the East African Middle and Later Stone Age. Journal of Anthropological Archaeology 16,189-225.

Marean, C. W., Spencer, L. M., Blumenschine, R. J., Capaldo, S. D. 1992. Captive Hyaena bone choice and destruction, the Schlepp effect and Olduvai archaeofaunas. Journal of Archaeological Science 19,101-121.

Pobiner, B. L., Blumenschine, R. J. 2003. A taphonomic perspective on Oldowan hominid encroachment on the carnivoran paleoguild. Journal of Taphonomy 1, 115-141.

Potts, R. 2003. Chapter 15: Early human predation. In: Kelley, P. H., Kowalewski, M., Hansen, T. A., (Eds.), Predator-Prey Interactions in the Fossil Record. Plenum, New York, pp. 359-376.

Shipman, P. 1981. Life History of a Fossil: An Introduction to Taphonomy and Paleoecology. Harvard University, Cambridge.

Tappen, M. 1995. Savanna ecology and natural bone deposition: Implications for early hominid site formation, hunting, and scavenging. Current Anthropology 36, 223-260.

Tappen, M. 2001. Deconstructing the Serengeti. In: Stanford, C. Bunn, H. T. (Eds.), Meat-Eating and Human Evolution. Oxford University, Oxford, pp. 13-32.

Trueman, C. N. G., Behrensmeyer, A. K., Tuross, N., Weiner, S. 2004. Mineralogical and compositional changes in bones exposed on soil surfaces in Amboseli National Park, Kenya: Diagenetic mechanisms and the role of sediment pore fluids. Journal of Archeological Science 31,721-739.

Tuross, N., Behrensmeyer, A. K., Eanes, E. D. 1989. Strontium increases and crystallinity changes in taphonomic and archaeological bone. Journal of Archeological Science 16, 661-672.

Weigelt, J. 1927. Resente Wirbeltierleichen und ihre Paläobiologische Bedeutung. Verlag von Max Weg, Leipzig.

Western, D. 1973. The structure, dynamics and changes of the Amboseli ecosystem. Ph.D. Dissertation, University of Nairobi, Nairobi, Kenya.

Western, D. 1980. Linking the ecology of past and present mammal communities. In: Behrensmeyer, A. K., Hill, A. (Eds.), Fossils in the Making. University of Chicago, Chicago, pp. 72-93.

Western, D. 2002. In the Dust of Kilimanjaro. Island Press, Washington, D.C.

Western, D. and C. van Praet. 1973. Cyclical changes in the habitat and climate of an East African ecosystem. Nature 241,104-106.

Western, D. and D. Maitumo. 2004. Woodland loss and restoration in a savanna park: a 20-year experiment. African Journal of Ecology 42,111-121.

Yellen, J. E. 1977. Cultural patterning in faunal remains: evidence from the !Kung Bushmen. In: Yellen, J E. (Ed.), Experimental Archaeology. Academic Press, New York, pp. 271-331.

# CHAPTER 9

# VERTEBRATE TAPHONOMIC PERSPECTIVES ON OLDOWAN HOMININ LAND USE IN THE PLIO-PLEISTOCENE OLDUVAI BASIN, TANZANIA

ROBERT J. BLUMENSCHINE, CHARLES R. PETERS, SALVATORE D. CAPALDO, PETER ANDREWS, JACKSON K. NJAU AND BRIANA L. POBINER

## ABSTRACT

We survey vertebrate taphonomic studies being conducted by the Olduvai Landscape Paleoanthropology Project (OLAPP) at Olduvai Gorge, Tanzania. These studies continue the tradition established by C.K. Brain of using taphonomy to reconstruct past environments and to identify the influence of carnivores on fossil bone assemblages, while extending it to establish some of the landscape ecological correlates of variability in Oldowan hominin activity traces in the Plio-Pleistocene Olduvai Basin. We use neotaphonomic findings in a variety of modern settings to interpret aspects of vertebrate fossil and stone artifact assemblages recovered from landscapes in the eastern lowermost Bed II Olduvai Basin by OLAPP since 1989, and to some of the Bed I and Lower Bed II bone assemblages recovered earlier by M.D. Leakey (1971). We address selective bone burial in lake-margin settings, crocodiles as taphonomic agents in wetland settings, small mammal assemblages as indicators of vegetation structure, tooth-marked and percussion-marked bone as indicators of the sequence of access by hominins and carnivores to carcass foods, and correlations of hyaenid ravaging of mammal long bones with patterns of stone artifact discard and loss by Oldowan hominins.

## INTRODUCTION

Fewer than 40 years after C.K. Brain's (1967, 1969) pioneering observations of differential bone survivorship, vertebrate taphonomy has come to play a pivotal role in interpretations of hominin eco-behavioral evolution. Brain showed that density-dependent destruction of goat bones by dogs in Hottentot villages produced a skeletal part profile similar to that attributed by Dart (e.g., 1949) to hunting and tool-use by *Australopithecus africanus* at the South African Pliocene site of Makapansgat. This pioneering work was later extended to include a series of other modern bone-modifying species, with applications to other South African fossil-bearing cave sites (compiled in Brain, 1981). Brain was one of the first to demonstrate that non-hominin species could be important contributors of bones to assemblages that include hominin fossils and artifacts. Of perhaps greater conceptual importance, Brain also showed that vertebrate fossil assemblages contain diverse information about animal behavior and ecology that can be interpreted reliably if they are evaluated neotaphonomically, using systematic observations of relevant processes in modern settings. While studies of modern physical processes had been shown by geological scientists to provide effective interpretive models of physical processes acting in the past (e.g., Emiliani's [1955] paleo-temperature studies), Brain's work was the first to show in a systematic manner for paleoanthropology that reconstructions of paleo-behavior were amenable to a similar approach. For archaeology, Binford (e.g., 1981) would later formalize this approach to behavioral reconstructions under the label of middle range research, but it was Brain's work that laid the foundation for paleoanthropology's growing and increasingly productive reliance on behaviorally-oriented neotaphonomic research.

Brain focused on skeletal part profiles to identify the biological agents involved in the formation of bone assemblages. He expanded his early studies of differential destruction of bones by dogs to include bone modification by leopards, cheetah, brown hyenas, spotted hyenas, porcupines, various owls, and black eagles. He invoked bone density and timing of bone fusion in growing animals as the osteological properties underlying differential destruction of bone elements and portions thereof, properties which were likely to have influenced the survivability of bones of ancient animals being consumed by prehistoric carnivores. He also used bone surface modification to identify the actions of ancient animals. This work includes most famously his demonstration that the punctures fossilized in the cranium of an australopithecine (SK 54) from Swartkrans were likely inflicted by an animal with the canine morphology and inter-canine spacing of the fossil leopards also found at the site (Brain, 1970). Today, skeletal part profiles and bone surface modifications remain as the categories of information about vertebrate fossil assemblages that are most widely used in hominin behavioral reconstructions.

In this chapter, we expand on the methodological tradition established by Brain, by applying neotaphonomic studies to an understanding of the Plio-Pleistocene landscapes and traces of hominin land use recorded at Olduvai Gorge, Tanzania. Brain's use of vertebrate taphonomy to infer aspects of prehistoric hominin land use was limited by the sparse distribution of the South African limestone caves and the poor temporal resolution of their fossil deposits. His inferences were focused necessarily on the extent to which hominins were responsible for accumulating single site assemblages, and on whether various cave sites were hominin living sites. Behrensmeyer's (e.g., 1975; 1985; Behrensmeyer et al., 1979) studies in Kenya's Amboseli Basin, with applications to Plio-Pleistocene surface fossil assemblages at East Turkana, Kenya, showed that landscape-scale variations in the identity of various biotic and abiotic agencies of vertebrate bone assemblage formation can help to resolve paleolandscape mosaics and aspects of prehistoric community structure. Blumenschine (1989) showed that the skeletal part and portion composition of modern surface bone assemblages in the Serengeti are sensitive to competition among spotted hyenas for fresh carcass foods from larger mammals.

Here, we preview some of the ongoing vertebrate taphonomic studies being conducted by the Olduvai Landscape Paleoanthropology Project (OLAPP). These studies continue the tradition of using taphonomy to reconstruct past environments and to identify the influence of carnivores on fossil bone assemblages, while extending it to explore some of the landscape ecological correlates of variability in Oldowan hominin activity traces in the Plio-Pleistocene Olduvai Basin. Our taphonomic studies integrate neotaphonomic observations with applications to the vertebrate fossil and stone artifact assemblages recovered from landscapes in the eastern lowermost Bed II Olduvai Basin by OLAPP since 1989, and to some of the Bed I and Lower Bed II bone assemblages recovered earlier by M.D. Leakey (1971). These studies include the following:

1. Selective burial of bone in lake-margin settings, based on our ongoing studies of the Wildebeest Graveyard (Capaldo and Peters, 1995) at Lake Masek, Tanzania.
2. The establishment of crocodiles as taphonomic agents in wetland settings in Bed I and Lower Bed II, and its implications for the setting of early hominin activities.
3. Indications from small mammal assemblages of changes in the vegetation structure of a part of the Olduvai Basin in Bed I times.
4. Interpretive applications to two Bed I sites (FLK 22 (*Zinjanthropus* level) and FLK-N levels 1 and 2) of simulations of the sequence of access by hominins and carnivores to carcass foods based on bone surface modifications.
5. Use of end to shaft ratios for medium-sized and larger mammal long bones as an index of hyaenid and possibly large canid bone ravaging, and the use of this index as a proxy for predation risk that correlates with patterns of stone artifact discard and loss by Oldowan hominins across ecostructurally distinct landscapes of the paleo-Olduvai Basin.

## Background to OLAPP

A long term goal of OLAPP is to understand the manner in which Oldowan hominins utilized the landscapes of the prehistoric Olduvai Lake Basin. The effort is designed to both reconstruct the landscape mosaics that existed in geologically exposed and unexposed parts of the paleo-Olduvai Basin, and to infer the nature of hominin activities in these landscapes. Landscape reconstruction is focused on variations in the basin's terrain, hydrology, and probable vegetation structure across space, and through time in response to short and long-term climate change as well as volcanically-induced landscape successions. Reconstructions of Oldowan hominin land use, focusing on diet, food acquisition, stone-tool technology and ranging patterns, are aimed at quantifying variability in hominin trace fossils (stone artifacts and butchered bones) across landscapes and through time, and relating these variations to contrasts in the resources and hazards that hominins are inferred to have encountered in the Olduvai Basin. Ultimately, our understanding of hominin interactions with these landscapes will be informative about the ecological circumstances surrounding the dependency of early *Homo* on flaked-stone tool use, their use of food resources from animals larger than the prey taken by non-human primates, and the evolution of the human central nervous system.

Vertebrate taphonomy is necessarily only one component of the larger multidisciplinary research program OLAPP is pursuing to achieve these goals. The project also emphasizes a variety of geological and paleobotanical approaches to landscape reconstruction. As with our vertebrate taphonomic work, these approaches integrate studies of the paleo-record with relevant observations of the natural environments of eastern and southern Africa. Although this chapter is restricted to OLAPP's vertebrate taphonomic studies, our multidisciplinary approach is an acknowledgement of the inferential complexity of landscape and land use reconstructions (cf. Gifford-Gonzalez, 1991).

We apply our neotaphonomic results to two sets of vertebrate fossil samples from Olduvai Gorge. One sample is derived from our selective re-analyses of bone assemblages recovered by Mary Leakey (1971) from excavations into Bed I and Lower Bed II. All of these assemblages are located in what Hay (1976) refers to as the Eastern Lake Margin (Figure 1a). The second sample is a subset of that recovered by OLAPP. It is restricted to the lowermost Bed II Eastern Lake Margin and distal Eastern Alluvial Fan (Figure 1). The sample consists of nearly 8000 vertebrate specimens (NISP) derived from approximately 465 $m^3$ of lowermost Bed II deposits excavated from 98 trenches. For the purposes of our analyses, the OLAPP trench sample is aggregated into a number of geographic locales depicted in Figure 1b.

## Differential Burial of Skeletal Parts in Lake-Margin Settings

Following Brain's lead, many studies have been conducted on the differential survivorship of bones subjected to a variety of destructive processes, particularly carnivore feeding. Very few studies have focused on the differential burial of bone (Behrensmeyer, 1983; Behrensmeyer and Dechant-Boaz, 1980) despite the fact that burial is prerequisite to fossilization. Burial potential of bone is apparently low in open-air and unvegetated settings, where exposure to subaerial weathering and trampling can lead to bone fragmentation and destruction beyond that inflicted by carnivores. Judging from modern shallow lake basins in East Africa, the lake-margin terrain that existed during Bed I and Lower Bed II times in the Olduvai Basin would have been exposed primarily as unvegetated mudflats or short grass and sedge pasturelands (Peters and Blumenschine, 1995, 1996). Bones in these setting would likely have been subjected to intense subaerial weathering and trampling except in well-vegetated (grass-covered) upper shore zones and in wet, low-lying areas (wet-mud lake shorelines, small channels, marshes). Observations of bones from drowned wildebeest at Lake Masek in northern Tanzania provide an opportunity to investigate the differential burial potential of skeletal parts.

Lake Masek is a small, relatively deep and narrow soda lake approximately 4 km long and 0.5 km wide. It is located in the southern Serengeti Plains adjoining Lake Ndutu, along the course of the drainage that subsequently flows eastward through Olduvai Gorge. Both lakes fluctuate in size with seasonal changes in rainfall. Compared to the very shallow and relatively broad Lake Ndutu, the flood zone along the north shore of Lake Masek is narrow, with a steeper gradient.

Both lakes are the site of episodic drownings of wildebeest that aggregate in the southeastern Serengeti Plains during the long rainy season. Mass drownings involving hundreds to thousands of individuals occur occasionally (for examples, see references in Capaldo and Peters (1995)). Drowning incidents that result in the death of only a few individuals are more common. Blumenschine (1986) reported on scavenging opportunities provided by a small drowning incident at Lake Masek in 1984. Capaldo and Peters (1996) reported on a small drowning incident, and on a bone assemblage along the north shore of the lake that resulted from a large drowning event in 1989 or 1990 (Capaldo and Peters, 1995). Njau (2000) later contrasted the surface and subsurface assemblages of wildebeest bones at both lakes. Since that time, we have continued to make at least once-yearly observations of the Capaldo and Peters study site. Referred to informally as the Wildebeest Graveyard, this study site lies along a 400 m stretch of the northern margin of Lake Masek, concentrated in a 30–40 m wide beach zone that has a slope of 4–6 degrees (Capaldo and Peters, 1995). The Wildebeest Graveyard is distinctive in containing a high density of relatively complete bones from a single species. Capaldo and Peters (1995) report a maximal minimum number of individuals (MNI) of 122 for a 100 m interval of the Graveyard area based on crania. The new observations reported briefly here are based on previously unpublished results.

Capaldo and Peters (1995) noted the dynamic burial potential of bone in the Wildebeest Graveyard. They reported a higher density of buried bone in the upper beach zone than in the lower beach zone, apparently due to the presence of clumps of grass on the upper beach. Here the grass traps bone and sediment, and resists the erosion and re-exposure of bone by the streamlets draining the grassy woodland above the beach. Bone was also buried in small channels that drain the upper beach. Overall, however, most of the burial was probably not permanent under current tectonic conditions. The area is in an erosional phase, with subsidence in the Ol'Balbal Depression at the downstream mouth of Olduvai Gorge determining the sub-regional drainage gradient. During years with low lake stands, the small channels cutting through the mid to upper beach have bone within them, some of which appears to be eroding out of previously grassy patches, and some of which appears to be undergoing burial in channel sediment. During exceptionally wet years, there is erosion of the beach at the high water line and drowning of the stoloniferous grass holding buried bone (now underwater on the mid to upper beach). Subsequent lowering of the lake level exposes the mid to upper beach to small-channel erosion once again. The grass succession and potential bone burial on the beach may begin again during this time. These local processes are still poorly understood, and we do not know what bone, if any, is buried in the deeper water of the lake.

Njau (2000) documented differential burial of skel-

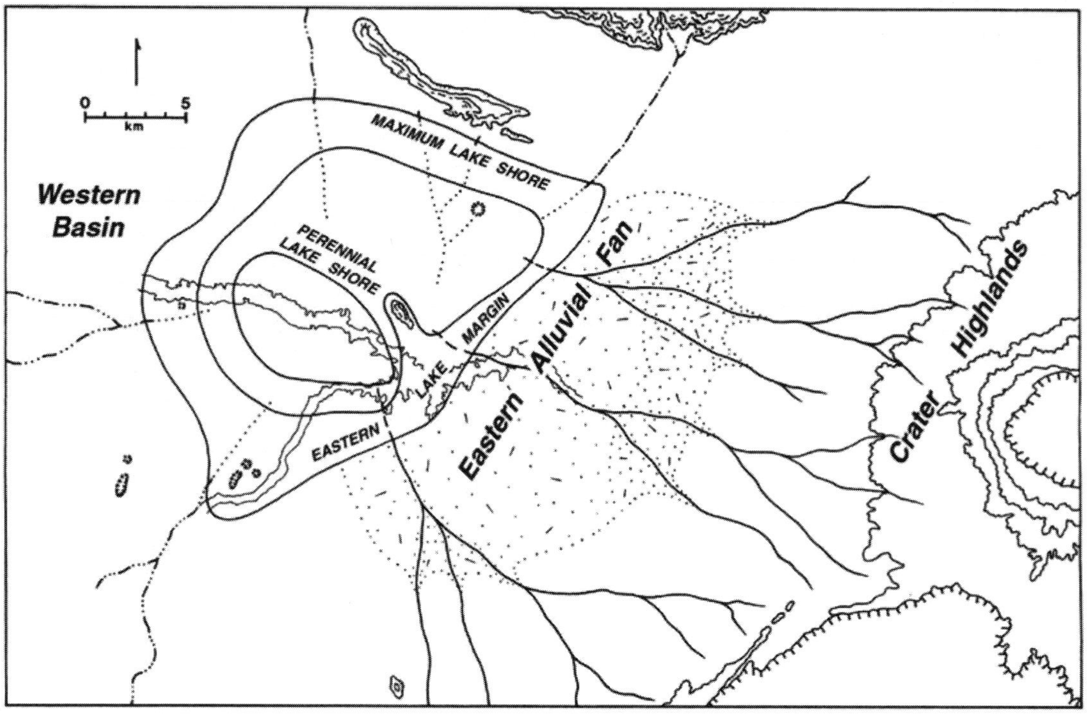


Figure 1 a. Paleogeography of the lowermost Bed II Olduvai Lake Basin (from Peters and Blumenschine, 1995, 1996; based in large part on Hay, 1976) superimposed on the outline of present-day Olduvai Gorge. The map shows the perennial (inner ring) and maximum (outer ring) paleo-lake shorelines, between which lies the lake-margin zone that was exposed during periods of relatively low lake level. The Eastern Alluvial Fan originates from streams draining the Crater Highlands on the east and southeast margins of the basin. The Western Basin extends into the Serengeti Plains. The paleogeographic zones are exposed to examination throughout the Gorge, and along fault escarpments (not shown) in the vicinity of the Gorge.

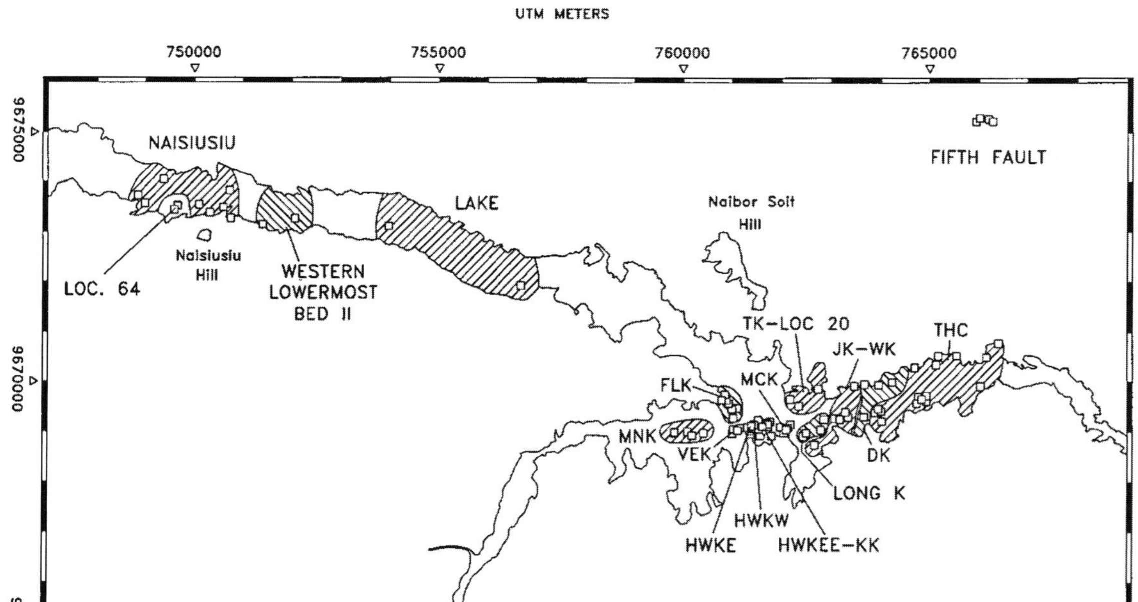


b. Olduvai Gorge, showing the location of OLAPP trenches (boxes) and the geographic locales into which they are allocated. The sample discussed in this paper derives from the Eastern Lake Margin and distal Eastern Alluvial Fan, and includes trenches in locales from MNK to THC.

etal elements in bone assemblages along the margins of Lakes Ndutu and Masek, including the Wildebeest Graveyard study site. He found buried bone in upper beach zones with small channels (Peters' observations). Relative to surface samples, vertebrae, ribs, and compact bones (tarsals, carpals, phalanges) are preferentially represented in partially and fully buried samples, while crania, scapulae, pelves, and to a lesser extent isolated teeth were underrepresented. The representation of mandibles and long bones is similar in buried and surface assemblages. The preferential burial of compact bones has been noted by Behrensmeyer (1983, Behrensmeyer and Dechant-Boaz, 1980), a phenomenon she attributes to the ease by which these bones can be pushed into a substrate by trampling. Trampling alone, however, cannot explain the preferential burial of vertebrae and ribs, given their larger size and surface area. Indeed, the preferential burial of ribs and vertebrae seems paradoxical given their relative low density and consequent lower resistance to carnivore ravaging and other subaerial destructive processes.

Figure 2 contrasts skeletal part profiles for two subsamples of the Wildebeest Graveyard. One is that provided by Capaldo and Peters (1995) for a 14 × 8 m focal plot on the upper beach. This sample represents the *in situ* remains of drowned wildebeest that had washed up on the beach during a rainy season when the lake was at a relatively high level. The second sample is provided by an unpublished inventory of bones occurring along the shoreline of the Graveyard during the long dry season in August 2001, when the lake level was lower than that when the first sample was deposited. The shoreline sample includes only those bones lying on the *c.* 1 m wide wave-wash zone of the beach, as well as those partially to fully submerged up to approximately 1 m from the shore.

The bone assemblage along the lakeshore is dominated by vertebrae and ribs. Phalanges, the unfused epiphyses of limb bone elements, and carpals/tarsals are also present, as is a single scapula and a single long bone. With the exception of the single long bone, all of these elements are easily transported by flowing water, falling into fluvial transport groups I and I/II of Voorhies (1969) and Behrensmeyer (1975).

Fluvial lag elements (transport groups II and III), including pelves, crania, and mandibles, are absent from

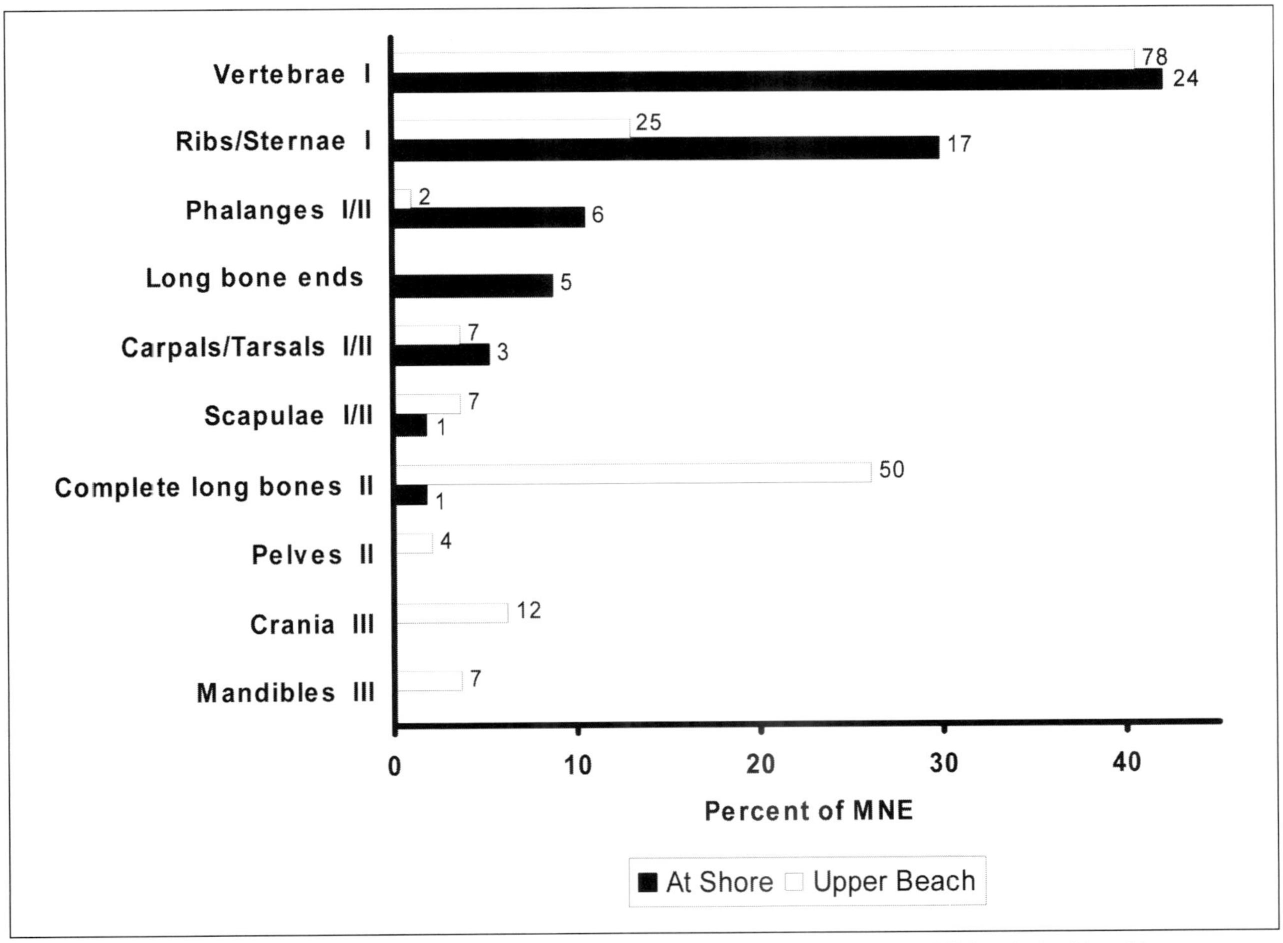


*Figure 2. Skeletal unit representation (percent of total minimum number of elements, MNE) at Lake Masek's Wildebeest Graveyard, contrasting bones at the shore (18 August 2001) with those recorded from a 14 X 8 m area in the upper beach zone by Capaldo and Peters (1995: Table 4). Fluvial transport groups (I, I/II, II, III) of Voorhies (1969) and Behrensmeyer (1975) are indicated for each skeletal unit. Long bone ends were not recorded separately by Capaldo and Peters (1995), and have had no transport group assigned.*

the shoreline sample. This pattern contrasts with the upper beach sample, which shows a more even skeletal part representation and no evidence of fluvial transport (Capaldo and Peters, 1995). Capaldo and Peters also noted that among partially buried bones, those along the lower beach are dominated by postcranial axial specimens, while those on the upper beach are dominated by long bones.

The Wildebeest Graveyard is transected by numerous streamlets that flow episodically during the rainy season. These streams apparently have a flow sufficient to entrain some bones and carry them toward the shoreline. Indeed, a number of these dry small channels contain bones transported part of the way to the shore. The shoreline assemblage contains bone that has been displaced from skeletons of animals that drowned during a rainy season when the shore was located higher on the beach. Transport mechanisms in addition to channelized flow in streamlets may also act to displace vertebrae, ribs and compact bones preferentially shoreward. These include sheetwash during heavy rainstorms, and disturbance by animals traversing the beach to drink at the lake. Regardless of the mechanism, bones that have been transported to the shore are those most likely to be pushed into mud by trampling, or buried at least partially during the next rise in lake level.

The survivorship of skeletal parts with low bone density on modern land surfaces is generally indicative of low degrees of ravaging of skeletons by bone-crunching carnivores. For example, Blumenschine (1989: Table 4) has shown that postcranial axial bones, particularly low-density ribs and vertebrae, show their highest MNE (minimum number of elements) per MNI values in Serengeti settings where carcass consumption is less complete, and bone destruction by spotted hyenas is less intense. On this basis, he proposed that relatively high proportions of postcranial axial bones in prehistoric settings may indicate that ravaging of carcasses by hyaenids was relatively low. Such prehistoric settings presumably offered greater scavenging opportunities for hominins.

For OLAPP's landscape sample from the eastern lowermost Bed II Olduvai Basin, variability in the proportion of postcranial axial to appendicular bones is evident among fossil samples from a number of geographic locales. This result might indicate that there was landscape heterogeneity in hominin scavenging opportunities. However, variation in postcranial axial skeletal part representation is strongly and positively correlated with the occurrence of bones from aquatic species (Figure 3). In light of our observations from Lake Masek, these results suggest that differential representation of skeletal elements in lowermost Bed II can be related not only to perimortem survivorship and carcass food availability to scavengers, but also to fluvial transport and burial potential of bones in lake-margin settings. Resolving this equifinality will require investigation of additional taphonomic features of the bone assemblages (e.g., degrees of fragmentation and tooth-marking), as well as detailed lithofacies analyses to identify depositional environments in which fluvial transport and differential burial of bone may have occurred.

## Crocodiles as Taphonomic Agents in Tropical Wetlands

Numerous studies have documented bone modification by a variety of mammalian carnivores in an effort to isolate evidence for hominin feeding behavior in prehistoric bone assemblages. However, bone modification by crocodilians has not been investigated systematically. Yet, crocodilian body fossils occur in many paleoanthropological assemblages from tropical lake and river basins, and modern crocodile predation on larger mammals is common. Davidson and Solomon (1990) provide the only published description of crocodilian damage to uningested bone, but their observations are limited to a single human victim killed by a saltwater crocodile (*Crocodylus porosus*). Crocodilian damage to fossil bones is inferred rarely and only anecdotally in the paleontological literature.

Our investigation of crocodilian damage to bone was motivated by the discovery among OLAPP's fossil collections from Bed I and lowermost Bed II of several larger mammal bone specimens that bear widespread and densely-packed tooth marks, many of which are very deep and broad (Figure 4). However, these specimens are complete, lacking even the minor gnawing that is produced commonly on lower-density portions of bone by mammalian carnivores. Recognizing this paradox, Blumenschine hypothesized that the bones had been modified by crocodiles during prey capture and/or consumption. The hypothesis was based on the co-occurrence of crocodile (*Crocodylus lloidi*) body fossils in the assemblages from which these specimens derive, as well as the distinctive feeding anatomy and behavior of crocodilians. Unlike mammalian carnivores, which gnaw and break bones of larger mammal carcasses to various degrees during consumption of flesh and within-bone tissues, crocodilians attempt to swallow whole carcass units—including bones—they detach from a prey item. Carcass units too large to ingest are abandoned.

To test Blumenschine's hypothesis, Njau observed Nile crocodiles (*C. niloticus*) feeding in captive settings. Njau also inventoried fresh bone in the vicinity of pools occupied or recently abandoned by crocodiles in the lower Grumeti River, Serengeti National Park, Tanzania. The results, which are reported in detail by Njau (2006) and Njau and Blumenschine (2006), are summarized qualitatively below.

In contrast to assemblages of larger mammal bones produced by mammalian carnivores, those produced by Nile crocodiles are composed of primarily complete elements, the majority of which lack tooth marks. Those bones that are tooth-marked lack gross gnawing and bear an often high density of shallow to deep, transversely to obliquely oriented tooth scores, along with shallow

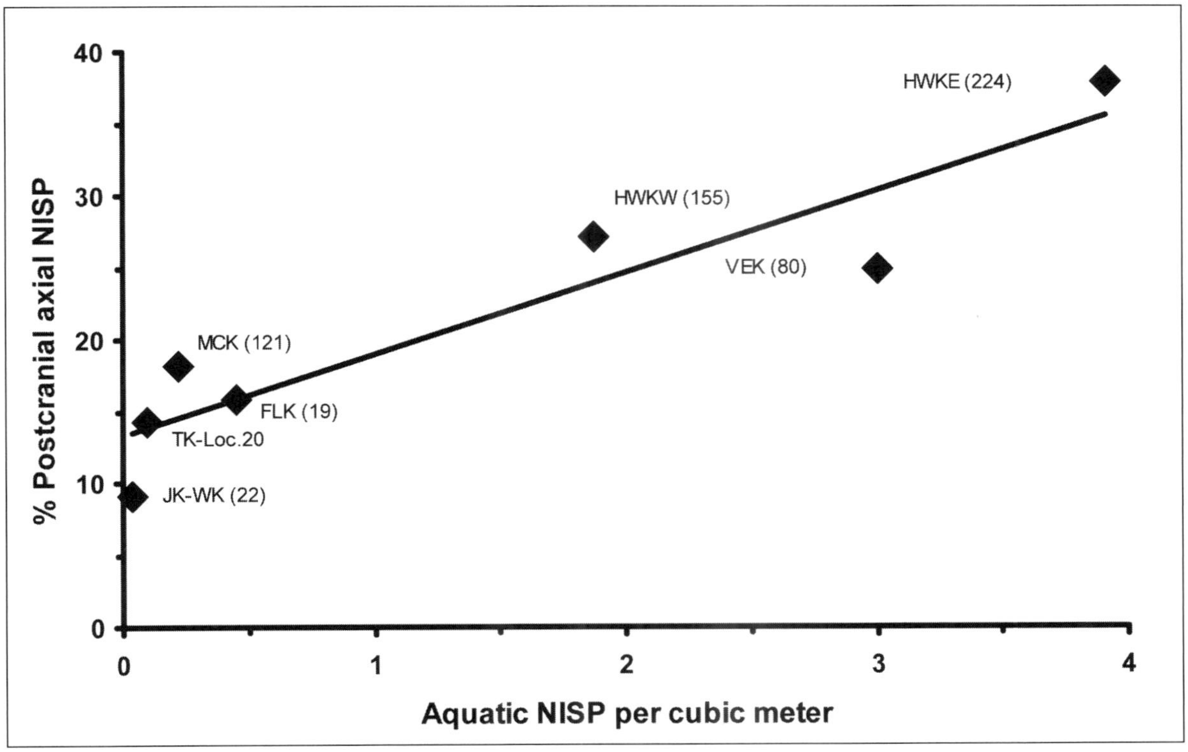


*Figure 3.* *The relationship between the abundance of specimens (NISP values) of fossil aquatic taxa (mainly hippopotamus and crocodile, rarely fish) and the proportionate occurrence of fossil postcranial axial specimens (as % of total non-tooth NISP; numbers in parentheses) of mammals the size of extant wildebeest and buffalo from trenches spanning the whole of lowermost Bed II grouped into seven geographic locales from the eastern Olduvai Basin (Figure 1b). MNK and Long K are excluded due to sample sizes under 10 (total NISPs = 5 and 8, respectively). DK and THC Complexes are excluded due to a lack of aquatic taxa. For the regression line $y = 13.2 + 5.7x$, $r^2 = 0.86$, one-tailed $p = 0.001$.*

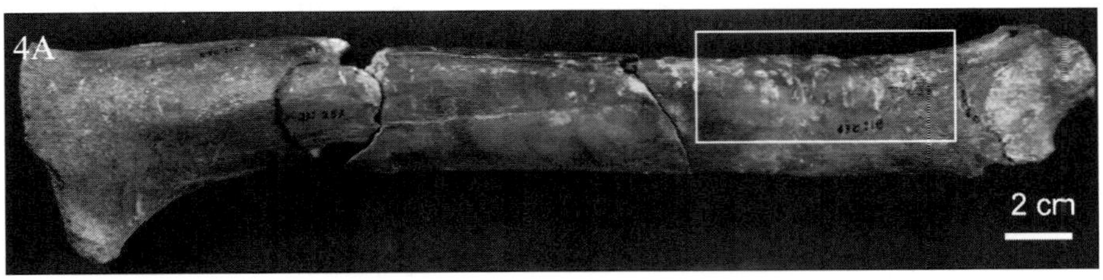


*Figure 4* *a. Posterior view of a complete tibia of a fossil juvenile bovid the size of the extant eland from OLAPP Trench 21. The specimen bears post-fossilization breaks.*

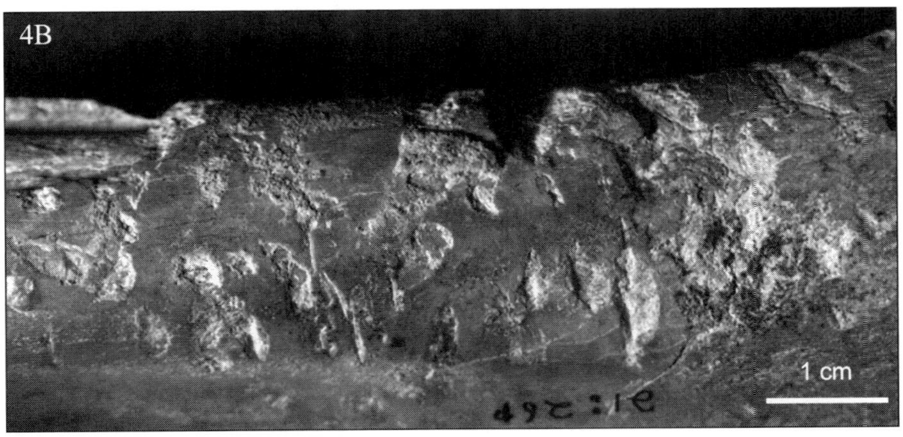


*b. Portion of the distal shaft of the tibia showing intense tooth-marking characteristic of damage by modern Nile crocodiles. The unfused proximal epiphjysis was not recovered, but the proximal end lacks gnawing damage.*

to deep pits and occasionally punctures, some of which have a morphology not observed among tooth marks produced by mammalian carnivores. The combination of intense tooth-marking and lack of gnawing and fragmentation is one of the most conspicuous traces of crocodile feeding that distinguishes it from bones fed on by mammalian carnivores. Assemblages produced by crocodiles are characterized by the retention of low-density bone portions and a number of associated bones that form articulating sets. In contrast, bone assemblages produced by mammalian carnivores are biased commonly against low-density elements and element portions, and contain fewer articulating specimens. The degree of such ravaging depends on the bone-destroying capabilities of the consumer species, and the competitiveness of the feeding episode (especially the number and hunger of consumers relative to carcass size and fat content of marrow and bone grease).

The distinctive types of feeding traces produced on fresh bones by Nile crocodiles are present on fossil specimens from Olduvai Gorge (Figure 4). We infer that these specimens were fed on by *C. lloidi,* the only crocodile known from this time period at Olduvai. We have recognized crocodile damage on a number of bone specimens recovered by OLAPP from stream and lake-margin contexts in Bed I and lowermost Bed II. Additionally, Njau (2006) reports damage consistent with that inflicted by crocodiles within all samples of larger mammal long bones he analyzed from assemblages excavated by Mary Leakey from lake-margin settings in Bed I and lower Bed II at Olduvai. These assemblages were investigated because they contain crocodile body fossils. They include DK levels 1–3, FLK 22, FLK-NN level 3, and HWK-E levels 1 and 2. In both sets of samples, damage inferred to have been inflicted by larger mammalian carnivores as well as by stone-tool-using hominins is also present, sometimes on the same bone specimen. This discovery increases the taphonomic complexity of the bone assemblages, specifically the range of bone accumulating and modifying agents that must be considered in models of site formation from localities where crocodile trace fossils are found. The great predation hazard likely posed by crocodiles to hominins (e.g., Peters and Blumenschine, 1995, 1996) also suggests that hominin activities in the vicinity of wetlands inhabited by crocodiles were more restricted than envisioned by the home base (Isaac, 1978) or living site (Leakey, 1971) models of land use. It also suggests that a significant portion of the hominin deaths represented by hominin body fossils in wetland contexts may have been the result of encounters with crocodiles.

## Paleoenvironmental Implications of Small Mammal Assemblages

To continue investigation of the environments of uppermost Bed I at Olduvai Gorge, a new but small collection of micromammals was collected by Andrews during OLAPP's 2002 field season from immediately below Tuff 1F at FLK-N. Identification of the causative agents responsible for this fauna is also considered here with reference to new data on recent owl pellet assemblages. Leakey (1971) identified six levels at FLK-N that correspond to the upper part of Bed I between Tuffs ID and IF. Levels 1 to 3, the younger of the six, were differentiated on the bases of sediment color and stone artifact density, but were also recognized as belonging to the same sedimentological level by Leakey (1971).

An earlier, more detailed study of the Bed I fossil rodent and shrew assemblages, conducted by Fernandez-Jalvo et al. (1998), documented faunal change from middle to upper Bed I. This study analyzed faunas from nine levels, six in FLK-N and three from middle Bed I in FLK-NN and FLK. The latter faunas from middle Bed I were found to contain small mammals indicative of sub-humid woodlands, with slightly drier conditions in the first part of upper Bed I in levels 4-6 in FLK-N. The top of the sequence in FLK-N levels 1-3, moreover, indicated considerably drier conditions. All three of these levels have low species richness relative to sample size, despite the fact that low rates of digestive wear on incisors suggest they were all accumulated by eagle owls (Fernandez-Jalvo et al., 1998). Eagle owls are noted for the wide range of prey species in their pellet accumulations (Andrews, 1990), both in terms of prey size and species composition. As such, the low species richness at these three levels cannot be attributed to a narrow range of predator prey-selection. Rather, it is more likely to reflect environmental conditions, with drier environments having low rodent species richness. This contrasts, for instance, with high species richness found in middle Bed I at FLK-NN level 2, where the small mammal fauna was probably accumulated by barn owls. Because barn owls are limited by their size to a lower range of prey, this indicates that the differences in species richness are a reflection of changes in the past ecosystem.

In 2002, a small excavation was made into the top of Bed I at FLK-N to confirm the stratigraphy of the Tuff IF channel cutting down into uppermost Bed I deposits (Leakey, 1971, Figure 31). The faunal sample obtained by screening is small, but it agrees in some essentials (Table 1) with the much larger ones collected by Mary Leakey from extensive excavations and described above (Fernandez-Jalvo et al., 1998). Six rodent species have been identified (Table 1), as well as one shrew and three unidentified birds. Digestion of the rodent incisors is light, with about 35% of teeth affected, and this includes a very light digestion category that has not been used previously (Andrews, 1990). This corroborates the previous conclusion that an eagle owl accumulated the small mammal fauna, most probably Verreaux eagle owl (Fernandez-Jalvo et al., 1998). On the other hand, the newly recovered sample shows how unreliable inter-taxa observations can be when derived from a small number of specimens.

Analyses of recent eagle owl and barn owl pellet assemblages are informative about the vegetation that existed in the vicinity of FLK-N during upper Bed I times. Three recent eagle owl (*Bubo lacteus*) pellet assemblages from the Serengeti were investigated earlier (Andrews, 1990), including two from kopjes in the short grass plains and one from within Olduvai Gorge. All were dominated by the gerbil *Tatera robusta*, a nocturnal species. Because of concerns that smaller owls such as barn owls could not predate this relatively large rodent, Andrews made several collections in the Serengeti in 1996 from both barn owl and eagle owl roosts. One such collection (Table 2) is from a barn owl nesting site on one of the kopjes in the short grass plains. The most abundant prey species was *Tatera robusta*, with other rodents and shrews being comparatively rare. Birds and lizards were also common, along with many remains of beetles. This collection compares well with a barn owl prey assemblage collected by Reed (2003), where gerbils were five times as abundant as murines. Finally, a collection of spotted eagle owl pellets (*Bubo africanus*) was made in 2002 within Olduvai Gorge at site MNK. It was dominated by gerbils, both large *(Tatera)* and small *(Gerbillus;* Table 3).

All of the modern prey assemblages considered above occurred in the middle of the short grass plains in the Serengeti. Those from Olduvai Gorge were accumulated by the spotted eagle owl and the Verreaux eagle owl, while those in kopjes away from the gorge were accumulated by the barn owl as well as the two eagle owls. These three predators each produced remarkably similar prey assemblages, with a super-abundance of gerbils and few if any murines. Although murines are known to be abundant in the Serengeti ecosystem (Andrews has seen large numbers of the diurnal grass mouse *Arvicanthis* in areas with high ground cover within the short grass plains), it would appear that they do not form part of the diet of any of the three owls in this part of the Serengeti. Therefore, the presence of eight other murine species in upper Bed 1 at FLK-N levels 1-3 (Table 1), as well as several non-murine and non-gerbilline species, in what is interpreted as an eagle owl prey assemblage, may indicate that the environment being sampled by the predators was not open grassland like that present today on the short grass plains. The gerbil/microtine ratio for FLK-N level 1-3 indicates a locally semi-arid environment, but it was probably one with considerable numbers of trees and/or bushes that can best be designated as wooded grassland.

## Sequencing the Actions of Multiple Consumer Species on Bone Assemblages

Again following Brain's lead, vertebrate taphonomists have made tremendous progress in describing the distinctive features of bone assemblages accumulated and modified by extant consumer species in addition to humans, including a variety of small to large mammalian carnivores, rodents and birds. These single species taphonomic descriptions have proved useful in identifying agents of bone accumulation and modification in fossil assemblages, often demonstrating that multiple consumer species were involved in an assemblage's formation. Bunn's (1981) and Potts and Shipman's (1981) identification of the presence of both stone tool cut marks and carnivore tooth marks on bones from Plio-Pleistocene assemblages is the original demonstration of this phenomenon.

Single species descriptions of assemblage formation are useful for isolating independent contributions of multiple species to a single bone accumulation. We can recognize, for example, bones deposited as owl pellets interspersed with bones of larger animals accumulated by hominins or carnivores in cave assemblages and in open-air settings that once supported roosting trees (e.g., Andrews, 1990).

Probably much more frequently than is recognized, other assemblages contain bones of animals that were fed upon by a series of consumer species, in some cases leaving feeding traces of two or more species on the same bone specimen. Such "dual-patterning" or multi-patterning (Blumenschine, 1988; Capaldo, 1997) is particularly likely to occur with assemblages containing the remains of larger mammal (> 10 kg) carcasses that once offered food surpluses to initial consumers, or within-bone nutrients they could not extract. Indeed, observations in both modern free-ranging and manipulated field settings have demonstrated the commonness by which a series of carnivore species feed on and modify the bones of the ever-diminishing remains of a single carcass. For such assemblages, single species descriptions of bone modification and accumulation are ill-suited for evaluating both the sequence of actions of multiple consumer species, and the types and amounts of carcass tissues consumed by each. This results from the fact that the parts consumed, and the associated bone modifications inflicted, by one consumer will alter both the nutritional attractiveness and the mechanical properties of bones remaining with food for a subsequent consumer (Blumenschine and Marean, 1993).

Blumenschine (1988) provided the first systematic and partially controlled field observations designed to help resolve the dual taphonomic effects of carnivores and hominins on bones of the same animal. The observations focused on the manner by which extant spotted hyenas alter the spatial distribution, long bone portion representation, and tooth and butchery mark frequencies of assemblages of long bones that had been defleshed and de-marrowed using stone tools. Using the frequencies of tooth-marked and tool-marked specimens, these observations were contrasted to simulations of a dual-patterned, "hominin-followed-by-carnivore" sequence of assemblage formation and to two single species models. The single species models include one that is "hominin only," affected only by metal-knife defleshing and hammerstone breakage of marrow cavities, and another that is "carnivore only," where lions and spotted hyenas

*Table 1. Fossil rodents recovered by M.D. Leakey (Fernandez-Jalvo et al., 1998) from FLK-N levels 1-3 combined, compared to those from OLAPP Trench 116 in levels correlative to Leakey's levels 1-3. MNI = minimum number of individuals.*

| Taxon | Leakey Assemblage MNI | OLAPP Assemblage MNI |
|---|---|---|
| *Arvicanthis primaveus* | 0 | 0 |
| *Aethomys lavocati* | 21 | 1 |
| *Mastomys minor* | 3 | 1 |
| *Mus petteri* | 5 | 0 |
| *Oenomys olduvaiensis* | 0 | 0 |
| *Pelomys dietrichi* | 0 | 0 |
| *Thallomys quadilobatus* | 3 | 0 |
| *Grammomys* spp. | 0 | 0 |
| *Zelotomys leakeyi* | 21 | 0 |
| **Total Murinae** | **53** | **2** |
| | | |
| *Gerbillus* spp. | 39 | 1 |
| *Tatera gentryi* | 35 | 2 |
| **Total Gerbillinae** | **74** | **3** |
| | | |
| *Steatomys* spp. | 37 | 1 |
| *Dendromus* spp. | 14 | 0 |
| *Saccostomus cf. mearnsi* | 22 | 0 |
| *Otomys petteri* | 63 | 1 |
| *Xerus cf. inauris* | 5 | 0 |
| *Heterocephalus jaegeri* | 2 | 0 |
| | | |
| **Total rodents (MNI)** | 270 | 7 |
| **Total number of species** | 11 | 6 |
| **Gerbillinae/Murinae MNI ratio** | 1.4 | 1.5 |

*Table 3. Small mammal species composition of recent owl pellets from a crevice at MNK, Olduvai Gorge, with data on relative amounts of damage on incisors by digestion. The first ten genera listed are rodents. Crocidura is a shrew. MNI = minimum number of individuals.*

| Taxon | MNI |
|---|---|
| *Tatera* | 24 |
| *Gerbillus* | 25 |
| *Steatomys* | 15 |
| *Saccostomus* | 2 |
| *Aethomys* | 1 |
| *Dendromus* | 0 |
| *Arvicanthi* | 0 |
| *Lemniscomys* | 0 |
| *Mastomys* | 0 |
| *Mus* | 0 |
| *Crocidura* | 0 |

| Incisor digestion | light | none | % light |
|---|---|---|---|
| *Tatera* | 5 | 34 | 12.8 |
| *Gerbillus* | 3 | 21 | 12.5 |
| isolated indet. | 2 | 15 | 11.8 |

*Table 2. Minimum numbers of individuals for recent rodents (*Tatera, Gerbillus, Steatomys, Otomys*), shrews (*Crocidura*), birds and reptiles from four levels of an excavation below a barn owl nesting site at Kopje 5 in the Serengeti short grass plains.*

| Level | *Tatera* | *Gerbillus* | *Steatomys* | *Otomys* | *Crocidura* | bird | lizard | chameleon | Total |
|---|---|---|---|---|---|---|---|---|---|
| level 1 | 13 | 2 | 0 | 1 | 1 | 4 | 2 | 1 | 24 |
| level 2 | 11 | 6 | 0 | 0 | 1 | 4 | 2 | 0 | 24 |
| level 3 | 49 | 6 | 3 | 0 | 2 | 13 | 6 | 0 | 79 |
| level 4 | 24 | 4 | 3 | 0 | 0 | 1 | 1 | 0 | 33 |
| | | | | | | | | | |
| Totals | 97 | 18 | 6 | 1 | 4 | 22 | 11 | 1 | 160 |

were the sole agents of defleshing and bone breakage. At about the same time, Binford et al. (1988) reported on a smaller, more casual series of observations on carnivore disturbance of butchered bones. Subsequently, several other researchers have expanded these simulations and observations to include additional consumer sequences and additional skeletal parts, and have applied their results to several Stone Age and Paleolithic zooarchaeological assemblages (e.g. Blumenschine, 1995; Capaldo, 1997, 1998; Dominguez-Rodrigo, 1997; Marean et al., 1992, 2000; Marean and Bertino, 1994; Marean and Kim, 1998; Selvaggio, 1994, 1998).

Brain's (1967, 1969) pioneering taphonomic work with Hottentot dogs also provided a human-followed-by-carnivore scenario of assemblage formation. The bones modified by dogs had been defleshed and at least partially de-marrowed by Hottentot villagers. While the results showed that carnivore ravaging could more parsimoniously explain the disproportionate fossil skeletal part profiles at Makapansgat than selection of bones for use as tools by australopithecines, they do not by themselves rule out a role, small or large, for australopithecines in assemblage formation prior to carnivore involvement.

Here, we compare results of independent observations of carnivore tooth-marking and hammerstone percussion-marking on the FLK 22 (*Zinjanthropus* level) assemblage from Bed I, Olduvai Gorge made by Blumenschine in 1989 and Capaldo in 1997. We also provide an initial report of Capaldo's observations of hammerstone percussion-marking and carnivore tooth-marking from the FLK-N levels 1 and 2 assemblage from upper Bed I, Olduvai Gorge. Although Capaldo analyzed all larger mammal skeletal elements from both assemblages (Capaldo et al., n.d.), the results presented here are restricted to long bones, as these were the sole focus of Blumenschine's earlier study. We focus on the issue of replicability of observations on frequencies of marked bones by different analysts using our independent results from FLK 22. We also address whether the sequence of consumers indicated at FLK 22 (the most intensely studied Early Stone Age bone assemblage) is similar to that at FLK-N levels 1 and 2.

The incidences of percussion-marked and tooth-marked larger mammal long bone fragments (NISP values) from FLK 22 and FLK-N levels 1 and 2 are shown in Figures 5a and 5b, respectively. The data are presented for individual long bone portions (epiphyseal, near-epiphyseal, and midshaft fragments) following conventions used by Blumenschine (1988; see also Figure 5). The fossil samples are compared to a series of modern samples derived from simulations and natural observations in the wild conducted by Blumenschine (1988, 1995).

The data on percussion marking are compared to a "hammerstone-only" simulation sample (327 fragmentary specimens from seven trials), and a "hammerstone-followed-by-hyena" simulation sample (598 fragmentary specimens from 12 trials). Bones in these samples, derived from wild animals, were defleshed with a steel knife and broken with a hammerstone-on-stone-anvil technique to remove all marrow from the medullary cavity. Those from the latter sample were used to simulate spotted hyena ravaging of hominin food refuse, where the hyenas are attracted to the grease contained in the trabecular bone of epiphyseal fragments.

The fossil tooth mark data are also compared to two samples. These include the same "hammerstone-followed-by-hyena" sample used above, and an additional "carnivore only" sample (231 fragmentary specimens from 9 carcass consumption episodes). This sample includes bones of wild animals observed to have been defleshed and broken by lions and/or spotted hyenas to access marrow and grease.

To make the fossil samples comparable to the modern samples, we excluded from the analyses several sets of fragments. These include a) those smaller than 2 cm, b) those bearing matrix or poorly preserved surfaces that potentially hid or erased marks, and/or c) those displaying a post-fossilization break in which an estimated 10% or more of the fragment is missing; green-bone fractures characterize the specimens retained in the analytical sample. These analytical conventions are discussed more thoroughly in Blumenschine (1995) and Capaldo (1997).

Our analysis excludes a consideration of cut-marking because of the inadequacy of the modern samples of bones that have been defleshed with stone implements and subsequently broken with a hammerstone. Bones in Blumenschine's and Capaldo's samples were defleshed and disarticulated with metal knives, and as such are of questionable applicability to prehistoric defleshing that employed stone knives. Selvaggio (1994, 1998) used stone tools to remove flesh remaining on long bones fed on earlier by carnivores, but she did not provide a comparative sample for which all flesh was removed using stone tools. Domínguez-Rodrigo (1997) produced modern samples that simulate four sequences of carnivore and hominin access to long bones. His analyses are based on samples that include at least some limbs subjected to both stone-tool defleshing and hammerstone breakage for marrow extraction. Including these limbs only, two of the four simulations are represented by long bones from one carcass each, producing 73 and 42 fragmentary specimens from a total of six elements. The two other simulations are represented by long bones from two carcasses each, producing a total of 88 and 133 fragmentary specimens from a total of 14 elements. These small samples do not afford an assessment of potential variability in the incidence of cut-marked specimens introduced by differences in carcass size, thoroughness of prior defleshing by carnivores, butchery (defleshing) techniques, and degrees of fragmentation, among other factors. Further, Domínguez-Rodrigo (1997) does not describe the procedures he used to deflesh and break bones, aside from stating that stone tools were used. The small samples and lack of procedural detail needed to replicate results leaves the interpretive reliability of his neotaphonomic observations in doubt.

One striking result of the analyses in Figure 5 is the close comparability between Blumenschine's and Capaldo's independent estimates of the proportion of bones that are marked in the FLK 22 assemblage. The result is obtained despite differences in sample size: Blumenschine's analysis was based on 731 specimens, while Capaldo's 1997 analysis included 1153 specimens, adding a large number of specimens that were not stored in the Kenya National Museums along with the FLK 22 assemblage when Blumenschine conducted his earlier analysis. This comparability between our results arises because we use the same techniques for mark identification that were set out in Blumenschine et al. (1996), and we both have considerable experience in generating and studying collections of modern bones where the agent of mark production is known. The substantially lower incidence of percussion marking and tooth marking reported on the same assemblage by Oliver (1994) is probably attributable to his relative inexperience with generating and studying relevant comparative modern collections. The interanalyst correspondence between Blumenschine and Capaldo is particularly tight for percussion-marking (Figure 5a): our independent determinations for the overall incidence of percussion-marked bone differs by 1%. Our estimates for epiphyseal and midshaft fragments are essentially identical, while those for near-epiphyseal fragments differ by less than three percentage points. This correspondence reinforces the similarity between percentages of percussion-marked bones at FLK 22 and in the modern samples, a similarity that was interpreted originally by Blumenschine (1995) to indicate that hominins broke the large majority of long bones at FLK 22. The new data presented here for FLK-N levels1 and 2 similarly indicates that hominins were the main agent of long bone diaphyseal fracture (and presumably marrow consumption) at this site.

Our results on the incidence of tooth-marked long bone specimens at FLK 22 show a lower interanalyst correspondence, differing by as much as 10% (Figure 5b). Still, both analyses place the rate of tooth-marking at FLK 22 clearly intermediate to and outside of the 95% confidence intervals for the modern carnivore only and modern hammerstone-followed-by-carnivore samples. This result is consistent with the percussion mark data in indicating that carnivores were not the primary agents of bone breakage and marrow consumption at FLK 22. The significantly higher rate of tooth-marking on midshaft fragments at FLK 22 in comparison to the modern hammerstone-followed-by-carnivore sample was suggested by Blumenschine (1995) to indicate that felids had defleshed at least some of the fossilized long bones prior to their acquisition in an unbroken state by

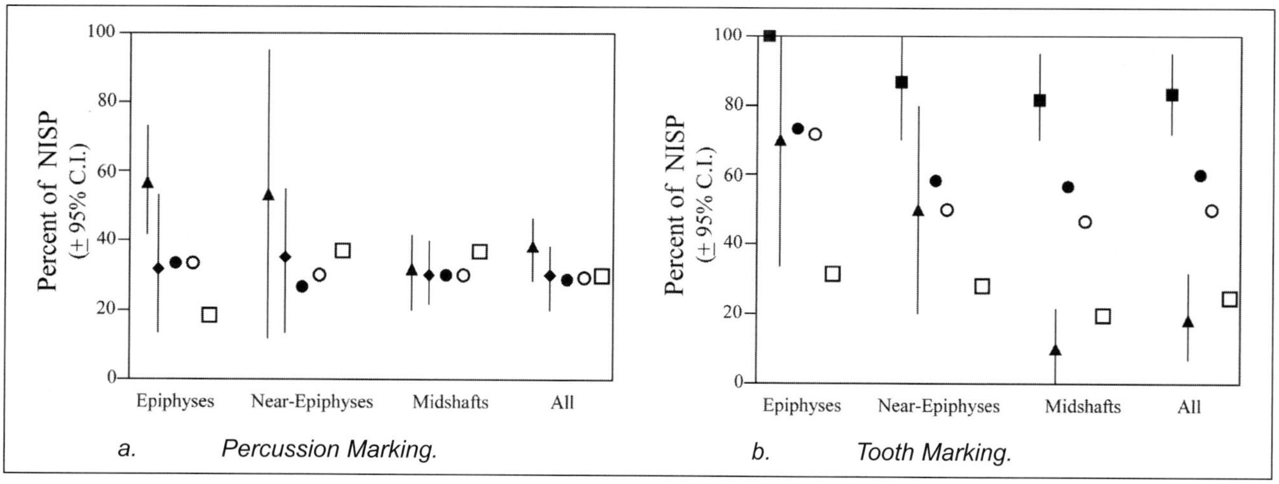


*Figure 5. Proportion of larger fossil mammal long bone specimens bearing at least one percussion mark (a) or tooth mark (b) at FLK 22 (Zinjanthropus level) and FLK-N, levels 1 and 2, compared with modern samples (see below). ● = FLK 22, based on Blumenschine's (1995) sample of 731 specimens. ○ = FLK 22, based on Capaldo's (1997) sample of 1153 specimens. □ = FLK-N, levels 1 and 2, based on a sample of 1067 specimens from Capaldo et al. (in prep.). Symbols with 95% confidence intervals represent mean values from three sets of observations of modern assemblages of fresh long bones defleshed with a steel knife and broken by hammerstone-on-stone-anvil technique, and/or by spotted hyenas (Blumenschine, 1988, 1995), as follows: ▲ = wild bovid bones defleshed with a steel knife and hammerstone-broken, followed by hyena ravaging; ◆ = wild bovid bones defleshed with a steel knife and hammerstone-broken only; ■ = wild bovid bones defleshed by lions and or spotted hyenas, and then broken by spotted hyenas only. Following the convention established by Blumenschine (1988), epiphyses are defined as any specimen bearing at least part of the proximal or distal end of the long bone; near-epiphyses are fragments derived at least in part from the proximal or distal shaft, identified on the basis of anatomical features on the cortical surface, or the presence of at least some trabecular bone on the medullary surface. Midshafts form the remainder of the fragments. Complete bones, which are few in number in both assemblages, are not included in the analysis*

hominins, inflicting in the process tooth marks on the midshafts. Selvaggio's (1998) observations of long bone shafts defleshed by lions and subsequently broken with a hammerstone strongly support this interpretation. Unlike FLK 22, the similarity in percentages of midshaft tooth-marking at FLK-N levels 1 and 2 to the modern hammerstone-followed-by-carnivore sample suggests that at this fossil site, hominins had access to not only all marrow, but access also to relatively large quantities of flesh.

## Hyaenid Bone Ravaging, Predation Risk, and Landscape Variation in Artifact Assemblages

Blumenschine and Peters (1998) used theoretical landscape variations in tree cover abundance and the inversely associated degrees of predator encounter risk for hominins to predict the landscape-specific character of stone artifact assemblages throughout the Plio-Pleistocene Olduvai Basin during lowermost Bed II times. They assumed that in settings with few trees and shrubs, carnivore density and potential competition among large carnivores for carcasses would be higher, as would hyaenid to carcass ratios. This is consistent with the reconstruction of habitat preferences for fossil hyaenids as open, relatively unwooded settings (Werdelin and Lewis, 2005).

The predictive model specified that in landscape settings with greater tree cover abundance (and therefore less predator encounter risk for hominins), the density of stone artifact discard and loss, and the functional diversity of artifact assemblages would be higher. On the other hand, detached pieces (flakes and flake fragments) were predicted to dominate assemblages in potentially more dangerous and competitive unwooded areas; here, hominins were modeled to transport still usable cores, hammerstones and manuports along with hastily acquired carcass parts to relatively safe (well wooded) areas where full butchery could take place.

Blumenschine et al. (2005) provide information on the extent to which proportions of long bone shafts to long bone ends of medium-sized (e.g., wildebeest) and large (e.g., buffalo) fossil mammals correlate with variations in Oldowan artifact density and diversity for 12 paleogeographic locales in the Olduvai Basin's lowermost Bed II Eastern Lake Margin and distal Eastern Alluvial Fan (Figure 1). The relative abundance of medium-sized and large fossil mammal long bone shafts compared to long bone ends (based on specimen counts) is used as the measure of the relative presence of hyaenids and possibly large canids in the paleo-setting. Restriction of the analysis to fossil long bones the size of those of extant wildebeest and buffalo is based on observations that show spotted hyenas to be the the only extant East African carnivore capable of fragmenting long bone diaphyses of this size (Blumenschine, 1988, 1989; Blumenschine and Marean, 1993; Marean and Spencer, 1991; Marean et al., 1992). For Plio-Pleistocene Olduvai, the relevant hyaenid was *Crocuta ultra*, a species with a similar lower third premolar (the main bone-cracking tooth) as *Crocuta crocuta*, the modern spotted hyena, and therefore similar bone-cracking capabilities (Lewis and Werdelin, 1999). *Canis africanus*, a wolf-sized canid (Werdelin, in Peters et al., 2007), was also present at Plio-Pleistocene Olduvai. This large canid may also have contributed to the fragmentation of medium-sized and large long bones judging from the bone fracturing capabilities of modern wolves (Haynes, 1980, 1983), but little published information on this fossil canid is available to evaluate this possibility. Therefore, relatively high proportions of long bone shafts to long bone ends (although difficult to translate into absolute carnivore abundance) is a reasonable index or relative measure of the increased presence of hyaenids and possibly large canids in paleo-settings. Landscape settings providing hominins with both relatively greater scavenging opportunities and a lower chance of encountering large predators theoretically would have a relatively high proportion of long bone ends.

The results of our analysis are reproduced in Figure 6. A moderately strong correlation exists between the relative abundance of long bone ends and artifact weight densities (Pearson's $r = 0.69$, $p = 0.01$). The relationship between the variables is positive, as predicted by the model, with relatively high artifact densities being associated with high proportions of long bone ends, suggesting a relative absence of hyaenids and large canids. Here, the presumed relative abundance of these carnivores apparently accounts for about half (48%) of the variability in artifact weight densities. On the other hand, the relative abundance of long bone ends shows the expected negative correlation with the proportion of whole detached pieces among all whole artifacts (Pearson's $r = -0.69$, $p = 0.01$), with the presumed relative abundance of hyaenids and possibly large canids apparently explaining about half of the variability in this measure of artifact assemblage composition. Here, hominins presumably discarded or lost a broader range of artifact functional types in locales with a relative absence of these carnivores.

Blumenschine's earlier interpretation of these results (Blumenschine et al., 2005) emphasized the role of potential competition among hominins and hyaenids for carcass resources as the major determinant of stone artifact assemblage density and functional diversity. This interpretation was based on neotaphonomic observations and conceptual frameworks developed earlier (Blumenschine, 1986, 1989, Blumenschine et al., 1994). The neotaphonomic studies suggest that the medium-sized and large mammal long bone end:shaft ratio is an indirect measure of the completeness of fresh carcass consumption by spotted hyenas, and therefore a measure of the general scale of scavenging opportunities for non-hyaenid consumers. However, emphasizing competition for carcasses assumes that traces of hominin land use were determined in large part by the availability of scavenge-

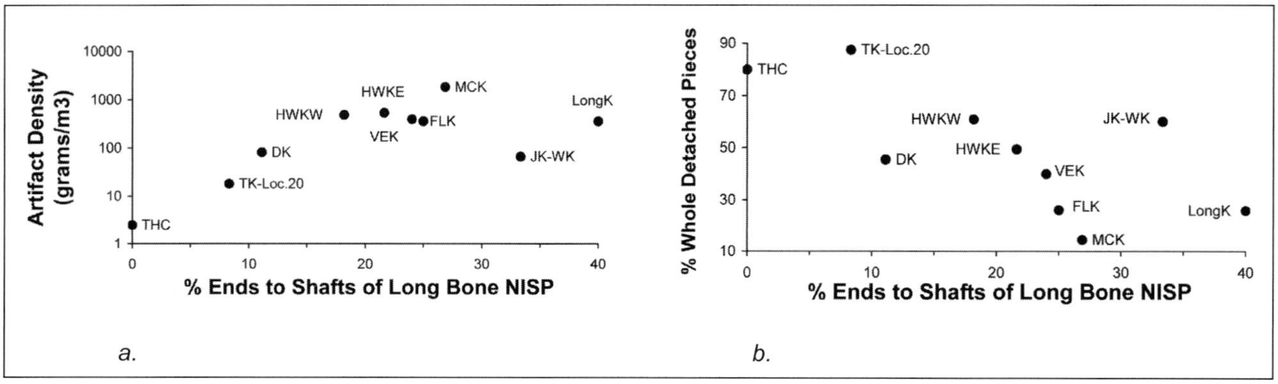


*Figure 6. The relationship between the hyaenid bone-ravaging index, expressed as the percentage of articular ends among all long bone specimens, and characteristics of the artifact assemblages for 10 geographic locales in the lowermost Bed II Olduvai Basin (see Figure 1b). Artifact assemblage characteristics are (a) weight density, and (b) the percentage of whole detached pieces (flakes) among all whole artifacts. HWKEE-KK and MNK are not plotted owing to their lack of larger mammal long bone shafts.*

able food. While Oldowan activity traces may largely record stone-tool butchery in landscape context (Blumenschine and Peters, 1998), constraints on utilizing other resources (water, refuge trees, plant foods, materials for tools), particularly predation risk, probably had a more regular influence on the composition and landscape distribution of stone artifact assemblages. Because of the above theoretical considerations, Werdelin and Lewis's work, and the fact that neotaphonomic studies demonstrate that hyaenid bone ravaging is greater in relatively open landscapes affording little to no arboreal refuge from predators, Peters has revised the conceptual framework to emphasize predation risk to hominins from possible encounters with large carnivores as a major determinant of Oldowan hominin land use. Now, the results of the analysis relating the relative presence of hyaenids and possibly large canids (bone-ravaging index) to the density and composition of Oldowan stone artifact assemblages are interpreted as strong evidence that differential use of the landscapes of the eastern Olduvai Basin by Oldowan hominins was conditioned in substantial part by the effects of potential predation risk. Following this interpretation, open landscape avoidance or rapid deployment and withdrawal by hominins in these settings can be related to the land use patterns of the bone-cracking hyaenids and other open-habitat large Plio-Pleistocene carnivores. One test of this interpretation will be provided by a planned analysis of carnivore tooth-marking on the fossil long bone specimens that is designed to assess the degree to which long bone fragmentation and destruction of long bone ends can be attributed to bone-cracking carnivores. If upheld, the relationships between the bone-ravaging index and stone artifact assemblage density and composition will constitute the first probable demonstration that community-level interactions involving large carnivores likely conditioned the land use patterns of prehistoric hominins.

Pobiner (2007) is currently conducting a study that may provide additional lines of evidence for evaluating the role of predator encounter risk as a possible influence on hominin land use. The study focuses on establishing taxon-specific patterns of bone damage through observations of modern carnivore feeding behavior. Taxon-specificity in bone damage is being sought at two levels: gross damage patterns to bones, and the morphology and anatomical patterning of tooth marks on bone surfaces. To date, efforts to establish taxon-specificity have been limited (Andrews and Fernandez-Jalvo, 1997; Domínguez-Rodrigo and Piquerus, 2003; Haynes, 1983; Pobiner and Blumenschine, 2003; Selvaggio and Wilder, 2001). Even family-level specificity would provide paleoecologically useful information, particularly if the relative dominance of flesh-specialist felids versus bone-cracking hyaenids in different paleo-landscape settings can be determined.

The results of Pobiner's study are not yet available, but analysis of data collected by Blumenschine in the Serengeti in 1983 and 1984 suggests that some species differences can be detected in gross damage patterns to bones. Figure 7 shows the contrasting degrees of bone damage and destruction caused by four extant carnivore species on fresh Size Class 1 (e.g., Thomson's gazelle) and Size Class 3 (e.g., wildebeest) carcasses. Figure 7 includes data on upper hindquarter parts (Pobiner and Blumenschine, 2003), plus previously unpublished data on upper forelimb parts. This figure shows a pattern within each carcass size group of progressively greater bone destruction that corresponds to increasing body size and/or jaw strength of the carnivores under consideration. Carnivores with greater bone destruction ability reduce and eventually destroy skeletal elements and skeletal element portions more intensely than do less capable bone-breakers. Destruction and marginal gnawing by lions on Size Class 3 carcasses is restricted to the same skeletal parts reduced by cheetah on Size Class 1 carcasses, with the addition of the caudal part of the sacrum and the olecranon process of the ulna. Destruction of Size Class 3 carcass parts by spotted hyenas is more intense, and includes virtually all parts destroyed or heavily fragmented by lions on Size Class 1 carcasses, with the major exception of the lumbar vertebral centra. In short, these observations demonstrate that lions damage the same bones to the same extent on Size Class 3 carcasses that cheetah damage on Size Class 1 carcasses, while spotted hyena damage and destruction of Size Class 3 carcasses mirrors the location and intensity of damage inflicted by lions on Size Class 1 carcasses. This patterning suggests the existence of a simple mechanism underlying the degree of skeletal reduction inflicted by modern carnivores when extracting flesh and within-bone edible tissues: the increased bone size and strength characteristic of larger mammalian carcasses imposes greater mechanical constraints to nutrient extraction that can be overcome only by carnivores with greater jaw strength and dental capabilities for breaking bone.

## Conclusions

OLAPP's vertebrate taphonomic work will continue to focus on ways in which paleolandscape ecology and hominin land use are revealed by analyses of fossil bone assemblages. Our work focuses on the development of reliable methods for identifying the feeding traces of individual carnivore taxa, and on resolving sequences of carnivore and hominin access to carcasses, including the types and amount of tissues consumed by each. It also focuses on measuring landscape variability in the differential traces of carnivore and hominin presence and their respective patterns of land use. Brain's contributions to vertebrate taphonomy encouraged the development of models of land use for individual larger carnivore taxa, inspiring our ongoing model-building for Oldowan hominins. Future work will also lead to new understandings of evolutionary changes in the larger carnivore guild, wherein hominins played an increasingly important role. We can also expect an increasingly refined role for taphonomy in the pursuit of paleoenvironmental reconstruction.

## Acknowledgements

We are grateful to a number of Tanzanian institutions for permitting and cooperating with our research. These include the Tanzania Commission for Science and Technology, the Antiquities Department, the National Museums of Tanzania, the Tanzania Wildlife Research Institute, the Ngorongoro Conservation Area Authority, Tanzania National Parks, and the Serengeti National Park. Funding for the research was provided by the National Science Foundation, the Wenner-Gren Foundation for Anthropological Research, the L.S.B. Leakey Foundation, and Rutgers University's Center for Human Evolutionary Studies. CRP also thanks the Department of Anthropology at the University of Georgia for continued support as Professor Emeritus. An anonymous reviewer provided useful comments on the manuscript. We are grateful to members of the OLAPP team who have contributed to the work reported here on modern and fossil bone assemblages at Olduvai. Finally, we thank the editors of this volume for organizing and hosting the conference upon which this volume is based.

## References

Andrews, P. 1990. Owls, Caves and Fossils. Natural History Museum Publications, London.

Andrews, P., Fernandez-Jalvo, Y. 1997. Surface modifications of the Sima de los Huesos fossil humans. Journal of Human Evolution 33, 191-217.

Behrensmeyer, A.K. 1975. The taphonomy and paleoecology of vertebrate assemblages east of Lake Rudolf, Kenya. Bulletin of the Museum of Comparative Zoology 146, 473-578.

Behrensmeyer, A.K. 1983. Patterns of natural bone distribution on recent land surfaces: implications for archaeological site formation. In: Clutton-Brock, J., Grigson, C. (Eds.), Animals and Archaeology, Volume 1: Hunters and Their Prey. British Archaeological Reports International Series 163, Oxford, pp. 93-106.

Behrensmeyer, A.K. 1985. Taphonomy and the paleoecologic reconstruction of hominid habitats in the Koobi Fora Formation. In: Coppens, Y. (Ed.), L'Environment des Hominides au Plio-Pleistocene. Masson, Paris, pp. 309-324.

Behrensmeyer, A.K., Dechant-Boaz, D.E. 1980. The recent bones of Amboseli National Park, Kenya, in relation to East African paleoecology. In: Behrensmeyer A.K., Hill, A. (Eds.), Fossils in the Making, University of Chicago, Chicago, pp. 72-92.

Behrensmeyer, A.K., Western, D., Dechant-Boaz, D. 1979. New perspectives in vertebrate paleoecology from a recent bone assemblage. Paleobiology 5, 12-21.

Binford, L. R. 1981. Bones: Ancient Men and Modern Myths. Academic Press, New York.

Binford, L. R., Mills, M. G. L., Stone, N. M. 1988. Hyena scavenging behavior and its implications for the interpretation of faunal assemblages from FLK 22 (the Zinj floor) at Olduvai Gorge. Journal of Anthropological Archaeology 7, 99-135.

Blumenschine, R. J. 1986. Early Hominid Scavenging Opportunities: Implications of Carcass Availability in the Serengeti and Ngorongoro Ecosystems. British Archaeological Reports International Series 283, Oxford.

Blumenschine, R. J. 1988. An experimental model of the timing of hominid and carnivore influence on archaeological bone assemblages. Journal of Archaeological Science 15, 483-502.

Blumenschine, R. J. 1989. A landscape taphonomic model of the scale of early prehistoric scavenging opportunities. Journal of Human Evolution 18, 345-371.

Blumenschine, R. J. 1995. Percussion marks, tooth marks and experimental determinations of the timing of hominid and carnivore access to long bones at FLK *Zinjanthropus*, Olduvai Gorge, Tanzania. Journal of Human Evolution 29, 21-51.

Blumenschine, R. J., Marean, C. W. 1993. A carnivore's view of archaeological bone assemblages. In: Hudson, J. (Ed.), From Bones to Behavior: Ethnoarchaeological and Experimental Contributions to the Interpretation of Faunal Remains. Center for Archaeological Investigations, University of Southern Illinois Press, Carbondale (IL), pp. 273-300.

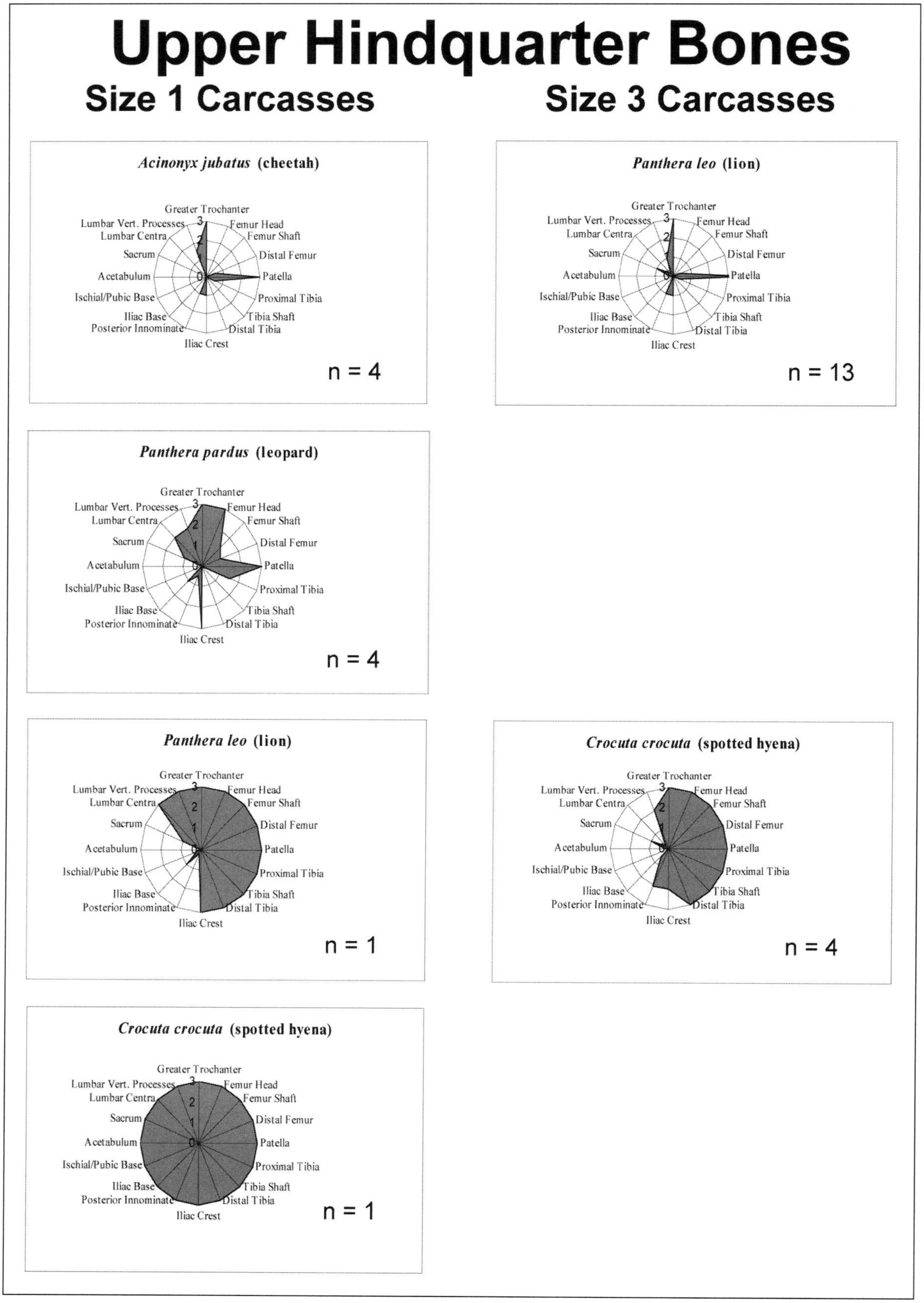


*Figure 7. Contrasting degrees of bone damage and destruction caused by four extant carnivore species to upper hindquarters and upper forelimbs of fresh Size Class 1 ungulate carcasses (e.g., Thomson's gazelle), and by two extant species to Size Class 3 carcasses (e.g., wildebeest). Modal values for damage categories are those observed by Blumenschine to have been inflicted on fresh carcasses by the specified carnivore as the exclusive consumer of the carcass in the Serengeti. Sample size refers to the number of carcasses included in each graph. Degrees of bone damage are categorized by numbers 0-3, in order of increasing damage: 0 = no gross damage, though tooth pits and scores are likely to occur; 1 = marginal gnawing (< 50% of portion is removed); 2 = heavy gnawing/fragmentation; 3 = destruction (may leave bone splinters).*

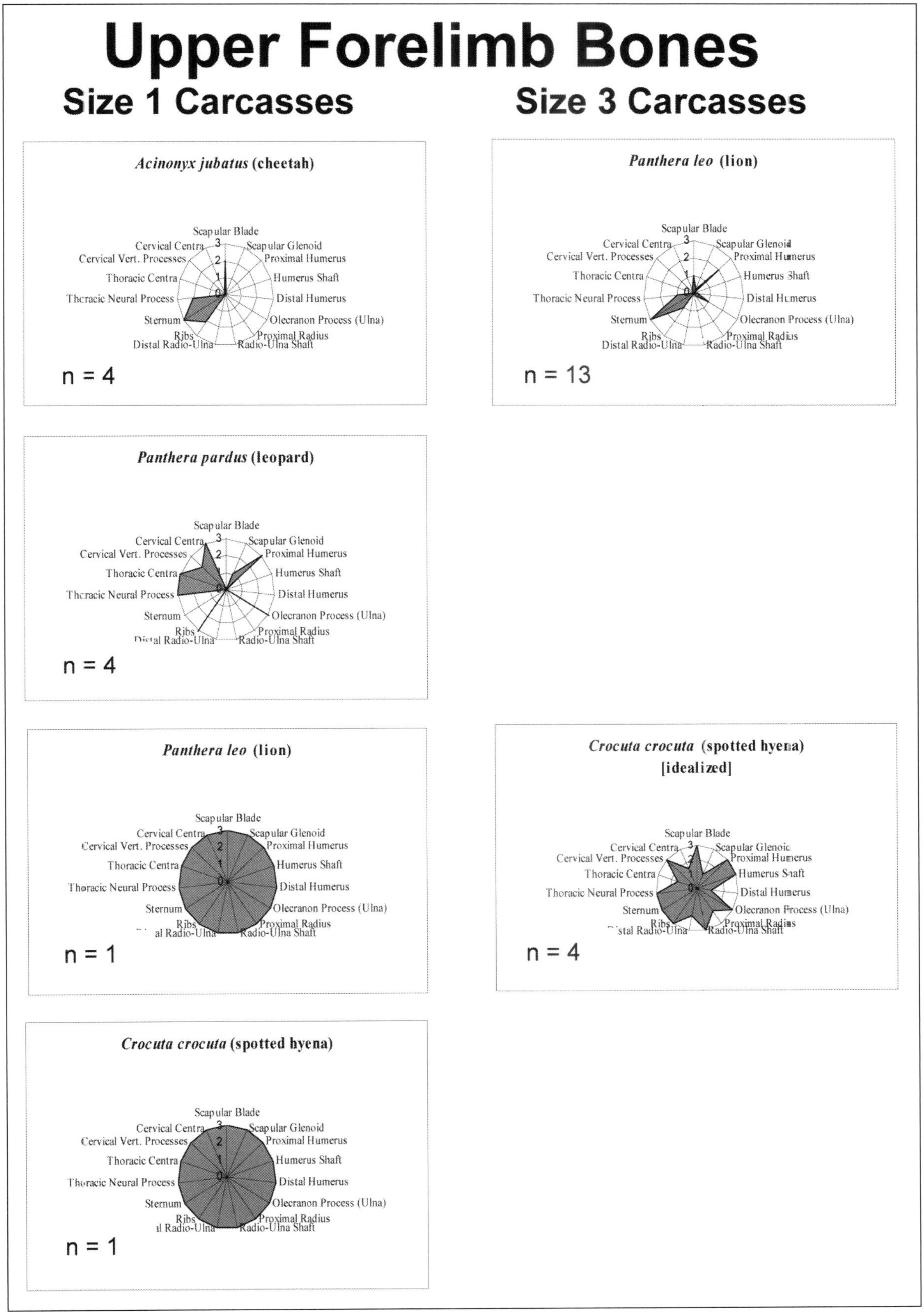
Upper Forelimb Bones
Size 1 Carcasses
Size 3 Carcasses
Acinonyx jubatus (cheetah)
n = 4
Panthera pardus (leopard)
n = 4
Panthera leo (lion)
n = 1
Crocuta crocuta (spotted hyena)
n = 1
Panthera leo (lion)
n = 13
Crocuta crocuta (spotted hyena) [idealized]
n = 4
Scapular Blade
Scapular Glenoid
Proximal Humerus
Humerus Shaft
Distal Humerus
Olecranon Process (Ulna)
Proximal Radius
Radio-Ulna Shaft
Distal Radio-Ulna
Ribs
Sternum
Thoracic Neural Process
Thoracic Centra
Cervical Vert. Processes
Cervical Centra
0
1
2
3

Blumenschine, R.J., Cavallo, J.A., Capaldo, S.D. 1994. Competition for carcasses and early hominid behavioral ecology: a case study and conceptual framework. Journal of Human Evolution 27, 197-213.

Blumenschine, R. J., Marean, C. W., Capaldo, S. D. 1996. Blind tests of interanalyst correspondence and accuracy in the identification of cut marks, percussion marks, and carnivore tooth marks on bone surfaces. Journal of Archaeological Science 23, 493-507.

Blumenschine, R. J., Peters, C. R., 1998. Archaeological predictions for hominid land use in the paleo-Olduvai Basin, Tanzania, during Lowermost Bed II times. Journal of Human Evolution 34, 565-607.

Blumenschine, R.J. Masao, F.T., Peters, C.R. 2005. Broad-scale landscape traces of Oldowan hominid land use at Olduvai Gorge, and the Olduvai Landscape Paleoanthropology Project. In: Mapunda, B.B.B., Msemwa, P. (Eds.), Salvaging the Cultural Heritage of Tanzania. University of Dar es Salaam, Dar es Salaam, pp. 158-189.

Brain, C. K. 1967. Hottentot food remains and their bearing on the interpretation of fossil bone assemblages. Scientific Papers of the Namib Desert Research Station 32, 1-11.

Brain, C. K. 1969. The contribution of Namib Desert Hottentots to an understanding of australopithecine bone accumulations. Scientific Papers of the Namib Desert Research Station 39, 13-22.

Brain, C.K., 1970. New finds at the Swartkrans australopithecine site. Nature 225, 1112-1119.

Brain, C. K. 1981. The Hunters or the Hunted? An Introduction to African Cave Taphonomy. University of Chicago, Chicago.

Bunn, H.T. 1981. Archaeological evidence for meat-eating by Plio-Pleistocene hominids from Koobi Fora and Olduvai Gorge. Nature 291, 547-577.

Capaldo, S. D. 1997. Experimental determinations of carcass processing by Plio-Pleistocene hominids at FLK 22 (*Zinjanthropus*) at Olduvai Gorge, Tanzania. Journal of Human Evolution 33, 555-597.

Capaldo, S. D. 1998. Simulating the formation of dual-patterned archaeofaunal assemblages with experimental control samples. Journal of Archaeological Science 25, 311-330.

Capaldo, S.D., Pante, M.C., Johnson, P.K. n.d. Experimental determinations of carcass processing by Plio-Pleistocene hominids and carnivores at, and inter-assemblage comparisons between FLKN 1-2, FLKN 3, FLK 22, and FLKNN 3, Olduvai Gorge, Tanzania. Submitted to Journal of Human Evolution.

Capaldo, S.D., Peters, C.R. 1995. Skeletal inventories from wildebeest drownings at lakes Masek and Ndutu in the Serengeti ecosystem of Tanzania. Journal of Archaeological Science 22, 385-408.

Capaldo, S.D., Peters, C.R. 1996. Observations of wildebeest *(Connochaetes taurinus)* crossing Lake Masek (Serengeti ecosystem, Tanzania) including one small drowning. Mammalia 60, 303-306.

Dart, R. A. 1949. The predatory implemental technique of *Australopithecus*. American Journal of Physical Anthropology 7, 1-38.

Davidson, I., Solomon, S. 1990. Was OH 7 the victim of a crocodile attack? In: Solomon, S., Davidson, I., Watson, D. (Eds.), Problem Solving in Taphonomy: Archaeological and Palaeontological Studies from Europe, Africa and Oceania, Vol. 2, Tempus, St. Lucia, Queensland, pp. 197-206.

Dominquez-Rodrigo, M. 1997. Meat-eating by early hominids at the FLK 22 *Zinjanthropus* site, Olduvai Gorge (Tanzania): an experimental approach using cut-mark data. Journal of Human Evolution 33, 669-690.

Dominguez-Rodrigo, M., Piqueras, A. 2003. The use of tooth pits to identify carnivore taxa in tooth-marked archaeofaunas and their relevance to reconstruct hominid processing behaviors. Journal of Archaeological Science 30, 1385-1391.

Emiliani, C. 1955. Pleistocene temperatures. Journal of Geology 63, 539-578.

Fernandez-Jalvo, Y., Denys, C., Andrews, P., Williams, T., Dauphin, Y., Humphrey, L. 1998. Taphonomy and palaeoecology of Olduvai Bed-I (Pleistocene, Tanzania). Journal of Human Evolution 34, 137-172.

Gifford-Gonzalez, D. 1991. Bones are not enough: Analogues, knowledge and interpretive strategies in zooarchaeology. Journal of Anthropological Archaeology 1, 355-381.

Hay, R.L., 1976. Geology of the Olduvai Gorge. University of California Press, Berkeley.

Haynes, G. 1980. Evidence of carnivore gnawing on Pleistocene and Recent mammalian bones. Paleobiology 6, 341-351.

Haynes, G. 1983. A guide for differentiating mammalian carnivore taxa responsible for gnaw damage to herbivore limb bones. Paleobiology 9, 164-172.

Isaac, G.Ll., 1978. Food sharing and human evolution: archaeological evidence from the Plio-Pleistocene of East Africa. Journal of Anthropological Research 34, 311-325.

Leakey, M.D., 1971. Olduvai Gorge: Excavations in Beds I and II, 1960-1963. Cambridge University, Cambridge.

Lewis, M.E., Werdelin, L. 1999. Implications of mosaic evolution in non-primate taxa for understanding changes in hominid behavioral ecology. Journal of Human Evolution 36, A10-A11.

Marean, C. W., Bertino, L. 1994. Intrasite spatial analysis of bone: subtracting the effect of secondary carnivore consumers. American Antiquity 59, 748-768.

Marean, C. W., Kim, S. Y. 1998. Mousterian large-mammal remains from Kobeh Cave: Behavioral implications for Neanderthals and early modern humans. Current Anthropology 39, S79-S113.

Marean, C. W., Spencer, L. M. 1991. Impact of carnivore ravaging on zooarchaeological measures of element abundance. American Antiquity 56, 645-658.

Marean, C. W., Abe, Y., Frey, C. J., Randall, R. C. 2000. Zooarchaeological and taphonomic analysis of the Die Kelders Cave 1 Layers 10 and 11 Middle Stone Age larger mammal fauna. Journal of Human Evolution 38, 197-233.

Marean, C. W., Spencer, L. M., Blumenschine, R. J., Capaldo, S. D. 1992. Captive hyena bone choice and destruction, the schlepp effect, and Olduvai archaeofaunas. Journal of Archaeological Science 19, 101-121.

Njau, J. K. 2000. Taphonomic relationships between subaerial and subsurface assemblages in recent lake margin environments: Its relevance to the formation of fossil bone assemblages. M.A. Thesis, Rutgers University.

Njau, J. K. 2006. The relevance of crocodiles to Oldowan hominin palaeoecology at Olduvai Gorge, Tanzania. Ph.D. Dissertation, Rutgers University.

Njau, J. K., Blumenschine, R. J. 2006. A diagnosis of crocodile feeding traces on larger mammal bone, with examples from the Plio-Pleistocene Olduvai Basin, Tanzania. Journal of Human Evolution 50, 142-162.

Oliver, J. S. 1994. Estimates of hominid and carnivore involvement in the FLK *Zinjanthropus* fossil assemblage: some socioeconomic implications. Journal of Human Evolution 27, 267-294.

Peters, C.R., Blumenschine, R.J. 1995. Landscape perspectives on possible land use patterns for Early Pleistocene hominids in the Olduvai Basin. Journal of Human Evolution 29, 321-362.

Peters, C.R., Blumenschine, R.J., 1996. Landscape perspectives on possible land use patterns for Early Pleistocene hominids in the Olduvai Basin, Tanzania: Part II, expanding the landscape models. In: Magori, C.C., Saanane, C.B., Schrenck, F. (Eds.), Four Million Years of Hominid Evolution in Africa: Papers in Honour of Dr. Mary Douglas Leakey's Outstanding Contribution in Paleoanthropology. Darmstadter Beitrage zur Naturgeschichte, Kaupia 6, Darmstadt, pp. 175-221.

Peters, C.R., Blumenschine, R.J., Hay, R.L., Livingstone, D.A., Marean, C.W., Harrison, T., Armour-Chelu, M., Andrews, P., Bernor, R.L., Bonnefille, R., Werdelin, L. 2007. Paleoecology of the Serengeti ecosystem. In: Sinclair, T., Packer, C., Coughenour, M., Mduma, S. (Eds.), Serengeti III: Humans in Complex Ecosystem Processes and the Conservation of Biodiversity. University of Chicago, Chicago, in press.

Pobiner, B.L. 2007. Hominin-carnivore interactions: Evidence from modern carnivore bone modification and Early Pleistocene archaeofaunas (Koobi Fora, Kenya; Olduvai Gorge, Tanzania). Ph.D. Dissertation, Rutgers University.

Pobiner, B.L., Blumenschine, R.J. 2003. A taphonomic perspective on Oldowan hominid encroachment on the carnivoran paleoguild. Journal of Taphonomy 1, 115-141.

Potts, R., Shipman, P. 1981. Cutmarks made by stone tools on bones from Olduvai Gorge, Tanzania. Nature 291, 577-580.

Reed, D. 2003. Micromammal paleoecology: Past and present relationships between African micromammals and their habitats. Ph.D. Dissertation, State University of New York at Stony Brook.

Selvaggio, M. M. 1994. Carnivore tooth marks and stone tool butchery marks on scavenged bones: archaeological implications. Journal of Human Evolution 27, 215-228.

Selvaggio, M. M. 1998. Evidence for a three-stage sequence of hominid and carnivore involvement with long bones at FLK *Zinjanthropus*, Olduvai Gorge, Tanzania. Journal of Archaeological Science 25, 191-202.

Selvaggio, M. M., Wilder, J. 2001. Identifying the involvement of multiple carnivore taxa with archaeological bone assemblages. Journal of Archaeological Science 28, 465-470.

Voorhies, M. 1969. Taphonomy and population dynamics of an Early Pliocene vertebrate fauna. Knox County, Nebraska. Contributions in Geology Special Paper No. 1, University of Wyoming.

Werdelin, L., Lewis, M.E. 2005. Plio-Pleistocene Carnivora of eastern Africa: Species richness and turnover patterns. Zoological Journal of the Linnean Society 144, 121-144.

# CHAPTER 10

# The Earlier Stone Age in South Africa: Site Context and the Influence of Cave Studies

Kathleen Kuman


## Abstract

Although South Africa lacks volcanic deposits and sites with fauna are limited, its Earlier Stone Age (ESA) provides a long record which extends from the Oldowan (*c* 1.9 million years ago, Mya), through all phases of the Acheulean, and to final ESA industries transitional to the Middle Stone Age (MSA). Until now, the earliest sites have been identified mainly from secondary deposits within underground dolomitic limestone caves, but this record is set to expand with new dating of alluvial deposits in the Vaal River basin. Apart from four later Acheulean cave occupation sites and the early sites in underground cave fills, all other sites are found in open-air contexts, frequently close to standing water. These contexts are variable—in large river basins, in seasonal lake basins or pans, in river terrace and colluvial basin deposits, in one spring deposit, in coastal aeolian deposits, and in deflated inland lag deposits that in one case preserves three stratified cultural units. This paper presents the data for the sequence of ESA sites in the South and discusses the variety of contexts in which these archaic sites are found.


## Introduction

The impact of Bob Brain's work on taphonomy is widespread and well appreciated. But perhaps less widely recognized is the profound influence that Bob has had on our understanding of cave systems and site formation processes. This is probably because there are fewer specialists who excavate dolomitic cave sites than there are taphonomists working with faunal collections. However, the understanding of cave infill formation and transformation is integral to taphonomic interpretations, and it is perhaps even more critical for archaeological interpretations made on such cave assemblages. The elegant explanations that Bob has developed on the cyclical nature of cave sedimentation and erosion processes have been of considerable influence in my own work in site formation analysis. For those prehistorians like myself who do not practice zooarchaeology, site formation is the sister discipline to taphonomy. It is the study of the processes by which sites form and become transformed through time and through geological forces. While these two methods are usually applied as independent specializations, practitioners of both disciplines have a tacit understanding of how well the two approaches complement each other in deciphering the history of any site that preserves both artifacts and fauna.

Site formation analysis has relied heavily on data generated by research on the open-air alluvial and lacustrine sites of East Africa (Isaac, 1967; Schick, 1987a, 1991, 1997; Morton, 1996). Experimental work has helped analysts to evaluate just how pristine (or primary) is the context of an open-air, stratified site. In addition to study of the sedimentary context, formation and disturbance processes can be assessed with the condition and size profile of an assemblage, as well as the orientations and dips of individual artifacts, and a site can then be ranked along a continuum from primary to secondary contexts. In contrast with East Africa, most of the earliest South African artifacts (*c*. 2-1 million years ago, Mya) occur in secondary context, re-deposited from surface occupations around shaft-like entrances that fed into underground cave infills. In thinking through interpretations of behaviour from these cave sites, I have also come to have a certain perspective on the overall South

African Earlier Stone Age (ESA). This paper reviews the contexts of ESA sites from 2 to 0.2 Mya and discusses the importance of site formation analysis to a general appreciation of prehistory in the southern sites.

## The ESA Record

The South African ESA is preserved in a variety of contexts, both as 'sites' in the traditional sense and as extensive surface and geological assemblages, and even as buried deflated assemblages. Artifacts of the ESA are so numerous in some parts of the country that the Abbé Breuil once commented that there were not only enough specimens to fill a museum (on Canteen Kopje) to overflowing but to build it of them (Clark, 1959: 127). The prolific nature of this record also led to its historical importance. In the first half of the 20th Century, the South African ESA was influential in establishing the antiquity of the *African* cultural record at a time when the European Palaeolithic was much better understood (Jansen, 1926; Breuil, 1930; Goodwin, 1928, 1933; Goodwin and Van Riet Lowe, 1929; Malan, 1947; and Van Riet Lowe, 1937, 1945, 1952a, 1952b). But it is not merely a rich record, it is also a record that is gradually becoming more complete, particularly in the earlier and later phases.

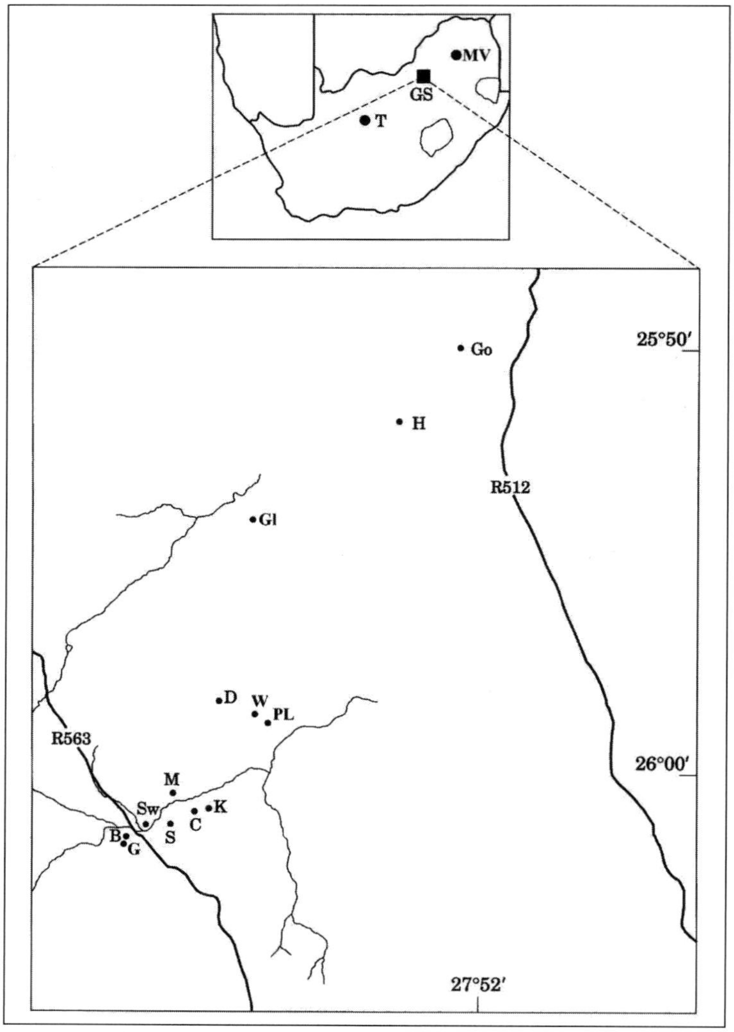


*Figure 1. The Cradle of Humankind World Heritage Site, Gauteng Province (GS): G: Goldsmith's; B: Bolt's Farm; Sw: Swartkrans; S: Sterkfontein; M: Minnaar's; C: Cooper's; K: Kromdraai; D: Drimolen; W: Wonder Cave; PL: Plover's Lake; Gl: Gladysvale; H: Haasgat; Go: Gondolin; T: Taung; MV: Makapans Valley.*

### Earliest ESA Sites

#### *Oldowan*

Thus far there are only two sites confirmed old enough to belong to an Oldowan phase of the ESA.

*Kromdraai*

Kromdraai B (KB) has thus far yielded only two certain artifacts, a core and a flake (Kuman et al., 1997). The KB site has produced *Paranthropus* fossils and at least part of it is placed by paleomagnetism at 1.9 Mya (Thackeray et al., 2002; and see Vrba, 1975). It seems that KB was not much used by Oldowan hominids because its physical setting was less than ideal, as there is evidence in the fauna and geology that the cave was locally quite wet at this time (Brain, 1958; Vrba, 1981). As KB lies within easy reach of the same gravels used by hominids at Swartkrans and Sterkfontein, such conditions may explain the avoidance of the site by hominids and therefore the paucity of stone tools. All three of these sites lie within 300m of gravels associated with the Blaaubank River (T.C. Partridge, unpublished maps of the gravels).

*Sterkfontein*

In the early 1990s, a large Oldowan assemblage was excavated from Sterkfontein in Member 5 East, associated with fauna estimated at 2 to 1.7 Mya (Clarke, 1994a; Kuman, 1994a). The only associated hominids in the Oldowan Infill are a few fragmentary fossils of *Paranthropus*. However, one molar closely resembles the KB specimen TM 1536 (Kuman and Clarke, 2000), and the 1.9 Mya palaeomagnetic relative date for KB fits within the faunal age estimate for the Sterkfontein infill. The Oldowan assemblage consists of 3,245 pieces, 84% of which is small flaking debris under 20mm size. Almost all of the material one would expect to find in a completely preserved assemblage is present, with mainly flakes <10mm size poorly represented. The catchment of artifacts around the cave entrance was good, probably because the entrance to the cave lay on top of the hill rather than on its slope. A moderately wooded environ-

*Table 1. The earliest occurrences of artifacts in South Africa. With the exception of the two alluvial site localities, all other occurrences are in underground dolomitic limestone cave infills. The age estimates are based on published sources and provide the range of possible dates. Although bifaces have been found in dump material, there are no provenanced examples thus far from the Swartkrans and Kromdraai assemblages, which are designated Early Acheulean on technological grounds.*

| Site Name | Age Estimate in Mya | Stratigraphic Context | Basis for Age Estimate | Cultural Industry |
|---|---|---|---|---|
| Sterkfontein | 2.0-1.7 | Oldowan Infill, Member 5 East | fauna | Oldowan |
| Kromdraai B | 1.9 | KB deposits | palaeomagnetism; fauna | Oldowan age |
| Swartkrans | 1.8 or 1.7 | Member 1, Lower Bank | fauna | undetermined |
| Sterkfontein | 1.7-1.4 | Member 5 West | fauna; artifacts | Early Acheulean |
| Swartkrans | 1.5 | Member 2 | fauna | Early Acheulean |
| Swartkrans | 1.0 | Member 3 | fauna | Early Acheulean |
| Kromdraai A | 1.5-1.0 | KA deposits and dump | fauna | Early Acheulean |
| Three Rivers and Klipplaatdrif | ? | river gravels | artifacts | Early Acheulean |
| Coopers | 1.6-1.9 | Coopers D | fauna | unpublished |
| Goldsmith's | 1.4 or more | Miners' dumps and associated infill | fauna | undetermined |
| Drimolen | 1.5-1.8 | Decalcified breccia | fauna | undetermined |
| Gladysvale | >0.780 | Decalcified breccia | ESR; fauna | Acheulean |
| Rietputs Formation | 1.3–1.7 | alluvial gravel and sand | cosmogenic burial $^{26}Al–^{9}Be$ | Early Acheulean |

ment is also reconstructed at this time (Luyt and Lee-Thorp, 2003), which probably helped to retain lithic debris around the cave entrance. Thus the land surface was stable and relatively little material was lost from surface occupations during the Oldowan. Fauna accumulated in the deposit as a death-trap assemblage, with only a minor contribution from slope wash around the cave entrance (Pickering, 1999).

Sixty-five percent of the Oldowan material is fresh, and 25% is weathered, showing that the bulk of the material entered the cave without lying on the surface for an extended period of time (Kuman, 1998; Field, 1999). The most logical scenario to explain the Oldowan accumulation envisions hominids sheltering under shade trees growing around a narrow cave opening. Debris around the cave entrance entered the cave with rainwater and gravitation. A smaller portion washed in from the surrounding landscape as an additional 10% of the artifacts are abraded. Some fauna from the slope-wash component were probably the result of hominid meals, but there is only one certain cut-marked bone and the bulk of the fauna seems not to be associated with hominid activities (Pickering 1999).

The top of the Oldowan Infill has been exposed to the surface by erosion of the cave roof. However, at the base of this infill lies a narrow shaft 12 m long which opens into the roof of an underlying cavern termed the Name Chamber. Beneath the shaft lies an enormous talus, which Clarke (1994a) suggests is likely to contain, within its heart, a collapsed breccia from an infill that preceded the Oldowan deposit. There are numerous artifacts and fauna in the outer, uncemented portion of this talus, which we have sampled. Thus far, the stone tools most resemble the Oldowan from the site, and both artifacts and fauna are currently under study. The size distribution of the artifacts will determine if the breccia sampled from this talus contains an earlier component of the Oldowan Infill, or alternatively, Oldowan material that has filtered down to a lower level through the narrow shaft.

### *Swartkrans?*

It is uncertain whether any artifacts from Swartkrans may be old enough to belong to the Oldowan. Member 1 has a broad faunal age published as *c.* 1.7 or 1.8 Mya (Brain et al., 1988; Brain, 1993)—dates that encompass both Oldowan and Developed Oldowan sites elsewhere in Africa. While Clark (1993) suggested that the assemblage could be Oldowan, the material is undiagnostic and his reasons were tenuous. The Lower Bank of Member 1 has 277 artifacts and 21 manuports (Field, 1999), which found their way into the cave in a sporadic manner through slope wash and gravitation from the surface. The incomplete capture of debris from surface occupations seems related to the fact that the cave opening was located on the side of the hill. Fifty-five percent of the assemblage is weathered, showing that the majority of material was exposed on the surface for long periods of time (Field, 1999). Interestingly though, artifacts in a range of sizes entered the cave, with 36% being small flaking debris under 20mm size. This may indicate that the immediate area was well vegetated enough to preserve some smaller material. There are some water dependent species in Swartkrans Member 1 that suggest the presence of a river or stream near the cave supporting riparian woodland and reed beds (Watson, 1993; Reed, 1997).

The associated hominids in Member 1 are *Paranthropus* and *Homo* sp., with *Homo ergaster* found thus far only in a younger section of Member 1 (the Hanging Remnant; Brain, 1981, 1993; Clarke 1994b). No *Homo* fossils diagnostic to species have yet been found in the Lower Bank of Member 1. *H. ergaster* dates from 1.78 until 1.49 Mya (Schwartz and Tattersall, 2003), a period which spans both the Oldowan and early Acheulean. The earliest dates for assemblages more complex than the Oldowan are 1.6 Mya at Koobi Fora for the Developed Oldowan (Isaac, 1997) and *c.* 1.65 for the early Acheulean in western Kenya (Roche and Kibunjia, 1996), both of which industries are widely considered today to belong to the same industrial complex. Hence we cannot resolve the affinity of the Swartkrans assemblage through chronological comparisons. At Sterkfontein, *H. ergaster* is associated with early Acheulean deposits (Kuman and Clarke, 2000). We hope that renewed work on both the Lower Bank and the Hanging Remnant of Member 1 begun in 2005 may produce better information on the archaeology, as fauna is not a sensitive enough time indicator. If a larger sample produced bifaces, larger flake sizes and an abundance of manuports, this would indicate a post-Oldowan industry. Without such elements, only better dates for Member 1 could support an Oldowan designation, as this is a simple core and flake industry lacking the kind of diagnostics that appear after 1.7 Mya. Thus far, only the Lower Bank has produced artefacts, while the important Hanging Remnant section of Member 1 has not.

## Early Acheulean

Slightly younger artifacts are more widespread, occurring in at least three sites. These assemblages have been designated Early Acheulean (Kuman, 1998; Field, 1999), but they would also fit the definition of the Developed Oldowan used by some researchers because bifaces are rare. Only a few bifaces occur in good context (at Sterkfontein), but there are almost two dozen additional specimens from poorer contexts. In East Africa, years of research at Koobi Fora have demonstrated the validity of a Developed Oldowan lacking bifaces in the Karari Industry, which has its own distinctive character (Isaac and Harris, 1997). The Gauteng industry, on the other hand, includes bifaces, but they are rare or absent at individual sites presumably because of small sample size or activity differences. This paper thus considers the Developed Oldowan and Early Acheulean as industries within the early Acheulean complex of sites. I support the view that even a single biface justifies calling an assemblage Acheulean. Sites may be termed Developed Oldowan for descriptive or diagnostic purposes, but the possible causes for such differences and their significance are a point for discussion.

### *Sterkfontein*

The largest assemblage is found at Sterkfontein in Member 5 West with 701 pieces (493 artifacts and 208 manuports). Vrba (1982) considers that one *Antidorcas recki* specimen compares well with Olduvai Lower Bed II specimens (1.7 Mya), and she states that the Sterkfontein bovids may either be similar in age to Swartkrans Member 1 or perhaps marginally younger. There are two bifaces—one cleaver and one handaxe—and the assemblage is heavily winnowed: only 4% of artifacts are under 20mm size. In contrast with the largely fresh Oldowan assemblage from Member 5 East at Sterkfontein, only 32% of artifacts are fresh, showing that occupation debris underwent long-term exposure, with erosion winnowing the site of most of the lighter material. The land surface by this time may have been more sloped, or the position of the cave entrance may have changed, resulting in a less stable catchment surface. Fauna and carbon isotopes both confirm that the environment was open savannah at this time (Vrba, 1975; Reed, 1997; Luyt and Lee-Thorp, 2003), and a drier habitat may have at times enhanced erosion on the Sterkfontein hill. A taphonomic study by Pickering (1999) shows that the fauna was accumulated by carnivores, and there are no cut-marked bones to indicate any obvious hominid involvement.

These Acheulean artifacts are in a secure context because the bulk of the Member 5 West breccia that we have excavated is well-cemented and undisturbed by decalcification. However, there are numerous artifacts excavated from adjacent breccias to the east and south of this area, and many of these may also be the work of Acheulean hominids. However, the site has been disturbed in places by solution pockets, and this could cause some mixing with younger material (Kuman and Clarke, 2000). Therefore only the Member 5 West assemblage with diagnostic artifacts and visible stratigraphy in calcified breccia has been used for the Early Acheulean description. From other locations in the site, there are in fact two additional cleavers and eight handaxes, most of which appear to be Early Acheulean. On its eastern side, the Member 5 West breccia is also truncated by a large solution channel that separates it from other Acheulean material in Member 5 East. An infill with some limited Middle Stone Age material subsequently filled the void and almost certainly connects with an underground deposit within the adjacent Lincoln Cave (Reynolds et al., 2003). An interesting feature of this latter deposit is that some ESA artifacts and two hominid teeth from Member

5 have been incorporated through erosion of older breccia (Reynolds et al., 2003). Such mixing of deposits can be visible if blocks of older breccia become incorporated in a younger infill, which Robinson (1962) noted in his excavation of western breccia. In the case of this younger infill, however, mixing cannot be detected in this manner, and the artifacts and hominid fossils tend to be much better indicators of mixing than the stratigraphy or the fauna.

In 2004, we discovered a new dump with over 100 Acheulean artifacts, several metres south of Member 5 West. It is undoubtedly one of the limeminers' dumps that went unnoticed because it was heavily obscured with vegetation. Although bifaces are absent, the artifacts appear to derive from decalcified areas of Member 5 West cleared by limeminers in search travertine. Such 'erosion channels' in solid breccia were noted by Robinson (1962) in his early work on the western breccia. The assemblage is highly consistent with the early Acheulean in Member 5 West in its typology, technology, and biased size profile, and it lacks mixing with younger tools that affects eastern portions of the site. This find has significantly enlarged the early Acheulean collection at Sterkfontein. The complicated nature of the Sterkfontein stratigraphy, with its mix of cemented and decalcified breccias, is a good illustration of how cautiously cave infill assemblages need to be approached by the analyst. In such contexts, artifacts generally prove to be more sensitive time-indicators than fauna. They can also alert one to mixing or complications in stratigraphy that may not be obvious in the sediments.

### *Swartkrans*

Field's (1999) comprehensive study of the Swartkrans artefacts suggests that Member 1 could possibly belong to the Acheulean (see above for details), but Members 2 and 3 are not in question despite the absence of *in situ* bifaces (see also Clark, 1993). This view is supported by four bifaces processed from limeminers' dumps by Brain, which Leakey (1970) described. One cleaver and one handaxe had enough adhering breccia to allow Brain (1981) to assign them to Member 2. Member 2 is estimated by fauna at *c.* 1.5 Mya and has *H. ergaster* and *Paranthropus* fossils (Brain et al., 1988; Brain, 1993). Sixty-eight percent of the artifacts are weathered, reflecting a long period of surface accumulation before deposition in the cave (Field, 1999). Member 3, containing only *Paranthropus* fossils, is estimated at 1.0 Mya (Brain, 1993), although the carnivores suggest it could be as old as 1.5 Mya (Turner, 1997). Fifty-one percent of these artifacts are also weathered (Field, 1999). Both of these infills show a wider range of flake sizes than the early Acheulean deposit at Sterkfontein, but small flaking debris <20 mm is still under-represented (16% in M2 and 26% in M3). As with Member 1, the capture of material in the catchment area of the shafts was sporadic and incomplete, with the most weathered assemblage showing the poorest capture of small flaking debris. Overall, however, the local cave environment at Swartkrans appears to have been more vegetated than at Sterkfontein during the early Acheulean because water-dependent species are present in the faunas, but none is found in Sterkfontein Member 5 West.

Swartkrans is thus far the only early site where hominids have accumulated a significant portion of the fauna. A high frequency of cut and percussion marks on upper and intermediate limb bone shafts indicates that by 1 Mya hominids in the valley had early access to hunted or scavenged meat (Pickering et al., 2004), and such marks are also now being quantified for Members 2 and 1 (Pickering, pers. comm.). Further evidence of hominid influence in the faunal assemblage at Swartkrans has been suggested for Member 3, where large numbers of burnt bones have been heated to temperatures that suggest hominids had controlled the use of fire (Brain and Sillen, 1988; Brain, 1993). The counter-argument is that bone may have been heated to very high temperatures within the cave entrance or shaft if accumulations of vegetation and wood were ignited by natural fires, continuing to burn for longer periods than in open bushfires. If, however, the hominids were the responsible party, their fires were probably tended on the surface, with burnt bone washing into the cave in a manner similar to the artifacts. Had hominids tended fires within the cave or its entrance area, it is likely that they would have also flaked stone in these spots, and their lithics would then show less weathering and more complete size profiles. Although the controlled use of fire has been proposed in East Africa *c.* 1.5 Mya (Bellomo, 1994), perhaps the earliest date for less controversial evidence comes from the Acheulean site of Gesher Benot Ya'akov in Israel, nearly 790,000 years old (Goren-Inbar et al., 2004).

### *Kromdraai*

Kromdraai A (KA) is a third early site with a small but significant collection of lithics. One hundred artifacts and manuports have been published (Kuman et al., 1997, Field, 1999), and some additional artifacts have been excavated in recent years by Thackeray (pers. comm.). The KA assemblage lacks bifaces but has some larger flakes more common in Developed Oldowan / Early Acheulean infills in the valley. Although the age of KA has only broadly been estimated at 2-1.0 Mya (Thackeray, pers. comm.), the artifacts are very similar to those excavated from Swartkrans and Sterkfontein Member 5 West, which suggests they date somewhere between 1.7 and 1 Mya. Like Swartkrans, the KA artifacts show a range of sizes reflecting a sporadic and incomplete capture of surface material.

### *Rietputs Formation, Vaal basin*

In 2006, the first absolute dates for alluvial deposits in the Vaal River basin, in the interior of the country, were achieved through cosmogenic burial dating for the Rietputs Formation (not far from Pniel and Canteen Kopje, Fig 2). Alluvial deposits with Acheulean artifacts consist of coarse gravel and sand, exposed in active diamond mining pits at depths up to 16 m. Ratios of $^{26}Al/^{9}Be$ were measured using accelerator mass spectrometry in quartz grains collected from the pits (Gibbon, Granger, Kuman and Partridge, in submission). Although dating results will be finalized during 2007, it is already clear that the Rietputs Formation contains artifacts of early Acheulean age (1.3 to 1.7 Mya, R. Gibbon, pers. comm.). Artifacts are made mainly on lavas and occasionally on finer grained material such as hornfels, in contrast with quartz, quartzite and chert used in the Gauteng early sites. These early Acheulean artifacts now expand the distribution of the industry beyond the early hominid sites and provide new technological information of considerable value.

## New Assemblages Not Yet Assigned to Industry

### *Coopers*

Several hundred metres west of Kromdraai A is Coopers, a site with a similar lithic pattern to KA. Two areas of this site have yielded fauna and *Paranthropus* fossils: Coopers B (Steininger and Berger, 2001) and Coopers D (Berger et al., 2003). Since excavations began at Coopers D in 2001, it has also produced over 50 artifacts similar to others in the valley (Hall, 2004). Bifaces are currently absent but the tools are associated with a large faunal assemblage estimated at *c.* 1.6-1.9 m.y. (Berger et al., 2003). The site is located between Kromdraai and Sterkfontein, and hominids had access to the same Blaaubank River gravels for their raw materials.

### *Goldsmith's*

Discovered in 2003, Goldsmith's is situated about 4 km southwest of Sterkfontein. R.J. Clarke and W. Mokokwe have processed faunal samples from breccia dumps prior to excavating undisturbed deposits, and a dozen stone tools have been recovered from these dumps (Mokokwe, 2005). Ten are similar to the 2-1 Mya old material from the other sites in the valley, while one flake and one core are MSA-like (pers. observation). The artifacts were found in loose earth within the dump and not in blocks of breccia. However, the fauna is also contained in both loose earth and breccia blocks within the dump, and we found no surface artifacts in a survey of the area. The ESA artifacts are thus most likely to derive from decalcified breccia, while the MSA pieces could come from a second decalcified breccia or from overburden. Generally speaking, ESA types found in surface or hillwash deposits are very rare in the Sterkfontein valley, and those found in dumps usually derive from decalcified site breccias. At present, the significance of these ten ESA artifacts is that they are found in the Blaaubank valley, in the vicinity of gravels and close to the more prolific artifact-bearing sites.

### *Drimolen and Gladysvale*

Only four early artifacts have thus far been excavated from sites outside of the Blaaubank River valley. Drimolen, which has been worked for over 10 years, lies 7 km from Sterkfontein. It has produced many fossils of *Paranthropus*, plus a small number of *Homo* specimens from deposits estimated at *c.* 1.5 to 1.8 Mya (Keyser, 2000; Keyser et al., 2000). It has to date yielded only three stone tools (one core and two flakes). Drimolen is not situated near a good source of raw materials. The same is true of Gladysvale, 10 km from Sterkfontein, where one handaxe was excavated from a deposit considered to be >780,000 years (Lacruz et al., 2002; Hall et al. 2006). Although there are quartzite boulders in the vicinity of Gladysvale (Hall et. al, 2006), gravels are absent. In the Sterkfontein valley assemblages, some artifacts were made on rocks obtained from the landscape around the caves, but there was a distinct preference for river cobbles in assemblages from 2 to 1 Mya. The Drimolen and Gladysvale finds are significant in showing that hominids transported stone tools some considerable distances around the landscape. Due to the rather different nature of the faunal accumulations in the early Gauteng sites, they currently lack the pattern of cutmarked bone occurring in the absence of stone tools which is noted from even the earliest East African sites. However, these rare finds of artifacts at distance from good raw material sources attest to the planning and transport abilities of the early southern tool-making hominids—a pattern which has been well demonstrated in East African contexts.

## Discussion

As excavations of sites like Drimolen and Gladysvale continue, we may anticipate additional artefacts, but the proximity of most early African sites to raw material sources (see Plummer, 2005) suggests that such finds will be small in number. It is striking that the sizeable Gauteng ESA collections lie within easy reach of a major source of cobbles in the Blaaubank River terrace. Boulders do occasionally occur in the Blaaubank gravels and may have been used to make the rare large flake or biface blank in the Early Acheulean, but river cobbles are dominant in all of the early assemblages (Field, 1999, Kuman, 2003).

Prior to the discovery of the Rietputs Formation early Acheulean, only one area outside of the dolomitic caves region has thus far been published as producing Early Acheulean artifacts. Mason (1962) excavated two series of artifacts from open alluvial sites at Three Rivers and Klipplaatdrif in southern Gauteng Province. Handaxes and cleavers are present, and the technology is indeed comparable to the artifacts at Sterkfontein. However, the material is excavated from river gravels and there is no associated fauna to support the age of the industry. Hence the assignment of the collection to the Early Acheulean is based only on technological grounds. The new dates for alluvial sediments of the Rietputs Formation is thus valuable new evidence that early Acheulean occupation occurred elsewhere in the country in a differing geologic context.

It is significant that the Sterkfontein assemblages are so large relative to the other Gauteng sites. Even compared with Swartkrans as the second largest collection, the Sterkfontein collection is many times larger. If all of the ESA and (unpublished) MSA artifacts from Swartkrans are tallied, there are close to 2,000 pieces. Sterkfontein, on the other hand, has about 9,000 ESA and MSA artifacts (Kuman, 1996). This figure includes over 4,000 pieces in good context and thousands more in disturbed or more complex stratigraphic situations that have not been published in detail. This fact remains, even though the three major sites all lie close to the same raw material source. The Blaaubank River terraces have been mapped by Partridge, who has found that gravel deposits lie within 300 m of each of the three major sites, Sterkfontein, Swartkrans and Kromdraai (T.C. Partridge, pers. comm.). The area under discussion is only about 4.5 km long, with sites situated on both sides of the Blaaubank River. Hence the extent of the accumulation at Sterkfontein suggests that the site had some physical properties that made it a favored location for hominid activities over long stretches of time. These features could have been good shade trees, low outcrops of dolomite that provided some shelter, dry ground with a good outlook, a setting safe from carnivores, or possibly some com-

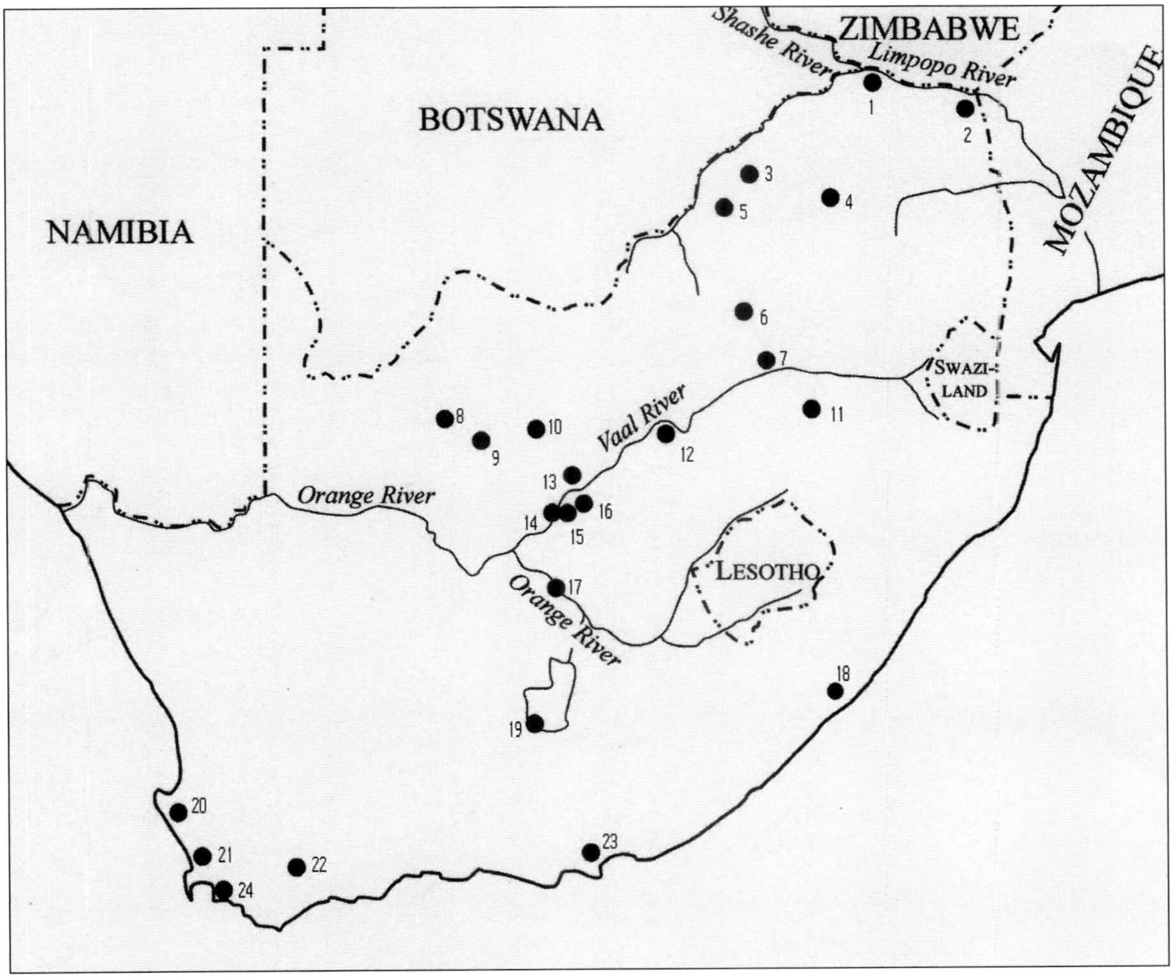


Figure 2. Approximate location of selected ESA sites mentioned in the text:

| | | | |
|---|---|---|---|
| 1. | Kudu Koppie, Hackthorne, Keratic Koppie | 13. | Pniel, Canteen Koppie |
| 2. | Northern Kruger Park | 14. | Muirton |
| 3. | Olieboomport Shelter | 15. | Rooidam |
| 4. | Cave of Hearths | 16. | Doornlagte |
| 5. | Blaaubank River | 17. | Orange River Scheme |
| 6. | Sterkfontein, Swartkrans, Kromdraai | 18. | Port Edward, Pondoland |
| 7. | Three Rivers, Klipplaatdrif | 19. | Seacow Valley Survey |
| 8. | Kathu Pan | 20. | Elandsfontein |
| 9. | Wonderwerk Cave | 21. | Duinefontein |
| 10. | Taung DB3 | 22. | Montagu Cave |
| 11. | Cornelia | 23. | Amanzi Springs |
| 12. | Munro | 24. | Cape Hangklip |

*Table 2. Earlier Stone Age sites of South Africa less than 1.0 Mya. The chronological placement of the sites is based on published statements (see references within text). In some cases, the literature may suggest a middle, later or late sub-stage within the Acheulean or ESA. Given the paucity of absolute dates and systematic comparative studies, however, these designations should be read as interpretations subject to revision.*

| Site Name | Broad Chronology in Mya | Sedimentary Context | Basis for Age Estimate |
|---|---|---|---|
| Cornelia | 1.0–0.7 | resorted hillside rubble and valley gravel deposits | fauna |
| Elandsfontein | 1.0.–0.6 | thin palimpsests in dune sands formed on land surface near waterhole | fauna possibly from two periods and later Acheulean bifaces |
| Powers Site | 1.0–>0.2 | disturbed alluvial sediments from different contexts | fauna mostly from Gravel C; Fauresmith artifacts from surface of Gravel C |
| Doornlaagte | middle Acheulean | pan margin sands: a lag of multiple occupations | artifacts |
| Kathu Pan | <ca1.0/later Acheulean | pan margin: silty to gravelly sands with spring vents | fauna |
| Cape Hangklip | later Acheulean | palimpsest exposed on raised beach | artifacts |
| Canteen Kopje: Stratum 1 | final Acheulean? | Hutton Sands with palaeosols | artifacts (Fauresmith & MSA, unpublished) |
| Stratum 2a (top 30-40cm) | final Acheulean | predominantly colluvial gravel | artifacts (Fauresmith) |
| Stratum 2a (below 30-40cm) | Acheulean | predominantly colluvial gravel | artifacts (Victorial West and sporadic Levallois) |
| Stratum 2b Upper Unit | Acheulean | colluvial sandy gravel | artifacts (sporadic Levallois) |
| Stratum 2b Lower Unit | Acheulean | predominantly colluvial gravel | unpublished |
| Pniel 6, Stratum 4 | Acheulean | colluvial rubble deposits | artifacts comparable with Pniel I with early Pleistocene fauna |
| Northcliff (Acacia Road) | Acheulean | hillslope rubble with artifacts in two conditions | artifacts |
| Northern Kruger Park | predominantly middle Acheulean | extensive landscape with alluvial and colluvial deposits (secondary contexts) | artifacts (some prepared core technology) |
| Haaskraal Pan | later Acheulean | pan floor and adjacent valley slope deposit within a 360,000 yr sequence | artifacts |
| Amanzi Springs | Acheulean | disturbed spring mound | artifacts |
| Montagu Cave | later Acheulean | cave strata | artifacts |
| Cave of Hearths Beds I–III | >0.4/later Acheulean | cave breccias | artifacts; fauna; ESR |
| Wonderwerk Cave: Major Unit 3 | ca 0.276–0.286 Late Fauresmith | cave sediments | Uranium Series |
| Major Unit 4 | >0.350, Middle Fauresmith | cave sediments | inferred age |
| Major Units 7–8 | >0.780, Acheulean | cave sediments | palaeomagnetism |

*Table 2. (continued)*

| Site Name | Broad Chronology in Mya | Sedimentary Context | Basis for Age Estimate |
|---|---|---|---|
| Taung DB3 | Acheulean | factory site on quartzite outcrop, only partially buried | Victoria West artifacts |
| Olieboompoort Shelter | Acheulean | rockshelter: palimpsest in basal rubble | artifacts |
| Wonderboom | later Acheulean | colluvial hill rubble | artifacts |
| Munro | later Acheulean | gravel and overlying calcretised colluvial and alluvial deposit | artifacts |
| Other Vaal Younger Gravels and Riverton Formation Sites | Acheulean | alluvial deposits | some faunal collections |
| Rooidam | >0.2/late Acheulean | deep pan sediments | Uranium Series minimum of 0.174 for higher stratum; Fauresmith artifacts |
| Duinefontein 2:<br>Horison 3<br>Horison 2 | 0.290/late Acheulean<br>0.270/late Acheulean | accumulations on two land surfaces near a marsh or large pond and now encased in dune sands | OSL dates on encasing sands, confirmed by fauna, artifacts and capping U Series date |
| Muirton | late Acheulean | in calcified silty sand overlying Vaal Younger Gravels with Acheulean | artifacts |
| Nooitgedacht 2 | late Acheulean | base of Hutton Sands and in underlying gravel | Fauresmith artifacts |
| Roseberry Plain 1 | late Acheulean | base of Hutton Sands and on underlying bedrock | Fauresmith artifacts |
| Orange River Scheme | Acheulean<br>late Acheulean | 3 surface sites<br>1 site in thin gravel deposit | artifacts<br>Fauresmith artifacts |
| Seacow Valley Survey | later and final Acheulean | quarry and surface sites in floodplain basin | artefact typology: Seven Acheulean and many Fauresmith assemblages |
| Nakop | Acheulean | surface exposure in gravel overlying aeolian sand | Victoria West artifacts |
| Blaaubank River | later ESA | surface of outwashed gravel | 'Earlier Sangoan' and later Acheulean artifacts |
| Geelhoutboom | Acheulean | deflated coastal dune plain | artifacts |
| Kudu Koppie | final ESA | deflated lag deposit on sandstone, overlain by MSA palimpsest in koppie rubble, overlain by late Pleistocene sandcover with LSA | Sangoan-like artifacts stratified under MSA |
| Hackthorne | final ESA | deflated lag deposit on calcretised Miocene terrace beneath late Pleistocene sandcover | Sangoan-like artifacts, mixed with MSA? |
| Keratic Koppie | final ESA | deflated lag deposit on sandstone beneath late Pleistocene sandcover | Sangoan-like artifacts, mixed with MSA? |
| Port Edward and Pondoland | final ESA | surface exposures in deflated coastal dune deposits | Sangoan artifacts |

bination of these factors (Kuman, 1994b). Significant numbers of manuports are found at the three main sites (Field, 1999), which shows that repeated visits occurred at some venues in the valley.

The presence of larger artifact collections at the three main sites is not merely a product of the intensity of excavations. The richest site, Sterkfontein, also has an abundance of artifacts in the overburden above intact breccia. This is a feature that is not so prominent at other sites and must result from certain favourable physical properties. The sporadic finds of artifacts at sites like Drimolen and Gladysvale, or the alluvial sites in southern Gauteng and the Vaal basin, attest to the fact that tool-using early hominids were more widely distributed than the best sites would indicate. Early artifacts in secondary context are also present elsewhere in southern Africa in Angola, the Democratic Republic of the Congo, and possibly Botswana (see Kuman, 1998 for references). Thus the best southern evidence for cultural behaviour *c.* 2 to 1 m.y. ago is found mainly at sites in special geographic settings near good sources of river cobbles. With the exception of Swartkrans, the lack of direct associations between early Pleistocene faunas and hominid activities also shows the limitations of the South African record. Nevertheless, it is a valuable early record because it complements the East African sites and shows that the distribution of early industries was quite widespread on the African continent. The increase in the numbers of manuports at Sterkfontein and other sites after 1.7 Mya also parallels the East African evidence (Potts, 1991; Schick, 1987b), reflecting the more habitual use of stone by hominids (probably *H. ergaster*) following the Oldowan period.

## Middle, Later and Late ESA Sites

### *Chronology*

After 1 Mya, ESA sites and surface occurrences become more numerous (Sampson, 1974). Table 2 provides a list of published sites. The list is not exhaustive as some published surface and gravel collections are not detailed, but it illustrates the variety of contexts in which South African ESA material is preserved. Unfortunately, datable volcanic sediments do not occur in South Africa, and only three sites have published radiometric ages based on other methods (**Wonderwerk Cave**, **Rooidam**, and **Duinefontein**. All are at the younger end of the ESA chronology, and for two of these sites, the dates provide only minimum ages. Fauna is preserved at only some of the sites. Although many of the associations are not horizon-specific, these fauna are valuable in illustrating the great antiquity of some of the sites.

Artifacts from a few of the sites may possibly be called middle Acheulean on technological or faunal grounds (**Cornelia**, **Canteen Kopje** in part, **Doornlaagte**, and **northern Kruger Park**). Because of the poor dating resolution, however, this distinction is not widely used, and most researchers prefer to lump all of the sites 1 Mya or younger into the 'Upper Acheulean,' in contrast with the 'Lower Acheulean' of the Gauteng cave sites (Volman, 1984; Klein, 2000). In this paper, the 'middle Acheulean' category is kept as potentially useful. As refinements in dating techniques develop, we may hope one day to test the idea of change within these industries over time, as has been suggested for a few sites by Mason (1962). In East Africa, for example, the lengthy sequence at Olduvai Gorge shows that the Acheulean is not totally static in that region. The overall evaluation of the assemblages through time shows that, in contrast with the heavier, thicker, less standardized handaxes of the early Acheulean in Bed II, the middle Acheulean Bed IV handaxes (from ca 1.2 Mya) show more regularity of shapes and improved flatness in section, while cleavers become more frequent and are often more elegantly made (Roe, 1994: 204). In the post-Bed IV or Masek sites (>0.4 but <ca 1.0 Mya), handaxes show even greater standardization of preferred shapes and a high degree of technological competence. Another East African site that should be re-evaluated for the concept of a middle Acheulean is Olorgesailie, where new dates place the majority of the beds between 992,000 and 601,000 years (Deino and Potts, 1990). Prior to the benefit of these older daters, Isaac (1977: 213) classified the site as Upper Acheulean because "reliable chronological distinctions…are not possible." However, he commented that the assemblages are somewhat less refined than those from other sites and that one might be tempted to designate them as middle Acheulean. Like Isaac, Noll (2000) found no trends within the Olorgesailie artefact sequence in his comparison of material from Member 1 (0.99 Mya) and Member 6/7 (0.97-0.75 Mya), relating differences only to intensity of flaking and raw materials Noll (2000). It is widely noted that only in the final phase of the ESA do significant changes in lithic character and regional differences appear (Noll, 2000; Klein, 2000), which Clark (1970, 2001a) attributed to the emergence of more modern humans.

A few of the sites in Table 2 have faunal age estimates that may shed light on such a distinction between the middle and later Acheulean. The highly refined bifaces of **Kathu Pan**, for example, are associated with elephant fossils that are more evolved than the form found in Olduvai Bed IV (Volman, 1984: 184), which translates to an age of <ca 1.0 Mya. Hence Kathu Pan is classed as later Acheulean on both technological and faunal grounds. The **Cave of Hearths** also has refined bifaces and a younger fauna, and has been published as later Acheulean (Mason, 1988a; see also Ogola, 2003; and McNabb et al., 2004). Efforts to date the site with ESR have recently been undertaken by Grun (2003), who announced a preliminary estimate of 0.4 to 0.6 Mya for the hominid mandible in the top Acheulean bed. Grun (pers. comm.) is now working to expand the sample of ESR readings to refine these preliminary results. For **Canteen Kopje**, Mason (1962) had designated an assemblage

from this site as middle Acheulean, but renewed work by Beaumont and McNabb (2000) indicates that a longer sequence is present.

Sites in Table 2 are termed 'Acheulean' if there is insufficient information to suggest a middle, later or late placement. Within the later to late Acheulean, a few authors have noted a trend toward smaller and broader handaxes over time, particularly at **Rooidam** and **Montagu Cave** (Fock, 1968; Keller, 1973), or in comparisons across sites (Mason, 1962). Late or final Acheulean assemblages in the table are designated either by dates that suggest an age of *c.* 0.3–0.2 Mya, or by a refined technology often termed **'Fauresmith'** in the literature. **Rooidam** is perhaps the best example of a Fauresmith site but has only a minimum Uranium Series date (Szabo and Butzer, 1979). The top Acheulean stratum at **Wonderwerk Cave** has been published as containing Fauresmith tools dated by amino acid racemisation at >0.2 Mya (Binneman and Beaumont, 1992). Recently, Beaumont and Vogel (2006) revised the age of the late Fauresmith unit to 0.276-0.286 Mya and also inferred an age of >0.35 Mya for a middle Fauresmith unit. Small numbers of Acheulean artefacts also derive from lower units >0.780 Mya based on palaeomagnetic readings. **Duinefontein 2** currently has the best late Acheulean dating result, but the tool assemblage is rather informal (Cruz-Uribe et al., 2003). Fauresmith assemblages are said to include small, refined handaxes, sometimes associated with points, blades and prepared cores. Based on the occurrence of facetted platforms and convergent flaking, Beaumont (Beaumont and Vogel 2006) assigns Fauresmith assemblages to the MSA, whereas I prefer to refer to them as a final ESA phase due to the persistence of handaxes and cleavers. However, there is clearly much variability in the late ESA, and patterns are more important than nomenclature in this period, which is increasingly understood to be a time of change, in terms of both technology and evolutionary biology. The origins of prepared core technology in southern Africa appear to extend back to the Victoria West industry. Its antiquity is not yet understood as sequences are limited, and detailed descriptions are currently lacking, except at sites that lack stratigraphy (e.g., **Taung DB3**, Kuman, 2001; or **Nakop**, Brain, and Mason, 1955). Although prepared core or Levallois technology becomes more widespread in the later Acheulean, it cannot be assumed that it originated in this phase. The earliest published dates for some form of this technology are *c.* 0.7–0.8 Mya at Gesher Benot Ya'aqov in Israel (Goren-Inbar and Saragusti, 1996; Goren-Inbar et al., 2000), but detailed descriptions are awaited to aid its assessment.

The **Fauresmith** has traditionally been seen as a late or final industry within the Acheulean, which equates with the end of the ESA. However, there is a second final ESA industry thought to be comparable in time to the Fauresmith, namely the **Sangoan** (or Sangoan Industrial Complex). Classic Sangoan assemblages have been described in Zambia and Uganda (Clark, 1970, 2001b), but a variant occurs in Kenya in the Sangoan-Lumpemban Industrial Complex (McBrearty, 1988, 1991). It has also been suggested to occur in Zimbabwe (Cooke, 1966), although here the assemblages are small and require further substantiation. In South Africa, surface finds of Sangoan-like tools occur along the KwaZulu-Natal and Eastern Cape Wild Coast (Davies, 1976). In 2006, we confirmed that some of these deflated surface sites (respectively at **Port Edward** and **Pondoland**) are Sangoan-like, and survey for intact deposits should begin in the near future. Deacon (1970) also suggests that the relatively heavy and unstandardized form of artifacts at **Amanzi Springs** could similarly suggest a late to final ESA placement of this near-coastal site.

In the Transvaal, Mason (1962) excavated a small collection of 'Sangoan' artifacts from a river gravel, but his sample was identified typologically from a mixed collection, and thus its integrity was questioned. Recently, however, the first evidence for the stratified occurrence of a Sangoan-like industry in South Africa has been found in the far north of the country, close to the borders with Botswana and Zimbabwe. This is at **Kudu Koppie**, where a large assemblage occurs in a deflated context, stratified below a palimpsest of MSA artifacts (Kuman et al., 2005). At two nearby sites, similar Sangoan-like material has also been excavated from deflated contexts at **Hackthorne** (Kuman et al., 2004) and **Keratic Koppie** (Kuman et al., 2005). However, these are single-component lag deposits buried by younger sands, and there is a stronger likelihood that mixing with younger (early? MSA) material may have occurred as stratification is absent. At Kudu Koppie, however, the integrity of the stratified occurrences is confirmed by differences in sedimentary matrix and in artefact typology and technology (Pollarolo, 2004; Kempson, 2005; Kuman et al., in prep.) Elsewhere in Africa, Sangoan occurrences are generally thought to date to *ca* 300,000 years (McBrearty and Brooks, 2000). Both the Sangoan and Fauresmith have long been considered to be the first signs of regional specialization at the end of the ESA (Clark, 1959).

## *Site Contexts*

The variety of contexts in which these southern ESA sites are preserved is interesting, and each has implications for the nature of the archaeological data.

### *Alluvial Sands and Gravels*

A majority of sites has deposits formed in alluvial sediments or exposed on alluvial surfaces: **Cornelia**, **Powers**, **Munro**, **Muirton**, **Nooitgedacht 2**, and other **Vaal basin sites** are prominent among them, along with **Nakop**, **Blaaubank River**, the **Orange River basin**, the adjacent **Seacow valley sites**, and extensive alluvial exposures in **Northern Kruger Park** (Clark, 1974; Humphreys, 1969; Mason, 1962, 1967, 1969; Sampson, 1972, 1974, 1985; Helgren, 1978; Butzer, 1984; and previously cited references). The quality of such sites is obviously least good on exposed surfaces or within gravels, the latter representing high energy conditions. Sands can also represent disturbed conditions, but finer sands and silts have better potential for preservation of a greater range of artifact sizes. Cornelia, Powers and some other Vaal basin sites preserve faunal collections, which are highly valuable for their age estimates. Overall, the value of this group of sites is the information it provides on the wide-ranging geographic distribution of Acheulean peoples.

### *Colluvial Deposits*

Colluvial deposits form through slope wash or the gravitation of sediment. Such processes often result in better preservation of assemblages as the associated gravitational and erosional processes can be less destructive than high-energy alluvial forces. During very arid periods, thinner ground covers enhance erosion, especially on hillslopes, and substantial deposits of coarse colluvial rubbles can result. While colluvium is often a component associated with sedimentation processes at alluvial, pan or other sites, colluvial rubble deposits are especially prominent at **Pniel**, **Northcliff**, **Wonderboom**, and in areas of **Northern Kruger Park** (Mason, 1962; Beaumont, 1969; Beaumont, 1990d; Gibbon, 2004). In rare circumstances, a stratified sequence may accumulate. At **Canteen Kopje**, for example, a stratigraphic sequence occurs with four components that appear to span a large part of the Acheulean (Beaumont and McNabb, 2000).

### *Pans*

Sites found in pan settings generally provide good preservational contexts because they form under lower energy depositional conditions and incorporate colluvial pan margin sediments. Many pan sites also tend to preserve long term, time-averaged accumulations because hominids repeatedly visited such venues for water and game. At **Doornlaagte**, preservation is good, with artifacts occurring as a vertically dispersed accumulation that is the lightly winnowed lag of multiple occupations (Mason, 1988b; Deacon, 1988). **Haaskraal Pan** is another deep site that may contain over 360,000 years of deposits (Partridge and Dalbey, 1986). **Rooidam** is a late Acheulean pan site that has deep sediments, preserving a lengthy accumulation of Fauresmith artifacts which shows some change in handaxe breadth over time (Fock, 1968). Finally, **Kathu Pan** preserves three components of later Acheulean artifacts; however, this site is complicated by its association with underground springs that have created dolines or collapsed areas (Beaumont, 1990a). The artifacts are often associated with spring vents in complex sedimentary contexts, but the collections are large and represent a lengthy sequence of refined, later Acheulean material fortunately associated with fauna.

### *Spring Mound*

The only ESA spring mound site in the country is **Amanzi Springs**. Two phases of accumulation preserve large numbers of ESA material in the lower deposits. The context of the artifacts has been disturbed by spring activity, but Amanzi is especially valuable as the only Acheulean site in South Africa to preserve wood and other botanical remains (Deacon, 1970).

### *Coastal Dune Sites*

Four sites are preserved in coastal dune fields. The first two were associated with standing water during moister periods. **Elandsfontein** is a later Acheulean site that formed on a land surface associated with a waterhole, created by large-scale deflation that scoured out sediments to a point at which the water table became exposed (Butzer, 1973; Klein, 1978; Klein and Cruz-Uribe, 1991; Deacon, 1998; Klein et al. 2006). Artifacts occur in thin palimpsests accumulated on an ancient landsurface associated with the waterhole. **Duinefontein 2** is a somewhat younger site in a similar context (J. Deacon, 1976; Klein et al., 1999; Cruz-Uribe et al., 2003). In this case, two ancient land surfaces are preserved as separate horizons associated with a large pond or marsh. Although deflation has played a role in concentrating material on the land surfaces at Elandsfontein and possibly at Duinefontein, these two sites provide extremely valuable records as both preserve fauna and land surface stratigraphy. The OSL and Uranium Series dates for Duinefontein are particularly useful for late ESA chronology. Two further sites occur in poorer contexts. **Cape Hangklip** is a later Acheulean palimpsest of artifacts that appears to have accumulated on an ancient shoreline, with the site now exposed through deflation (Sampson, 1972). **Geelhoutboom** is a lesser known coastal site with artifacts in a series of deflation platforms in coastal dunes (Laidler, 1945; Deacon, 1970). The **Port Edward** and **Pondoland** Sangoan sites thus far surveyed contain artefacts deflated onto red dune sands from overlying deposits that no longer exist. These sands are intermittently exposed along a stretch of coast over 20 km long from Pondoland to Port Edward, and deposits east of Port Edward are likely to be similarly extensive. Weathering has produced the red coloration of the sand, and aeolian activity has resulted in concentrations of heavy minerals. It is on these more resistant, mineral-rich horizons that stone tools have come to rest. We hope that a thorough survey of more inland areas may potentially locate Sangoan sites that are less altered by deflation.

### *Cave Deposits*

Although each site has some limitations, four later Acheulean sites are fortunately preserved in cave deposits. **Montagu Cave** in the Cape has a lengthy sequence of artifacts with extensive factory site debris (Keller, 1973). It is a site rich in artifacts but unfortunately lacks fauna. **Cave of Hearths** is the next richest site and has a sequence of later Acheulean in three beds, with artifacts that have been very well published (Mason, 1962, 1988a; McNabb et al., 2004). It also has fauna, accumulated and modified by a number of agents, including both humans and carnivores (Ogola, 2003). Its limitation is that the deposits consist of cemented cave breccias that complicate the stratigraphy, with some areas having undergone subsidence and collapse. Nevertheless, the Acheulean sequence has been carefully ascribed to the original three beds (Mason, 1988a), and fauna from the most reliable positions have been studied, showing the complex nature of its accumulation processes (Ogola, 2003). **Wonderwerk Cave** is close to the Vaal basin sites and has deep stratified deposits, including Fauresmith (from ca 0.27 to >0.35 Mya) (Beaumont and Vogel 2006). An inferred age of ca 0.5 Mya has been suggested for the lowest Fauresmith levels, but more detailed assemblage descriptions or dating resolution would aid assessment. Underlying Acheulean material said to be >0.78 Mya is present but unfotunately sparse. Fauna is present but limited, with some deposits disturbed by diggers for bat guano. (Malan and Wells, 1943; Beaumont, 1990b; Binneman and Beaumont, 1992). However, the stratified sequence and dating potential make this site especially important for future excavation, especially as it may assist with understanding the rich regional record of open-air sites in the Vaal basin. Finally, **Olieboompoort Shelter** is a rock-shelter with a small sample of Acheulean artifacts noted to occur in a basal rubble deposit (Mason, 1962). There are some indications for the controlled use of fire at Cave of Hearths, Montagu Cave, and Wonderwerk Cave.

*Aeolian Sands and Underlying Deposits (Vaal Basin)*

Two late Acheulean sites in the Vaal basin are **Nooitgedacht 2** and **Roseberry Plain 1**, both of which contain Fauresmith artifacts in the lowest levels of an aeolian sand cover known as Hutton Sands (Beaumont, 1990c). At Nooitgedacht the artifacts continue down into an underlying gravel, but at Roseberry Plain they continue downwards to lie on bedrock. These sites have not been published in detail and are thus more difficult to discuss, but they are apparently just two examples of widespread occurrences around Kimberley that are "discontinuously mantled by up to *c.* 3m depth of Hutton sands, in the lowest levels of which occur a normally low density of Fauresmith artifacts" (Beaumont, 1990c; van Riet Lowe, 1927). The nature of the underlying deposits in which the artifacts from these two sites continue requires further clarification. **Taung DB3** is a third site preserved in a related circumstance (Kuman, 2001). It is a factory site located on a quartzite outcrop, only partially buried in a thin veneer of aeolian sediment weathered from the parent rock on which the site is located. The site is located on a high escarpment in the upper Vaal basin, and its interest lies in the presence of Victoria West technology in a factory context.

*Aeolian Sands and Underlying Lag Deposits (Limpopo basin)*

The lag deposits at **Kudu Koppie** were originally accumulated as colluvium and rock rubble associated with an adjacent outcrop of sandstone that provided shade and shelter for hominid occupations during more mesic regional climatic conditions. These occupations were followed by ensuing phases of large-scale, intense deflation episodes during xeric periods that transformed the original sediments into lag deposits (Pollarolo, 2004; Kuman et al., 2005; Kempson, 2005). The basal unit is a closely packed deflated lag deposit with a weathered, small particle size matrix (Kuman et al., in prep). Above this is a thick unit of deflated koppie rubble, less weathered and with a larger particle size matrix. A thick sand unit of predominantly late Pleistocene age directly overlies the MSA rubble. It contains LSA with pottery in the upper levels, and sparse LSA material throughout, which has filtered down through the unconsolidated sand. Two other deflated sites occur within 1-2 km of Kudu Koppie. **Hackthorne** is an unstratified site at which Sangoan-like artifacts have been deflated onto a calcretised Miocene river terrace, with pockets of artifact-bearing sediments also contained in numerous solution cavities within the calcrete (Kuman et al., 2004). **Keratic Koppie** is a similar deflated site, but in this case the artifacts lie within a deflated basal horison on sandstone bedrock (Le Baron, 2004; Kuman et al., 2005). At both of these single component sites, late ESA artefact types occur. To determine the degree of mixing, Kempson (2007) has compared these two assemblages with the stratified sequence at Kudu Koppie, where there are clear differences between the ESA and MSA assemblages. Like Kudu Koppie, Hackthorne and Keratic Koppie are buried under aeolian sand. OSL dating at all three sites has produced age estimates of *c.* 15–23,000 years for this regional sandcover (Kuman et al. in prep.).

## Conclusion

For many years, the earliest phases of the ESA have been best known from the dolomitic limestone caves of Gauteng, in deposits *c.* 2-1 Mya. If Oldowan hominids initially favored the semi-tropical sheltered valleys in which these early hominid sites are found (Bamford, 1999), by early Acheulean times they had certainly become more visible in southern Africa (Kuman, 1998). Absolute dating of the Rietputs Formation deposits (1.7 to 1.3 Mya, Vaal River basin) is now set to provide a new suite of sites, significantly expanding the geological context in which such early material is found. The prominence of erosion is undoubtedly responsible for our limited record of earliest ESA sites, and the restricted preservation of fauna and limited dating methods have been responsible for our difficulty in recognizing the true antiquity of some alluvial occurrences.

ESA sites in a variety of contexts fill the gap between these earliest sites and the early MSA, which appears about 260,000 years ago in South Africa (Grun et al., 1996; McBrearty and Brooks, 2000). South African landscapes are dominated by erosion and planation (Deacon, 1975; Klein, 2000), which is evidenced in the widespread distribution of surface finds in poor context. Assemblages in river gravels are common, but those in alluvial and colluvial sediments with lower energy sedimentary processes preserve a better record. Pan sites, which are remnants of older, more extensive drainage systems, provide a few of the best sedimentary contexts. Aeolian activity is prominent in coastal areas and in the drought-prone far north of the country, and despite their extreme conditions, such contexts have preserved a few stratified sites of much significance. Cave occupation sites are generally limited to the later Acheulean, but they preserve a key record, especially in terms of dating potential, stratified sequences, and evidence for the controlled use of fire.

The overwhelming impression one has of the South African ESA is that sites are most visible close to good raw material sources. This is certainly the case for the early Gauteng cave infills, where only small numbers of tools are found at sites lying more than 500 m from good gravels (Kuman, 2003). It is also a clear pattern at all but one of the Acheulean sites—only Elandsfontein has exotic raw materials carried in from 20–30 km distance, while all other sites lie either close to rock sources, or within several km (Klein, 2000). This is also true for the Sangoan-like sites. Miocene gravels of the Limpopo River deposits provided raw materials close to Kudu Koppie, Hackthorne and Keratic Koppie, while beach and stream cobbles were readily available at the Port Edward and Pondoland sites. We cannot know, however, whether this pattern of proximity to raw materials reflects hominid ranging patterns, or merely 'cultural visibility.' The strong association of ESA sites with standing water in one form or another has often been noted, and even linked to the inability of Acheulean hominids to transport water in containers (Deacon, 1975; Deacon and Deacon, 1999; Klein, 2000). However, standing water, raw materials and good sedimentation processes are often physically associated, which suggests that the location of ESA sites may not be the most reliable evidence for cultural conservatism. However, it seems that the artifacts themselves provide strong clues to this conservatism. Detailed studies of biface technology and form indicate that ESA hominids did not impose stylistic differences in their toolkits, whether these are considered regionally within Africa or across the Old World. Differences that 'appear' to be stylistic have been shown to relate only to raw material size and shape (McPherron, 2000). This view has recently been confirmed by McNabb et al. (2004) for the later Acheulean bifaces at Cave of Hearths. They argue that the most important determinant in the shaping of bifaces is not style, but the least-effort approach to the creation of functional cutting edges.

The South African ESA record has its limitations as many sites lack stratigraphy, primary contexts, or good materials for absolute dating. However, it is a rich record, nonetheless, and one which now includes all of the stages of cultural development, from the earliest to the latest industries. The long-term, persevering research by Bob Brain at Swartkrans over three decades has served as a true model for the kind of patient and insightful approach needed to understand site formation and transformation through the ESA. This work should serve as an inspiration to archaeologists, as much as it does to taphonomists.

## Acknowledgments

I am most grateful to the three conference organizers for giving me the opportunity to participate in this very special workshop in honour of a scientist whom I have long admired and held as a role model. The work of both Nick Toth and Kathy Schick has also been of particular influence in my outlook on archaeology, and it was a pleasure to be part of their first workshop at the Stone Age Institute. Travis Pickering is an important fellow traveler in both my professional and personal life. I highly value my collaborations with him and time spent with his family, and I look forward to much more of this in the future. Ron Clarke, Phillip Tobias and Tim Partridge have been excellent colleagues in my endeavours in the early archaeology of South Africa. Finally I am grateful to the postgraduate students who have been part of my Limpopo valley research programme. Over the last several years, we have both learned from one another, and without their interest and enthusiasm the new discoveries noted here in the far north would not have happened. Primary research cited in this paper has been supported by the Palaeoanthropological Scientific Trust, the De Beers Fund Educational Trust, the L.S.B. Leakey Foundation, The National Research Foundation, and the University of the Witwatersrand.

## References

Bamford, M. 1999. Pliocene fossil woods from an early hominid cave deposit, Sterkfontein, South Africa. South African Journal of Science 95, 231-237.

Beaumont, P. 1969. A very early Stone Age site at Northcliff Extension 2, Johannesburg. South African Journal of Science 65, 65-71.

Beaumont, P. 1990a. Kathu. In: Beaumont, P., Morris, D. (Eds). Guide to Archaeological Sites in the Northern Cape. McGregor Museum, Kimberley (South Africa), pp. 75-100.

Beaumont, P. 1990b. Wonderwerk Cave. In: Beaumont, P., Morris, D. (Eds). Guide to Archaeological Sites in the Northern Cape. McGregor Museum, Kimberley (South Africa), pp. 101-134.

Beaumont, P. 1990c. Nooitgedacht 2 and Roseberry Plain 1. In: Beaumont, P., Morris D. (Eds). Guide to Archaeological Sites in the Northern Cape. McGregor Museum, Kimberley (South Africa), pp. 4-6.

Beaumont, P. 1990d. Pniel 1, Pniel 6. In: Beaumont, P., Morris, D. (Eds). Guide to Archaeological Sites in the Northern Cape. McGregor Museum, Kimberley (South Africa), pp. 7-13.

Beaumont, P., McNabb, J. 2000. Canteen Kopje – the recent excavations. The Digging Stick 17, 3-7.

Bellomo, R.V. 1994. Methods of determining early hominid behavioral activities associated with the controlled use of fire at FxJj 20 Main, Koobi Fora, Kenya. Journal of Human Evolution 27, 173-195.

Berger, L.R., de Ruiter, D.J., Steininger, C.M., Hancox, J. 2003. Preliminary results of excavations at the newly investigated Coopers D deposit, Gauteng, South Africa. South African Journal of Science 99, 276-278.

Binneman, J., Beaumont, P. 1992. Use-wear analysis of two Acheulean handaxes from Wonderwerk Cave, Northern Cape. Southern African Field Archaeology 1, 92-97.

Brain, C.K. 1958. The Transvaal Ape-Man-Bearing Cave Deposits. Transvaal Museum, Pretoria.

Brain, C.K. 1981. The Hunters or the Hunted? An Introduction to African Cave Taphonomy. University of Chicago Press, Chicago.

Brain, C.K. (Ed.). 1993. Swartkrans: A Cave's Chronicle of Early Man. Transvaal Museum , Pretoria.

Brain, C.K., Churcher, C.S., Clark, J.D., Grine, F.E., Shipman, P., Susman, R.L., Turner, A., Watson, V. 1988. New evidence of early hominids, their culture and environment from the Swartkrans cave. South African Journal of Science 84, 828-835.

Brain, C.K. Mason, R.J. 1955. A later African Chelles-Acheul site near Nakop, South Salahari. South African Archaeological Bulletin 10, 22-25.

Brain, C.K., Sillen, A. 1988. Evidence from the Swartkrans cave for the earliest use of fire. Nature 336, 464-466.

Breuil, H. 1930. Premières impressions de voyage sur la préhistoire Sud-Africaine. L'Anthropologie XL.

Butzer, K.W. 1973. Re-evaluation of the geology of the Elandsfontein (Hopefield) site, south-western Cape, South Africa. South African Journal of Science 69, 234-238.

Butzer, K.W. 1984. Archeogeology and Quaternary environment in the interior of southern Africa. In: Klein, R.G. (Ed.). Southern African Prehistory and Paleoenvironments. A.A. Balkema, Rotterdam and New York, pp.1-64.

Clark, J.D. 1959. The Prehistory of Southern Africa. Penguin Books, London.

Clark, J.D. 1970. The Prehistory of Africa. Thames and Hudson, London.

Clark, J.D. 1974. The stone artifacts from Cornelia, O.F.S., South Africa. In: Butzer, K.W., Clark, J.D., Cooke, H.B.S. (Eds). The Geology, Archaeology and Fossil Mammals of the Cornelia Beds, O.F.S. Memoirs of the National Museum, Bloemfontein, No. 9.

Clark, J.D. 1993. Stone artifact assemblages from Members 1-3, Swartkrans Cave. In: Brain, C.K. (Ed.). Swartkrans: A Cave's Chronicle of Early Man. Transvaal Museum, Pretoria, pp. 167-194.

Clark, J.D. 2001a. Variability in primary and secondary technologies of the Later Acheulian in Africa. In: Milliken, S., Cook J. (Eds.), A Very Remote Period Indeed. Oxbow Books, Oxford, pp. 1-18.

Clark, J.D. 2001b. Kalambo Falls Prehistoric Site, Vol. III. Cambridge University Press, London.

Clarke, R.J. 1994a. On some new interpretations of Sterkfontein stratigraphy. South African Journal of Science 90, 211-214.

Clarke, R.J. 1994b. The significance of the Swartkrans *Homo* to the *Homo erectus* problem. Courier Forschungs-Institut Senckenberg 171, 185-193.

Cooke, C.K., 1966. Re-appraisal of the industry hitherto named the Proto-Stillbay. Arnoldia (Rhodesia) 2, 1-14.

Cruz-Uribe, K., Klein, R.G. Avery, G., Avery, M., Halkett, D., Hart, T., Milo, R.G., Sampson, C.G., Volman, T.P. 2003. Excavation of buried Late Acheulean (Mid-Quaternary) land surfaces at Duinefontein 2, Western Cape Province, South Africa. Journal of Archaeological Science 30, 559-575.

Davies, O. 1976. The 'Sangoan' industries. Annals of the Natal Museum 22, 885-911.

Deacon, H.J. 1970. The Acheulean occupation at Amanzi Springs, Uitenhage district, Cape Province. Annals of the Cape Provincial Museums (Natural History) 8, 89-189.

Deacon, H.J. 1975. Demography, subsistence and culture during the Acheulean in southern Africa. In: Butzer, K.W., Isaac, G. (Eds.), After the Australopithecines. Mouton, The Hague, pp. 543-569.

Deacon, H.J. 1988. Geology of the Doornlaagte deposit. In: Mason, R.J. (Ed.). Cave of Hearths, Makapansgat, Transvaal. Occasional Paper No. 21, Archaeological Research Unit, University of the Witwatersrand, pp. 643-647.

Deacon, H.J. 1998. Elandsfontein and Klasies River revisited. In: Ashton, N.M., Healy F., Pettitt, P. (Eds). A Master of His Craft: Papers in Stone Age Archaeology Presented to John Wymer. OxBow, Oxford, pp. 23-28.

Deacon, H.J. Deacon, J. 1999. Human Beginnings in South Africa. David Philip, Johannesburg and Cape Town.

Deacon, J. 1976. Report on stone artifacts from Duinefontein 2, Melkbosstrand. South African Archaeological Bulletin 31, 21-25.

Deino, A., Potts, R. 1990. Single crystal $^{40}Ar/^{39}Ar$ dating of the Olorgesailie Formation, southern Kenya Rift. Journal of Geophysical Research 95, 8453-8470.

Deino, A., Potts, R. 1992. Age-probability spectra for examination of single-crystal $^{40}Ar/^{39}Ar$ dating results: examples from Olorgesailie, southern Kenya rift valley. Quaternary International 7/8, 81-89.

Delson, E., Van Coevering, J.A. 2000. Composite stratigraphic chart of Olduvai Gorge. In: Delson, E., Tattersall, I., Van Coevering, J.A., Brooks, A.S. (Eds.). Encyclopedia of Human Evolution and Prehistory, Garland, New York, p. 488.

Field, A.S. 1999. An analytical and comparative study of the Earlier Stone Age archaeology of the Sterkfontein Valley. M.Sc. Dissertation, University of the Witwatersrand.

Fock, G.J. 1968. Rooidam, a sealed site of the First Intermediate. South African Journal of Science 64, 153-159.

Gibbon, R.J. 2004. Earlier Stone Age geoarchaeology of the northern Kruger Park. MSc Dissertation, University of the Witwatersrand.

Gibbon, R.J., Granger, D.E., Kuman, K , Partridge, T.C. in submission. Early Acheulean technology in the Vaal River Gravels, South Africa, dated with cosmogenic nuclides.

Goodwin, A.J.H. 1928. Archaeology of the Vaal River gravels. Transactions of the Royal Society of South Africa 16, 77–102.

Goodwin, A.J.H. 1933. Some developments in technique during the Earlier Stone Age. Transactions of the Royal Society of South Africa 21, 109–123.

Goodwin, A.J.H., Van Riet Lowe, C. 1929. The Stone Age cultures of South Africa. Annals of the South African Museum 27, 1–289.

Goren-Inbar, N., Alperson, N., Kilev, M., Simchoni, O., Melamed, Y., Ben-Nun, A., Werker, E. 2004. Evidence of hominin control of fire at Gesher Benot Ya'aqov, Israel. Science 304, 725-727.

Goren-Inbar, N., Saragusti, I. 1996. An Acheulian biface assemblage from Gesher Benot Ya'aqov, Israel: indications of African affinities. Journal of Field Archaeology 23, 15-30.

Goren-Inbar, N., Feibel, C.S., Verosub, K.L., Melamed, Y., Kislev, M.E., Tchernov, E., Saragusti, I. 2000. Pleistocene milestones on the Out-of-Africa Corridor at Gesher Benot Ya'aqov, Israel. Science 289, 944-947.

Grun, R. 2003. Direct dating of South African human fossils. Paper presented at the XVth Biennial Conference of the Southern African Society for Quaternary Research. March-April, University of the Witwatersrand, Johannesburg.

Grun, R., Brink, J.S., Spooner, N.A., Taylor, L., Stringer, C.B., Franciscus, R.G., Murray, A.S. 1996. Direct dating of Florisbad hominid. Nature 382, 500-501.

Hall, G. 2004. New archaeological material from the Plio-Pleistocene sites of Coopers Cave, Plovers Lake and Gladysvale, Witwatersrand, South Africa. Paper delivered at the annual conference of the Palaeontological Society of South Africa, University of the Witwatersrand, Johannesburg.

Hall, G., Pickering, R., Lacruz, R., Hancox, J., Berger, L.R., Schmid, P. 2006. An Acheulean handaxe from Gladysvale Cave Site, Gauteng, South Africa. South African Journal of Science 102, 103-105.

Helgren, D.M. 1978. Acheulian settlement along the lower Vaal River, South Africa. Journal of Archaeological Science 5, 39–60.

Humphreys, A.J.B. 1969. Later Acheulian or Fauresmith? Acontribution. Annals of the Cape Provincial Museum (Natural History) 6, 87–101.

Isaac, G. Ll. 1967. Towards the interpretation of occupation debris: some experiments and observations. Kroeber Anthropological Society Papers 37, 31-57.

Isaac, G. Ll. 1977. Olorgesailie. University of Chicago Press, Chicago.

Isaac, G. Ll., Isaac, B. (Eds.) 1997. Koobi Fora Research Project, Vol. 5: Plio-Pleistocene Archaeology. Clarendon Press, Oxford.

Isaac, G,Ll., Harris, J.W.K. Kroll, E.M. 1997. The stone artefact assemblages: A comparative study. In Isaac G.Ll., Isaac, B. (Eds.), Koobi Fora Research Project, Vol. 5: Plio-Pleistocene Archaeology. Clarendon Press, Oxford, pp. 262-362.

Jansen, F.J. 1926. A new type of stone implement from Victoria West. South African Journal of Science 23, 818–825.

Keller, C. 1973. Montagu Cave in prehistory. University of California Anthropological Records 28, 1-150.

Kempson, H. 2007. Late Earlier Stone Age sites in the Mapungubwe National Park, South Africa: A technological study. MSc Dissertation, University of the Witwatersrand.

Kempson, H. 2005. Earliest archaeology of the Limpopo Basin, South Africa: Final ESA and early MSA assemblages. Paper presented at the 12th Congress of the Pan African Archaeological Association for Prehistory and Related Studies. Gaborone, Botswana, 3 to 10 July 2005.

Keyser, A.W. 2000. The Drimolen skull: The most complete australopithecine cranium and mandible to date. South African Journal of Science 96, 189-193.

Keyser A.W., Menter, C.G., Moggi-Cecchi, J., Pickering T.R., Berger, L.R. 2000. Drimolen: A new hominid-bearing site in Gauteng, South Africa. South African Journal of Science 96, 193-197.

Klein, R.G. 1978. The fauna and overall interpretation of the "Cutting 10" Acheulean site at Elandsfontein (Hopefield), Southwestern Cape Province, South Africa. Quaternary Research 10, 69-83.

Klein, R.G. 2000. The Earlier Stone Age of southern Africa. South African Archaeological Bulletin 55, 107-122.

Klein, R.G., Avery, G., Cruz-Uribe, K., Halkett, D., Hart, T., Milo, R.G., Volman, T.P. 1999. Duinefontein 2: An Acheulean site in the western Cape Province of South Africa. Journal of Human Evolution 37, 153-190.

Klein, R.G., Avery, G., Cruz-Uribe, Steele, T.E. in press. The mammalian fauna associated with an archaic hominin skullcap and later Acheulean artifacts at Elandsfontein, Western Cape Province, South Africa. Journal of Human Evolution.

Klein, R.G., Cruz-Uribe, K. 1991. The bovids from Elandsfontein, South Africa, and their implications for the age, palaeoenvironment, and origins of the site. The African Archaeological Review 9, 21-79.

Kuman, K. 1994a. The archaeology of Sterkfontein: Past and present. Journal of Human Evolution 27, 471–495.

Kuman, K. 1994b. The archaeology of Sterkfontein: Preliminary findings on site formation and cultural change. South African Journal of Science 90, 215-219.

Kuman, K. 1996. Recent findings on the archaeology of Sterkfontein. Kaupia, Damstadtler Beitrage zur Naturgeschichte 6, 31-36.

Kuman, K. 1998. The earliest South African industries. In: Petraglia, M., Korisettar (Eds.), Early Human Behavior in Global Context: The Rise and Diversity of the Lower Palaeolithic Record. Routledge Press, London, pp. 151-186.

Kuman, K. 2001. An Acheulean factory site with prepared core technology near Taung, South Africa. South African Archaeological Bulletin 55, 107-122.

Kuman, K. 2003. Site formation in the early South African Stone Age sites and its influence on the archaeological record. South African Journal of Science 99, 251-254.

Kuman, K., Clarke, R.J. 2000. Stratigraphy, artifact industries and hominid associations for Sterkfontein, Member 5. Journal of Human Evolution 38, 827-847.

Kuman, K., Field, A.S., Thackeray, J.F. 1997. Discovery of new artifacts at Kromdraai. South African Journal of Science 93, 187-193.

Kuman, K., Le Baron, J.C., Gibbon, R.J. 2004. Earlier Stone Age archaeology of the Vhembe-Dongola National Park (South Africa) and vicinity. Quaternary International: 129, 23-32.

Kuman, K., Gibbon, R.J., Kempson, H., Langejans, G., Le Baron, J.C.. Pollarolo, L., Sutton, M. 2005. Stone Age signatures in northernmost South Africa: early archaeology of the Mapungubwe National Park and vicinity. In: D'Errico, F., Backwell, L. (Eds): From Tools to Symbols, From Early Hominids to Modern Humans. Johannesburg: Witwatersrand University Press, pp 163-182.

Kuman, K., Pollarolo, L., Kempson, H., Jacobs, Z., Woodbourne, S., Kekana, M., Wilkins, J., Le Baron, J., Tooth, S. in preparation. Kudu Koppie: A stratified Sangoan-MSA open-air site in northern Limpopo Province, South Africa.

Laidler, P.W. 1945. The evolution of Middle Palaeolithic technique at Geelhout, near Kareedouw, in the southern Cape. Transactions of the Royal Society of South Africa 31, 283-313.

Leakey, M.D. 1971. Olduvai Gorge, Vol. 3: Excavations in Beds I and II, 1960-1963. Cambridge University Press, Cambridge.

Le Baron, J.C. 2004. A landscape of Stone Age archaeology: a primary investigation near Hackthorne 1, Limpopo Province, South Africa. Paper presented at the Annual Meeting of the Paleoanthopology Society, McGill University, March, Montreal.

Luyt, C.J., Lee-Thorp, J.A. 2003. Carbon isotope ratios of Sterkfontein fossils indicate a marked shift to open environments *c.* 1.7 Myr ago. South African Journal of Science 99, 271-273.

Malan, B.D. 1947. Flake tools and artifacts in the Stellenbosch-Fauresmith transition in the Vaal River Valley. South African Archaeological Bulletin 56, 8–22.

Malan, B.D., Wells, L.H. 1943. A further report on the Wonderwerk Cave, Kuruman. South African Journal of Science 15, 258-270.

Mason, R.J. 1962. Prehistory of the Transvaal. Witwatersrand University Press, Johannesburg.

Mason, R.J. 1957. The archaeology of the earliest superficial deposits in the lower Vaal basin near Holpan, Windsorton District. South African Geographical Journal 49, 39–56.

Mason, R.J. 1969. The Oppermansdrif Dam archaeological project, Vaal basin. South African Archaeological Bulletin 24, 182-192.

Mason, R.J. (Ed.). 1988a. Cave of Hearths, Makapansgat, Transvaal. Occasional Paper No. 21, Archaeological Research Unit, University of the Witwatersrand.

Mason, R.J. 1988b. The Doornlaagte Early Stone Age. In: Mason, R.J. (Ed.). Cave of Hearths, Makapansgat, Transvaal. Occasional Paper No. 21, Archaeological Research Unit, University of the Witwatersrand, pp. 623-630.

McBrearty, S., 1988. The Sangoan-Lupemban and Middle Stone Age sequence at the Muguruk site, western Kenya. World Archaeology 19, 388-420.

McBrearty, S., 1991. Recent research in western Kenya and its implications for the status of the Sangoan industry. In: Clark, J.D. (Ed.). Cultural Beginnings. Dr. Rudolf Habelt GMBH, Bonn, pp 159-176.

McBrearty, S., Brooks, A.S. 2000. The revolution that wasn't: a new interpretation of the origin of human behaviour. Journal of Human Evolution 39, 453-563.

McNabb, J., Binyon, F., Hazelwood, L. 2004. The large cutting tools from the South African Acheulean and the question of social traditions. Current Anthropology 45, 653-677.

McPherron, S. P. 2000. Handaxes as a measure of the mental capabilities of early hominids. Journal of Archaeological Science 27, 655-663.

Mokokwe, W. 2005. Goldsmith's Site: A newly discovered Pleistocene fossil and stone tool site near Sterkfontein. MSc Research Report, University of the Witwatersrand.

Morton, A.G.T. 1996. Archaeological site formation: Experiments in lake margin processes. PhD Dissertation, University of Cambridge.

Noll, M. 2000. Components of Acheulean lithic assemblage variability at Olorgesailie, Kenya. Ph.D. Dissertation, University of Illinois at Urbana-Champaign.

Ogola, C. 2003. Taphonomy of the Cave of Hearths Acheulean fauna. Research Report, University of the Witwatersrand.

Partridge, T.C., Dalbe, T.S. 1986. Geoarchaeology of the Haaskraal Pan: A preliminary palaeoenvironmental model. Palaeoecology of Africa and the Surrounding Islands 17, 69-78.

Pickering, T.R. 1999. Taphonomic interpretations of the Sterkfontein early hominid site (Gauteng, South Africa) reconsidered in light of recent evidence. PhD Dissertation, University of Wisconsin-Madison.

Pickering, T.R., Domínguez-Rodrigo, M., Egeland, C.P., Brain, C.K. 2004. New data and ideas on the foraging behaviour of Early Stone Age hominids at Swartkrans Cave, South Africa. South African Journal of Science 100, 215-219.

Plummer, T. 2005. Discord after discard: Reconstructing aspects of Oldowan hominin behaviour. In Stahl, A.B. (Ed.). African Archaeology: A Critical Introduction. Blackwell Publishing, Oxford, pp. 55-92.

Pollarolo, L. 2004. Archaeological investigations at Kudu Koppie, A Stone Age site in Limpopo Province, South Africa. PhD Dissertation, University of Florence.

Potts, R. 1991. Why the Oldowan? Plio-Pleistocene toolmaking and the transport of resources. Journal of Anthropological Research 47, 153-177.

Reed, K. 1997. Early hominid evolution and ecological change through the African Plio-Pleistocene. Journal of Human Evolution 32, 289-322.

Reynolds, S.C., Vogel, J.C., Clarke, R.J , Kuman, K. 2003. Preliminary results of excavations at Lincoln Cave, Sterkfontein, South Africa. Sterkfontein. South African Journal of Science 99, 286-288.

Robinson, J.T. 1962. Sterkfontein stratigraphy and the significance of the Extension Site. South African Archaeological Bulletin 17, 87-107.

Roche, H., Kibunjia, M. 1996. Contribution of the West Turkana Plio-Pleistocene sites to the archaeology of the Lower Omo/Turkana Basin. Kaupia, Darmstadter Beitrage zur Naturgeschichte 6: 27-30.

Roe, D.A. 1994. A metrical analysis of selected sets of handaxes and cleavers from Olduvai Gorge. In: Leakey, M.D. (Ed.). Olduvai Gorge, Vol. 5: Excavations in Beds III, IV and the Masek Beds, 1968-1971. Cambridge University Press, Cambridge, pp.

Sampson, C.G., 1972. The Stone Age Industries of the Orange River Scheme and South Africa. Memoir No. 6, National Museum, Bloemfontein.

Sampson, C.G., 1974. The Stone Age Archaeology of Southern Africa. Academic Press, New York.

Sampson, C.G. 1985. Atlas of Stone Age Settlement in the Central and Upper Seacow Valley. Memoirs of the National Museum (Bloemfontein), No. 20.

Schick, K.D. 1987a. Stone Age Sites in the Making: Experiments in the Formation and Transformation of Archaeological Occurrences. British Archaeological Reports, International Series 319, Oxford.

Schick, K. 1987b. Modelling the formation of Early Stone Age artifact concentrations. Journal of Human Evolution 16, 789-807.

Schick, K.D. 1991. On making behavioral inferences from early archaeological sites. In: Clark, J.D. (Ed.). Cultural Beginnings. Dr. Rudolf Habelt GMBH, Bonn, pp.

Schick, K.D. 1997. Experimental studies of site-formation processes. In: Isaac, G.Ll., Isaac, B. (Eds.). Koobi Fora Research Project, Vol. 5: Plio-Pleistocene Archaeology. Clarendon Press, Oxford.

Schwartz, J.H., Tattersall, I. 2003. The Human Fossil Record, Vol. II: Craniodental Morphology of Genus *Homo* (Africa and Asia). Wiley-Liss, New York.

Steininger C.M., Berger L.R. 2001. Taxonomic affinity of a new specimen from Cooper's, South Africa. American Journal of Physical Anthropology Supplement 30, 291.

Szabo, B.J. Butzer, K.W. 1979. Uranium-series dating of lacustrine limestones from pan deposits with Final Acheulean assemblages at Rooidam, Kimberley District, South Africa. Quaternary Research 11, 257-260.

Thackeray J.F., Kirschvink, J.L., Raub, T.D. 2002. Palaeomagnetic analyses of calcified deposits from the Plio-Pleistocene hominid site of Kromdraai, South Africa. South African Journal of Science 98, 537-539.

Turner, A. 1997. Further remains of Carnivora (Mammalia) from the Sterkfontein hominid site. Palaeontologia Africana 34, 115-126.

Van Riet Lowe, C. 1927. The Fauresmith Coup de Poing. South African Journal of Science 24, 502–505.

Van Riet Lowe, C. 1937. The archeology of the Vaal River basin. In: Söhnge, P.G., Visser, D.J.L., Van Riet Lowe, C. (Eds.). The Geology and Archaeology of the Vaal River Basin. Geological Survey (South Africa), Memoir 35, pp. 61–164.

Van Riet Lowe, C. 1945. The evolution of the Levallois Technique in South Africa. Man XLV, 37–51.

Van Riet Lowe, C. 1952a. The development of the Hand-Axe Culture in South Africa. In: Leakey, L.S.B., Cole, S. (Eds.). Proceedings of the Pan-African Congress on Prehistory, 1947. Basil Blackwell, Oxford, pp. 167-177.

Van Riet Lowe, C. 1952b. The Vaal River chronology—an up-to-date summary. South African Archaeological Bulletin 7, 135–149.

Volman, T.P. 1984. Early prehistory of Southern Africa. In: Klein, R. G. (Ed.). Southern African Prehistory and Paleoenvironments. A.A. Balkema, Rotterdam and New York, pp. 169-220.

Vrba, E.S. 1975. Some evidence of chronology and paleoecology of Sterkfontein, Swartkrans and Kromdraai from the fossil Bovidae. Nature 254, 301-304.

Vrba, E.S. 1981. The Kromdraai australopithecine site revisited in 1980: Recent excavations and results. Annals of the Transvaal Museum 33, 17-60.

Vrba, E.S. 1982. Biostratigraphy and chronology, based particularly on Bovidae, of southern hominid-associated assemblages: Makapansgat, Sterkfontein, Taung, Kromdraai, Swartkrans; also Elandsfontein (Saldanha), Broken Hill (now Kabwe) and Cave of Hearths. Congrès International de Paleontologie Humaine, 1er Congrès, Tome. 2. CNRS, Nice.

Watson, V. 1993. Composition of the Swartkrans bone accumulations, in terms of skeletal parts and animals represented. In: Brain, C.K. (Ed.). Swartkrans: A Cave's Chronocle of Early Man. Transvaal Museum, Pretoria, pp. 35-73.

# CHAPTER 11

# TAPHONOMY OF STERKFONTEIN *AUSTRALOPITHECUS* SKELETONS

RON J. CLARKE

## ABSTRACT

The rarity of complete or partial skeletons of Plio-Pleistocene hominids in open-air sites is due to the scattering and breakage of bones by carnivores and scavengers. In caves, it is also due to those factors, plus the scattering and breakage caused by falling rocks and movement in a talus slope. At the Sterkfontein Cave site, however, there are two partial skeletons (Sts14 and StW431), consisting of torsos that appear to have resulted from feeding by a particular predator or scavenger. There is also a virtually complete skeleton that resulted from an individual falling into a shaft, being apparently mummified, and then sealed in by flowstone before the bones could be crushed and scattered by rock movement on the steep talus slope. This paper discusses the factors involved in the preservation and distribution of hominid skeletal remains at Sterkfontein, including those hominid remains less complete than these three skeletons.

## INTRODUCTION

Under the right conditions it is possible for complete animal skeletons and even soft tissues and stomach contents to be preserved as fossils for millions of years as they were in the 50 million year old Eocene oil shale deposits of Messel, near Frankfurt, Germany (Schaal and Ziegler, 1992). Thus, theoretically, there is no reason why, with correct conditions for preservation, complete skeletons and soft tissues of early hominids should not somewhere have been preserved. However, except in the case of deliberate burials by Neanderthals and Cro-Magnons of relatively recent times, complete or partially complete skeletons of early hominids are very rare. Natural preservation of soft tissues has only occurred in very recent mummified bodies such as those from the Bronze Age in the Alps and the Incas in the Andes that were preserved by freezing or from the Iron Age of Denmark and England preserved in peat bogs (Bahn, 1996).

In the case of the Plio-Pleistocene sites that have yielded fossil remains of early hominids it is rare to find complete or partial skeletons. Although in East Africa the sedimentary conditions for preservation are good (volcanic deposits in ancient lake beds), most hominid skeletons were damaged and disturbed either before burial by predators and scavengers and/or after burial by water disturbance, erosion, weathering and animal trampling. Classic examples of such post-burial disturbance are the skeletons of "Lucy" and Olduvai Hominid 62.

The preservation conditions in the dolomite caves of South Africa, while very different to those of the East African lake beds, were affected by their own set of modifying factors. The degree of completeness of a fossil specimen at Sterkfontein Caves, be it cranial or postcranial bone or whole or partial skeleton, is dependent upon the following factors:

1. Mode of introduction to the cave, for example, by big cat, hyena, porcupine, bird of prey, death trap or slope wash.
2. Position within the cave, such as under protective roof overhang, beneath open shaft, on talus slope which is subject to rock fall and displacement, in water, beneath stalactites, or on floors which might be subject to collapse into lower chambers.
3. Decalcification of cracked and crushed specimens that had been held together by the natural consolidant calcium carbonate.

4. Mode of recovery of fossils can also damage or separate parts of what was a more complete specimen. These recovery methods are dynamite blasting by the lime miners and palaeontologists, drilling and breaking of fossil-bearing rock, pick-axe and shovel in decalcified cave infill, chiselling away of breccia encasing fossil bones and acetic acid cleaning of cracked and crushed specimens.

As with much of the other fauna from the Sterkfontein caves, the *Australopithecus* specimens usually consist of either partial or almost complete cranial material or fragments of cranial and post-cranial material. There are only two partial skeletons, Sts 14 and StW 431 from member 4 (dating to between 2 and 3 million years ago) and one virtually complete skeleton, StW 573 from member 2 dated to at least 3.3 million years ago. It is of interest to examine these skeletons in order to determine whether there are any clues as to how the individuals came to be in the cave and why it is that all other hominid specimens are so fragmentary.

## *Australopithecus* Partial Skeleton Sts 14

This individual is represented by nine thoracic vertebrae with some rib fragments, six lumbar vertebrae, a sacrum and both os coxae forming a virtually complete pelvis, superior shaft and neck of left femur, and a tibia fragment (Oakley et al., 1977). These elements, with the exception of the tibia and ribs are described by Robinson (1972) and the partial skeleton is illustrated in Howell (1965) and Reader (1981).

These bones were present in a relatively small block of breccia that according to Broom et al. (1950) "was blasted out on 1st August 1947." As it was Broom and Robinson who were blasting there at that time we can assume that they would have recovered any associated blocks from the blast. Unless other parts of the skeleton had been in adjacent breccia blasted out prior to 1939 by the lime miners we might also assume that the specimen Broom and Robinson recovered represented all that had been preserved of the individual in the cave. Broom et al. (1950) described only the right side of the pelvis (ilium, ischium and pubis) illustrating it from both lateral and medial sides. They also described and illustrated the left femur. Robinson (1972) referred to the right "innominate" having been separated from the block and that a "natural cast of the lateral face of the right innominate was intact in the block". He clearly meant the medial face of the innominate because one published photograph (Findlay, 1972) shows Broom holding the block with the lateral face of the right ilium exposed on the surface and the medial face still buried in the block. Two other photographs (Place, 1957) show the same view of the block in close-up and also John Robinson cleaning the ilium out of the block with the aid of hand-held tools. Robinson (1972) explained that because of the natural cast of the right "innominate" and "the fragmentary, damaged, and delicate appearance of the remaining pieces in the block," Broom decreed that no preparation was to be done on the block. Only after Broom's death did Robinson use acetic acid to extract the remaining parts of the pelvis, the vertebrae and ribs from the block.

Broom et al. (1950) mentioned part of a badly crushed skull being associated with the skeleton. This specimen must have been Sts 13 recovered on 29th July 1947 (i.e., two days before the pelvis) and consisting of an elderly adult partial face with most of right dentition and part of left dentition. Unfortunately the specimen was never described and is listed as missing by Oakley et al. (1977). It would anyway have been impossible to demonstrate that such a crushed partial maxilla belonged with the Sts 14 partial skeleton even had it been found in the same block. Firstly, there are many other adult *Australopithecus* cranial fossils in the surrounding member 4 breccia and any of them could potentially have come from the same individual as Sts 14. Secondly, as so much of the Sts 14 individual is missing (cervical vertebrae, shoulder girdle, arms, hands, feet and both legs apart from the proximal femur fragment and tibia fragment) there is no reason to suppose that the torso had an associated skull when it reached its final resting place in the cave.

The big question concerns what happened to those missing parts and how is it that just the central part of the body is intact? Theoretically the lime-mining of the 1930s could have removed those missing portions but, if so, then at least some parts should have been found during our processing of the lime miners breccia dumps. Many hominid fragments have been discovered during that processing and I found that some of them fitted with specimens recovered by Broom (Clarke, 1990). One of the most important of such associations occurred on 29th of August 2002 when I identified a hominid tooth just cleaned out of breccia from the miners dump (D18) as being the missing left upper third molar from Broom's 17 August 1936 discovery of the first *Australopithecus* adult cranium TM1511. It then occurred to me that if there was one tooth from the dump, perhaps there had been others and so on 2nd September I checked our collection of fossil hominids from D18 and identified the missing right upper third premolar that had been recovered on 24 March 1981 and catalogued as StW 91 (Figure 1). Although we have a large amount of the dump still to process it seems most probable that, had there been any other parts of skeleton belonging to Sts 14 we should by now have found at least one bone and we have not. Thus I am drawn to the conclusion that possibly the fact that mainly the spinal column, pelvis, and an attached partial limb-bone are preserved in Sts 14 suggests the action of a particular predator or scavenger. Support for such a conclusion comes from another partial skeleton of *Australopithecus*, StW 431 which will now be discussed.

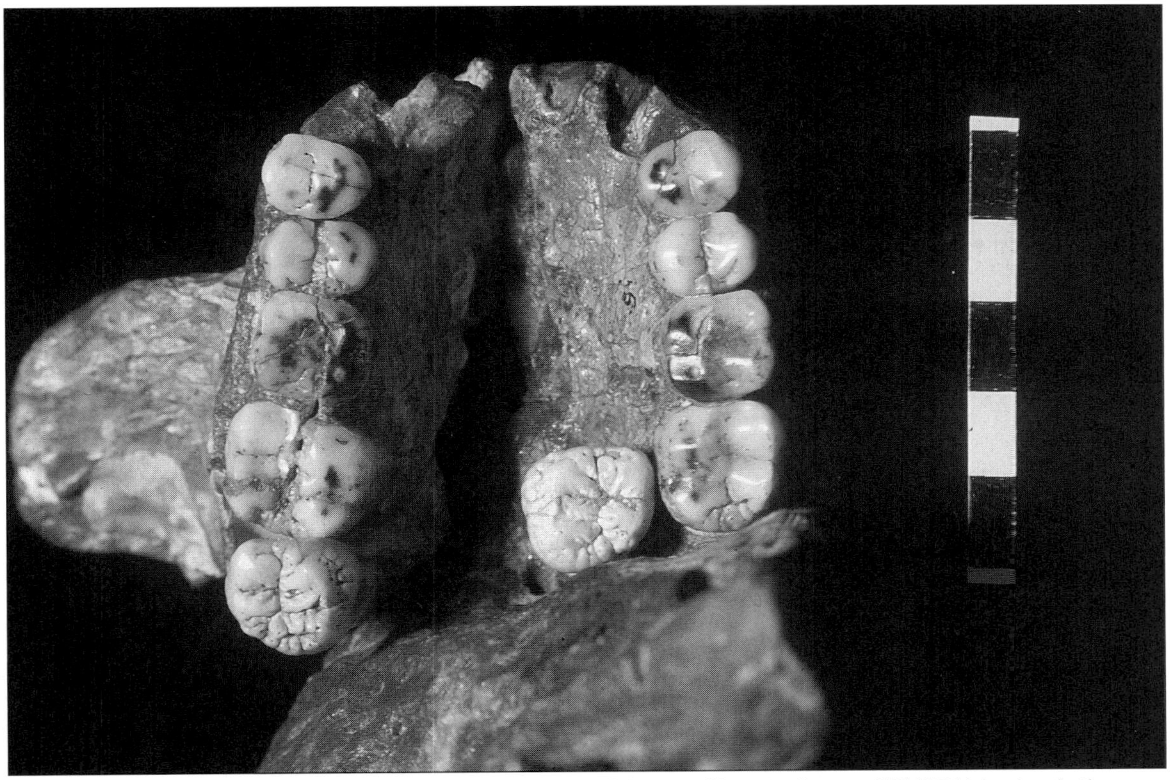

*Figure 1. A left upper third molar set on the lingual side of the maxillary toothrow of TM1511 is clearly the antimere of TM1511's right upper third molar, as discovered and identified by the author in breccia from a lime miners dump. He also identified the right upper third premolar from that dump.*

## *Australopithecus* Partial Skeleton StW 431

This individual is represented by nine consecutive thoracic and lumbar vertebrae, sacrum that joins to left and right incomplete os coxae, one right rib fragment, partial right scapula, partial right clavicle, distal half of right humerus and the articulating proximal halves of right radius and ulna (Toussaint et al., 2003, Kibii and Clarke, 2003) (Figure 2). These fossils were recovered close together during excavation of an area of decalcified Member 4 breccia by A. R. Hughes in February and March 1987.

As with Sts 14, the specimen consists largely of the pelvis and spinal column (minus the cervical vertebrae) and parts of one limb. Again the question arises as to what happened to the skull and other missing parts of the skeleton. In this case all the surrounding areas of decalcified breccia were excavated and sieved by Hughes and his excavation team. Thus, if there had been other parts they should have been recovered. Job Kibii analysed all of the faunal remains from the excavated decalcified Member 4 area and did recover a crucial fragment of the ilium that provided a link with the sacrum. This piece had previously been misidentified as bovid. A subsequent thorough search by Kibii and Clarke of all the bags of bone fragments from that area resulted in the recovery of four more small but important fragments of the pelvis (Kibii and Clarke, 2003). If there had been other

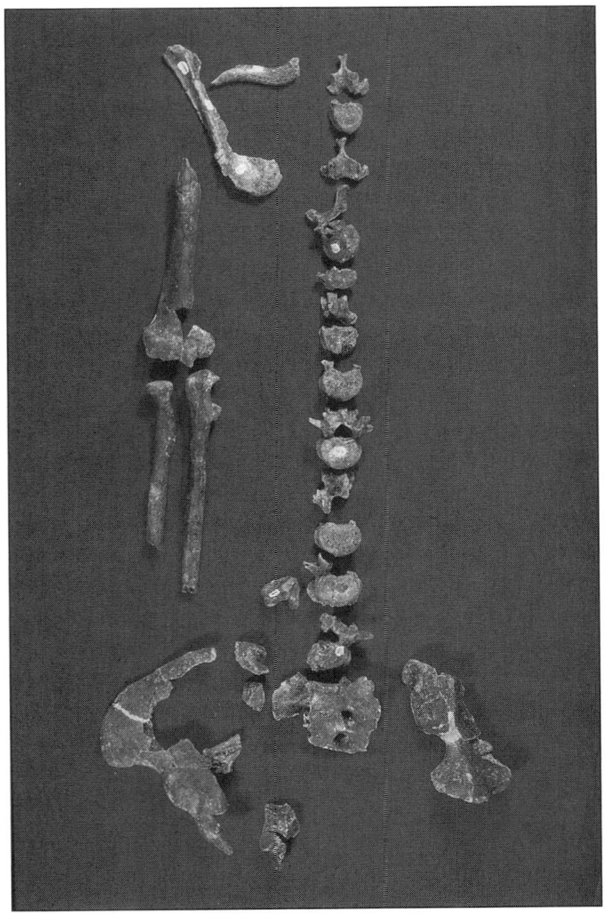

*Figure 2. StW431, a partial* Australopithecus *skeleton from Sterkfontein Member 4.*

substantial elements of the same skeleton present they would have been identified by Kibii during his detailed analysis. He did indeed identify eleven other hominid post-cranial fossils of different individuals from various parts of the Hughes excavation (Tobias et al., 2003).

Thus the fact that the preserved skeletal portion of the StW 431 individual follows a similar pattern to that of the Sts 14 individual suggests that both could have been subjected to the same kind of predation or scavenging that left only the thoraco-lumbar vertebrae and pelvis with remnants of ribs and of one limb.

## Other Sterkfontein *Australopithecus* Pelvic and Vertebral Fossils

It is of interest to note that from 68 years of fossil recovery and excavation of the Sterkfontein member 4 breccia and out of a substantial collection of hominid post-cranial fossils, only two other catalogued pelvic fossils and three catalogued vertebral fossils have been recovered. These are Sts 65 right ilium associated with a vertebral fragment (Robinson, 1972), StW 611 left ischium found by Job Kibii (Tobias et al., 2003), Sts 73 thoracic vertebra (Robinson, 1972) and StW 8 four lumbar vertebrae conjoined with StW 41 two thoracic vertebrae (Tobias, 1973) (Figure 3). These conjoining vertebrae are of significance because they suggest that they could have been part of a torso similar to that of Sts 14 and StW 431. The four lumbar vertebrae (StW8) were identified by P.V. Tobias in November 1969 in a display cabinet in the Sterkfontein tea room, and the two thoracic vertebrae (StW 41) were cleaned out of breccia from the limeminers' Dump 18 on 10th of January, 1975. Thus there is a strong possibility that other parts of the same skeleton were blasted out by the lime miners and taken or sold as souvenirs. Although the ilium of Sts 65 could potentially have belonged with these vertebrae there is a size difference suggesting that they are probably from two different individuals. The StW 8 lumbar vertebrae are considerably larger than those of Sts 14 whereas the Sts 65 ilium is of similar size to Sts 14. The Sts 73 thoracic vertebra is however of similar size to the StW 41 thoracic vertebrae and therefore could possibly have come from the same individual.

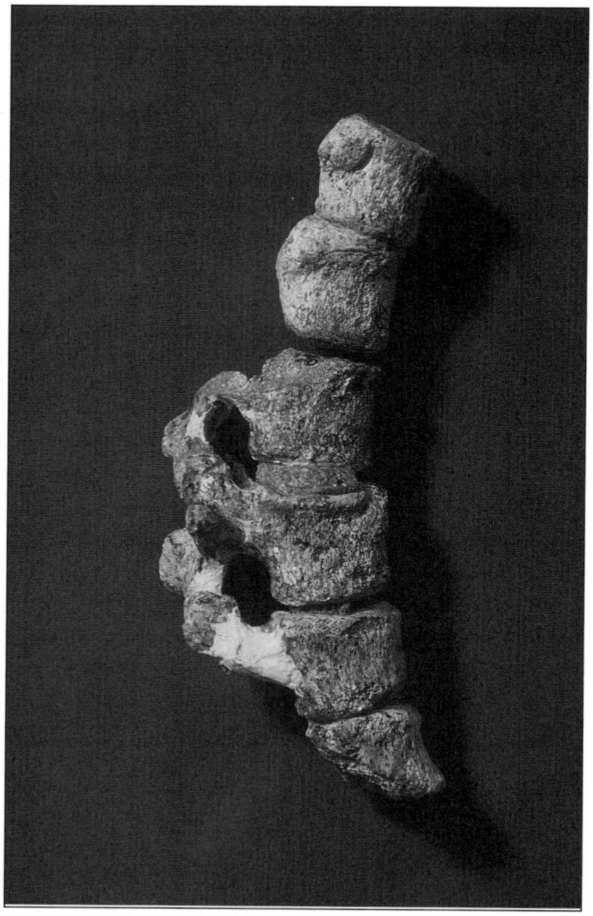

*Figure 3. The conjoining hominid thoracic (StW 41) and lumbar (StW 8) vertebral series from Sterkfontein Member 4.*

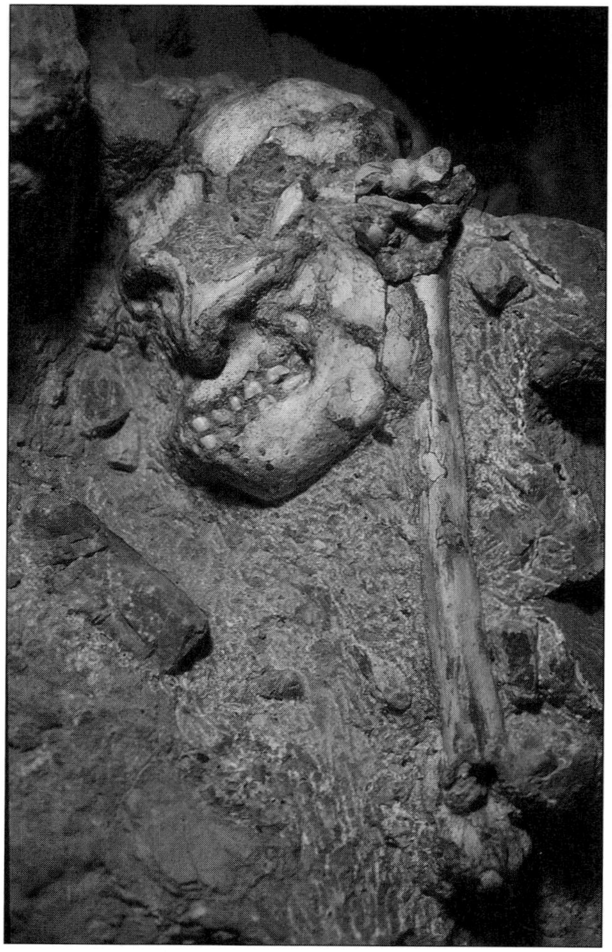

*Figure 4. The skull and left humerus of StW 573, a complete* Australopithecus *skeleton from Sterkfontein Member 2.*

## *Australopithecus* Complete Skeleton StW 573

This individual is represented by virtually a whole skeleton including the skull, currently being excavated from the Member 2 breccia within the Silberberg grotto (Clarke, 1998, 1999, 2002) (Figure 4). Its significance lies in the following factors: 1) it is complete apart from most of the foot bones that were blasted away by lime miners and were never recovered. No other such complete skeleton of *Australopithecus* has ever been found. 2) it has an age of at least 3.3 million years old. (Partridge et al., 1999) and has also been dated by another method to *c*. 4 million years old (Partridge et a., 2003). It is thus the oldest *Australopithecus* in South Africa. 3) its total anatomy is revealing information for the first time about a complete *Australopithecus* individual including stature, limb ratios and clues to its mode of locomotion. 4) it provides for the first time an opportunity to analyse the taphonomic history of a complete *Australopithecus* skeleton.

The fact that the skeleton is complete and does not show any signs of carnivore damage rules out carnivore as the agent of accumulation. The location of the skeleton near the base of a steep talus cone suggests that it was probably an individual that either fell into a shaft by accident or climbed in but was unable to climb out. The pose of the skeleton with left leg crossed over right and one arm stretched above its head is suggestive of a body that rolled down the talus slope. The skull is intact with the mandible still in its closed position and the other skeletal elements are generally in correct anatomical relationship to each other. The left hand is particularly informative as it is not only complete but is clenched with the thumb across the palm (Figure 5). This indicates that the body possibly mummified in dry conditions in the cave and that the mummified skin and tendons held the bones in their correct anatomical relationships. It seems that the corpse was not accessed by dermestid beetles because, had they cleaned the flesh from the bones, then many of the skeletal parts, especially the hand bones, would have been disarticulated and scattered on the stony talus slope.

Following the mummification during a dry phase there was a change in climate to wet conditions. This resulted in water removing sediment beneath part of the surface of the talus slope and forming a cavity into which the infill supporting the mid-section of the skeleton collapsed. This collapse had the effect of wrenching the left humerus away from the forearm at the elbow joint and breaking the femurs in mid-shaft. The disturbance caused to the central part of the skeleton by this collapse in the rock-filled matrix also resulted in breaking and scattering of ribs and of the right forearm, wrist and hand as well as crushing and partial disintegration of the pelvis. Some of this disruption could also have been exacerbated by the water still partially flowing through that collapsed area. Following this there was the build-up of a massive stalagmite within the cavern and part of it consisted of a thick flowstone that covered the Member 2 talus slope and sealed in the upper part of the skeleton down to the broken ends of the upper femur shafts. At this point the flowstone filled the cavity beneath the

*Figure 5. View of the left hand bones of StW 573.*

lower femurs, tibiae, fibulae and feet.

Thus one can see in this one individual skeleton three different conditions of preservation resulting from different post-depositional influences. Firstly, there is the intact preservation of the left hand skeleton as a consequence of apparent mummification followed by burial with no disturbance. Secondly, there is the broken up nature and fragmentation of the central part of the skeleton and right arm due to collapse into a cavity. Thirdly, there is the crushing and disintegration of parts of the pelvis due to rock fall and pressure and possibly to the water action from the flow beneath the talus surface.

One can thus see that in a talus deposit within dolomite caves a complete skeleton could be broken, fragmented and scattered by rock-fall, collapse, water-flow and filtering of small elements through holes in the talus slope. However, if the skeleton of StW 573, was to be subjected to decalcification there would still remain some substantial identifiable post-cranial and cranial portions and the end result would be very different to that of the decalcified StW 431 or the breccia-encased Sts 14. Those two cases do appear to have resulted from predation or scavenging in which the skull, cervical vertebrae, most of the limb bones, hands and feet were separated from the torso before it reached its final resting place in the talus slope. There are no signs of carnivore gnawing on any of the elements of these torsos but there are definite carnivore gnaw marks on several of the isolated limb bones from other parts of the member 4 talus deposit. One very good example is that of the heavily gnawed distal right humerus StW 339 which was recovered from an adjacent square, just above the right radius shaft StW 348 and right ulna shaft StW 349. These two bones both display apparent carnivore tooth crushing on their proximal and distal ends and I consider them to belong to the same individual as StW 339 humerus. As a unit this mid portion of arm contrasts markedly with the StW 431 mid portion of right arm which displays no carnivore marks (Figure 6).

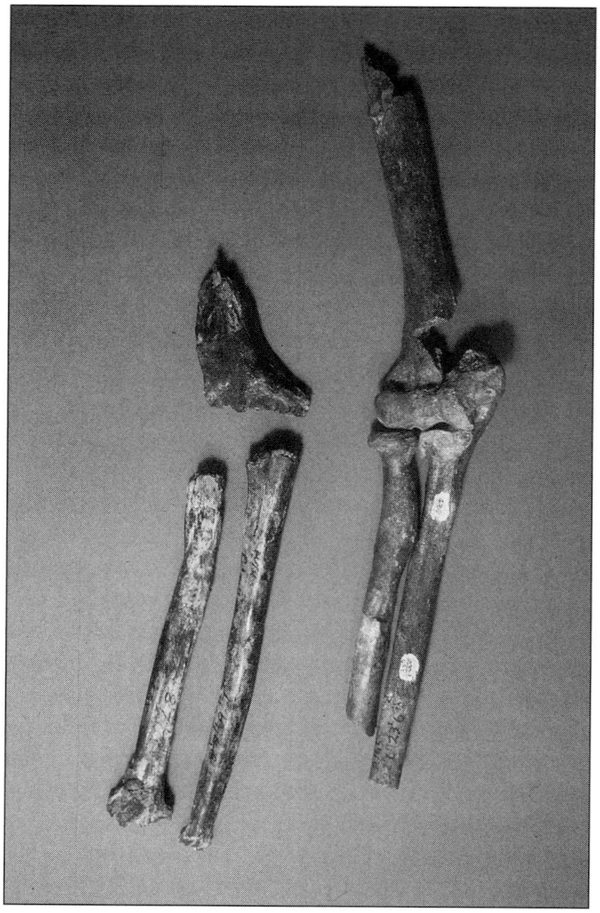

*Figure 6. Comparison of a hominid forelimb set heavily gnawed by carnivores at the elbow joint (note the missing distal end of humerus StW 339 and damage to proximal ends of radius StW 348 and ulna StW 349; left side of photograph) and an undamaged forelimb from the partial* Australopithecus *skeleton StW 431 (right).*

## Summary

From the complete articulated skeleton and skull of StW 573 it is apparent that it was an individual that fell into the cave, was not attacked by carnivores or fed upon by scavengers, was apparently mummified and buried before the bones could be scattered. The cranial fossils from a large number of *Australopithecus* individuals in the Member 4 breccia suggest that there should also have been, in the vicinity, an equally large number of *Australopithecus* skeletons. Yet only two partial skeletons have been recovered and they have no associated cranial fossils. These two torsos seem to be what was left after the individuals had been fed upon by a particular type of carnivore or scavenger that had removed the skull, cervical vertebrae and most of the limbs.

The postcranial bones from the other *Australopithecus* individuals represented by skulls could have been broken up and scattered by predators, scavengers, rock-fall and filtering of the fragments through the talus cone. Indications of all these taphonomic agents are shown by the condition of the bones and the widespread distribution of elements assigned to single individuals within the Member 4 talus.

## References


Bahn, P. (Ed.) 1996. Tombs, Graves and Mummies. Weidenfield and Nicolson, London.

Broom, R., Robinson, J. T., Schepers, G.W.H. 1950. Sterkfontein Ape-Man *Plesianthropus*. Transvaal Museum, Pretoria.

Clarke, R.J. 1990. Observations on some restored hominid specimens in the Transvaal Museum, Pretoria. In: Sperber, G.H. (Ed.), From Apes to Angels, Essays in Anthropology in Honor of Phillip V. Tobias. Wiley-Liss, New York, pp. 135-151.

Clarke, R.J. 1998. First ever discovery of a well-preserved skull and associated skeleton of *Australopithecus*. South African Journal of Science 94, 460-463.

Clarke, R.J. 1999. Discovery of complete arm and hand of the 3.3 million-year-old *Australopithecus* skeleton from Sterkfontein. South African Journal of Science 95, 477-480.

Clarke, R.J. 2002. Newly revealed information on the Sterkfontein Member 2 *Australopithecus* skeleton. South African Journal of Science 98, 523-526.

Findlay, G. 1972. Dr Robert Broom. Balkema, Cape Town.

Howell, F.C. 1965. Early Man. Time-Life Books, New York.

Kibii, J.M., Clarke, R.J. 2003. A reconstruction of the Stw 431 *Australopithecus* pelvis based on newly discovered fragments. South African Journal of Science 99, 225-226.

Oakley, K.P., Campbell, B.G., Molleson, T.I. 1977. Catalogue of Fossil Hominids, Part 1: Africa. British Museum, London.

Partridge, T.C., Shaw, J., Heslop, D., Clarke, R.J. 1999. The new hominid skeleton from Sterkfontein, South Africa: age and preliminary assessment. Journal of Quaternary Science 14, 293-298.

Partridge, T.C., Granger, D.E., Caffee, M.W., Clarke, R.J. 2003. Lower Pliocene hominid remains from Sterkfontein. Science 300, 607-612.

Place, R. 1957. Finding Fossil Man. Rockliff, London.

Reader, J. 1981. Missing Links. Little, Brown and Co., Boston.

Robinson, J.T. 1972. Early Hominid Posture and Locomotion. University of Chicago Press, Chicago.

Schaal, S., Ziegler, W. 1992. Messel. Clarendon Press, Oxford.

Tobias, P.V. 1973. A new chapter in the history of the Sterkfontein early hominid site. Journal of the South African Biological Society 14, 30-44.

Tobias, P.V., Clarke, R.J., White, H. 2003. Sterkfontein Research Unit 2nd Report, April 2001–June 2003. University of the Witwatersrand, Johannesburg.

Toussaint, M., Macho, G.A., Tobias, P.V. Partridge, T.C., Hughes, A.R. 2003. The third partial skeleton of a late Pliocene hominin (Stw 431) from Sterkfontein, South Africa. South African Journal of Science 99, 215-223.

# CHAPTER 12

# Taphonomy of Immature Hominid Skulls and the Taung, Mojokerto, and Herto Specimens


Gail E. Krovitz and Pat Shipman



## Abstract

This study explored quantitative and qualitative methods for deducing the taphonomic history of immature hominid crania, in the hopes of developing models for use in diagnosing the timing and pattern of breakage in specimens of unknown history. First, two sets of cranial inventory data (of modern human archaeological samples) were used to develop a model of the taphonomic vulnerability of the different parts of the immature hominid cranium. We found that the nasal bones, vomer, basilar and lateral elements of the occipital, and zygomatic arches were highly vulnerable to breakage or loss, while the orbital rim of the frontal bone was rarely missing. In fact, the frontal bone plays an important role as a keystone that keeps crania articulated. Second, we reviewed the pertinent medical and forensic data on cranial damage, and discussed three temporal stages of cranial breakage: stage 1 (wet or fresh bone breakage), stage 2 (dry bone or postmortem breakage), and stage 3 (post-fossilization breakage). Patterns of breakage and disarticulation in 20 immature fossil hominid crania were also included in this discussion. Several new fracture types were observed in the fossil hominids, including temporal line, perpendicular, and metopic or para-metopic fractures, and a mosaic fracture pattern. Finally, the models discussed above were used to deduce the breakage histories of three immature fossil specimens that were exposed to different taphonomic influences: Taung 1 (*Australopithecus africanus*), Mojokerto (or Perning 1, *Homo erectus*), and Herto BOU-VP-16/5 (*Homo sapiens idaltu*).


## Introduction

Although the tremendous importance of immature crania for documenting the growth, development, and evolution of various hominid species is widely recognized, their taphonomy has not been systematically addressed. That immature crania are rarer in the fossil record than adult crania is generally attributed to their greater vulnerability to damage because immature sutures are unfused and immature bones are thinner than adult ones (see discussion in Saunders, 2000). Here, we report a first step toward deducing the taphonomic history of immature hominid crania, which may prove useful in determining the timing, causes, and implications of damage to such specimens.

The primary aim of our project was to establish the expected pattern of breakage and destruction to immature hominid crania that have been subjected to minimal taphonomic disturbance. To this end, we carried out both quantitative and qualitative studies. We relied upon two sets of cranial inventory data. One of us (G.K.) conducted an inventory of the immature cranial remains of 272 recent humans in six cemetery populations from four broad geographic regions. Here, we refer to these crania as the Krovitz sample. For each cranium, she recorded the presence or absence of anatomical landmarks as an indicator of the loss or breakage of cranial elements. From these inventory data we developed a model of the taphonomic vulnerability of the different parts of the immature hominid cranium. We also analyzed bone inventory data on a sample of 81 modern human crania from a cemetery population in England that was excavated and studied by the Sedgeford Historical and Archaeological Project (SHARP); unpublished data were kindly provided to us by Patricia Reid of SHARP. The SHARP sample had been scored by the original researchers for the presence or absence of eight major cranial bones; age at death and sex had also been assigned. Only three of 82 individuals in the SHARP sample were immature, so this sample must be taken as representative of the cranial taphonomy of adults. Description and comparison of the results of the inventory studies constitute Part I of this paper.

In Part II, we first review the pertinent medical and forensic data on cranial damage and then describe and discuss the results of qualitative and quantitative observations conducted on photographs, casts, and occasionally upon originals of 20 immature fossil crania of *Homo erectus, Australopithecus africanus,* Neandertals, and anatomically modern *Homo sapiens* (including *H. sapiens idaltu*). Following forensic practice, we recognize three temporal stages of cranial or bone breakage: *stage 1, wet or fresh bone breakage*, which incorporates the stages forensic scientists recognize as antemortem breakage, which shows signs of healing on the broken edges, and perimortem breakage, which does not; *stage 2, dry bone or postmortem breakage*; and *stage 3, post-fossilization (fossilized bone) damage*. These three stages grade into one another along a temporal continuum and the placement of a specimen along this continuum has a marked impact on its response to potentially damaging agents. From our observations of modern and fossil crania, we summarize the types, frequency, and morphology of breaks seen in crania in each of these stages for potential use in diagnosing the timing of breakage in specimens of unknown history.

Finally, in Part III, we use the quantitative and qualitative results from the previous two sections to deduce the breakage histories of three immature fossil specimens that were exposed to different taphonomic influences. These are: Taung 1, *Australopithecus africanus,* which was dropped into a cave, probably by a leopard, while it was both fresh and fleshed (McKee, 2001; McKee, 2004, personal communication to P.S.); Mojokerto, *Homo erectus,* which was deposited in fluvial sediments (Huffman, 2001; Huffman and Zaim, 2003) and subjected to breakage and plastic deformation (Anton, 2003, personal communication to P.S.) at an unknown time; and Herto BOU-VP-16/5, *Homo sapiens idaltu,* which was modified and curated by hominids after the death of the individual (Clark et al., 2003; White et al., 2003) and underwent primarily post-fossilization damage (White, 2004, personal communication to P.S.). We use the taphonomic vulnerability model developed from the inventory data and the patterns of breakage and taphonomic destruction observed on casts and high resolution photographs of these specimens to deduce the specific taphonomic history of each of these specimens.

## Quantitative Studies

### Materials and methods

To determine how immature crania were preserved and damaged in specimens that were exposed to relatively few taphonomic forces, we used two sources of information.

The Krovitz sample consists of a landmark inventory of 272 immature crania of recent humans from archaeological samples conducted by one of us (Krovitz, 2000). Although the inventory was not designed for a taphonomic study, the results are useful here. The samples represent cemetery populations from England (Christ Church Spitalfields, 18$^{th}$ and 19$^{th}$ centuries; Adams and Reeve, 1987; Molleson and Cox, 1993), Medieval Denmark (A.D.1000-1500 ), Nubia (A.D. 0-1500; Vagn Nielsen, 1970), Edo Period Japan (A. D.1603-1867; Mizoguchi, 1997), St. Lawrence Island Yupik Eskimo (A.D. 1800; Collins, 1937; Utermohle, 1984; Heathcote, 1986), and Indian Knoll (2500-2000 B.C.; Snow, 1948). For each cranium, G.K. recorded the presence or absence of a set of 39 anatomical landmarks (Figure 1, Table 1). This provides fine-grained data on the location of breakage or complete loss of cranial elements.

Specimens were only included if they were undistorted, non-pathological, and had at least one anatomical region (face or neurocranium) that was largely articulated. Since disarticulated cranial bones, no matter how complete, were not included in the Krovitz sample, this inventory provides a conservative estimate of breakage for these samples. Recent human samples where completeness of the crania was a primary criterion for collection were excluded from the inventory.

Individuals in the Krovitz sample were divided into four developmental age groups based on tooth formation and eruption sequences: 0 – 3.0 years (Age Group 1), 3.1 – 6.0 years (Age Group 2), 6.1 – 9.0 years (Age Group 3), and 9.1 – 13.5 years (Age Group 4). Tooth formation was the primary method for dental age estimation (using data from Thoma and Goldman, 1960; Moorrees et al., 1963; Smith, 1991), although tooth eruption was also used when necessary (see discussion in Krovitz, 2000). These developmental age groups roughly coincide with the following developmental criteria (after Minugh-Purvis, 1988): 1) infancy (birth to completion of deciduous tooth eruption and development), 2) early childhood (period between deciduous tooth development and permanent tooth eruption), 3) mid-childhood (eruption of the first permanent teeth), and 4) late childhood (completion of permanent tooth eruption and development, except for the third molar).

The SHARP data, which were made available to us but were not collected by us, consist of bone inventories of 82 individuals represented by articulated remains from the Anglo Saxon cemetery at Sedgeford buried between 662 and 881 A.D. (Stillwell, 2002; Sedgeford Historical and Archaeological Research Project or SHARP,

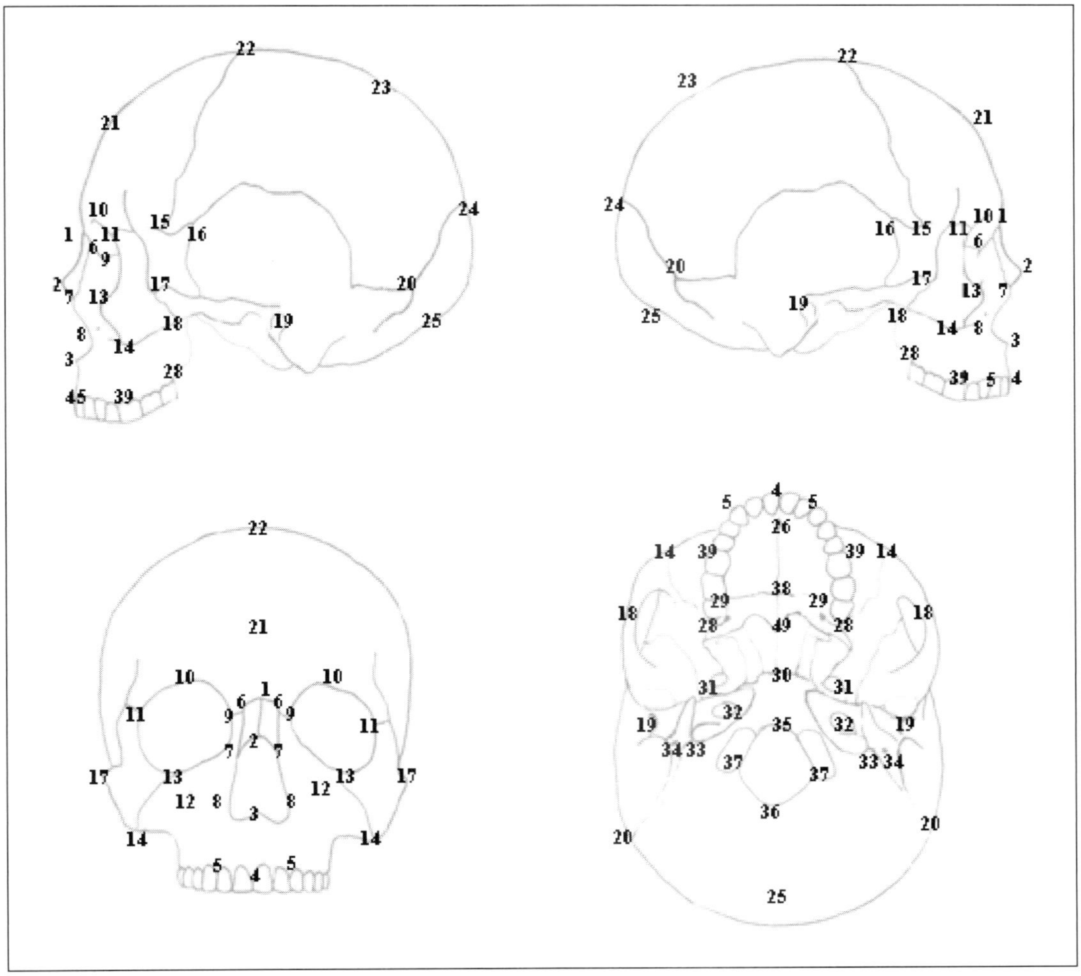

*Figure 1. Landmarks used in this study. See Table 1 for landmark descriptions.*

*Table 1. Description of landmarks used in this analysis; see Figure 1 for location of landmarks. All landmarks described with L/R were collected from either the right or left side, depending on which side was better preserved in that individual.*

| LANDMARK DESCRIPTION | | | |
|---|---|---|---|
| 1. NAS (Nasion) | 11. L/R FZJ (Frontal-zygomatic junction at orbital rim) | 21. N-B (1/2 way between nasion and bregma in midline) | 31. L/R FOV (Foramen ovale, posterolateral point) |
| 2. NAL (Nasale) | 12. L/R IF (Infraorbital foramen, marked inferolaterally, most superior if many) | 22. BRG (Bregma, coronal-sagittal suture intersection in midline) | 32. L/R CAR (Carotid canal, posterolateral point) |
| 3. ANS (Anterior nasal spine) | 13. L/R ZYS (Top zygomatic-maxillary suture, at orbital rim) | 23. B-L (1/2 way between bregma and lambda in midline) | 33. L/R JUG (Jugular process, anterior point) |
| 4. IDS (Intradentale superior, between central incisors) | 14. L/R ZYI (Bottom zygomatic-maxillary suture) | 24. LAM (Lambda, sagittal-lambdoid suture intersection, in midline) | 34. L/R STY (Stylomastoid foramen) |
| 5. L/R PMM (Premaxilla-maxilla junction at alveolar border, between I2 and C) | 15. L/R PTN (Pterion, intersection of frontal-parietal and sphenoid) | 25. L-O (1/2 way between lambda and opisthion, in midline) | 35. BAS (Basion) |
| 6. L/R NMT (Top nasal-maxillary suture at frontal bone) | 16. L/R SPH (Squamous temporal-parietal-greater wing of sphenoid) | 26. ICF (Incisive foramen, marked posteriorly) | 36. OPI (Opisthion) |
| 7. L/R NMB (Bottom nasal-maxillary suture at nasal aperture) | 17. L/R SZA (Superior temporal-zygomatic junction on the arch) | 27. PNS (Posterior nasal spine) | 37. L/R CFM (Posterior border of the occipital condyle with foramen magnum) |
| 8. L/R ALA (Alare) | 18. L/R IZA (Inferior temporal-zygomatic junction on the arch) | 28. L/R MXT (Maxillary tuberosity = junction maxilla and palatine bones on alveolus) | 38. MXP (Junction palatine/maxilla in midline) |
| 9. L/R FMO (Frontal-maxillary suture at orbital rim) | 19. L/R EAM (External auditory meatus, uppermost lateral point) | 29. L/R PAL (Junction on palatine suture with edges/curve of palate) | 39. L/R DPM (Behind DM2/P4 on exterior alveolus) |
| 10. L/R ORB (Top of orbit, 1/2 way between NAS-FZJ) | 20. L/R AST (Asterion = parietal-temporal-occipital) | 30. VSJ (Vomer sphenoid junction, taken on vomer) | |

unpublished data; Reid, 2004, personal communication to P.S.). For each individual, the presence or absence of the frontal, maxilla, palatine, zygomatic, sphenoid, parietal, temporal, and occipital was scored separately for the right and left sides; the values for both sides were averaged for our purposes. If present, each bone was scored in completeness categories consisting of: < 25% complete; 25-50% complete; 51-74% complete; and 75-100% complete.

Unlike the Krovitz sample (which consisted entirely of immature individuals), 79 out of 82 (95%) individuals in the SHARP sample were adult. Cranial fragments and isolated bones that could not be associated with a particular burial were excluded from the database provided to us; however, the SHARP sample did include disarticulated cranial bones that could be associated with a burial (for example, one individual's cranial remains consist solely of an occipital bone). This is an important difference from the Krovitz sample that only considered relatively articulated crania and no disarticulated bones. Although healed lesions were observed on five individuals in the SHARP sample (Stillwell, 2002), no crania were grossly pathological. The SHARP bone inventory provided coarser-grained data on the preservation of various cranial elements in a predominantly adult sample of buried modern humans.

We selected deliberately buried specimens because this greatly simplified the potential range of taphonomic histories exhibited by the samples. Further, most modern humans are buried in some fashion, so most sizeable samples of modern human crania are derived from cemetery populations. Since interment occurred shortly after death, the possibility of lengthy surface weathering, significant waterborne transport, or substantial carnivore damage was eliminated. What we observed on these specimens represents a generalized or baseline pattern of destruction, damage, and preservation of immature crania undergoing deposition rapidly after death. Therefore, the condition of crania in our samples reflects the taphonomic vulnerability of various parts of the cranium based primarily on their mechanical resistance to breakage and on the structural integrity of various sutures between bones. The ways in which specimens of unknown taphonomic history, such as the fossil crania considered in Part III, deviate from this baseline pattern should provide clues to their exposure to other destructive agents.

## Results and discussion

From the landmark inventory, we calculated the percentage of specimens in which each landmark was absent in each age group, and in all age groups averaged. The presence or absence of many landmarks was highly correlated with that of other nearby landmarks, with three obvious clusters of covarying landmarks (face, neurocranium, and basicranium). These groupings were undoubtedly caused by the close spatial relationships of the landmarks within each anatomical region and the general similarity in terms of robustness and/or geometry of bones within each region.

We could not derive a single predicted sequence of disarticulation and damage from these landmark inventory data because the variability among specimens was too great. Instead, we identified clusters of landmarks that exhibited high taphonomic vulnerability, intermediate taphonomic vulnerability, and low taphonomic vulnerability (Figures 2 and 3; Table 2). The high and low vulnerability clusters together comprise 16 of the 39 landmarks (41%). The remaining 23 landmarks (59%) are of intermediate vulnerability.

We argue that: (1) the primary factor determining the taphonomic vulnerability of a landmark is its structural resistance to breakage and destruction, which is a function of the density of the skeletal element and of its placement within or projection from the cranium as a whole; (2) the breakage of immature hominid crania is intimately related to the placement and physical nature of sutures; (3) although crania of younger individuals were generally less complete, the taphonomic vulnerability grouping of most cranial landmarks does not change dramatically between the ages of 0-13 years.

### *High taphonomic vulnerability*

Landmarks with high taphonomic vulnerability are illustrated in Figure 3 and listed in Table 2a. Several landmarks in the facial region are missing with remarkably high frequency in the Krovitz sample. These most vulnerable landmarks are missing in almost 70% of individuals in the youngest age group and are absent in 40-47% of the specimens across all age groups. All of the most vulnerable facial landmarks are associated with the nasal bones (NAL), the zygomatic arch (SZA, IZA), and the vomer (VSJ). Each of these bones has sutures with other bones that cover small linear distances and which therefore probably break or separate more easily than do more extensive sutures. The nasals and the vomer are thin and fragile bones prone to damage, and both have an edge projecting into open space. In contrast, the zygomatic is not a particularly fragile bone but the zygomatic arch projects from the generally ovoid shape of the cranium, which makes it vulnerable to breakage. The anterior portion of the zygomatic arch is also a thin strut of bone that is very susceptible to crushing. In cadaver experiments, McElhaney and colleagues found that the zygomatic arch will break under as little as 130 psi (McElhaney et al., 1976; Mackey, 1984), whereas the pressure required to fracture the cranial vault is much greater: 450 to 750 psi (Cox et al., 1987). On dry crania, the zygomatic arch encloses empty space and requires even less force to break.

Out of the 272 individuals inventoried, 163 (60%) had a face judged to be in good condition, while preservation of the face was judged to be fair or poor in the remaining specimens. The entire face was missing in 25 specimens (9% of the sample), suggesting that loss of the entire face, usually from nasion downwards, is only moderately common in archaeological remains. This

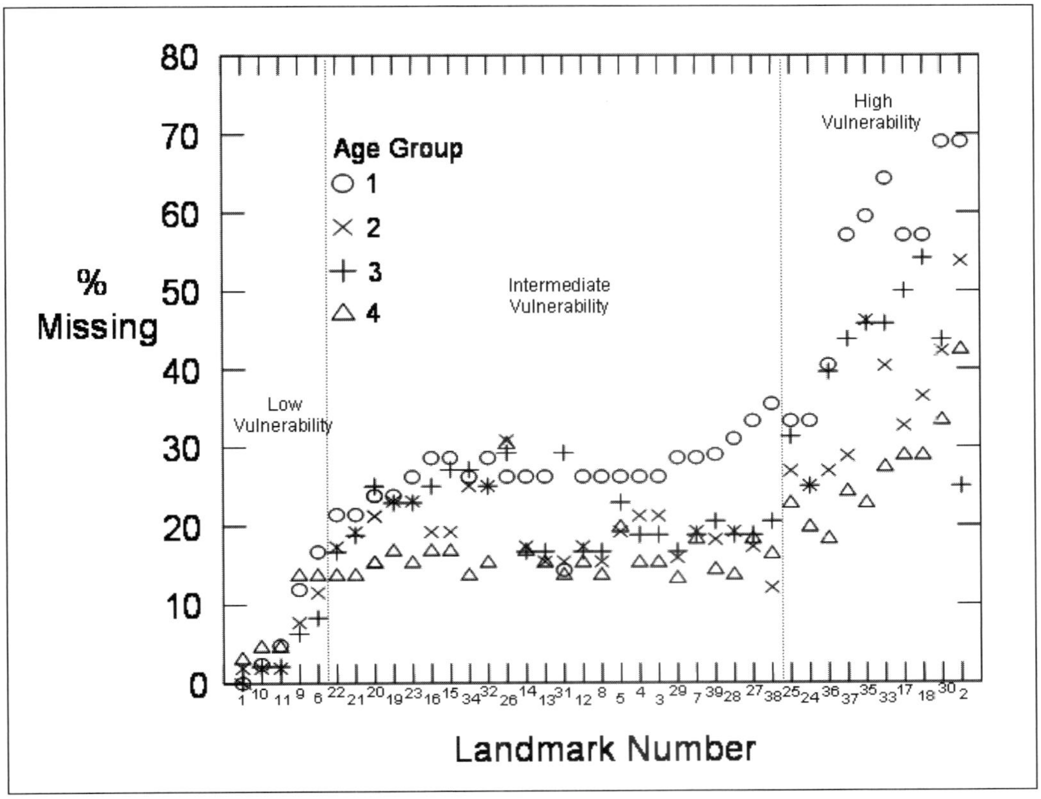


Figure 2. Percentage of landmarks missing for each age group, ordered into low, intermediate and high taphonomic vulnerability. Landmark numbers as in Table 1.

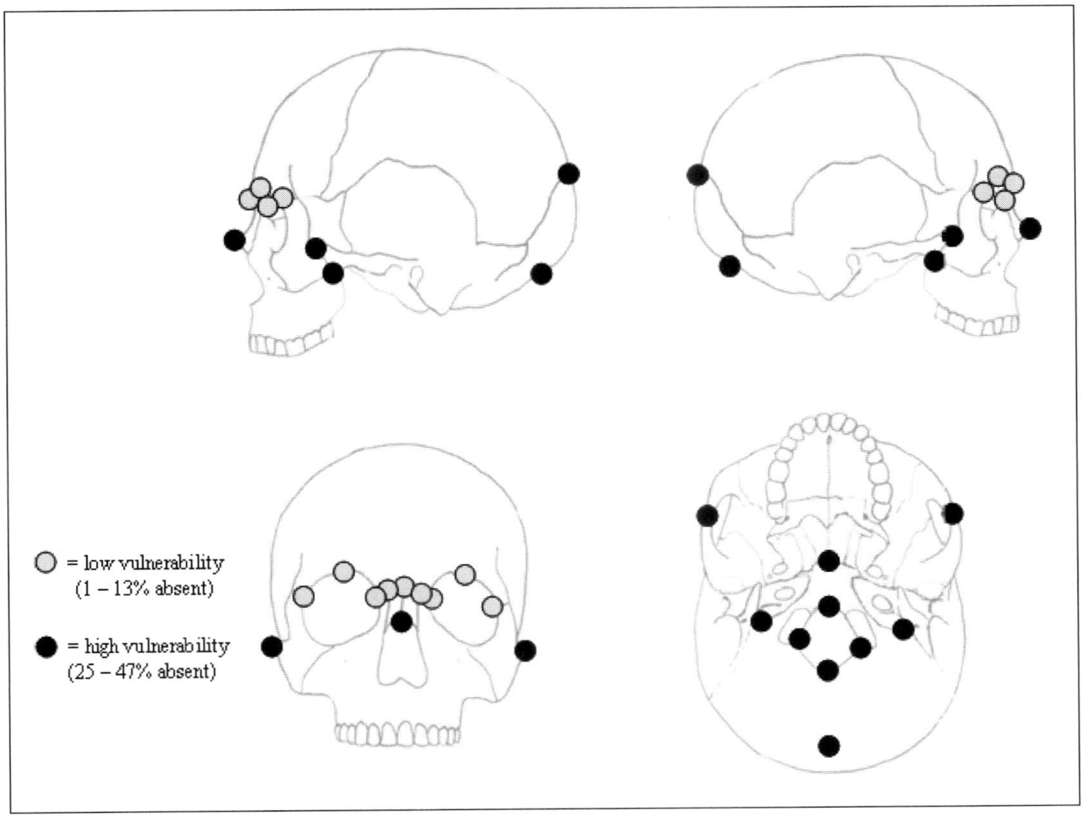


Figure 3. Landmarks with high and low taphonomic vulnerability (see Table 2).

*Table 2. Taphonomic vulnerability of landmarks, separated into High (a), Intermediate (b) and Low (c) taphonomic vulnerability. Landmark abbreviations and numbers as in Table 1.*

| 2a. High Taphonomic Vulnerability | | | | | | |
|---|---|---|---|---|---|---|
| | Landmark (number) | % Absent 0-3 years | % Absent 3-6 years | % Absent 6-9 years | % Absent 9-13 years | Mean % All ages |
| Facial: | NAL(2) | 69 | 53.8 | 25 | 42.2 | 47 |
| | VSJ(30) | 69 | 42.3 | 43.8 | 33.3 | 45 |
| | SZA(17) | 57.1 | 32.7 | 50 | 28.8 | 40 |
| | IZA(18) | 57.1 | 36.5 | 54.2 | 28.8 | 42 |
| Basal: | JUG(33) | 64.3 | 40.4 | 45.8 | 27.3 | 42 |
| | BAS(35) | 59.5 | 46.2 | 45.8 | 22.7 | 41 |
| | CFM(37) | 57.1 | 28.8 | 43.8 | 24.2 | 37 |
| | OPI(36) | 40.5 | 26.9 | 36.9 | 18.2 | 30 |
| Vault: | LAM(24) | 33.3 | 25 | 25 | 19.7 | 28 |
| | L-O(25) | 33.3 | 26.9 | 31.3 | 22.7 | 25 |
| **2b. Intermediate Taphonomic Vulnerability** | | | | | | |
| | Landmark (number) | % Absent 0-3 years | % Absent 3-6 years | % Absent 6-9 years | % Absent 9-13 years | Mean % All ages |
| Facial: | ICF(26) | 26.2 | 30.8 | 29.2 | 30.3 | 29 |
| | PNS(27) | 33.3 | 17.3 | 18.8 | 18.2 | 22 |
| | PMM(5) | 26.2 | 19.2 | 22.9 | 19.7 | 22 |
| | MXP(38) | 35.5 | 12.1 | 20.5 | 16.3 | 21 |
| | MXT(28) | 31 | 19.2 | 18.8 | 13.6 | 21 |
| | NMB(7) | 8.6 | 19.2 | 18.8 | 18.2 | 21 |
| | ANS(3) | 26.2 | 21.2 | 18.8 | 15.2 | 21 |
| | IDS(4) | 26.2 | 21.2 | 18.8 | 15.2 | 21 |
| | ZYI(14) | 26.2 | 17.3 | 16.7 | 16.7 | 19 |
| | DPM(39) | 29 | 18.2 | 20.5 | 14.3 | 19 |
| | PAL(29) | 8.7 | 15.9 | 16.7 | 13.2 | 18 |
| | IF(12) | 26.2 | 17.3 | 16.7 | 15.2 | 18 |
| | ZYS(13) | 26.2 | 15.4 | 16.7 | 15.2 | 18 |
| | ALA(8) | 26.2 | 15.4 | 16.7 | 13.6 | 17 |
| Basal: | CAR(32) | 28.6 | 25 | 25 | 15.2 | 23 |
| | STY(34) | 26.2 | 25 | 27.1 | 13.6 | 22 |
| | FOV(31) | 14.3 | 15.4 | 29.2 | 13.6 | 18 |
| Vault: | PTN(15) | 28.6 | 19.2 | 27.1 | 16.7 | 22 |
| | SPH(16) | 28.6 | 19.2 | 25 | 16.7 | 22 |
| | B-L(23) | 26.2 | 23.1 | 22.9 | 15.2 | 22 |
| | EAM(19) | 23.8 | 23.1 | 22.9 | 16.7 | 22 |
| | AST(20) | 23.8 | 21.2 | 25 | 15.2 | 21 |
| | N-B(21) | 21.4 | 19.2 | 18.8 | 13.6 | 18 |
| | BRG(22) | 21.4 | 16.7 | 13.6 | 13.6 | 17 |
| **2c. Low Taphonomic Vulnerability** | | | | | | |
| | Landmark (number) | % Absent 0-3 years | % Absent 3-6 years | % Absent 6-9 years | % Absent 9-13 years | Mean % All ages |
| Facial: | NMT(6) | 16.7 | 11.5 | 8.3 | 13.6 | 13 |
| | FMO(9) | 11.9 | 7.7 | 6.3 | 13.6 | 10 |
| | FZJ(11) | 4.8 | 1.9 | 2.1 | 4.5 | 3 |
| | ORB(10) | 2.4 | 1.9 | 2.1 | 4.5 | 3 |
| | NAS(1) | 0 | 1.9 | 0 | 3 | 1 |
| Basal: None | | | | | | |
| Vault: None | | | | | | |

pattern of damage is known in the forensic literature as a LeFort III fracture (described below, after Rogers, 1992) and occurs in about 10% of patients seeking medical attention for cranial fractures (Richardson, 2000). A LeFort III fracture leaves the cranium in two pieces, the face and the skullcap (neurocranium plus basicranium). The thinness and vulnerability of the facial bones to taphonomic forces means that once these two portions of the cranium have separated, the skullcap is much more likely to survive and be recovered than the face. Additionally, the sphenoid is extremely likely to be broken or missing in crania missing the face.

Three landmarks in the basicranial region are also consistently missing in high frequencies (25-42% across all age groups): the anterior part of the jugular process (JUG); the anterior point of the foramen magnum on the midline or basion (BAS); and the posterior border of the occipital condyle with the foramen magnum (CFM). All are clustered spatially and are intimately related to the basilar and lateral parts of the occipital bone. The sutures between these ossification centers and the squamous occipital are short and vulnerable to separation; the squamous fuses to the basilar portion at about five years, and the lateral and basilar parts fuse in the sixth year (Byers, 2002). The basilar suture is vulnerable to separation until it fuses to the sphenoid between 18-21 years.

Three landmarks on the vault are also missing in moderately high frequencies (25-30% across all age groups). These are: lambda (LAM) at the junction of the lambdoid and sagittal sutures; a point (L-O) halfway between lambda and opisthion on the midline; and opisthion (OPI), the posterior midline point of the foramen magnum. All are associated with the squamous occipital. One of the most likely types of taphonomic damage to occur to an immature cranium is the loss of the occipital bone due to separation of the lambdoid suture.

The high frequency with which both basilar and squamous occipital landmarks are missing shows that loss of part or all of the occipital must be considered one of the most common types of cranial damage among immature individuals, as is loss of the vomer, nasals, and zygomatic arches. Separation at the coronal suture seemed common in the Krovitz sample but was not specifically quantified.

### *Intermediate taphonomic vulnerability*

This grouping includes most of the neurocranial landmarks and a mixture of facial and basicranial landmarks (Table 2b). Because these landmarks exhibit the greatest variability, both within and between age groups, we suggest that the preservation of and damage to these landmarks tends to reflect particular differences in individual growth and taphonomic history.

Because the squamous temporal suture is beveled and not interdigitated, we expected that this suture would be more likely to open and fall apart than other neurocranial sutures. Contrary to our expectations, the landmarks along this suture (AST, PTN, and SPH) fell into the intermediate vulnerability group; just over 20% of the specimens (regardless of age group) were missing at least one of the landmarks on the squamous temporal suture.

### *Low taphonomic vulnerability*

Landmarks with low taphonomic vulnerability are illustrated in Figure 3 and listed in Table 2c. These landmarks were present in nearly all of the specimens in the samples examined here regardless of the age at death of the individual. All of the landmarks with the lowest taphonomic vulnerability are facial and all are located on the sturdy orbital rim of the frontal bone. They are nasion (NAS), orbitale (ORB), the top of the nasomaxillary suture at the frontal bone (NMT), the frontomaxillary suture at the orbit (FMO), and the frontozygomatic junction at the orbit (FZJ). It is apparent that the robustness and structural strength of the orbital rim of the frontal bone has a substantial impact on the frequency of preservation of this region of the cranium.

We also believe that the frontal bone acts as a physical keystone in holding the cranium together and is critical in determining how an immature cranium will break. The frontal has sutural connections with the parietals, temporals, sphenoid, maxillae, zygomatics, nasals, and ethmoid. The frontal bone plays a key role in hafting the face onto the neurocranium and in reinforcing and strengthening sutural connections (such as the sagittal suture) within the neurocranium. If the coronal suture opens and the face and frontal separate as a unit from the rest of the cranium, then the face has an improved chance of survival, although the rest of the cranium will almost certainly disarticulate. However, if the face breaks off below nasion in a LeFort II or III fracture (discussed below), then the neurocranium has a better chance of survival, but the facial bones will almost certainly disarticulate.

In general, once a cranial bone is isolated its individual chance of survival is lessened. However, isolated cranial elements vary in their likelihood of survival due to their structural or mechanical properties. Sturdier bones (or bone parts, such as the petrous temporal, suitably named for its rocklike properties) almost always survive in higher frequencies and with less damage than delicate bones (such as the nasals or vomer), thin bones with a complex shape (such as the sphenoid), or bones with projecting processes (such as the zygomatic).

The Krovitz sample considered only fairly well articulated crania; thus most individuals preserved the frontal bone, which explains the apparent low taphonomic vulnerability of the landmarks on the frontal bone. Individuals with a missing or badly broken frontal bone probably did not survive to be included in the Krovitz sample. Because of its general robustness and its many sutural attachments to other bones, the frontal has an unusually large effect on the taphonomic survival of the cranium.

## *Application of taphonomic vulnerability categories*

The categories described above represent three levels of taphonomic vulnerability. In general terms, knowing which cranial parts are missing from a specimen can be used with the taphonomic vulnerability categories to judge the intensity of the taphonomic damage to which a specimen was exposed.

Specimens showing only a few fractures or absences in the high vulnerability category have probably been subjected to *minimal* taphonomic destruction. Those showing breaks in high and intermediate categories (but not necessarily the loss of all landmarks in those categories) have been subjected to *moderate* taphonomic destruction. Those crania showing some fractures to or loss of some landmarks in all three groupings have been subjected to *extensive* taphonomic destruction. The categories of taphonomic vulnerability are nested, so that specimens subjected to any level of taphonomic destruction have also, by definition, been subjected to the less severe types of damage as well. Especially indicative of intense taphonomic destruction is the breakage and partial loss of the orbital margin on the frontal bone or separation of the frontal bone from other parts of the cranium. As argued above, once the frontal is absent, the cranium as a whole is less likely to be preserved in the archaeological or fossil record. However, it is important to note that the response of a cranium to taphonomic agents is highly variable according to the particular circumstances of burial. Further, these categories do not necessarily represent three widely separated points in the time (relative to death) at which damage occurred. For example, extensive damage can and does occur early in a specimen's taphonomic history while the bone is fresh and the cranium is fully fleshed; conversely, minimal damage can and does occur after fossilization.

If the age at death of an immature cranium can be determined with some precision, then a more refined deduction can be made using both age at death and taphonomic vulnerability data. Within the Krovitz sample, most landmarks show decreasing vulnerability as age increases (Table 2), although landmarks rarely shift from one vulnerability category to another. Only four landmarks change taphonomic vulnerability categories with increased age: the juncture of the palatine suture with the edges or curve of the palate (PAL); the top of the zygomatic-maxillary suture at the inferior orbital rim (ZYS); the foramen ovale (FOV); and the incisive foramen (ICF).

The taphonomic vulnerability of PAL is intermediate in the youngest individuals and steadily diminishes as age increases, until PAL eventually ranks among the landmarks showing the lowest vulnerability in individuals aged 9-13 years. Between the ages of 0-13 years, the palate lengthens and strengthens considerably and, between ages of 5-12 years, the first and second permanent molars erupt. We hypothesize that the presence of these molars may partially shield the palatine suture from destruction in older individuals. A similar drop in vulnerability is shown in ZYS, which is in the intermediate vulnerability group in the youngest group of individuals and drops to the low vulnerability group in individuals older than 3 years. This change may be related to the increasing strength and buttressing in the face as adulthood is approached. We can offer no hypotheses for the change in taphonomic vulnerability of FOV and ICF in the different age groups other than individual variability or some inadvertent sampling bias.

## *Comparison of breakage in the Krovitz and SHARP samples*

The results of the Krovitz landmark inventory were compared with those of the SHARP bone inventory to see if these patterns of survival were consistent between immature and adult crania. To calculate frequency of damage/absence in the SHARP sample, we added the numbers of bones that were entirely missing to those in which a significant portion (25% or more) of the bone was missing (Table 3). As there are only three immature crania in the sample (Reid, 2004, pers. comm. to P.S.), we did not subdivide the sample into age groups.

Generally, the frequency and location of cranial damage is similar between the two samples. The Krovitz data showed that the category of high taphonomic vulnerability included facial landmarks associated with the maxilla, nasals, zygomatic arches, and vomer. Although

*Table 3. Completeness of bones in the SHARP sample (N=82; 79 adults and 3 immature crania). Completeness data represent an average of lefts and rights for each bone. Bone abbreviations as follows: Front = frontal, Pariet = parietal, Occipit = occipital, Temp = temporal, Sphen = sphenoid, Zygo = zygomatic, Max = maxilla, and Pal = palatine.*

| Completeness | Front | Pariet | Occipit | Temp | Sphen | Zygo | Max | Pal |
|---|---|---|---|---|---|---|---|---|
| 75–100% | 55 | 54 | 55 | 36.5 | 31.5 | 30 | 38.5 | 42 |
| 26–74% | 11 | 12 | 10 | 16 | 10.5 | 10.5 | 13.5 | 6.5 |
| 1–25% | 7.5 | 7 | 7 | 10 | 8 | 11.5 | 8.5 | 6 |
| 0% | 8 | 9 | 8 | 19.5 | 32 | 30 | 21.5 | 27.5 |
| 0–74% combined | 25 | 27.5 | 26 | 45.5 | 50.5 | 43.5 | 43.5 | 42 |
| % damaged or absent | 33% | 34% | 32% | 55% | 62% | 63% | 53% | 51% |

data on the nasals and vomer were not available for the SHARP sample, facial bones (zygomatics and maxillae) were damaged or absent in a high number of individuals (63% and 53% respectively). Thus, in both the immature sample inventoried by Krovitz and in the largely adult SHARP sample, aspects of the facial bones were among the most highly vulnerable to taphonomic destruction. Landmarks in the category of lowest taphonomic vulnerability, based on the Krovitz sample, were related to the sturdy superior margin of the orbit and the crucial role of structural keystone that the frontal bone plays within the cranium. This is consistent with the observation that only 33% of individuals in the SHARP sample had seriously damaged or absent frontals, making the frontal the second least vulnerable bone after the occipital.

The two samples differed strikingly in the vulnerability of the occipital bone. Landmarks in this region were among the most highly vulnerable in immature individuals in the Krovitz sample. In the largely adult SHARP sample, the occipital bone was the least vulnerable bone, being missing or seriously damaged in only 32% of individuals. We hypothesize that this difference is related to the adult nature of the SHARP sample. Despite considerable variability in the timing of the closure of cranial sutures in adults (Todd and Lyon, 1924; Todd and Lyon, 1925a, b, c; McKern and Stewart, 1957; Meindl and Lovejoy, 1985), the lambdoid sutures in adults of the SHARP sample would have fused to some degree, and sutures between the basilar and lateral occipital elements would be completely fused. Stronger bony attachments between the occipital and other cranial bones would make the occipital bone itself less likely to separate from the rest of the cranium and less vulnerable to damage.

Another marked difference in the survival of cranial bones in these two samples involves the sphenoid and the temporal bones. In the Krovitz sample, landmarks associated with the sphenoid and temporal bones fall into the intermediate category. In the SHARP sample, these bones are highly vulnerable, showing serious damage or destruction in 62% and 55% of the individuals respectively. Thus the sphenoid is the second most vulnerable bone and the temporal is the third most vulnerable bone in the SHARP sample. We hypothesize that the beveled nature of the squamous temporal suture is a more important source of vulnerability in adults because the other endocranial sutures are partially or wholly fused. In contrast, in immature individuals, the squamous temporal suture is not distinctly more vulnerable to separation than the other cranial sutures. Similarly, the sphenoid is relatively more vulnerable in adults than in immature individuals. The sphenoid is a very thin and fragile bone in adults compared to the other vault bones, which have thickened and acquired greater robustness with age; in immature crania, the sphenoid is not so markedly different in robustness from the other cranial bones. This difference could also be due to sampling differences between the Krovitz and SHARP samples, as the SHARP sample included disarticulated bones and the Krovitz sample did not. If a large number of the sphenoids and temporals contained in the SHARP sample were from disarticulated crania then they would likely be less well preserved than those from the more articulated crania in the Krovitz sample.

In summary, cranial breakage and survival in the immature sample studied by Krovitz and the adult SHARP sample show a generally similar pattern. There are important exceptions pertaining to the survival of the occipital, the sphenoid, and the temporal bones.

## Qualitative Breakage Patterns in Immature Fossil Crania

Qualitative observations on patterns of fracture location and morphology provided additional tools with which to deduce the approximate timing of damage in an immature cranium. We examined some original specimens but more usually photographs and casts of 20 fossilized immature hominid crania that were relatively complete and thus might be comparable to the Krovitz sample (Appendix I). These data were compared with similar observations on crania in the forensic and medical literature (primarily representing damage to living or recently dead individuals) and with archaeological specimens from the inventory sample (representing primarily post-burial damage).

For each fossil specimen we observed, we noted the general frequency of damage, the location of fractures, and the attributes (length, course, texture, and type) of fractures and of the fractured surfaces. We used standardized terminology from the forensic literature, where possible, to describe fracture type and morphology and to deduce the time of fracture relative to death. Appendix I summarizes our observations for each of the 20 immature fossil specimens considered.

### Chronology of taphonomic damage

We divide the taphonomic history of an immature cranium into three phases at which breakage can occur. In chronological order, these are:

*Stage 1) Wet or fresh bone breakage,* incorporating antemortem and perimortem breakage, generally takes place while the bone is partially or wholly fleshed. Antemortem breakage is recognized by the fact that healing began before death; under controlled conditions, grossly detectible healing may be evident in as little as one to two weeks after the time of injury (Sauer, 1998; Galloway, 1999: 15, citing Murphy et al., 1990 and Rogers, 1992). Identifying perimortem breakage (occurring at the time of death) can be less straightforward than antemortem damage, but still incorporates fresh bone fracture patterns (Sauer, 1998). Although antemortem and perimortem fractures can be differentiated by evidence of healing, they both occur on wet, fresh bone and both have similar fracture characteristics; therefore, they are considered together in stage 1.

Fresh or living bone is a composite tissue comprised of flexible protein (mostly collagen) and brittle hydroxyapatite. In both antemortem and perimortem breakage, the bony tissue and the sutures between bones contain intact collagen and other organic components that give bone its elasticity, meaning that it is able to bend or deform under load before failure (breaking) occurs. The flexible collagen and membranes surrounding unfused sutures stop cracks from propagating through the bony tissue by deforming and dissipating force (Currey, 1984).

The mechanical properties of one square inch samples of fresh tissue from long bones has been measured. In tension, such a sample of whole fresh bone fractures or fails at only about two-thirds the pressure that is required to fracture bone under compression (Gordon, 1968: 42-44). This is why, in fresh bone subjected to blows, fractures are initiated not at the point of impact where the bone is compressed but in the surrounding bone which is placed under tension (Rogers, 1992). Whole, fresh bone has a low modulus of elasticity, which means it has a tendency to bend without breaking (Gordon, 1968: 42-44). Younger individuals with bones that are more cartilaginous and less mineralized will have an even lower modulus of elasticity than adults.

*Stage 2) Dry bone breakage* occurs postmortem after the bone has lost much of its organic content, and is usually defleshed, although dried or mummified flesh can be found on stage 2 crania. Dry bone breakage may occur during a lengthy time span ranging from shortly after death to centuries later, depending on specific preservation conditions; the organic component of bone will also vary according to the time since death and preservation conditions. The archaeological sample we inventoried had been subjected primarily or exclusively to stage 2 damage.

Immature crania in stage 2 are substantially more vulnerable to separation along unfused sutures than stage 1 bones since the connective tissue holding the sutures together is degraded or decayed in stage 2. Therefore, the modulus of elasticity of the bone tissue is compromised as is the crack-stopping ability of the flexible components of bone. Thus breakage will occur in stage 2 bones at lower loads than in stage 1 bones (Lyman, 1994; Galloway, 1999). The loss of elastic tissue from the bone not only lowers the force needed to produce failure in bone but also alters the morphology of the resulting fracture. As elastic tissue degrades, the fracture surface becomes progressively flatter, more planar, and less likely to splinter or bevel. The course of vault fractures is more likely to be curvilinear and longer in fresher crania and straighter and shorter in dried crania.

*Stage 3) Post-fossilization breakage* occurs after the bone has been mineralized, but fossilization is not an all-or-nothing event. Bones at a single site may range from fully mineralized to a condition close to that of dry bones with only minimal geochemical changes. What typifies this stage is that the bone has no significant elastic or soft tissue left and behaves more like a brittle, inorganic material, such as stone or ceramic, than like fresh bone. We do not know of anyone who has measured of mechanical strength of fossilized bone; strength would presumably vary with the degree of mineralization. As a crude approximation, we might expect fossilized bone to have the very low tensile strength, very low modulus of elasticity, and high compressive strength of stony substances such as brick or concrete.

Another issue in breakage is microstructure. Living bone is riddled with osteons and other spaces that house bone cells; these "holes" can and do serve as the sites of fracture origin because they are inherently weak places in the tissue. Similarly, foramina act as stress concentrators where fractures often initiate (Currey, 1984). In fossilized bone, these "holes" are more or less completely filled with mineral. The increased strength caused by the absence of holes seems to be more than offset by the complete lack of flexibility.

Both the diminished flexibility and the geometry or morphology of the bone(s) assumes greater importance in determining breakage in stages 2 and 3 than in stage 1. It is also important to appreciate that breakage of stage 1 crania is more likely to be the result of accident or human violence than breakage in older crania. Fractures caused by violence are more likely to be directed at the face than fractures caused in other ways during stages 2 or 3.

## Review of the forensic and medical terminology: fracture types

Six basic types of cranial fractures, differentiated on the basis of location and morphology, are frequently discussed in the medical and forensic literature (e.g., Galloway, 1999; Byers, 2000; Richardson, 2000). Three combined fracture patterns are frequently identified in the face (i.e., LeFort fractures), and disarticulation of cranial sutures is also noted. These types of cranial damage are reviewed here with particular focus on immature crania.

(1) *Linear fractures* are elongated, single breaks that go through the outer table of the bone, the diploe, and the inner table of the cranial vault bones. Linear fractures comprise 70-80% of observed fractures in the forensic context (Gurdjian, 1975; Rogers, 1992), which usually involves stage 1 breakage. Linear fractures are often the result of impact with objects having a large mass, such as heavy weapons or automobiles. In forensic cases, linear fractures occur less commonly in children than in adults due to the greater elasticity of immature bone (Duncan, 1993) but are known to occur in cases of child abuse, especially in children under the age of three years (Naim-Ur-Rahman et al., 1994). In fresh crania, linear fractures occur as a result of forces between 450 to 750 psi (Cox et al., 1987) although there is considerable individual variation.

(2) *Diastatic fractures* are linear fractures that follow the course of sutures, fused or unfused, in stage 1 crania. In antemortem or perimortem circumstances, diastatic fractures cause traumatic interruptions of sutures,

sometimes leading to the springing outward of the vault bone on one side of the fracture and the evulsion of brain tissue through the crack. This springing out is a result of the release of the inherent tension of the intact cranial vault by a fracture. This phenomenon is very unlikely to occur in dry or fossilized crania because elasticity of the bones is so diminished that fracture is more likely to occur than a rebounding outward of part of the cranium.

In the forensic context, diastatic fractures constitute about 5% of all fractures and occur most commonly in the coronal and lambdoid sutures (Galloway, 1999). Blount (1955, 1977) observed that true diastatic fractures are rare in children in stage 1. He postulated that linear fractures made on fresh or wet bone very rarely cross a suture because the area of the suture has different mechanical properties from the bone surrounding it. The greater flexibility of connective tissue in and near sutures in immature crania acts to stop cracks by dissipating force through deformation.

(3) *Depressed fractures*, together with comminuted and stellate fractures (discussed below), comprise 15% of fractures in forensic contexts (Gurdjian, 1975). Depressed fractures involve deformation of the cranial vault in response to impact, usually of high velocity by a blunt object of small or moderate diameter. The bone at the point of impact is pushed inward while the area immediately surrounding the impact is bent outward, placing the bone under tension and initiating fractures (Gurdjian et al., 1953; Gurdjian, 1975; Rogers, 1992). The fragments of bone pushed inward by the impact in stage 1 remain attached to the cranium; Byers (2002) refers to this phenomenon as *hinging* and regards it as diagnostic of stage 1 breakage from depressed fractures. We have observed a specimen that received a depressed fracture in stage 1 with the fragments still in place some 200 years later, well after the cranium had reached stage 2 (G. Milner, personal communication to authors, 2003). Depending on the strength of the blow, a depressed fracture may be surrounded by a number of linear breaks that radiate outward from the depressed area known as *radiating fractures*. A depressed fracture surrounded by radiating fractures is one typical result of blunt force trauma under stage 1 conditions. Radiating fractures are unlikely to occur in dry stage 2 bone (Byers, 2002: 270) or in fossilized bone.

Depressed fractures are 3.5 times more common in stage 1 children than in stage 1 adults (Zimmerman and Bilaniuk, 1981). Even though an immature cranium is more flexible than an adult's, the absolute thinness of the cranial vault bones makes immature crania more prone to fracture. Sometimes depressed fractures occurring in young individuals do not break through both inner and outer bony tables of the vault but may simply dimple the surface; this stage 1 phenomenon is known as a *ping-pong fracture* because similar depressions occur on ping-pong balls. Among the immature fossil crania we examined through photographs, Qafzeh 11 shows a clear depressed fracture on the frontal just above the left orbit (Tillier, 1999), but this does not appear to be a ping-pong fracture and the individual is older than those who typically incur such fractures. Close-up photographs in Tillier (1999) support her suggestion that some healing had occurred at the time of death, proving that this fracture was antemortem by at least a few weeks.

(4) *Stellate fractures* are a set of linear fractures radiating in a star-shaped pattern from a single point where impact occurred. Gurdjian (1975) found that stellate fractures are typical of heavy loads of relatively low velocity on stage 1 crania and are somewhat more common on upper parietals than elsewhere. Where stellate fractures are centered on a depressed fracture, they are functionally identical to radiating fractures.

Although radiating fractures do not usually occur later than stage 1, we observed stellate fractures without depressed fractures on crania in stages 2 and 3. We inferred that these were caused by the slow crushing or flattening of curved bones or parts of bones probably under sedimentary load. Examples of stellate fractures centered at inion but not involving a depressed fracture, which probably occurred in stage 2 or 3, can be seen on the immature fossil crania from Engis, Pech de l'Azé, or Roc de Marsal.

(5) *Communited fractures* of the vault involve large numbers of small fragments, usually produced by low velocity/heavy impact force. Crushing incidents are one common cause of comminuted fractures of fresh (stage 1) crania. We suggest that comminuted fractures may also result from sedimentary pressures acting on stage 2 or 3 (dry or fossilized) crania. The fossil crania Dederiyeh 1 and 2, Qafzeh 11, and KNM-WT 15000, among others, exemplify an overall comminution of the cranial vault (which we call a mosaic fracture pattern, see below). We discuss below ways in which stage 1 and stage 2 comminuted fractures may be distinguished.

(6) The *tripod* or *zygomatic-maxillary fracture* is one of the most common cranial fractures observed in medical and forensic circumstances (Rogers, 1992; Richardson, 2000). Frequently caused by a blow to the malar eminence, the tripod fracture separates the zygomatic bone from the rest of the cranium by breaks in the zygomatic arch, at or near the zygomatic-maxillary suture, and at or near the zygomatic-frontal suture (Rogers, 1992).

In addition to the six types of cranial fracture described above, forensic experts distinguish three combined fracture patterns involving the face called LeFort fractures (see Figure 4) (Galloway, 1999; Byers, 2002). These fractures may occur in combination as well as separately. A *LeFort I fracture* is an approximately horizontal break above the alveolar processes of the maxillae and below the nasal aperture. The typical cause of a LeFort I fracture is a blow to the lower face from the front or side. A *LeFort II fracture* isolates the midface from the vault, with breakage passing through the maxilla, the infraorbital foramen, and nasion; these fractures typically result from a blow to the midface at midline. A *LeFort III*

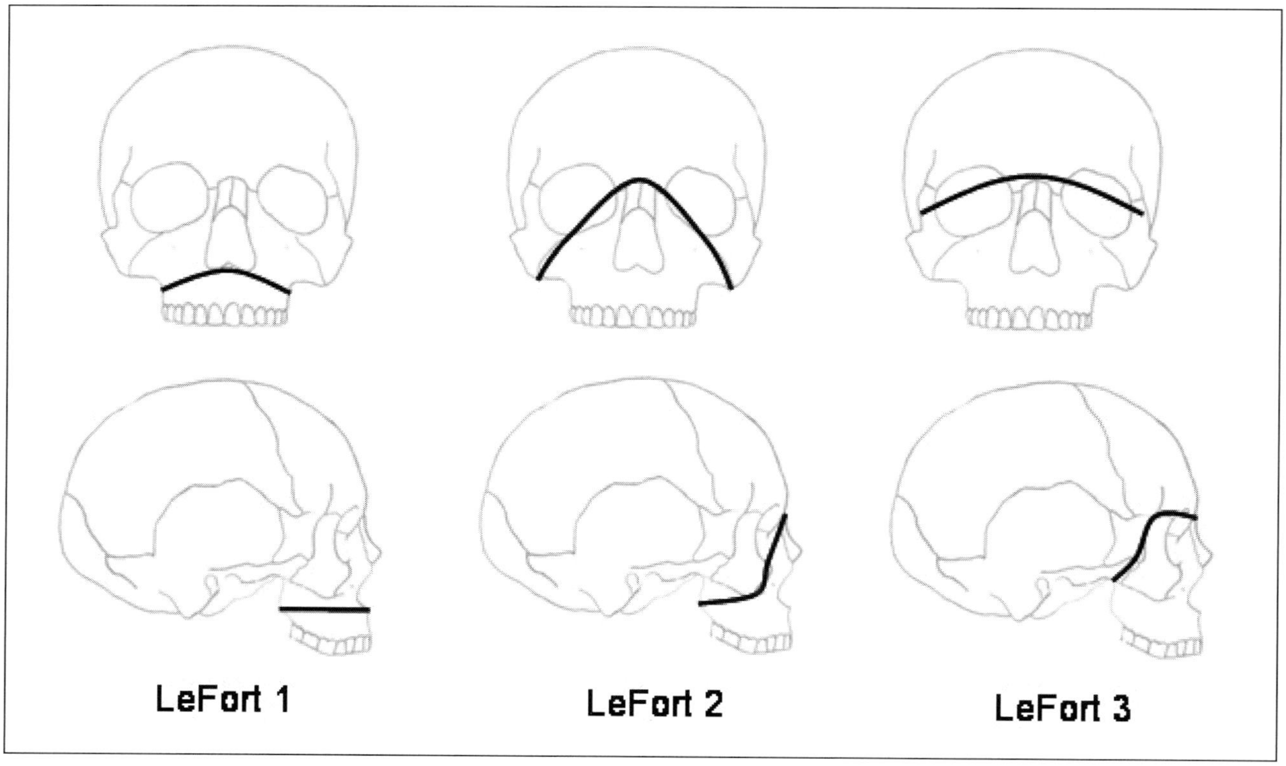


*Figure 4: Illustration of LeFort fractures (after Byers, 2002, and Galloway, 1999).*

*fracture* passes through nasion, through the bony orbits below the superciliary ridge, back through the sphenoid and zygomatic arch. This fracture separates the cranium into a skullcap and a facial portion and occurs in forensic circumstances when a blow is struck to the upper central part of the face.

In the Krovitz sample, 9% of the crania showed breaks that appeared to be LeFort III fractures, i.e., the specimens were neurocrania without faces. Eight percent of the SHARP sample showed LeFort III fractures, as judged by the simultaneous absence of right and left maxillae, zygomatics, and palatines. However, because we did not have data on isolated cranial bones (those not associated with skeletons) in the SHARP sample, the actual frequency of LeFort III fractures may have been higher.

Breakage that appears as a LeFort III fracture seems to be relatively common among fossil hominids. Numerous adult crania of *Homo erectus* show this pattern as do immature crania including Engis 2, Subalyuk 2, Mojokerto 1, and Skhul 1. We hypothesize that during burial and taphonomic destruction, crania with LeFort I and II fractures (or other facial damage) are so weakened that they tend to deteriorate further and present as LeFort III fractures by the time of excavation and study. It is uncommon to find fossil crania with LeFort I or II types of damage.

Finally, *sutural separations* are another type of cranial "damage" that is commonly observed in immature crania. A sutural separation involves the falling apart of an unfused or open suture due to the loss or degradation of the collagen component of bone tissue and of the sutural membranes themselves. Although diastatic fractures are breaks that follow the course of a suture, they differ from sutural separations in that the application of force is necessary to produce a diastatic fracture; while sutural separations occur naturally as a by-product of decompositional changes in an immature cranium. Sutural separations occur in stage 2 rather than stage 1 unless there is some processing of the fresh cranium (such as cooking or chemical treatment) that destroys or degrades the organic component of the bone. Sutural separations yield cranial fragments that may be entirely unbroken but which are separated from the rest of the cranium. The vast majority of crania from individuals who die under the age of one year (and many older immature individuals as well) will come apart along sutural lines once the collagen and membrane at the sutures has broken down (i.e., once stage 2 is reached). However, since the exposed sutural edges involve many small, protruding points that are vulnerable to further breakage, not all cranial bones isolated by sutural separation will be complete. Close inspection of the sutural edge should reveal whether the breakage is a single diastatic fracture or a series of fractures subsequent to a sutural separation.

The Taung 1 cranium shows a separation of the coronal suture, with subsequent minor breakage of the frontal. Gibraltar 2, the Devil's Tower Neandertal, also experienced a separation of the coronal, sagittal, and temporal sutures; subsequent to that separation, there has been some breakage of the exposed sutural edges. The Amud 7 specimen preserves the occipital bone of a young Neandertal which was isolated by sutural separation; the rest of the cranium was not recovered.

In summary, six types of fracture, three patterns of facial fracture, and sutural separations are distinguished in the forensic and medical literature. Many but not all of these were recognized in the archaeological or fossil samples we surveyed.

## Influence of chronological stage on fracture morphology

### *Stage 1*

To assess stage 1 fractures, we turned again to the medical and forensic literature. Richardson (2000), a radiologist, reports the frequency of different fractures for a hospital population. Less than 10% of the injuries in his sample were incurred by children. Most probably, his sample was biased toward adults and toward young males, since in many medical reports young males are found to sustain a higher frequency of facial fractures than other age and sex categories (Barker et al., 2003). In Richardson's sample, automobile accidents and assaults were the two most common causes of the facial fractures.

Richardson found that the most common type of midfacial fracture is a tripod fracture of the zygomatic and maxilla (40%). LeFort I fractures comprise 15%; LeFort II, III, simple zygomatic arch fractures, and comminuted fractures each comprise another 10%; and alveolar fractures, often associated with fractured teeth, make up the last 5%. Richardson points out that 60-70% of all facial fractures (the tripod fractures plus LeFort II and III) involve the orbit.

Any of the cranial fractures described above may occur on stage 1 crania, with the exception of sutural separations (see discussion on sutural separations above). Since most stage 1 crania do not sustain trauma prior to death, Richardson's sample is not expected to parallel the incidence of facial fractures in a general cemetery population. Prior to the invention of mechanized forms of transport, an individual cranium probably experienced few breaks during stage 1 except in cases of attack, warfare, or a major accident. Even then, the point or points of impact may be evident. The effects and sequence of a series of blows or impacts can be deduced by a skilled examiner in forensic or archaeological cases (e.g., Sauer, 1998; Marks et al., 1999).

The morphology of fractures is helpful in deducing the chronological stage of the cranium at the time of breakage. Linear fractures in stage 1 crania are usually long and curved rather than straight; their course relates more to the direction and magnitude of force applied than to the geometry of the cranium per se. The outline of a depressed fracture is also often curved into a rounded or oval shape, reflecting the shape of the object which impacted the cranium. Concentric rings of fracture may encircle a depressed fracture. Stage 1 fractures of crania are often beveled, with the direction of the bevel indicating the direction of movement of the force causing the fracture (Berryman and Symes, 1998).

Burial per se may protect the cranium from further damage by many taphonomic agents. Although crania buried in stage 1 may undergo non-traumatic sutural separations, especially on very young (<1 year old) individuals, it is important to remember that these sutural separations do not normally occur at the time of burial but after burial, during stage 2, when the organic components of the cranial bone have decayed. Stage 1 crania do not generally show sutural separations unless there is some processing of the fresh cranium that destroys or degrades the organic component of the bone.

In summary, stage 1 fractures are usually few in number per specimen, except in cases of warfare, violent attack, or accidents. Stage 1 depressed fractures frequently show a rounded outline that reflects the shape of the object responsible for the damage; depressed fractures often show hinged pieces or retention of the pushed-in fragments. Linear fractures occur and are elongated and curving in course; where they intersect sutures, the course may change abruptly to follow the plane of weakness represented by the suture, becoming a diastatic fracture. Stage 1 breaks are often beveled and the fracture surface itself will show irregularities caused by microscopic variations in the amount of elastic tissue in the bone. The fracture itself is sharp-edged and appears crisp and cleanly defined. Sutural separations are generally absent.

### *Stage 2*

Breakage occurring after the bony tissue has dried and its organic component has decayed differs from that in stage 1 because the material properties of the bone tissue have changed (as discussed above). Generally, the edges of fragments fractured during stage 2 are flatter and more planar than those resulting from stage 1 fractures, indicating that the bone is responding more uniformly to pressure since it is no longer elastic. Stage 2 fractures of the cranial vault may or may not be beveled.

Synthesizing our original observations of immature archaeological and fossil crania with those reported in the literature, we observed meaningful differences in the general frequency of the different fracture types on crania in different temporal stages. Long, linear fractures with a curving course across the cranial vault are rare in crania that were dry (stage 2 or 3) when broken, as are true diastatic fractures. In contrast, sutural separations commonly occur in immature crania during stage 2, judging from the Krovitz sample. Fully or partially separated sutures, especially interdigitated ones, are very vulnerable to further breakage because of the irregularity of the exposed edge. One common consequence of this post-separation breakage is the loss of an angular piece of vault bone at bregma, as in the specimen from Le Figuier or Qafzeh 11. Another is the loss of small fragments along the course of separated coronal, sagittal, or lambdoid sutures, such as in the Mojokerto specimen. The lack of a linear course in such sutural fragments distinguishes them from true diastatic fractures.

Dry crania (stage 2) show fewer discrete depressed fractures than wet ones (stage 1). Oval or curving holes where pieces are missing—holes which might represent depressed fractures that have lost their small fragments—are uncommon in stage 2 breakage. An exemplar might be the rounded hole on the right side of the Roc de Marsal cranium. However, if this was originally a depressed fracture, there is no hinging nor are the depressed fragments still attached. Holes in stage 2 are much more commonly geometric or angular in outline. Comminuted fractures in stage 2 crania may separate into fragments. Their recovery and recognition depends largely on excavation and curation techniques. Cases where large numbers of fragments have been reassembled into partial fossil crania include Dederiyeh 1 and 2, KNM-WT 15000, and Herto BOU-VP-16/5.

A comparison of the frequency of various fracture types in stage 1 and stage 2 crania is instructive. It is important to remember, however, that breaks occurring in stage 1 crania are likely to weaken the specimen and make it more vulnerable to further breakage. This means that the initial fractures may be obscured by later damage, and if stage 1 damage is significant, the specimens may not survive to be examined as archaeological (stage 2) or fossil (stage 3) specimens.

The frequency of tripod and zygomatic arch fractures (together 50%) in stage 1 crania reported by Richardson matches closely with the high frequency of fractures to the zygomatic arch, 40-47% across all age groups in the Krovitz stage 2 crania (Table 2a). The zygomatic fractures observed in the Krovitz sample were not sutural separations but fractures of the zygomatic arch. In the SHARP sample, 27% of the individuals showed zygomatic breakage and another 37% were missing the zygomatic bones altogether.

In contrast, the high frequency of orbital fractures (60-70%) cited by Richardson is not paralleled by the frequency of orbital fractures in stage 2 crania. Richardson does not specify where the orbits are fractured, simply that fractures involving the orbit are very common in patients seeking medical attention for cranial trauma. In the Krovitz sample, four landmarks (ORB, FZJ, ZYS, and FMO) reflect damage to the superior, lateral, inferior, and medial parts of the orbital rim respectively. Three of these landmarks (ORB, FZJ, and FMO) are so rarely missing in the Krovitz sample that they are in the low taphonomic vulnerability category. The fourth landmark, ZYS, falls into the intermediate taphonomic vulnerability category, being missing in 18% of the 272 specimens in the landmark sample. Clearly, then, either orbital fractures are substantially less common in stage 2 than in stage 1 immature crania or, because of the important structural role the frontal bone plays in protecting crania from breakage (as discussed above), crania with frontal or orbital fractures that occurred during stage 1 did not survive to be inventoried in stage 2.

This conclusion is supported by the SHARP sample data which do not show high levels of breakage on bones involved in the orbital rim. The frequency of orbital fractures in this sample can be assessed only from data on the completeness of the frontal and maxilla. Twenty-two percent of the SHARP individuals show breakage to some part of the frontal bone and an additional 11% are missing their frontal bones. Twenty-seven percent of the SHARP individuals show maxillary breakage, with an additional 26% percent missing the maxillae entirely.

Pure LeFort I fractures, which comprised 15% of Richardson's sample, were not observed in the Krovitz sample and LeFort II fractures appeared to be rare, suggesting that both types probably progressed to LeFort III fractures by the time they were observed in archaeological or fossil samples. In other words, specimens missing the landmarks near the maxillary teeth were invariably also missing much more of the face. LeFort III fractures occurred in 9% of the Krovitz specimens versus 10% in Richardson's sample. Our inventory data show that the nasal sutures and the zygomatic arches are especially likely to open or break in dry immature crania. Loss of both of these regions would encourage the separation of the face and vault in a LeFort III pattern. The number of LeFort I and II fractures could not be readily estimated from the SHARP data. LeFort III fractures, as judged by the simultaneous absence of maxillae, palatines, and zygomatics, apparently occurred in 8.5% (N=7 out of 82) of the SHARP sample. Among the immature fossilized crania, there was only one LeFort I fracture (Herto BOU-VP-16/5). LeFort II or III fractures occurred in 8 of 20 (40%) of the specimens. Because the original specimens were not examined in most cases and the sample size is small, we do not know if this difference between the archaeological and fossil samples is meaningful.

In the Krovitz sample, common sutural separations in stage 2 crania occur at the basilar suture (37-64% of the specimens according to age) and, less frequently, along the lambdoid suture. Separation along the basilar suture contributes to further breakage of the basicranium; separation along the basilar and lambdoid sutures results in the isolation of the occipital from the rest of the cranium. Separation at the coronal suture occurred, separating the face and frontal bone from the rest of the cranium in 38 specimens (14% of the Krovitz sample); note that in these individuals the rest of the cranium did not survive intact once the face and frontal bones separated. Other observed (though not quantified) sutural separations include the zygomatic-temporal suture (contributing to zygomatic arch breakage), zygomatic-maxillary suture (contributing to loss of the maxilla or zygomatic), and springing out of the beveled parietal-temporal suture (contributing to loss of the temporal bone, especially if the occipital bone is also missing).

We observed two new forms of linear fracture regularly in stage 2 crania although we did not quantify their occurrence. We call the first of these a *temporal line fracture*. This fracture differs from an elongated, curving linear fracture by its anatomical placement. A true linear fracture occurring on a stage 1 cranium curves across the vault of the cranium with no characteristic placement. A temporal line fracture is found on one or both sides of a cranium or skullcap that has been subjected to a compressive force after burial and drying. Dry crania can be considered ovoids that may be hollow, incompletely filled with sediments, or filled with unconsolidated sediments. Slow compression, such as the weight of accumulating sediments overlying the cranium, will tend to flatten the cranial vault from side to side. Such compression will produce a linear fracture that follows the approximate course of the temporal line on one or both sides of the specimen. It is not the location of this muscle marking but the more acute curvature of the cranial vault in this region that renders it especially vulnerable to fracture. This acuteness of curvature is especially evident in posterior view, where modern human crania show parietal bosses or the typical "en maison" shape. Temporal line fractures typically pass from the superior margin of the orbit (or from the coronal suture) through the parietal and stop when the fracture encounters the lambdoid suture, as in the La Quina H18 and Teshik Tash specimens.

The second new type of fracture we observed in stage 2 crania is a *perpendicular fracture* in our terminology. These fractures run perpendicular to the sagittal suture inferiorly from that suture until they encounter the temporal suture or a temporal line fracture. Single specimens often show two or more perpendicular fractures, which are the natural result of diffuse pressure applied to the ovoid cranium. Perpendicular fractures are common in archaeological specimens (G. Milner, 2004, personal communication to G.K. and P.S.). Longitudinal bending stress tends to flatten the curvature toward the front and back of the cranium, causing the perpendicular fractures, as in the Teshik Tash, Engis 2, Grotte des Enfants 6, and Skhul 1 specimens, among others.

The intersection of temporal line and perpendicular fractures effectively breaks the parietal into large, roughly rectangular or trapezoidal fragments. We hypothesize that continued sedimentary pressure (or some other type of diffuse compressive load) will break these rectangular fragments further until most or all of the cranial vault surface is broken into triangular or irregularly geometric fragments in what we call a *mosaic fracture pattern.* This pattern differs from a comminuted fracture in that the mosaic fracture pattern covers a large area of the cranial vault and has no clear outline or point of impact. The mosaic fracture pattern can be observed in KNM-WT 15000, Mojokerto, Herto BOU-VP-16/5, Dederiyeh 1 and 2, Subalyuk 2, and Qafzeh 10 and 11.

Another new type of fracture we saw in stage 2 crania is the *metopic* or *parametopic fracture.* Younger individuals, under the age of about four years, may show a vertical fracture of the frontal bone either along the metopic suture or parallel to it. Although Cobain et al. (2002) report that the metopic suture is fused in most individuals by the time of birth, the site of the former suture may be weaker than the adjacent bone over the orbits, which is reinforced by the superciliary ridge. Additionally, the cause might be due to the change in curvature which becomes more acute at or near the midline of the frontal bone, thus rendering this region especially vulnerable to breakage. The Le Figuier and Qafzeh 10 specimens show metopic or parametopic breaks, which are distinct from patent metopic sutures such as in Pech de l'Azé.

During our reading of descriptions of fossil crania, we noted a disturbing tendency for any fracture along the midline of the frontal bone to be described as a patent metopic suture, even when the edge of the break was planar and not interdigitated. Similarly, missing fragments at bregma were sometimes labeled as patent anterior fontanelles without anatomical evidence. Great caution is needed in concluding that missing bone indicates a natural anatomical consequence of immaturity rather than breakage.

A *stellate fracture* of the squamous occipital, centering on inion, occurred in a number of fossil crania although the bone tissue is thicker at inion. This pattern of breakage is common on stage 2 specimens. The cause of such breaks would seem to be the geometry of the occipital, which is effectively a very blunt cone the point of which is at inion. Virtually any diffuse pressure on such a structure will tend to flatten the cone, producing a stellate fracture at inion, as in Engis 2 and Amud 7.

We may make some broad generalizations in comparing stage 1 and stage 2 fractures. Breakage in stage 2 typically involves the orbit much less often than in stage 1, and in stage 2 the lateral or inferior orbital margins are more often damaged than the superior margin. Temporal line, perpendicular, and metopic/parametopic fractures are typical stage 2 breaks. Sedimentary pressure during stage 2 may cause a widespread mosaic fracture pattern comprised of numerous geometric fragments lacking a discrete area of impact, as distinct from a circumscribed area with a comminuted or depressed fracture, which is more typical of stage 1. The surfaces of stage 2 fractures are usually planar with blunt edges lacking beveling or hinging. The course of stage 2 fractures is not curved or rounded. Immature crania damaged during stage 2 sustain more breaks per specimen and frequently show nontraumatic sutural separations. Overall, stage 2 breaks rarely show a discrete point of impact and instead result from more diffuse pressure.

### *Stage 3*

Post-fossilization breakage occurs after the bone has been mineralized. Though rarely of concern in forensics, post-fossilization breakage is important to paleoanthropologists as it may yield clues to the circumstances under which a hominid died and became buried in sediments. In this stage, the cranium acts very much like a ceramic vessel of similar shape. The primary influences on breakage of a stage 3 cranium seem to be the geometry of the cranium or parts of the cranium in question, as in stage 2, and the lack of bony elasticity. Bone density per se is a less important issue in post-fossilization breakage than in earlier breakage because stage 3 bone tissue is mineralized.

Determining the timing of postmortem fractures relative to the time of death can be difficult; we are not confident of our ability to distinguish between stage 2 and stage 3 breakage in most cases. Stage 3 breakage may reveal internal surfaces of the fossil that are different (often lighter) in color than those that have been exposed longer; thus fresh breaks on fossils are often readily recognizable. Some fossil specimens we examined showed mosaic fracture patterns generally similar to those produced in stage 2 with many discrete, short, and straight fractures yielding numerous geometric fragments. These specimens generally lacked either a long temporal fracture or perpendicular fracture, which we believe are more common in stage 2 crania. We hypothesize that crania subjected primarily to stage 3 damage are more likely to show an overall diffuse shattering rather than the creation of longer, more linear fractures along geometric planes of weakness. As a subjective impression, we believe that the average fragment size is somewhat smaller in stage 3 mosaic fracture patterns than in stage 2 patterns, but it will be difficult to test this because we cannot reliably separate stage 2 and 3 breakage. Fragment edges, where visible, are generally planar in both stage 2 and 3 mosaic fracture patterns.

Another taphonomic factor that has a considerable influence upon the breakage of stage 3 crania is the presence or absence of a consolidated natural endocast that was formed while the cranium or a portion of the cranium was still intact. A cranium that is largely intact when it becomes a sedimentary particle in an open air or cave site is likely to become filled with sediment which will eventually harden into an endocast. At least some deliberately buried crania also develop natural endocasts (i.e., Skhul 1, see McCown and Keith 1939: 299-301). Because these sedimentary infillings obscure the internal surfaces of the fossilized bones, they are usually removed during preparation. The exceptions are cases where the sedimentary infilling (endocast) preserves an impression of bones which were not recovered or where the endocast separates naturally from the fossilized bone without causing damage, as in Taung 1.

We have few examples to generalize from in which endocasts have been recovered intact or where their extent and placement at the time of discovery is documented. However it is intuitively obvious that the presence and nature of a consolidated natural endocast effectively transforms a fossilized cranium from a hollow ovoid, structurally similar to a ceramic vessel in terms of potential for breakage, into a solid comprised of a dense center (the endocast) and a relatively thin outer covering (the fossilized bone). We suggest that a consolidated, natural endocast will not necessarily prevent fragmentation of a fossilized cranium as pressure is exerted but will act to keep fragments together and in or near their original anatomical position. Although a fossilized cranium with a complete endocast may be more vulnerable to breakage than the surrounding rock, the endocast helps the cranium resist flattening with the result that the vault bone shatters but the pieces are not destroyed. A partial endocast, as in Taung 1 or Skhul 1, obviously leaves the portion of the cranium without the endocast extremely vulnerable to fragmentation and loss.

## Three Case Studies

### Taung 1

Taung 1, the type specimen of *Australopithecus africanus,* was collected during quarrying of a South African limestone cave and was recognized by Raymond Dart as a previously unknown species of hominid (Dart, 1925). The specimen is that of a young individual about 3-4 years old (Bromage, 1985); only the partial cranium and mandible have been recovered. There are no other hominid remains to date from the site. Subsequent studies of the remaining portions of the Taung deposits and similar caves nearby suggest a complex taphonomic history for the bones preserved in those caves. It is most probable that the Taung 1 skull was washed or dropped into a cave in the tufa by a leopard or other mammalian predator, while the skull was both fresh and fleshed (Brain, 1981; McKee and Tobias, 1994; McKee, 2001; McKee, 2004, personal communication to P.S.). An alternative interpretation made by Berger and Clarke (1995), that the Taung individual was preyed upon by a large avian raptor, is less likely (McKee 2001). Whatever the precise cause of death, the skull became a sedimentary particle while it was largely or completely intact and the mandible was attached to the cranium by soft tissue.

The Taung skull is one of those rare specimens with a natural endocast, which preserves the impression of the entire right side of the cranial vault and occiput although those parts of the cranium are now missing. The endocast did not fill the skull completely and does not preserve the left side of the cranium, which was not recovered. The face, frontal bone, and mandible of the individual were intact and in anatomical position when found. Because the venous markings and the sulci and gyri of the interior surface of the right side of the braincase are clearly preserved on the endocast, it is obvious that the skeletal elements of this side of the cranium were also present in the rock. The vault fragments from the right side and occiput were either destroyed by the blast that exposed the skull or were not collected by the workmen who retrieved the skull.

Remarkably, the face and frontal bone show no weathering and no fractures; very fragile regions of the skull with high taphonomic vulnerability are preserved (such as the nasal bones and zygomatic arch). The coronal suture has separated neatly and there is only minimal additional breakage on the frontal. The right zygomatic arch is intact. The mandible was found in place, attached to the maxilla by sediment (Dart, 1925). This is strong evidence that the cranium and mandible were deposited shortly after death while the bone was in stage 1 condition and held together by soft tissue. Burial in alkaline sediments and the partial infilling of the skull protected it from further damage until the specimen was exposed during a mining operation.

There is one area where a small fragment, probably of parietal, was pushed into the then-unconsolidated endocast, to which the fragment still adheres although the surrounding bone is missing. This damage could not have occurred when the specimen was fresh or the fragment would have been pushed into the brain tissue and would not now adhere to the endocast (McKee, 2001). A pointed rock or other object probably caused this small fracture before the endocast was fully consolidated.

We conclude that the Taung skull came into the ancient tufa cave during stage 1 when its flesh was still intact, as has been proposed before (e.g., Brain, 1981; McKee and Tobias, 1994). A natural endocast was formed. The specimen was subjected to minimal taphonomic destruction thereafter except for sutural separation during stage 2. Possible crushing or destruction of left side of cranium occurred during late stage 2 or early stage 3, after drying of the cranium and before consolidation of the endocast.

## Mojokerto (Perning 1)

Mojokerto is an immature *Homo erectus* specimen (Anton, 1997), approximately 4-6 years old, that was discovered in Java in 1936 by Andoyo, an Indonesian geological assistant (Duyfjes, 1936; von Koenigswald, 1936a, b). The specimen was deposited in fluvial sediments (Huffman, 2001; Huffman and Zaim, 2003). To our knowledge, no one has attempted to reconstruct the taphonomic history of the Mojokerto skull as a bony specimen, although its taphonomic history as a geological and sedimentary particle has been discussed (Huffman and Zaim, 2003).

The Mojokerto cranium appears to have suffered a LeFort III fracture; the face and much of the basicranium of the specimen was lost or destroyed. Transport of the specimen after this fracture may have occurred but was probably not extensive, judging from the preservation of fragile edges of the broken right parietal, the occipital, and the frontal where it articulates with the ethmoid. Von Koenigswald (1936a) perhaps overstated the fragility of the parts that remain intact, writing: "It is in fact a miracle that such a fragile object has been so well preserved under these circumstances." Later he wrote (1937: 25): "we are certain that it [the cranium] was found in situ, because the bone is so thin that it would have been destroyed by any movement or rewashing."

The Mojokerto cranial vault is broken into many angular fragments. A piece of the frontal is missing at midline, and a fracture which runs from the edge of the missing section to bregma suggests that there was probably a metopic fracture. Small pieces of bone are missing at bregma and at various points along the coronal, sagittal, and occipital sutures. However, the main cranial sutures did not separate. Most of the small fractures along the sutures have beveled edges with the inner table being more extensive; they may represent bending and fracturing of the specimen in situ that caused small fragments to separate from the cranium. Alternatively, it is possible that this damage occurred during excavation or preparation, procedures which are not well documented. The cranium shows a possible temporal line fracture, several perpendicular fractures on the cranial vault and a stellate fracture at inion. A number of fragments of the cranial vault have beveled edges, suggesting that these fractures occurred before all organic tissue and flexibility of the bones was lost. Most of the occipital portion of the basicranium is missing although the (damaged) petrous portions of both temporals are preserved.

Once buried, the Mojokerto skullcap filled with sediment which became a natural endocast. Venous markings are visible on the better preserved (and exposed) left side of this endocast, showing that additional vault fragments were present in the rock. If these fragments survived until the moment of discovery, they were unfortunately not collected. It is important to note that the fossil was collected as an aid in geologic mapping and biostratigraphy, not for paleontological studies (Duyfjes, 1936). An alternative interpretation is based upon the fact that at least one credible report of the discovery of the cranium mentions that there were fossil fragments lying on the surface, which prompted Andoyo to excavate there and discover the cranium (Duyfjes, 1936). Possibly the now-missing fragments of the left side of the cranial vault were the surface fragments seen by Andoyo and presumably judged too small to be useful. If so, the discovery occurred after these pieces became separated from the rest of the fossilized specimen but before weathering and erosion could destroy the impression of interior surface of the parietal and temporal fragments on the endocast.

Several pieces of bone from the right side of the vault and from the occipital bone are pushed sharply into the endocast, which is not complete in this area, and there are sizeable areas where there is no preserved bone at all but only endocast. The placement of bevels and pushed-in fragments suggests that, prior to the complete consolidation of the endocast, sedimentary pressure produced numerous fractures and forced some of the resulting fragments inward.

The Mojokerto cranium is subtly but markedly deformed (Anton, 2003, pers. comm. to P.S.); symmetry could not be restored even if all of the pieces were separated from the matrix endocast. The remaining portion of the left temporal, bearing the zygomatic process, has been moved in an anterior direction and rotated in a clockwise direction from lateral view. It is possible that the plastic deformation and warping of the specimen occurred in stage 1, while the bone was still somewhat elastic. However, we cannot judge with certainty when in the taphonomic history of the specimen this plastic deformation and warping of the bony tissue occurred, since sedimentary pressures are capable of warping consolidated rock.

Fractures of the Mojokerto cranium typical of postmortem stage 2 include: the deduced metopic fracture, the perpendicular fractures, and the mosaic fracture pattern of the cranial vault. The number of angular fragments are neither as numerous nor as small as those in, for example, Herto BOU-VP-16/5 (see below). The Mojokerto specimen also shows various fragments of bone pushed into the endocast, a possible temporal line fracture, and a rotation of the temporal, all of which must have occurred after the bone had dried but before the endocast was consolidated.

Many areas of the cranium in the high and intermediate taphonomic vulnerability groups have been broken or are missing in this specimen: the entire face, the zygomatic arches, and the basilar occipital. Much of the squamous occipital is preserved, as is most of the neurocranium. The superior margin of the right orbit is broken, even though this is an area of the cranium most likely to be intact in archaeological specimens (see discussion above).

In summary, the Mojokerto cranium was probably subjected to late stage 1–early stage 2 breakage. Damage most likely occurred after sedimentary burial but before consolidation of the endocast and while the cranial bones were sufficiently elastic to warp and deform as well as break with beveled edges. Without the natural endocast, it seems likely that many of the individual fragments would have separated from one another along fracture lines. There is little or no evidence of separation along sutures. The face was broken off in a LeFort III pattern. The preservation of this cranium suggests exposure to moderate taphonomic destruction.

## Herto BOU-VP-16/5

Herto BOU-VP-16/5, an immature cranium of *Homo sapiens idaltu,* was discovered in 1997 in the Herto Bouri region of Ethiopia (Clark et al., 2003; White et al., 2003). The specimen was recovered in over 180 pieces, which were found on the surface after eroding out of an indurated sandstone. In the view of the discoverers, the cranium was modified and curated by hominids after the death of the individual (Clark et al., 2003; White et al., 2003). The pieces of the Herto cranium are numerous, angular, and appear to be planar on the edges (White et al., 2003; White, 2003, personal communication to P.S.). The middle part of the face is missing but the maxillary alveoli are preserved; this is the only example of a LeFort I fracture we observed among the fossil crania. Much of the basicranium is damaged or missing with the exception of the petrous temporals. Metopic or parametopic, temporal line, and perpendicular fractures are absent as are elongated linear and diastatic fractures. Fractures of the cranial vault are numerous, short, and straight, but do not follow the sutures. Thus the entire neurocranium is comprised of angular fragments in a mosaic fracture pattern. While the coronal, sagittal, and lambdoid sutures have not separated, the temporal sutures apparently opened and portions of the temporal bones were not recovered. Only the squamous occipital is preserved; the basicranial part of the occipital and the sphenoid are both missing. These features might be expected in either in dry-bone damage or in post-fossilization fractures (stages 2 or 3).

White (2004, personal communication to P.S.) concluded that the fractures were primarily or wholly post-fossilization based on three observations. First, matrix-filled cracks between pieces and ectocranial matrix that bridged adjacent pieces indicate that the cranium was embedded whole. Second, the filling of various voids (such as sinuses, diploe spaces, etc.) with matrix shows in many cases that anatomically adjacent fragments were in place when the matrix hardened. Finally, there were plant rootcasts on the endocranial and ectocranial surfaces but none on the fracture surfaces, showing that breakage occurred well after sedimentary burial and probably after erosional exposure.

The Herto BOU-VP-16/5 cranium shows defleshing cutmarks around the perimeter of the glenoid fossa and polishing of the broken edges of the occipital and temporal bones. These alterations are taken as evidence of postmortem treatment of the cranium by hominids, perhaps as part of a mortuary ritual (Clark et al., 2003: 751). This damage was most probably inflicted during stages 1 (the defleshing) and 2 (the polishing of broken edges).

Despite the extensive fragmentation of Herto BOU-VP-16/5, which bespeaks intense exposure to taphonomic agents of destruction, large parts of the fragile facial bones are preserved. Both nasals are present; the left orbital rim is intact as is most of the left zygomatic arch; substantial parts of both maxillae are present. The survival of some (but not all) of the elements in the most taphonomically vulnerable category combined with extensive fragmentation suggests that breakage occurred after fossilization had enhanced the structural strength of elements that are fragile in stages 1 and 2. The presence of a LeFort I type fracture is very rare in immature fossils. If the specimen had been subjected to more extensive taphonomic destruction, the LeFort I fracture would have probably progressed to a LeFort III fracture. If efforts to recover fragmentary pieces of the cranium had been less intensive, the specimen might well appear to have had a LeFort III fracture.

We find no evidence that would lead us to question the interpretation that the cranium was defleshed and curated (during stage 1), resulting in polishing of edges around the broken-out basicranium (probably during stage 2). From the observations and data presented above, we deduce that most of the mosaic fragmentation and fracturing of the vault and face of Herto BOU-VP-16/5 occurred during stage 3, the post-fossilization period (White et al., 2003).

# Conclusions

We have summarized and integrated quantitative and qualitative data from medical, forensic, archaeological, and paleontological sources in an attempt to characterize the taphonomic attributes of immature hominid crania. From these diverse observations, we have created a set of expectations that relate fracture patterns to taphonomic vulnerability and that describe fracture morphology and placement in relation to the time of breakage relative to the death of the individual. Data on the breakage and preservation of individuals from the Krovitz and Sedgeford samples have been used to identify key differences in breakage between immature and adult crania, respectively.

We tried to show how these expectations might be used in practical terms by re-analyzing three immature fossil crania from Taung, Mojokerto, and Herto. We regard the work reported here as a first approximation and still speculative. We encourage further research along these lines in order to produce more refined and useful diagnostic tools for the taphonomist, paleontologist, and forensic anthropologist.

## Acknowledgements

We wish to gratefully acknowledge receiving information and assistance from Susan Anton, Berhane Asfaw, Frank Huffman, Patricia Reid, Alan Mann, Jeffrey McKee, George Milner, Erik Trinkaus, and Tim White on this paper. G.K. would like to thank those who gave her access to the skeletal collections in their care, or who helped her obtain or interpret the dental x-rays for the recent human samples: Hisao Baba, Pia Bennike, Jodie Blodgett, Luca Bondioli, Jennifer Clark, Kevin Conley, Jean-Marie Cordy, Kate Hesseldenz, Louise Humphrey, David Hunt, Robert Kruszynski, Helen Liversidge, Niels Lynnerup, Roberto Macchiarelli, Giorgio Manzi, Yuji Mizoguchi, Theya Molleson, Søren Nørby, Nancy O'Malley, Rosine Orban, Doug Owsley, Ildiko Pap, Rick Potts, Edourard Poty, Mary Powell, Patrick Semal, Ib Sewerin, Gabriella Spedini, Chris Stringer, and Erik Trinkaus. G.K.'s research was supported by grants from the L.S.B. Leakey Society, the National Science Foundation, and the Japanese Society for the Promotion of Science. P. S. especially wishes to thank Bob Brain for being an inspiration throughout her entire career.

## References

Adams, M., Reeve, J. 1987. Excavations at Christ Church, Spitalfields 1984-6. Antiquity 61, 247-256.

Anton, S. 1997. Developmental age and taxonomic affinity of the Mojokerto child, Java, Indonesia. American Journal of Physical Anthropology 102, 497-514.

Akazawa, T., Muhesen, S., Dodo, Y, Kondo, O., Mizoguchi, Y. 1995. Neanderthal infant burial. Nature 377, 585-586.

Barker, R., Hockey, R., Spinks, D., Miles, E. 2003. Facial Injury. Injury Bulletin, Queensland Injury Surveillance Unit, no. 79 (September), p. 1.

Berger, L.R., Clarke, R.J. 1995. Eagle involvement in the accumulation of the Taung child fauna. Journal of Human Evolution 29, 275-299.

Berryman, H.E., Symes, S.A. 1998. Recognizing gunshot and blunt cranial trauma through fracture interpretation. In: Reichs, K.J. (Ed), Forensic Osteology, Advances in the Identification of Human Remains, 2nd Edition. Charles C. Thomas, Springfield (IL), pp. 333-352

Billy, G. 1979. L'enfant Magdalenien de la Grotte due Figuier. L'Anthropologie 83, 223-252.

Blount, W.P. 1955. Fractures in Children. Williams and Wilkins, Baltimore.

Blount, W.P. 1977. Fractures in Children. Robert E. Krieger Publishing Co, Huntington (NY).

Brain, C.K. 1981. The Hunters or the Hunted? An Introduction to African Cave Taphonomy. University of Chicago, Chicago.

Bromage, T.G. 1985. Taung facial remodeling: A growth and development study. In: Tobias, P.V. (Ed), Hominid Evolution: Past, Present and Future. Alan R. Liss, New York, pp. 239-245.

Byers, S.N. 2002. Introduction to Forensic Anthropology. Allyn and Bacon, Boston.

Clark, J.D., Yonas Beyene, Y., WoldeGabriel, G., Hart, W.K., Renne, P.R., Gilbert, H., Defleur, A., Suwa, G., Katoh, S., Ludwig, K.R., Boisserie, J.-R., Asfaw, B., White, T.D. 2003. Stratigraphic, chronological and behavioural contexts of Pleistocene Homo sapiens from Middle Awash, Ethiopia. Nature 423, 747 - 752.

Cobain, L.R., Nabipour, S., Richards, G. 2002. The role of sutures in frontofacial growth: evidence from the metopic suture. Paper delivered at the 80th General Session of the International Association for Dental Research, the American association for Dental Research, and the California Association for Dental Research, San Diego, CA.

Collins, H.B.J. 1937. Archaeology of St. Lawrence Island, Alaska. Smithsonian Miscellaneous Collections 96(1). Smithsonian Institution Press, Washington, DC.

Cox, T.C., Buchilz, D.J., Wolf, D.J. 1987. Blunt force trauma from police impact weapons: some skeletal and neuropsychological considerations. Journal of Police Science Administration 15, 56-62.

Currey, J. 1984. The Mechanical Adaptations of Bones. Princeton University, Princeton.

Dart, R.A. 1925. *Australopithecus africanus*: The man-ape of South Africa. Nature 115, 195-199

Dean, C., Leakey, M.G., Reid, D., Schrenk, F., Schwartz, G.T., Stringer, C., Walker, A. 2001. Growth processes in teeth distinguish modern humans from *Homo erectus* and earlier hominins. Nature 414, 628-631.

Dodo, Y., Kondo, O., Muhesen, S., Akazawa, T. 1998. Anatomy of the Neandertal infant skeleton from Dederiyeh Cave, Syria. In: Akazawa, T., Aoki, K., Bar-Yosef, O. (Eds), Neandertals and Modern Humans in Western Asia. Plenum Press, New York, pp. 323-338.

Duncan, C.C. 1993. Skull fractures in infancy and childhood. Pediatric Review 14, 389-390.

Duyfjes, J. 1936. Zur Geologie und Stratigraphie des Kendenggebietes zwischen Trinil und Soerabaja (Java). De Ingenieur in Nederlandsch-Indië, Mijnbouw & Geologie, DeMijningingenieur 4, 136-149.

Ferembach, D. 1970. Le crane de l'enfant du Pech-de-l'Azé. Archives de l'Institut de Paleontologie Humaine 33, 13-51.

Fraipont, C. 1936. Les hommes fossils d'Engis. Archives de l'Institut de Paleontologie Humaine 16, 1-52.

Galloway, A. 1999. Fracture patterns and skeletal morphology: Introduction and the Skull. In: Galloway, A. (Ed), Broken Bones: Anthropological Analysis of Blunt Force Trauma. Charles C. Thomas, Springfield (IL), , pp. 63-79.

Gordon, J.E. 1968. The New Science of Strong Materials or Why You Don't Fall Through the Floor. Penguin, Middlesex.

Gurdjian, E.S., Webster, J.E., Lissner, H.R. 1953. Observation on prediction of fracture site: Head injury. Radiology 60, 226-235.

Gurdjian, E.S. 1975. Impact Head Injury, Mechanistic, Clinical and Preventive Correlations. Charles C. Thomas, Springfield (IL).

Heathcote, G.M. 1986. Exploratory Human Craniometry of Recent Eskaleutian Regional Groups from the Western Arctic and Subarctic of North America. British Archaeological Reports International Series 301, Oxford.

Huffman, O.F., Zaim, Y. 2003. Mojokerto Delta, East Java; Paleoenvironment of Homo modjokertensis—first results. Journal of Mineral Technology 10, 1-9.

Huffman, O.F. 2001. Geologic context and age of the Perning/Mojokerto *Homo erectus*, East Java. Journal of Human Evolution 40, 353-362.

Ishida, H., Kondo, O., Muhesen, S., Akazawa, T. 2000. A new Neandertal child recovered at Dederiyeh Cave, Syria in 1997-1998. American Journal of Physical Anthropology 30, 186-187.

Krovitz, G.E. 2000. Three-dimensional comparisons of craniofacial morphology and growth patterns in Neandertals and modern humans. Ph.D. Dissertation, The Johns Hopkins University.

Lyman, R.L. 1994. Vertebrate Taphonomy. Cambridge University, Cambridge.

Mackey, M. 1984. Injuries to the face and to skin. In: Aldman, B., Chapon, A. (Eds), Biomechanics of Impact Trauma. Elsevier Science Publisher, Amsterdam, pp. 335-340.

Madre-Dupouy, M. 1992. L'Enfant du Roc de Marsal: Etude Analytique et Comparative. Editions CNRS, Paris.

Marks, M.K., Hudson, J.W., Elkins, S.K. 1999. Craniofacial fractures: Collaboration spells success. In: Galloway, A. (Ed), Broken Bones: Anthropological Analysis of Blunt Force Trauma. Charles C. Thomas, Springfield (IL), pp. 258-286.

McCown, T.D., Keith, A. 1939. The Stone Age of Mount Carmel: Volume II. Clarendon University, Oxford.

McElhaney, J.E., Reynolds, V.L., Hilyard, J.F. 1976. Handbook of Human Tolerance. Japan Automobile Research Institute, Inc, Tokyo.

McKee, J.K. 2001. The Taung raptor hypothesis: Caveats and evidence. American Journal of Physical Anthropology 32, 107.

McKee, J.K., Tobias, P.V. 1994. Taung stratigraphy and taphonomy: Preliminary results based on the 1988-93 excavations. South African Journal of Science 90, 233-235.

McKern, T.W., Stewart, T.D. 1957. Skeletal Age Changes in Young American Males. Quartermaster Research and Development Center, Matick (MA).

Meindl, R.S., Lovejoy, C.O. 1985. Ectocranial suture closure: A revised method for the determination of skeletal age at death based on the lateral-anterior suture. American Journal of Physical Anthropology 68, 57-66.

Minugh-Purvis, N. 1988. Patterns of craniofacial growth and fevelopment in Upper Pleistocene hominids. Ph.D. Dissertation, University of Pennsylvania.

Mizoguchi, Y. 1997. Human Bones from the Ikenohata-Shichikencho Relics (in Japanese). Ikenohata-Shichikencho Iseki (Keianji-Ato). Taito-ku Ikenohata-Shichikencho Iseki Chosakai, Tokyo.

Molleson, T., Cox, M. 1993. The Spitalfields Project, Vol. 2: The middling sort. Council for British Archaeology Research Report 86.

Moorrees, C.F.A., Fanning, E.A., Hunt, E.E. 1963. Age variation of formation stages for ten permanent teeth. Journal of Dental Research 42, 1490-1502.

Murphy, S.P., Sledzik, P.S., Mann, R.W., Kelley, M.A. 1990. Macroscopic bone remodeling following trauma: Reconsidering the term perimortem. Paper presented at the annual meeting of the American Association of Forensic Scientists, Cincinnati.

Naim-Ur-Rahman, Jamjoon, Z., Jamjoon, A., Murshid, W.R. 1994. Growing skull fractures: Classification and management British Journal of Neurosurgery 8, 667-679.

Pap, I., Tillier, A.-M., Arensburg, B., Chech, M. 1996. The Subalyuk Neandertal remains (Hungary): A re-examination. Annales Historico-Naturales Musei Nationalis Hungarici 88, 233-270.

Patté, E. 1957. L'Enfant Néandertalien du Peche de l'Azé. Maison et Cie, Paris.

Rak, Y., Kimbel, W.H., Hovers, E. 1994. A Neandertal infant from Amud Cave, Israel. Journal of Human Evolution 26, 313-324.

Rak, Y., Kimbel, W.H., Hovers, E. 1996. On Neandertal autapomorphies discernible in Neandertal infants: A response to Creed-Miles et al. Journal of Human Evolution30, 155-258.

Richardson, M. L. 2000. Approaches to Differential Diagnosis in Musculoskeletal Imaging. <http://www.rad.washington.edu/mskbook/facialfx.html>

Rogers, L. 1992. Radiology of Skeletal Trauma. Churchill Livingstone, New York.

Saunders, S. 2000. Subadult skeletons and growth-related studies. In: Katzenberg, M.A., Saunders, S. (Eds), Biological Anthropology of the Human Skeleton. Wiley-Liss, New York, pp. 135-161.

Sauer, N. 1998. The timing of injuries and manner of death: Distinguishing among antemortem, perimortem and postmortem trauma. In: Reichs, K.J. (Ed), Forensic Osteology: Advances in the Identification of Human Remains. Charles C. Thomas, Springfield (IL), pp. 321-332.

Schwartz, J.H., Tattersall, I. 2002. The Human Fossil Record, Volume I: Terminology and Craniodental Morphology of Genus Homo (Europe). Wiley Liss, New York.

Schwartz, J.H., Tattersall, I. 2003. The Human Fossil Record, Volume II: Craniodental Morphology of Genus Homo (Africa and Asia). Wiley Liss, New York.

Smith, B.H. 1991. Standards of human tooth formation and dental age assessment. In Kelley, M.A., Larsen, C.S. (Eds), Advances in Dental Anthropology. Wiley-Liss, New York, pp. 143-168.

Smith, B.H. 1993. The physiological age of KNM-WT 15000. In: Walker, A.C., Leakey, R. (Eds), The Nariokotome *Homo erectus* Skeleton. Harvard University Press, Cambridge, pp. 196-220.

Snow, C.E. 1948. Indian Knoll Skeletons. The University of Kentucky Reports in Anthropology, Volume IV, Number 3, Part II. Lexington, Department of Anthropology, University of Kentucky.

Stillwell, B. 2002. An investigation into cranial trauma from the Anglo Saxon cemetery at Sedgeford, Norfolk, U.K. Undergraduate Dissertation, Bournemouth University.

Thoma, H.K., Goldman, H.G. 1960. Oral Pathology, 5th Edition. The C.V. Mosby Company, St. Louis.

Tillier, A-M. 1982. Les enfants Neanderthaliens de Devil's Tower (Gibraltar). Zeitschrift fur Morphologie und Anthropologie 3, 125-148.

Tillier, A-M. 1983. L'enfant Neanderthalien du Roc de Marsal (Campagne du Bugue, Dordogne): Le squelette facial. Annales de Paleontologie 69, 137-149.

Tillier, A-M. 1984. L'enfant *Homo* 11 de Qafzeh (Israel) et son rapport à la comprehension des modalités de la croissance des squelettes Moustériens. Paleorient 10, 7-48.

Tillier, A-M. 1999. Les Enfants Mousteriens de Qafzeh. CNRS, Paris.

Todd, T.W., Lyon, D.W. 1924. Endocranial suture closure, its progress and age relationship, Part I: Adult males of the while stock. American Journal of Physical Anthropology 7, 325-384.

Todd, T.W., Lyon, D. 1925a. Cranial suture closure, its progress and age relationship, Part II: Ectocranial closure in adult males of the white stock. American Journal of Physical Anthropology 8, 23-45.

Todd, T.W., Lyon, D. 1925b. Cranial suture closure, its progress and age relationship, Part III: Endocranial closure in adult males of Negro stock. American Journal of Physical Anthropology 8, 47-71.

Todd, T.W., Lyon, D. 1925c. Cranial suture closure, its progress and age relationship, Part IV: Ectocranial closure in adult males of Negro stock. American Journal of Physical Anthropology 8, 149-168.

Utermohle, C.J. 1984. From Barrow eastward: Cranial variation of the eastern Eskimo Ph.D. Dissertation, Arizona State University.

Vagn Nielsen, O. 1970. Human Remains: Metrical and Non-Metrical Anatomical Variations. The Scandinavian Joint Expedition to Sudanese Nubia Vol. 9. Scandinavian University Books, Copenhagen.

von Koenigswald, G.H.R. 1936a. Een Nieuwe Pithecanthropus Ontdekt. [A New Pithecanthropus Found.] De Preangerbode March 28, 1936. (translated from the Dutch by J. Volker).

von Koenigswald, G.H.R. 1936b. Ein fossiler Hominide aus dem Altpleistocãn Ostjavas [A Fossil Hominid from the Lower Pleistocene of Java.]. De Ingenieur in Nederlandsch-Indië, Mijnbouw and Geologie, De Mijningingenieur 4, 8, 149-157.

von Koenigswald, G.H.R. 1937. A review of the stratigraphy of Java and its relations to early man. In: MacCurdy, G.G. (Ed), Early Man. J. B. Lippincott, Philadelphia, pp. 23-32.

Walker, A., Leakey, R. (Eds). 1993. The Nariokotome *Homo erectus* Skeleton. Harvard University, Cambridge.

White, T.D., Asfaw, B., DeGusta, D., Gilbert, H., Richards, G., Suwa, G., Howell, F.C. 2003. Pleistocene *Homo sapiens* from Middle Awash, Ethiopia. Nature 423, 742-747.

Zimmerman, R.A., Bilaniuk, L.T. 1981. Computed tomography in pediatric head trauma. Neuroradiology 8, 257-271.

## Appendix

**Observations of breakage and preservation on immature fossil crania.** Specimens are listed in alphabetical order. We list the sources consulted for our observations, then each specimen is briefly described, and finally a tentative diagnosis of the taphonomic history of each specimen is given.

### Amud 7

Descriptions and pictures in: Rak et al. (1994, 1996); Schwartz and Tattersall (2003). Neandertal aged 10 months (Rak et al., 1994). The specimen primarily consists of the occipital bone with a tiny piece of the sphenoid at basion, and a piece of the left petrous attached. Complete foramen magnum, condyles unfused and missing. Diagnosis: Possible mosaic fracture pattern (specimen broken into >20 fragments) but specimen too incomplete to be sure. Large, irregular, angular pieces are missing where the basilar occipital joins the squamous occipital. Possible radiating fractures centering on inion. Probably there was a separation at the lambdoid suture, isolating the occipital during stage 2, and later the occipital was flattened with resultant mosaic fractures.

### Dederiyeh 1

Descriptions and pictures in: Akazawa et al. (1995); Dodo et al.(1998); Web site: http://www.nichibun.ac.jp/dederiyeh. Neandertal aged ~2 years (Akazawa et al., 1995). Many small fragments reconstructed into a cranium. Mostly neurocranium preserved from coronal suture posteriorly; few fragments of frontal and face; incomplete basicranium. Diagnosis: Extensive crushing resulting in mosaic fractures with many small angular pieces, and the loss or destruction of many pieces. Sutural separations at coronal and temporal sutures resulting in loss of frontal and temporal bones. Separation of zygomatic-maxillary sutures and then presumed damage to facial bones. All breakage probably stage 2 (sutural separations) or 3 (possible crushing).

### Dederiyeh 2

Descriptions and pictures in: Ishida et al. (2000); Web site: http://www.nichibun.ac.jp/dederiyeh. Neandertal aged ~ 2 years (Ishida et al., 2000). Crushed and heavily fragmented cranium; most of frontal preserved; superior and lateral orbital margins pieced back together from many fragments; orbital portion of right zygomatic preserved but neither zygomatic arch complete. Parts of both parietals preserved, left more complete than right. Right temporal present. Some facial fragments present including parts of both nasals. Most of occipital missing; no basicranium. Diagnosis: Most vault bones fragmented with mosaic crushing, although with larger angular pieces and more beveled edges than seen in Dederiyeh 1. Many possible sutural separations (sagittal, lambdoid, temporal, zygomatic-maxillary, and coronal) before crushing. Small hole at bregma, judged to be an open anterior fontanelle because of irregularities in bone texture. Piece of alveolar maxilla preserved with teeth. Zygomatic-maxillary sutures separated and/or LeFort I or II fractures. The cranium is extensively damaged and crushed, possibly during stages 1 and 2.

### Devil's Tower/ Gibraltar 2

Descriptions and pictures in: Tillier (1982); Schwartz and Tattersall (2003); inspection of original. Neandertal aged 4.5-5 years (Minugh-Purvis, 1988). The specimen consists of isolated, disarticulated bones (frontal, maxilla, parietal, temporal, occipital) and is not an intact cranium. Diagnosis: Probable separation at all sutures, but certainly separation occurred at the coronal, sagittal, and squamous temporal sutures, followed by mosaic cracking. Frontal largely intact but showing many small angular fragments due to cracking. Left parietal is complete, with some angular cracking but no bone loss at bregma; broken at lambdoid suture. Right temporal largely preserved. The left partial maxilla is preserved but its median palatine suture is open. It is impossible to know if the face broke off with a LeFort I, II, or III fracture or if the maxilla separated at the zygomatic-maxillary suture and then broke further. Primarily stage 2 damage.

### Engis 2

Descriptions and pictures in: Fraipont (1936); Schwartz and Tattersall (2002). Neandertal aged 4-5 years (Minugh-Purvis, 1988); inspection of cast and original. The specimen is a partial neuro- and basicranium without a face. The rim of the foramen magnum is intact. Diagnosis: The face is missing in a LeFort III pattern except for a small piece of the right zygomatic which is preserved to form the lateral rim of the right orbit. The frontal is broken into several small fragments over the right orbit; the left orbital rim is broken in the middle and the adjacent left parietal and sphenoid are missing, as is the vomer. The alveolar portions of both maxillae are preserved separately. The vault is comprised of large angular fragments in a mosaic breakage pattern. A large angular piece is missing at bregma on the right side, accompanied by breakage along the coronal and sagittal sutures. Perpendicular fractures of the left and right parietals are present but do not cross the sagittal suture. Most of the right temporal is present but the squamous suture appears to be open; both squamous and petrous portions of the temporal are present on the right side only. The left half of the squamous occipital is present; a straight, planar fracture runs vertically through the occipital to the lambdoid suture and the right half of the squamous occipital is missing. A stellate fracture can be observed at inion. Damage occurred during stage 2 or 3.

### Le Figuier

Descriptions and pictures in: Billy (1979). *Homo sapiens* aged ~ 3 years (Billy, 1979). The face and a fragmentary vault are present; some parts of the parietals are preserved; the temporals, occipital, and basicranium are missing. Diagnosis: The frontal is broken in a metopic fracture and a piece missing at midline. A piece is missing at bregma, but this is not judged to be an open fontanelle as its edges are planar. The orbits are relatively complete and the zygomatics and maxilla are complete to the midline. A LeFort III fracture separated the lower face from the vault. The vault bones are broken into large, rectangular fragments by fractures which include several perpendicular fractures and a possible temporal line fracture on the left. Separations occurred along the coronal, sagittal and lambdoid sutures with subsequent damage to the lambdoid suture. Damage occurred during stages 2 or 3.

## Grotte des Enfants 6

Descriptions and pictures in: Schwartz and Tattersall (2002). *Homo sapiens* aged 12-14 years based upon presence of erupted M2s and unerupted M3s. This is a cracked and very fragmented cranium with all bones present and relatively few pieces missing. Both zygomatic arches are broken. Diagnosis: The frontal is intact, with the nasal bones present, although the frontal-zygomatic suture opened. The face is largely intact except for the left inferior orbital margin where most of the zygomatic is missing. The left side of the vault and face were pushed inward and cannot be articulated properly. A large piece is missing from the left parietal towards the squamous temporal suture, but bregma is intact. The sphenoid is missing inferiorly but some lateral pieces are still present. There is a large break around the foramen magnum on the left side and up to opisthion. Although fundamentally intact, the right side of the cranium is broken into large angular pieces that do not quite fit back together, possibly due to plastic deformation. The squamous temporal suture separated and then sustained some damage. Perpendicular fractures on the parietals are very clear, as are planar and angular fractures, especially around asterion. The squamous occipital appears undamaged but the basilar portion is missing angular pieces. The cranium was probably intact when it was squashed from side to side during late stage 1, causing plastic deformation, and the cranium was then crushed during stage 2 or 3.

## Herto BOU-VP-16/5

Descriptions and pictures in: Clark et al. (2003); White et al. (2003). *Homo sapiens idaltu* aged 6-7 years (White et al., 2003). The specimen is a highly fragmented cranium with significant pieces missing from the basicranium, temporals, parietals, and face. Diagnosis: Mosaic fractures producing numerous angular fragments with clear planar edges cover the vault. All bones, including the frontal and occipital, are broken into many fragments. There is a possible stellate fracture at inion. The coronal, sagittal, and lambdoid sutures did not apparently open, although squamous temporal suture separated and the squamous temporal is missing. One perpendicular fracture across the vault crosses the sagittal suture. An apparent LeFort I (alveolar fracture) occurred; the maxillae and alveoli are preserved separately. Most fractures appear to be post-fossilization (stage 3) damage. See further discussion in Part III of this paper.

## La Quina 18

Descriptions and pictures in: Schwartz and Tattersall (2002); inspection of cast. Neandertal aged 7.5-8 years (Minugh-Purvis, 1988). The cranium is basically fragmented but intact with large portions of the basicranium and most of sphenoid missing. Both zygomatic arches are broken. Diagnosis: The cranium was mostly intact when it was crushed. The face is in very good condition and even the fragile nasal bones are intact; however the face shows some distortion and minor fracturing. Separations occurred along the coronal, sagittal, squamous temporal, and lambdoid sutures. Small missing pieces show that damage occurred after the sutural separation along the coronal suture and there are larger breaks along the sagittal suture. A right temporal line fracture extends from the coronal to the lambdoid suture and possibly into the occipital. There are perpendicular fractures on the parietals, one of which crosses the sagittal suture, resulting in very large angular pieces with planar edges. Separation of the lambdoid suture caused or contributed to the loss of the occipital and the basicranium. Most of the edges surrounding the missing pieces show straight, planar fractures not simple sutural separations. The squamous occipital is heavily fragmented and shows the most crushing and reconstruction. Probably this specimen was subjected to anterior-posterior crushing that detached or destroyed most of the inferior and posterior parts of the cranium. Most of damage probably occurred during stage 2, either before the sutural separations happened or after the separations but before displacement of the constituent bones occurred.

## Mojokerto (Perning 1)

Descriptions and pictures in: von Koenigswald, (1936a, b); Anton (1997); Schwartz and Tattersall (2003); inspection of cast. *Homo erectus* aged 4-6 years (Anton, 1997). The specimen is a skullcap lacking a face and most of the basicranium. Diagnosis: A LeFort III fracture separated or destroyed the face. The only parts of the orbits that remain are the superior margins and those are incomplete, with pieces missing; none of the lateral or inferior orbital margins are preserved. The frontal is cracked into angular fragments and is also missing a piece or pieces at midline. The vault sutures show minimal or no separation; small pieces along the sutures are missing, whether due to preparation, excavation, or natural causes is unknown. All of the vault bones show mosaic cracking into small angular pieces. Among the fracture edges that are visible, some are planar and others are clearly beveled. There is at least one perpendicular fracture and there may have been temporal line fractures on both right and left sides but missing pieces of the parietal make this uncertain. There is a stellate fracture centered on inion. The occipital was slightly "folded" along a horizontal plane, bending the basioccipital under at an unnatural angle. The basilar occipital is broken off posterior to the foramen magnum. A natural endocast held the cranial pieces together during the crushing and cracking process. This endocast preserved impressions of the meningeal vessels of the left parietal and temporal where there is no longer any bone. The fracture edges are distinctly beveled on the right side of the vault and fragments of the right parietal were pushed into the unconsolidated sedimentary infill, which later hardened into an endocast. There is some displacement of the petrous temporal on each side and the left temporal is rotated and pushed forwards. Very little of either zygomatic is preserved. Sides of the neurocranium are more heavily fragmented than is top of the vault. The beveling of some fracture edges and plastic deformation suggest that most damage occurred in late stage 1-early stage 2. See further description in Part III of this paper.

## KNM-WT 15000

Descriptions and pictures in: Walker and Leakey (1993); inspection of cast and original. *Homo erectus* aged 8-11 years (Smith, 1993; Dean et al., 2001). A largely complete cranium fragmented into many pieces and glued back together. Diagnosis: The frontal is missing a fragment at midline and shows planar fractures but the superior part of the nasals is intact. The face is essentially complete but the zygomatic-maxillary sutures opened and sustained some damage thereafter. Both zygomatic arches are broken and missing, though the body of each zygomatic bone is preserved to form the inferior and lateral orbital margins. The median palatine suture opened and subsequently a few small fragments became lost. All of the vault sutures separated and the exposed edges were then slightly damaged by erosion or weathering. The vault is crushed into many angular mosaic fragments, especially the parietals and frontal. Many fractures have planar edges and rarely if ever cross sutures. A large angular piece of the right parietal is missing where the sagittal and lambdoid sutures intersect. Inferiorly, the sphenoid and parts of the petrous temporals are broken away. Probably the sutures separated and then the individual bones were fragmented and weathered mostly or entirely during stage 3 without being transported significant distances.

## Pech de l'Azé

Descriptions and pictures in: Ferembach (1970); Patté (1957); Schwartz and Tattersall (2002); inspection of cast. Neandertal aged 2.5-3 years (Minugh-Purvis, 1988). The specimen is a fragmented cranium missing the posterior surface and the base of the neurocranium. The nasal processes and most of the maxillary body are missing, and only alveolar parts of the maxilla remain. The zygomatics still articulate with the frontal to form the lateral side of the orbits but both zygomatic arches are broken and missing. Diagnosis: The alveolar/palatal region was isolated from the neurocranium by a Le-Fort II fracture which may have started as separations at the zygomatic-maxillary sutures. The frontal is largely complete though it is fragmented and missing pieces on the left side. The frontal bears a patent metopic suture, judging from radiographs in Patté (1957), although the individual is unusually old for this condition. There are angular fragments missing along the coronal and sagittal sutures which opened and separated. A large fragment is missing at bregma; this hole is said to encompass a late-closing fontanelle (Tillier, 1999) which cannot be verified from the available photographs. The left side of the neurocranium shows many large angular fragments, some missing, from the left sphenoid suture to lambda. Both temporal squamous sutures appear to have separated but the temporal bones are partly preserved. The anterior half of the right parietal is complete; a perpendicular fracture with a clear planar edge splits the bone approximately in half and the posterior portion is missing. Most of the squamous and basilar occipital is missing. There is a consistent pattern of sutural separation followed by mosaic breakage, probably in stage 2, resulting in many angular fragments.

## Qafzeh 10

Descriptions and pictures in: Tillier (1999); inspection of cast. *Homo sapiens* aged 6 years (Tillier, 1999). This largely complete cranium was found lying on its left side in situ, crushed into numerous pieces. Diagnosis: The face, especially the left side, is preserved but broken into fragments, and the nasals are missing. The face is somewhat asymmetrical suggesting possible plastic deformation. Most of the frontal has been reconstructed or glued back together from many fragments, which include fractures to the superior orbital margins. Where they are visible the fracture edges look planar. Both zygomatic arches are broken but the bodies of the zygomatic bones are preserved to form the inferior and lateral margins of the orbit. The vault shows mosaic cracking caused by crushing, which was followed by erosion and weathering; fracture edges are more rounded and less crisp than usually observed. The vault fragments are relatively small. There are no clear signs of separations on the coronal and sagittal sutures. In fact Tillier (1999: 78, 165) suggested premature closure of the coronal suture and the sagittal suture is offset from the midline. There is probably a right side temporal line fracture with some pieces missing; pieces are also missing along the lambdoid suture, which probably separated. The basilar occipital is largely intact and the rim of the foramen magnum is complete, but the sphenoid and petrous temporals are damaged. Sutural separations followed by crushing, probably in stage 3.

## Qafzeh 11

Descriptions and pictures in: Tillier (1984, 1999); inspection of cast. *Homo sapiens* aged 12 years (Minugh-Purvis, 1988; Tillier, 1999). Overall the neurocranium is relatively complete save for the lower face and part of the base. The alveolar portion of the maxilla is preserved separately. The sphenoid, zygomatic, vomer and palatine bones are missing; there is no cranial base anterior to the occipital. The petrous temporals are heavily damaged. Diagnosis: Most of the face is missing, possibly due to a LeFort III and/or tripod fracture. Alveolar portions of the maxilla survive but there are no zygomatic bones. The frontal lacks a few small pieces above the left orbit; there are other small cracks in the orbits but no fragments are missing. The nasal region between the orbits is intact. A large angular fragment or fragments are missing from the right parietal at the coronal suture, and there is a hole at bregma. If this began as a coronal separation, then additional breakage occurred subsequently; however the left side of the coronal suture appears to be intact. There is a depressed fracture to the left frontal bone which possibly shows healing (our observations of Figure 63 in Tillier, 1999: 165), indicating it was an antemortem fracture. Parietal pieces are missing along the open sagittal, lambdoid, and squamous temporal sutures. At least one and possibly more fractures of the parietals cross the sagittal suture, showing that the bones were in anatomical position when the crushing and fragmentation occurred. The mosaic cracking of the vault was more intensive on the right side. The squamous occipital is heavily cracked, with subsequent damage to the fracture edges. The basilar occipital is fragmented and missing many pieces on left side, although almost the entire rim of the foramen magnum is preserved. Sutural separation occurred in stage 2 but the bones remained in approximate anatomical position before they were subjected to crushing.

## Qafzeh 12

Descriptions and pictures in: Tillier (1999). *Homo sapiens* aged 3-4 years (Tillier, 1999). This very fragmentary cranium consists of a partial superior surface of the vault with separate occipital and petrous temporal bones and alveolar fragments of the maxilla accompanying some of the right teeth. Diagnosis: Very little of the frontal survives as several angular fragments; nothing of the orbital margins remains. The patent metopic, coronal and sagittal sutures apparently separated, although a patent metopic suture at 3-4 years is unusual. Tillier (1999: 165) suggests this individual was also possibly hydrocephalic. The anterior fontanelle at bregma was clearly still patent: another unusual feature for a juvenile of this age. The vault bones are highly fragmented into angular fragments due to crushing following sutural separations in stage 2. Perpendicular fractures are also seen on the parietals, with some planar fracture edges visible. The petrous temporals survive separately in damaged form, as do the zygomatic bodies, but the zygomatic-maxillary sutures apparently opened and only very small alveolar portions of the maxillae are preserved. The lambdoid suture separated although a few fragments of parietal remain attached to occipital fragments; in other places, breakage removed pieces of the occipital subsequent to the opening of the lambdoid suture. The squamous occipital is broken into large, angular fragments, probably because of flattening after it separated from the rest of the neurocranium. Most damage probably occurred in stage 2.

## Roc de Marsal 1

Descriptions and pictures in: Tillier (1983); Madre-Dupouy (1992). Neandertal aged ~3 years (Minugh-Purvis, 1988; Madre-Dupouy, 1992). This specimen was part of a skeleton found lying on its right side with the cranium crushed almost flat. The neurocranium is fragmentary; the lower face is preserved and is less damaged on the right side. Both zygomatic arches are broken and incomplete although the articulation between the zygomatic and maxilla is intact on both sides. Most of the superior surface and left side of the vault are missing on the left side, as is the left side of the basicranium. Diagnosis: The face is nearly undamaged including the nasals and fragile processes of the maxillae. The median palatine suture is open with some small breaks that occurred after separation. The face was apparently separated from the neurocranium by a LeFort II fracture or by sutural separations followed by damage to approximate a LeFort II fracture. The frontal is broken into several large fragments and is missing a large piece at midline. Two parametopic fractures break the superior rim of each orbit and each intersects a hole produced by roughly horizontal fractures above the superciliary ridges and the subsequent loss of pieces. The left parametopic fracture joins with a massive hole encompassing bregma and most of the superior surface of the vault. Tillier (1999) indicates that the anterior fontanelle was open; although this cannot be verified from photographs and Madre-Dupouy (1992) indicates uncertainty on this point. Cracks radiate from the missing area of the parietals anteriorly, inferiorly, and posteriorly; these fractures divide the right parietal into angular fragments but do not appear to cross sutures. There is a discrete round hole in the right sphenoid at the intersection with the coronal suture. The edges of this hole look planar and not beveled; a few cracks surrounding the hole do not suggest radiating cracks from a stage 1 fracture but breakage later in the specimen's taphonomic history. The vault is comprised of many small angular fragments now held together with wax. The neurocranium was obviously flattened or crushed, probably after many or most of the sutures had opened but while the bones were essentially still in something approximating anatomical position. The cranial damage is progressively worse from the frontal, which is fragmented but largely preserved, posteriorly to what remains of the pieces of the occipital. Most damage probably occurred during stages 2 or 3.

## Skhul 1

Descriptions and pictures in: McCown and Keith (1939); Schwartz and Tattersall (2003); inspection of cast. *Homo sapiens* aged 4.5 years (McCown and Keith, 1939; Minugh-Purvis, 1988). The specimen is a cranial vault lacking most of the face, although isolated maxillary teeth were recovered. The specimen was part of a skeleton found lying on its left side (McCown and Keith, 1939). As excavated, the left radius protruded into the interior of the braincase. McCown and Keith suggest (p. 301) postmortem disturbance occurred prior to consolidation of the endocast. Diagnosis: No facial bones below the superior orbital margins are preserved due to an apparent LeFort III fracture. Pieces of the superior orbital margin are missing from the frontal at the midline and the fractures run posteriorly through the frontal. The frontal shows extensive mosaic cracking with small fragments now joined together by plaster. McCown and Keith suggest (1939) that there was a depressed fracture 30 mm by 13 mm on the frontal at midline that must have occurred when the bone was fresh, saying that "the bone on one side of the gap—the bone over the inner part of the right orbit—has 'sprung' or been lifted forwards its own thickness" (p. 309-310). In the cranium's current state, these observations are impossible to verify and what is visible does not convincingly support their hypothesis. The coronal suture opened and the frontal bone was discovered separated from the parietals with its interior surface uppermost, lying near the rest of the cranial vault. There was breakage along the coronal suture and at bregma subsequent to the separation. There are perpendicular cracks in the parietals and the right parietal is missing substantial pieces inferiorly. Only the right squamous temporal is present; the right petrous is broken. The sagittal and lambdoid sutures are intact and the squamous occipital is little broken; the basilar portion of the occipital shows much cracking and loss of the bone surrounding the foramen magnum. McCown and Keith (p. 299, Figs. 214, 215) discuss a natural endocast that filled the posterior part of the skull. This endocast preserved impressions of right parietal pieces that are now missing; the endocast was destroyed during preparation of the interior vault surface. The endocast probably accounts for the fact that the posterior skull stayed intact despite extensive cracking into mosaic fragments. The protection offered by the endocast did not prevent the shattering of the anterior, basilar and lateral portions of the skull probably caused by sedimentary pressure in stages 2 or 3.

## Subalyuk 2

Descriptions and pictures in: Pap et al. (1996); Schwartz and Tattersall (2002); inspection of original. Neandertal aged 2.5 years (Minugh-Purvis, 1988). The specimen consists of a neurocranium lacking most of its base and a separately preserved maxilla. Diagnosis: An apparent LeFort III fracture separated the maxilla from the neurocranium. The coronal suture did not separate and the sagittal suture is intact to lambda; there is some separation along the lambdoid suture and loss of large pieces from the left parietal. The frontal is missing a sizeable fragment at midline, however, the right frontonasal suture is intact. The metopic suture was probably patent (Tillier 1999). Both parietals show perpendicular fractures and the right parietal is missing several large, angular pieces. There is a possible left temporal line fracture from the orbit to the lambdoid suture. The occipital is cracked into many small pieces, some of which are missing, suggesting that the face and the occipital bore the brunt of the crushing force in stage 2 or 3.

Descriptions and pictures in: Dart (1925); inspection of cast. *Australopithecus africanus* aged 3-4 years (Bromage, 1985). The specimen is a face articulated with a mandible and frontal and a natural endocast to which is attached much of the basicranium. The endocast also preserves the impressions of the internal surface of the right parietal and temporal bones. The right parietal and temporal were not collected or were destroyed during the blasting process that revealed the specimen. A coronal separation apparently isolated the articulated frontal, face, and mandible from the rest of the skull. The natural face of the endocast shows that the skull did not fill with sediments completely as it lay in situ on its right side. The articulation of face and mandible and the complete absence of cracking or fragmentation of the preserved bones shows that the skull entered the tufa cave with soft tissues holding it together and protecting it from damage. The articulated position of the bones was preserved after the soft tissue decayed and while the cranial vault was infilled with sediments, which consolidated to form the natural endocast. The specimen became a sedimentary particle in stage 1 and suffered little damage thereafter. See further discussion in Part III of paper

## Teshik Tash

Descriptions and pictures in: Schwartz and Tattersall (2002). Neandertal aged 9-11 years (Minugh-Purvis, 1988). The cranium is largely complete as restored but the zygomatic arches are incomplete. Diagnosis: The cranium is comprised of large angular fragments produced by mosaic cracking. The frontal is cracked into large angular fragments but no pieces are missing; some fractures originate at the orbital rims. Bregma is intact but the parietals are fragmented. Small pieces are missing along the coronal, sagittal, and lambdoid sutures. Some fractures cross sutures uninterrupted, showing that the vault bones remained in articulated position or nearly so after the soft tissue deteriorated. There is one perpendicular fracture of the left parietal, and a clear left temporal line fracture from the coronal to the lambdoid suture. The face is largely intact including the nasal bones, which show only minor breakage. The zygomatic-maxillary suture apparently opened and then some additional breakage occurred. The median palatal suture is open but not separated. Most or all fractures occurred in stage 2 or 3.

# CHAPTER 13

# Carcass Foraging by Early Hominids at Swartkrans Cave (South Africa): A New Investigation of the Zooarchaeology and Taphonomy of Member 3

Travis Rayne Pickering, Manuel Domínguez-Rodrigo, Charles P. Egeland and C.K. Brain

## Abstract

While the Plio-Pleistocene paleontology of South African cave faunas is abundant and well-known, the zooarchaeology of these same assemblages is sparser and less appreciated. Most reconstructions of carcass foraging by Early Stone Age hominids are based largely on East African datasets. Here we take steps to remedy that situation by providing zooarchaeological and taphonomic data on the important *c.* 1.8 – 1.0 million year old archaeofauna from Swartkrans Member 3. Because most actualistic models of the interaction between hominids and carnivores over prey carcasses are focused on limb bones, we concentrated our study on the limb bone mid-shaft sub-assemblage from Member 3. Results indicate that tooth-marked specimens are approximately three and a half times as common as hominid-modified specimens in the limb bone shaft subassemblage as a whole. However, when taking into account diagenetic breakage, cortical surface preservation and differential fragmentation, hominids and carnivores seem to have contributed similarly to the formation of the Member 3 limb bone shaft subassemblage. Based on the anatomical distribution of stone tool cutmarks, Swartkrans hominids appear to have been capable carcass foragers during Member 3 times, gaining access to muscled carcass parts that are usually defleshed early and entirely by feeding carnivores. A similar pattern of cutmark distribution also characterizes broadly contemporary assemblages from East Africa, suggesting that hominids throughout the continent were capable acquirers of preferred parts from large animal carcasses.

## Introduction

Several specific models of subsistence behavior and sociality in Plio-Pleistocene hominids have been presented in the past 25 years (reviewed most recently in, Domínguez-Rodrigo, 2002; Domínguez-Rodrigo and Pickering, 2003; Pickering and Domínguez-Rodrigo, in press). We believe that most of those models fall into one of two major groups. The first of these asserts that hominids regularly acquired whole or substantial portions of large mammal carcasses that they then transported to favored locales to process, consume and possibly share with others group members (e.g., Bunn, 1981, 1982, 1983, 1986, 1991: Bunn and Ezzo, 1993; Bunn and Kroll; 1986; Isaac, 1978, 1981a,b, 1983, 1984). These models, which indicate early access to carcasses by hominids, also imply by extension that hunting and/or aggressive scavenging was a prominent feature of their carcass-foraging repertoire. In contrast, the second group of models posits very limited access to fleshed carcasses by hominids (e.g., Binford, 1981, 1985, 1988; Blumenschine, 1986, 1987, 1988, 1991, 1995; Blumenschine et al., 1994). According to this view, even those carcass parts that hominids infrequently secured were already picked-over by carnivores, leaving no appreciable "surplus" resources for hominid scavengers to share.

The vigorous debate that has emerged between advocates of these competing views is particularly fascinating when one considers that the relevant faunal database derives largely from just one archaeological site, FLK 22 *Zinjanthropus* (FLK *Zinj*), Olduvai Gorge, Tanzania (*c.* 1.75 million years old [Ma]). It is true that data from other important sites at Olduvai Gorge (site BK, *c.* 1.2 Ma), Peninj, Tanzania (the ST site complex, *c.* 1.5 Ma), and Koobi Fora, Kenya (various sites *c.* 1.88 – 1.6 Ma, but in particular, FxJj 50) have entered the debate (e.g., Bunn, 1994, 1997; Bunn et al., 1980; Domínguez-Rodrigo et al., 2002; Monahan, 1996), but more tangentially than those from FLK *Zinj*. At the very least, it can be said there is a geographic bias (i.e., the East African Rift Valley) in this dataset, with important information available from comparably aged South African sites rarely incorporated into overviews of the topic (for exceptions, see Pickering and Domínguez-Rodrigo, in press; Egeland et al., 2004). In addition to some socio-historical reasons for this bias (e.g., the *perception* that South African cave sites only inform about how early hominids died and not how they lived; the world community's relegation of South African science during the apartheid years), we also believe there several scientifically legitimate reasons for it.

First, the combination of topographic placement, unique ecological context and geomorphological form of South African hominid caves resulted in non-hominid "taphonomic overprints" on their faunas that are sometimes more complex than those from East African sites (e.g., Brain, 1981). Related to this point is the fact that all of the numerous actualistic models constructed since the 1980s to investigate early hominid foraging have been formulated in and with regard to the formation of open air sites in savanna mosaic habitats. Last, aside from purported bone tools from Sterkfontein, Swartkrans and Drimolen (Robinson, 1959; Brain and Shipman, 1993; Keyser, 2000) and indications of hominid-controlled fire from Swartkrans (Brain and Sillen, 1988), there is a paucity of evidence for other types of hominid-imparted bone modification reported for relevant South African sites. Until the results presented here, a *total* of only 15 cutmarked bone specimens (one from Sterkfontein, Pickering, 1999, and 14 from Swartkrans, Brain, 1993) and three chopmarked pieces (one from Sterkfontein, Brain, 1981, and two from Swartkrans, Brain, 1993) had been reported from the whole of Plio-Pleistocene South Africa.

In an effort to remedy this situation, we report here on 163 limb bone specimens from Swartkrans Member 3 with newly identified cutmarks and hammerstone percussion damage, and discuss the implications of these findings for the reconstruction of early hominid behavior in the Sterkfontein Valley, and beyond, *c.* 1.8 – 1.0 Ma. Our findings now rank the Swartkrans Member 3 archaeofauna as second only to FLK *Zinj* in number of hominid-modified bones from the Plio-Pleistocene and thus asserts the importance of this assemblage and the South African zooarchaeological record in general discussions of early hominid carcass foraging.

## Materials and Methods

A systematic zooarchaeological analysis of the complete limb bone shaft fragment subassemblage from Swartkrans Member 3 (1979 – 1986 excavations) was conducted. This 12,505 (number of identified specimens, NISP) piece sample derives from the larger 108,098 NISP fossil assemblage described initially by Brain (1993), Watson (1993) and Newman (1993) and also encompasses the subset of limb bone shaft specimens reported by Bishop and Blumenschine (1994). Limb bone shaft fragments were chosen as the analytical sample because most current actualistic models of hominid carcass use focus in large part on limb elements (e.g., Blumenschine, 1988, 1995; Blumenschine and Marean, 1995; Blumenschine and Selvaggio, 1991; Capaldo, 1995, 1997, 1998; Cleghorn and Marean, 2004 and this volume; Domínguez-Rodrigo, 1999a, 1999b, 2001; Marean and Cleghorn, 2003; Marean et al., 1992, 2004; Pickering et al., 2003; Selvaggio, 1994, 1998; Selvaggio and Wilder, 2001). Limb bone shaft specimens are defined here as pieces from ungulate humeri, radioulnae, metacarpals, femora, tibiae and metatarsals that preserve less than their complete, original diaphyseal circumferences and do not possess their articular ends (modified from Pickering, 1999; see also Pickering et al., 2003, 2005).

We isolated two sub-samples from the complete limb bone subassemblage for more in-depth analysis. Analytical Set I is comprised of every specimen ≥5 cm in maximum dimension plus every specimen <5 cm in maximum dimension that also preserves prehistoric bone surface modifications. This analytical set, with a NISP of 1466, is "unadjusted." In other words, it is not comparable to modern, actualistically derived samples of human butchered and carnivore ravaged bones (see discussion below), but it does provide "maximum" information on the frequency and distribution of hominid and carnivore bone surface modifications.

In addition to Analytical Set I, we created an adjusted sample, Analytical Set II, which is more comparable to actualistic samples that model the carcass-focused interactions of hominids and carnivores (see also, Blumenschine, 1995: 28, 33-39). Analytical Set II was assembled by beginning with the original limb bone subassemblage of 12,505 pieces and then taking the following steps. First, because the experimental control samples (i.e., Blumenschine, 1988, 1995; Blumenschine and Selvaggio, 1988; Capaldo, 1995, 1997, 1998; Selvaggio, 1994, 1998) do not consider specimens <2 cm in maximum dimension, specimens in the Member 3 fossil assemblage <2 cm were eliminated from consideration for comparative analyses—even if they bear prehistoric bone surface modifications. This resulted in a modified NISP of 8352. Second, processes of diagenetic fragmentation and cortical surface degradation not operant in the modern control samples had to be controlled in the fossil assemblage. Because of the assemblage's large size and time constraints, we were forced to adjust for these factors through a sampling procedure, rather than examining every specimen. This procedure is summarized thusly:

1. First, we sampled randomly 1,009 specimens from three size-range categories (2 – 3 cm, 3 – 4 cm, 4 – 5 cm).
2. Within each size-range category, we calculated the percentage of specimens with good cortical surface preservation and green versus dry breakage planes.[1]
3. We then averaged these percentages, which resulted in an average of 48.3 % of specimens <5 cm displaying good surface preservation and 65.0 % with dry breakage.
4. Next we applied these percentages from the sample back to the starting NISP of 8,352. Starting with the projection of well-preserved specimens, this is 8,352 × 0.483 = 4,034.
5. Adjusting for dry breakage was accomplished by multiplying 4,034 by the projected percentage of dry-broken specimens (4,034 × 0.65 = 2,622).
6. In order to reach a NISP estimate adjusted for green breakage, however, we first considered that the dry-broken NISP (2,622) is inflated by the fact that each originally deposited bone was broken into at least two pieces, at least doubling the dry-broken NISP. Thus, the most conservative approach divides the dry-broken NISP by two (2,622/2 = 1311). That estimate was then added to the green broken NISP (4,034 × 0.35 = 1,412), resulting in a new NISP of 2,723 of well-preserved and green-broken pieces.
7. Because all specimens in Analytical Set I (Member 3 NISP = 1,466) were coded individually for surface preservation and breakage, there was no need to following the sampling procedure outline in steps 1–6. Instead, we simply added the adjusted non-Analytical Set I NISP from above (2,723) to the adjusted Analytical Set I data to obtain a total adjusted NISP for Analytical Set II. From Analytical Set I there are 428 specimens >2 cm that display good cortical surfaces and green breaks. The resulting grand total of well-preserved, green-broken specimens is thus 3,151 (2,723 + 428), the final NISP for the Member 3 Analytical Set II.

We collected data on the following zooarchaeological and taphonomic attributes in both analytical sets.

## Skeletal element and element portion

When possible, specimens were identified to skeletal element (humerus, radioulna, metacarpal, femur, tibia, metatarsal, metapodial). Using the system of Domínguez-Rodrigo (1997, 1999a, Barba Egido and Domínguez-Rodrigo, 2005), we were able to categorize many of those specimens not identifiable to a specific skeletal element to a limb segment, as an upper (humerus or femur), intermediate (radioulna or tibia) or lower (metapodials) limb fragment. Specimens that remained unidentified after these steps were then simply entered into the database as limb bone shaft fragments. In addition, due to time constraints, no specimen <5 cm was identified beyond the level of limb bone shaft fragment; however, it is worth noting that a significant portion of these fragments are probably identifiable to skeletal part and will be considered in future analyses. Finally, following Blumenschine's (1988) bone portion classification system and in order to facilitate comparisons with experimental control samples (e.g., Blumenschine, 1988, 1995; Capaldo, 1995, 1997; Selvaggio, 1994, 1998), every specimen >2 cm in maximum dimension (n = 8,352) was identified as a near-epiphyseal or midshaft fragment.

## Animal body size

Each specimen was assigned to an animal body size group, following the size class system constructed for antelope by Brain (1974, 1981). For some analyses, individual body size groups were combined into three broad categories: small (corresponding to Brain's Size Class 1); medium (the combined remains of Size Classes 2 and 3); large (the combined remains of Size Class 4 and larger).

## Maximum linear dimension

Maximum length of each specimen, irrespective of orientation, was measured to the nearest centimeter.

## Circumference

In a modification of Bunn's (1983) system, the cross-sectional completeness of each specimen was recorded in increments of 25 %: <25 % of the original diaphyseal circumference preserved along a specimen's length; <50 % but >25 % of the original circumference preserved; <75 % but >50 % of the original circumference preserved; <100 % but >75 % of the original circumference preserved.

## Fracture patterns

Recent experimental results indicate that combined

[1] For assessment of cortical surface preservation, each fossil specimen was assigned to a subaerial weathering stage (Behrensmeyer, 1978). In addition, our observations suggest that bone surface preservation on specimens from Member 3 was also affected by various diagenetic processes, including water action, manganese formation and soil leaching. Thus, to account for overall surface condition, a subjective score of poor, moderate or good was assigned to each specimen (e.g., Pickering, 1999; Pickering et al., 2000). This is a qualitative assessment used to convey the relative "fidelity" of current bone surfaces for continuing to preserve prehistoric bone surface modifications. Distinguishing green- from dry-broken fracture edges is relatively simple. Green fractures occur on bone before loss of its organic fraction and are associated with smooth release surfaces and possess fracture angles (i.e., the "angle formed by the fracture surface [of a broken bone and its] cortical surface" [Villa and Mahieu, 1991: 34]) <85° or >95° (Pickering et al., 2005). In contrast, dry fractures occur after loss of a bone's organic content and are characterized by fracture angles closer to 90°.

fracture plane and angle data are useful for sorting dynamic (e.g., hammerstone percussion) and static (e.g., carnivore chewing) loading events on green bones (i.e., bones without significant loss of their organic fractions and desiccation) at the assemblage level (Alcántara García et al., in press; see also Capaldo and Blumenschine, 1994; Pickering et al., 2005). Thus, we conducted a detailed analysis of these features on all green fractures in the Member 3 assemblage (diagenetic and other "dry" break surfaces were ignored). Each green fracture plane ≥5 cm in length was recorded in relation to the long axis of the specimen: longitudinal (parallel) to the long axis, transverse (perpendicular) to the long axis or oblique (diagonal) to the long axis. Midpoint angles of those fracture planes were then measured to the nearest degree using a goniometer (Pickering et al., 2005).

## Bone surface modifications

Identification of bone surface modifications was undertaken using criteria and methods reviewed by Blumenschine et al. (1996). Each specimen was inspected under a strong oblique light source with the aid of at least 10 x magnification, as recommended by several analysts (e.g., Bunn, 1981, 1991; Bunn and Kroll, 1986; Blumenschine, 1995; Blumenschine and Marean, 1993; Blumenschine and Selvaggio, 1988, 1991; Blumenschine et al., 1996). During examination of each specimen, the bone surface was continuously repositioned in relation to the light source in order to discern modifications of any appreciable depth. Although other classes of bone surface modification (e.g., "random" striae, rodent gnaw marks, burning, alteration by gastric acids) were observed and noted, only carnivore tooth marks, stone tool cutmarks and hammerstone percussion marks were searched for and recorded systematically.

Several researchers have stressed the potential of various abiotic processes to mimic hominid-imparted bone surface damage, complicating inferential associations of particular marks and hominid butchery activity (e.g., Behrensmeyer et al., 1986, 1989; Fiorillo, 1989; Potts and Shipman, 1981; Oliver, 1989; Shipman and Rose, 1983). Thus, *all* specimens asserted to preserve hominid-imparted damage were subsequently examined by each researcher, and only after an unanimous decision was a specimen accepted and recorded as preserving the appropriate surface modification. Although time-consuming, this procedure was ultimately necessary for secure determinations. A prominent presence of abiotically derived linear striae (sometimes closely resembling stone tool cutmarks) was indicated by our many hours of experience with the curated collection and corroborated by observations of the sedimentary matrix from which the assemblage derives. As illustrated in Figure 1, the Member 3 deposit is a complex karstic coluvium, consisting

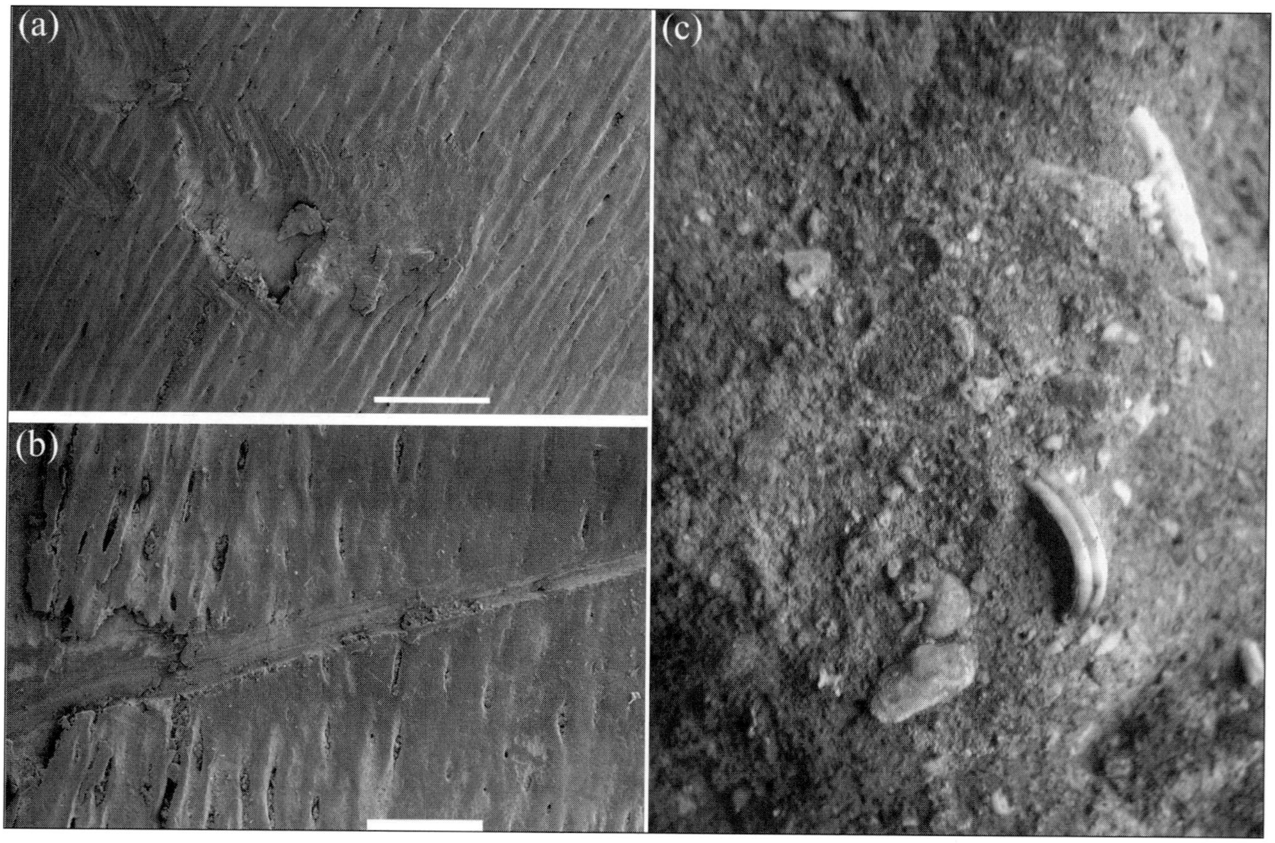


*Figure 1. Scanning electron microscope micrographs showing representative examples of hammerstone percussion marks (including pits and emanating patches of striae) (a) and a cutmark (with internal microstriations)(b). The other image (c) shows a close-up of* in situ *Member 3 sediment, which includes large angular clasts that held the potential to impart cutmark mimics on bone specimens.*

of materials ranging from clays to large angular clasts, which certainly held the potential to create abundant polish, abrasion and cutmark mimics on the Member 3 fossils. Thus, a configurational approach to cutmark identification, in which we considered anatomical placement as well as mark morphology, was absolutely necessary in this archaeofauna (see Binford, 1981; Domínguez-Rodrigo et al. 2005; Pickering et al., 2000).

## Results and discussion

### Skeletal element and taxonomic representation

Skeletal part representation of the Member 3 limb bone shaft assemblage is summarized by animal body size group in Table 1. At least two results emerge from consideration of these data that are generally relevant to the analysis of limb bone shaft specimens in zooarchaeology. While these are not the focus of this paper, we want to mention them and note that we are currently investigating them experimentally.

As is apparent from Table 2, identification of limb bone shaft specimens to specific element was accomplished most successfully for the remains of medium-sized animals (Size Classes 2 and 3), followed next by large animal specimens (Size Class 4 and larger) and last by small animal specimens (Size Class 1). The difference in proportion of specimens identified to skeletal element is statistically significant between Size Classes 1 and 2 ($X^2 = 51.238$, 1 d.f., $p<0.001$) and between Size Classes 1 and 3 ($X^2 = 52.867$, 1 d.f., $p<0.001$), but not between Size Classes 2 and 4 and larger ($X^2 = 0.502$, 1.d.f., $p<0.5$) or Size Classes 3 and 4 and larger ($X^2 = 1.068$, 1 d.f., $p<0.5$). Further, the differences between Size Classes 1 and 4 and larger are statistically significant ($X^2 = 17.276$, 1 d.f., $p<0.001$), while those between Size Classes 2 and 3 are not ($X^2 = 0.616$, 1 d.f., $p<0.5$). The lesser potential of shaft fragments from small ungulates to be identified to specific element is probably due to the absolutely smaller size of the preserved bone fragments from these diminutive animals.

Although shaft fragments from large-sized animals are only minimally less identifiable (i.e., statistically non-significant) than those from medium-sized animals, we noted that much more effort, time and consultation among analysts was involved in the assignment of large animal specimens to skeletal element than that for specimens from medium-sized animals. We hypothesize that this difference is at least in part because more of the total surface area of an absolutely larger bone is "featureless" than is that of an absolutely smaller bone. Thus, we

*Table 1. Skeletal element representation of Swartkrans Member 3 limb bone shaft fragments (number of identified specimens, NISP)[1]*

| | SMALL | MEDIUM | | LARGE | |
|---|---|---|---|---|---|
| Skeletal element | Size Class 1 | Size Class 2 | Size Class 3 | Size Class 4 | Size Class 5 |
| Humerus | 7 | 42 | 35 | 11 | |
| Radioulna | 13 | 32 | 20 | 5 | |
| Metacarpal | 14 | 22 | 15 | 4 | |
| Femur | 15 | 54 | 28 | 12 | |
| Tibia | 27 | 120 | 85 | 17 | 1 |
| Metatarsal | 12 | 44 | 31 | 5 | |
| Metapodial | 19 | 59 | 55 | 20 | |
| Upper | 23 | 30 | 24 | 9 | |
| Intermediate | 17 | 47 | 14 | 5 | |
| Limb bone shaft fragment | 152 | 124 | 66 | 16 | 1 |
| Total | 299 | 574 | 373 | 104 | 2 |

*1. Animal size classes are based on Brain's (1981) well-known system for antelope. Three separate categorizations of specimen identification are provided: (1) those specimens that could be identified to a specific element (above the third horizontal line); (2) those specimens that could be identified to a limb segment, as an upper (humerus or femur) or intermediate (radioulna or tibia) specimen but no further (between the third and fourth horizontal lines); (3) those specimens that could be identified as limb bone shaft fragments only (between the fourth and fifth horizontal lines). No lower limb bone specimens are listed in the second category, because they (by definition; see Domínguez-Rodrigo, 1997, 1999a) can be assigned more specifically as metapodials.*

*Table 2. Relative identifiability of Swartkrans Member 3 limb bone shaft fragments in different animal body size groups*[1]

| Size Class | Number identified to specific element | Number identified to limb segment only | Number identified to limb bone shaft only | Total |
|---|---|---|---|---|
| 1 | 88 (*29.4 %*) | 59 (*19.7 %*) | 152 (*50.8 %*) | 299 (*100.0 %*) |
| 2 | 314 (*54.6 %*) | 136 (*23.7 %*) | 124 (*21.6 %*) | 574 (*100.0 %*) |
| 3 | 214 (*57.4%*) | 93 (*24.9 %*) | 66 (*17.6 %*) | 373 (*100.0 %*) |
| 4 and 5 | 55 (*51.9 %*) | 34 (*32.1 %*) | 17 (*16.0 %*) | 106 (*100.0 %*) |
| Total | 671 (*49.6 %*) | 322 (*23.8 %*) | 359 (*26.6 %*) | 1352 (*100.0 %*) |

*1. Animal size classes are based on Brain's (1981) well-known system for antelope; in this study, Size Class 1 are considered small-sized animals, Size Classes 2 and 3 are considered medium-sized animals and Size Class 4 and larger are considered large animals. The second column lists the number of identified specimens (NISP) identified as humeri, radioulnae, metacarpals, femora, tibiae and metatarsals in each body size class. The third column lists the NISP for those fragments identified as upper limb pieces (humerus or femur), intermediate limb pieces (radioulna or tibia) or metapodial pieces. The fourth column lists the NISP for those fragments identified only as limb bone shaft pieces.*

predict that when a large-sized animal limb bone and a medium-sized animal limb bone are each comminuted heavily and equivalently, any given fragment of the large animal bone is less likely to preserve a landmark, or part thereof, useful for skeletal part identification than is any given fragment from the small animal bone. In addition, our observations indicate that any given shaft fragment from a large bone is more likely to be relatively "straight" (i.e., without clear indication of extrapolated cross-sectional shape and incipient curvature) than that of a fragment from a smaller bone. This is pertinent because cross-sectional shape and curvature are element-specific and thus useful features for distinguishing different limb bones.

The second point relevant to identification analysis in limb bone shaft studies concerns the high proportion of hindlimb specimens that we identified to specific skeletal element relative to forelimb specimens (Table 3). More particularly, the tibia possesses the highest NISP values of any specific limb bone across all body size classes. Femur NISP counts are ranked second in three of four cases. In contrast, radioulna NISP counts are second-to-last in rank in three of four cases and metacarpals score last in three of four cases. Perusal of some other well-known Pleistocene archaeofaunas reveals a similar pattern of tibia-highest representation, based on NISP and minimum number of elements (MNE) at FLK *Zinj* and FLKN levels 1 – 2 (Olduvai, Tanzania) (Bunn, 1986; Bunn and Kroll, 1986) and Kobeh Cave (Iran) (Marean and Kim, 1998), among other sites (Pickering et al., 2006). As we contend above that cross-sectional shape is the likely determining factor in the differential success of identifying small versus large animal limb shafts, our initial experimental results suggest the same for hindlimb versus forelimb elements (Pickering et al., 2006).

## Bone surface modifications

Table 4 provides a summary of prehistoric bone surface modifications in the Member 3 limb bone shaft subassemblage (see also Appendix). These data, analyzed in

*Table 3. Number of identified specimens (NISP) rank order for Swartkrans Member 3 limb bone shaft fragments identified to specific skeletal element*[1]

| Rank | Size Class 1 | Size Class 2 | Size Class 3 | Size Class 4 | Size Class 5 |
|---|---|---|---|---|---|
| 1 | Tibia | Tibia | Tibia | Tibia | Tibia |
| 2 | Femur | Femur | Humerus | Femur | |
| 3 | Metacarpal | Metatarsal | Metatarsal | Humerus | |
| 4 | Radioulna | Humerus | Femur | Metatarsal | }Tied |
| 5 | Metatarsal | Radioulna | Radioulna | Radioulna | |
| 6 | Humerus | Metacarpal | Metacarpal | Metacarpal | |

*1. Animal size classes are based on Brain's (1981) well-known system for antelope.*

combination with skeletal part data, can usefully inform about the relative contribution of hominids and large carnivores to the formation of the Member 3 fauna. In addition, consideration of the hominid-imparted modifications in isolation allows for specific inferences of the carcass-acquiring abilities of hominids.

### *Assessing the relative contribution of hominids and carnivores to assemblage formation*

When the limb bone shaft subassemblage is viewed as a whole (NISP = 12,505), both hominid-modified specimens (n = 163; 1.3 % of the total NISP) and carnivore tooth-marked specimens (n = 532; 4.3 % of the total NISP) are present at very low frequencies. A majority of the hominid-modified specimens bear damage inferred to be indicative of hammerstone percussion: 53 specimens are classified as impact flakes and an additional 50 preserve percussion pits, striae and/or notches (Table 4; Appendix). Collectively, these percussion-created specimens account for just 0.8 % of the total limb bone subassemblage NISP. A smaller number of 60 specimens (0.5 % of the total limb bone subassemblage) preserve cutmarks (Pickering et al., 2004a) (Table 4; Appendix).

The role of hominids and carnivores in the formation of the Member 3 fauna can be characterized more specifically when the data are examined by Size Class (Table 4). Carnivores were clearly the most active modifiers of Size Class 1 carcasses; 73.6 % of Size Class 1 limb bone specimens exhibit tooth marks. Tooth pit dimensions implicate leopards as one likely modifier of these small carcasses (Pickering et al., 2004b). Although tooth-marked specimens continue to appear in higher frequencies than hominid-modified specimens on Size Class 2, 3 and 4 remains, they are tooth-marked in lower frequencies compared to specimens from Size Class 1 carcasses. This coincides with an increase in hominid damage within these larger Size Classes, particularly in the frequency of cutmarked specimens, suggesting that hominids were a more active, though certainly not major, collector of especially Size Class 2 and 3 carcasses (Figure 2). Finally, limb bone fracture patterns support the suggestion that carnivores played a more important role in bone accumulation relative to hominids; fracture angle data indicate that a significant portion of green breakage in the Member 3 fauna was initiated through static loading characteristic of carnivore feeding (Pickering et al., 2005).

In order to more accurately assess the relative contributions of hominids and carnivores in assemblage formation we examined the Member 3 data within a comparative framework based on actualistic datasets of known derivation. The Member 3 bone surface modification percentages fall far short of experimental standards for both human- and carnivore-processed limb bones. For example, Blumenschine (1995) found that in his "carnivore-only" experiments on average 83.9 % of limb bone specimens are tooth-marked. Blumenschine and Selvaggio (1988) report that ~30 % of the total specimens in their experimental sample of hammerstone-broken limb bones bear at least one percussion mark. Finally, ethnoarchaeological and experimental studies indicate that hominid tool-assisted defleshing results in 15 – 30% of specimens bearing cutmarks (Bunn, 1982; Domínguez-Rodrigo, 1997, 1999a; Lupo and O'Connell, 2002).

However, as discussed above in the **Materials and Methods**, we made several adjustments to the Swartkrans data in Analytical Set II that renders them more comparable to the actualistic data and changes the compared fossil NISP to 3,151. This adjustment to the compared NISP value slightly alters bone surface mark percentages. When controlled for diagenetic fragmentation and cortical surface preservation, the frequencies of hominid- and carnivore-modified specimens are broadly similar (Table 5).

Examination of Table 5 reveals that when compared to the actualistic controls even the adjusted values in Analytical Set II are inconsistent with scenarios of intense hominid or carnivore involvement in the formation of the Member 3 accumulation. This is not surprising considering the depositional nature and time depth of the Member 3 fauna: like most other South African cave assemblages, the Member 3 assemblage was formed, at least in part, by secondarily deposited material derived from the cave's surface catchment. Over long periods of time it is likely that abiotic processes in addition to biotic actors not dealt with systematically in this analysis (e.g., rodents) contributed significantly to assemblage formation. Regardless, the adjusted bone surface damage frequencies suggest that hominids and carnivores contributed similarly to assemblage formation. Differential fragmentation supports this contention. A higher proportion of carnivore-modified specimens is comprised of pieces <2 cm in maximum dimension (68/532 = 12.8 %) relative to hominid-modified specimens (4/163 = 2.5 %), indicating that carnivore-modified specimens are more heavily fragmented than their hominid-modified counterparts. As Bartram (1993) has pointed out, intense fragmentation can artificially increase bone modification values based on NISP, in this case carnivore-modified specimens relative to hominid-modified specimens. Only an expanded analysis of the remaining skeletal parts will tell whether this suggestion can be applied to the Member 3 fauna as a whole.

In summary, there are low frequencies of both hominid and carnivore damage in the Member 3 fauna. Stratifying the sample by Size Class reveals that carnivores were the major modifiers of Size Class 1 carcasses, while hominids played their most significant role in Size Class 2 and 3 carcass modification. Thus, the bone surface damage evidence presented in this study supplements Brain's (1993) earlier arguments by indicating an important hominid contribution to assemblage formation in Member 3 times.

*Table 4. Summary of prehistoric bone surface modifications in the Swartkrans Member 3 limb bone shaft subassemblage[1]*

| Size Class | Skeletal element | NISP | Cutmarks | Percussion marks[2] | Tooth marks |
|---|---|---|---|---|---|
| 1 | | | | | |
| | Humerus | 7 | 1 (14.2) | | 3 (42.9) |
| | Radioulna | 13 | | | 7 (53.9) |
| | Metacarpal | 14 | | 3 (21.4) | 9 (64.3) |
| | Femur | 15 | | 3 (20.0) | 4 (26.7) |
| | Tibia | 27 | 1 (3.7) | 2 (7.4) | 14 (51.9) |
| | Metatarsal | 12 | 1 (8.3) | | 5 (41.7) |
| | Upper | 23 | 1 (4.3) | | 22 (95.7) |
| | Intermediate | 17 | | | 15 (88.2) |
| | Metapodial | 19 | 1 (5.3) | | 15 (79.0) |
| | Limb bone shaft | 152 | | 2 (1.3) | 127 (83.6) |
| | Total | 299 | 5 (1.7) | 9 (3.0) | 217 (73.6) |
| 2 | | | | | |
| | Humerus | 42 | 2 (4.8) | 3 (7.1) | 4 (9.5) |
| | Radioulna | 32 | 2 (6.3) | 3 (9.4) | 5 (15.6) |
| | Metacarpal | 22 | 2 (9.1) | 2 (9.1) | 10 (45.5) |
| | Femur | 54 | 5 (9.3) | 3 (5.7) | 13 (24.5) |
| | Tibia | 120 | 6 (5.0) | 4 (3.3) | 21 (17.5) |
| | Metatarsal | 44 | 2 (4.6) | 1 (2.3) | 8 (18.2) |
| | Upper | 30 | 4 (13.3) | | 14 (46.7) |
| | Intermediate | 47 | 1 (2.1) | | 15 (32.6) |
| | Metapodial | 59 | 2 (3.4) | | 14 (23.7) |
| | Limb bone shaft | 124 | 6 (4.8) | 4 (3.2) | 22 (17.7) |
| | Total | 573 | 32 (5.6) | 20 (3.5) | 126 (22.1) |
| 3 | | | | | |
| | Humerus | 35 | 2 (5.7) | 3 (8.6) | 8 (22.9) |
| | Radioulna | 20 | 2 (10.0) | | 6 (30.0) |
| | Metacarpal | 15 | 1 (6.7) | 1 (6.7) | 4 (26.7) |
| | Femur | 28 | 3 (10.7) | 2 (7.1) | 10 (35.7) |
| | Tibia | 85 | 3 (3.5) | 4 (4.7) | 29 (34.1) |
| | Metatarsal | 31 | 3 (9.7) | 1 (3.2) | 9 (29.0) |
| | Upper | 24 | 3 (12.5) | | 7 (29.2) |
| | Intermediate | 14 | | | 6 (42.9) |
| | Metapodial | 55 | 1 (1.8) | 1 (1.8) | 17 (30.9) |
| | Limb bone shaft | 66 | 1 (1.5) | 1 (1.5) | 20 (30.3) |
| | Total | 373 | 19 (5.1) | 13 (3.5) | 116 (31.1) |
| 4 | | | | | |
| | Humerus | 11 | 2 (18.2) | | |
| | Radioulna | 5 | 1 (20.0) | | 1 (20.0) |
| | Metacarpal | 4 | | 1 (25.0) | |
| | Femur | 12 | | | 2 (16.7) |
| | Tibia | 17 | | | 7 (41.2) |
| | Metatarsal | 5 | | | 1 (20.0) |
| | Upper | 9 | | | |
| | Intermediate | 5 | | | |
| | Metapodial | 20 | 1 (5.0) | 2 (10.0) | 1 (5.0) |
| | Limb bone shaft | 16 | | | |
| | Total | 104 | 4 (3.9) | 3 (2.9) | 12 (11.5) |

*1. Animal size classes are based on Brain's (1981) well-known system for antelopes. Parenthetical values in the fourth through sixth columns are percentages of the total number of identified specimens (NISP) for any row. Indeterminately identified pieces are those specimens that could be identified to a limb segment, as an upper (humerus or femur) or intermediate (radioulna or tibia) specimen but no further.*

*2. Percussion marks = pits and striae, in some cases associated with impact notches. Five additional specimens preserve notches only and a separate total of 53 impact flake specimens have been recovered from Swartkrans Member 3.*

*Table 4. Summary of prehistoric bone surface modifications in the Swartkrans Member 3 limb bone shaft subassemblage (continued)*[1]

| Size Class | Skeletal element | NISP | Cutmarks | Percussion marks[2] | Tooth marks |
|---|---|---|---|---|---|
| 1 – 4 (total) | Upper total | 290 | 23 (8.0) | 13 (4.5) | 87 (30.0) |
| | Humerus | 95 | 7 (7.4) | 6 (6.3) | 15 (15.8) |
| | Femur | 109 | 8 (7.3) | 7 (6.4) | 29 (26.6) |
| | Indeterminate | 86 | 8 (9.3) | | 43 (50.0) |
| | Intermediate total | 402 | 16 (4.0) | 13 (3.2) | 126 (31.3) |
| | Radioulna | 70 | 5 (7.1) | 3 (4.3) | 19 (27.1) |
| | Tibia | 249 | 10 (4.0) | 10 (4.0) | 71 (28.5) |
| | Indeterminate | 83 | 1 (1.2) | | 36 (43.4) |
| | Lower total | 300 | 14 (4.7) | 12 (4.0) | 93 (31.0) |
| | Metacarpal | 55 | 3 (5.5) | 7 (12.7) | 23 (41.8) |
| | Metatarsal | 92 | 6 (6.5) | 2 (2.8) | 23 (25.0) |
| | Metapodial | 153 | 5 (3.3) | 3 (2.0) | 47 (30.7) |

1. *Animal size classes are based on Brain's (1981) well-known system for antelopes. Parenthetical values in the fourth through sixth columns are percentages of the total number of identified specimens (NISP) for any row. Indeterminately identified pieces are those specimens that could be identified to a limb segment, as an upper (humerus or femur) or intermediate (radioulna or tibia) specimen but no further.*

2. *Percussion marks = pits and striae, in some cases associated with impact notches. Five additional specimens preserve notches only and a separate total of 53 impact flake specimens have been recovered from Swartkrans Member 3.*

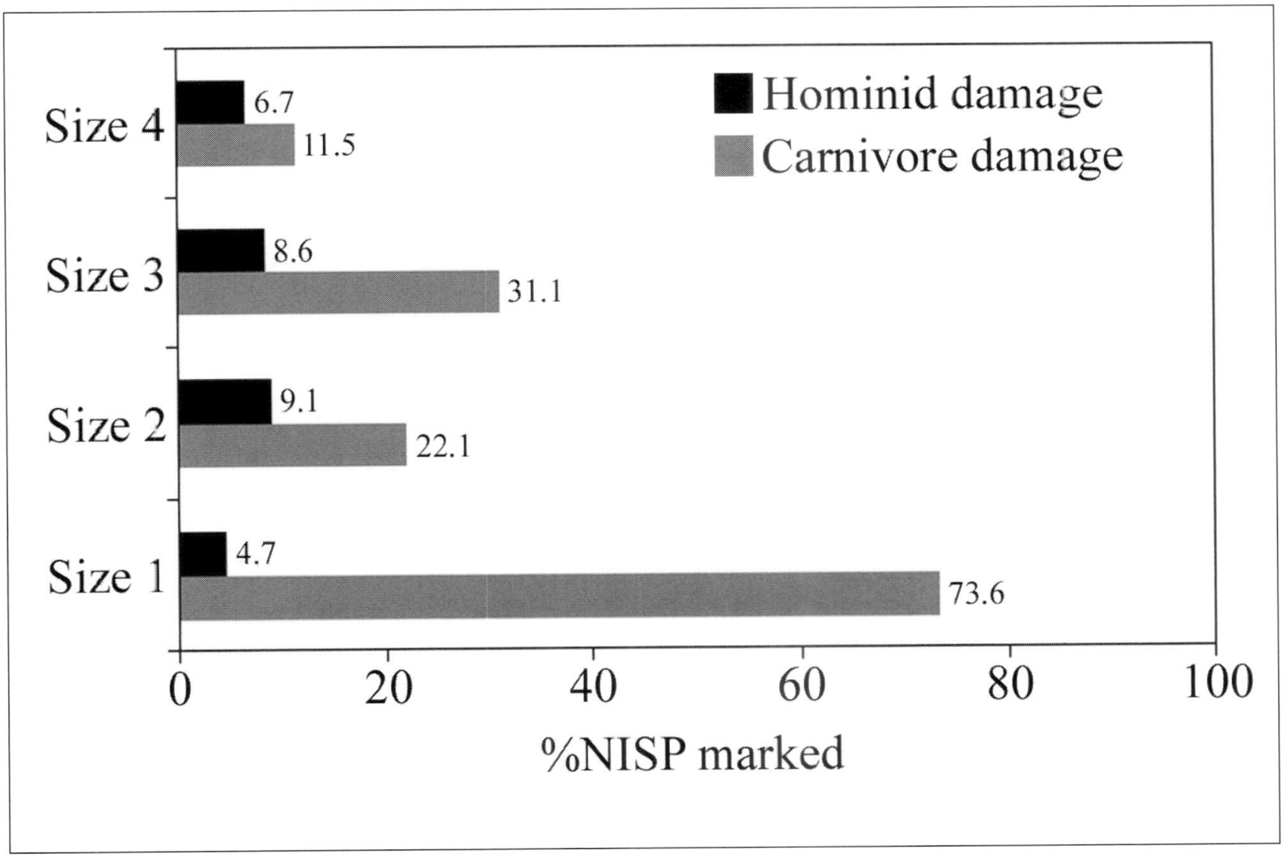


*Figure 2. Frequencies of modified limb bone shaft specimens summarized by animal size class (see Brain, 1974, 1981 for animal body size classes). Hominid damage = cutmarked plus percussed specimens; Carnivore damage = tooth marked specimens; %NISP marked = percentage of total number of identified specimens modified. Note that except for the high percentage of tooth marked Size Class 1 specimens, there is a relatively low proportion of both types of damage preserved across animal body sizes.*

*Table 5. Summary of prehistoric bone surface modifications in Analytical Set II (the adjusted Swartkrans Member 3 limb bone shaft sample)[1]*

| | Cutmarks | Percussion marks[2] | Percussion flakes | Total hominid modified | Total carnivore modified |
|---|---|---|---|---|---|
| N | 19 | 28 | 50 | 97 | 135 |
| % of total adjusted NISP | 0.6 | 0.9 | 1.6 | 1.5 | 4.3 |

*1. Data adjusted to make the sample comparable to modern actualistic samples; adjustments modified the total number of identified specimens (NISP) in the Swartkrans assemblage to 3,151.*

*2. Percussion marks = pits and striae, in some cases associated with impact notches.*

## *Assessing the carcass acquisition and exploitation abilities of hominids*

Data on the anatomical locations of stone tool cutmarks, on both the intra-skeletal and intra-bone levels, are the most convincing and direct indications of the timing of hominid access to large animal carcasses (e.g., Bunn, 1982, Bunn and Kroll, 1986; Domínguez-Rodrigo, 1997, 1999a, 2002; Domínguez-Rodrigo and Pickering, 2003; Pickering and Domínguez-Rodrigo, in press) (Figure 3). Obviously, the Member 3 sample discussed here is biased because limb bone shafts are the only type of specimen that we examined. However, we did study *all* shaft fragments, regardless of element, and also stratified the sample by animal body size. Thus, the restricted analysis is more informative than might be supposed initially.

Considering the remains of all size classes from Member 3 combined, 8.0 % of all upper limb bone specimens are cutmarked, while 4.0 % of intermediate specimens are cutmarked and 4.7 % of lower specimens are cutmarked (Figure 4; Table 4). These differences in cutmarked percentages approach more closely statistical significance between upper and intermediate specimens ($X^2 = 4.95$, 1 d.f., $p<0.05$) and between upper and lower specimens ($X^2 = 2.57$, 1 d.f., $p<0.2$), than between intermediate and lower specimens ($X^2 = 0.18$, 1 d.f., $p<0.5$).

We believe, however, that a more elucidating comparison is that of cutmarked percentages for upper and intermediate specimens *combined* (n = 39, 5.6 % of the total upper plus intermediate NISP) with that of lower specimens. With regard to the distribution of overlying meat, the distinction between these two limb segment groups is profound: midshaft portions of ungulate upper and intermediate limb bones are heavily muscled, while no appreciable meat covers these portions on metapodials. Thus, the differential distribution of cutmarks on the midshafts of upper and intermediate limb segments compared to metapodials is informative behaviorally. Interestingly, in the total Member 3 sample, there is no statistically significant difference in cutmark percentages between these grouped limb segments (i.e., upper plus intermediate versus lower: $X^2 = 0.41$, 1 d.f., $p<0.5$), indicating that hominids were removing overlying soft tissues from *all* classes of limb bones at nearly equivalent frequencies. This suggests that hominids may have, at least occasionally, gained access to fleshed whole limb units of various sized ungulates that they then processed completely for overlying soft tissues, from humerus-to-metacarpal and femur-to-metatarsal. Statistically non-significant differences in cutmark percentages between each limb segment (upper, intermediate, lower) supports this contention for ungulate remains of every size: small ($X^2 = 0.72$, 2 d.f., $p<0.5$); medium ($X^2 = 5.26$, 2 d.f., $p<0.1$); large ($X^2 = 0.37$, 2 d.f., $p<0.5$) (Figure 5).

Cutmarks on metapodial midshafts indicate the removal of skin and/or tendons. In addition to the anatomical fact that a paucity of meat is available on these bones, numerous ethnoarchaeological and experimental observations (e.g., Bartram, 1993; Binford, 1978, 1981; Binford and Bertram, 1977; Bunn, 2001; Domínguez-Rodrigo, 1997, 1999a; Nilssen, 2000) also corroborate this inference. Whether skin and/or tendons were the *actual object* of hominid butchery directed at the Swartkrans metapodials is difficult to infer. In ethnoarchaeological and experimental contexts, metapodial skinning is an important initial step in at least two fundamental butchery scenarios, which are usually not mutually exclusive. The first is to simply remove skin from the whole limb unit (or, in most cases, whole carcass) for eventual defleshing. Second, metapodials are often skinned to prepare bones (i.e., expose cortical surfaces) for subsequent marrow extraction by hammerstone percussion.

Very few specimens in the Member 3 sample preserve the co-occurrence of cutmarks and hammerstone percussion marks (n = 4; only one of which is a metapodial specimen). Such a co-occurrence might be predicted on metapodial specimens had cutmarking activities been conducted to simply prepare bones for hammerstone percussion. However, we note that no refitting of specimens was attempted. Thus, whole limb bones may have been processed for overlying tissues and then broken open, resulting in currently disassociated fragments from the same original element, some of which preserve cutmarks and others that preserve percussion marks. Within the pooled sub-sample of hominid-modified specimens, percussion mark frequencies by individual element and

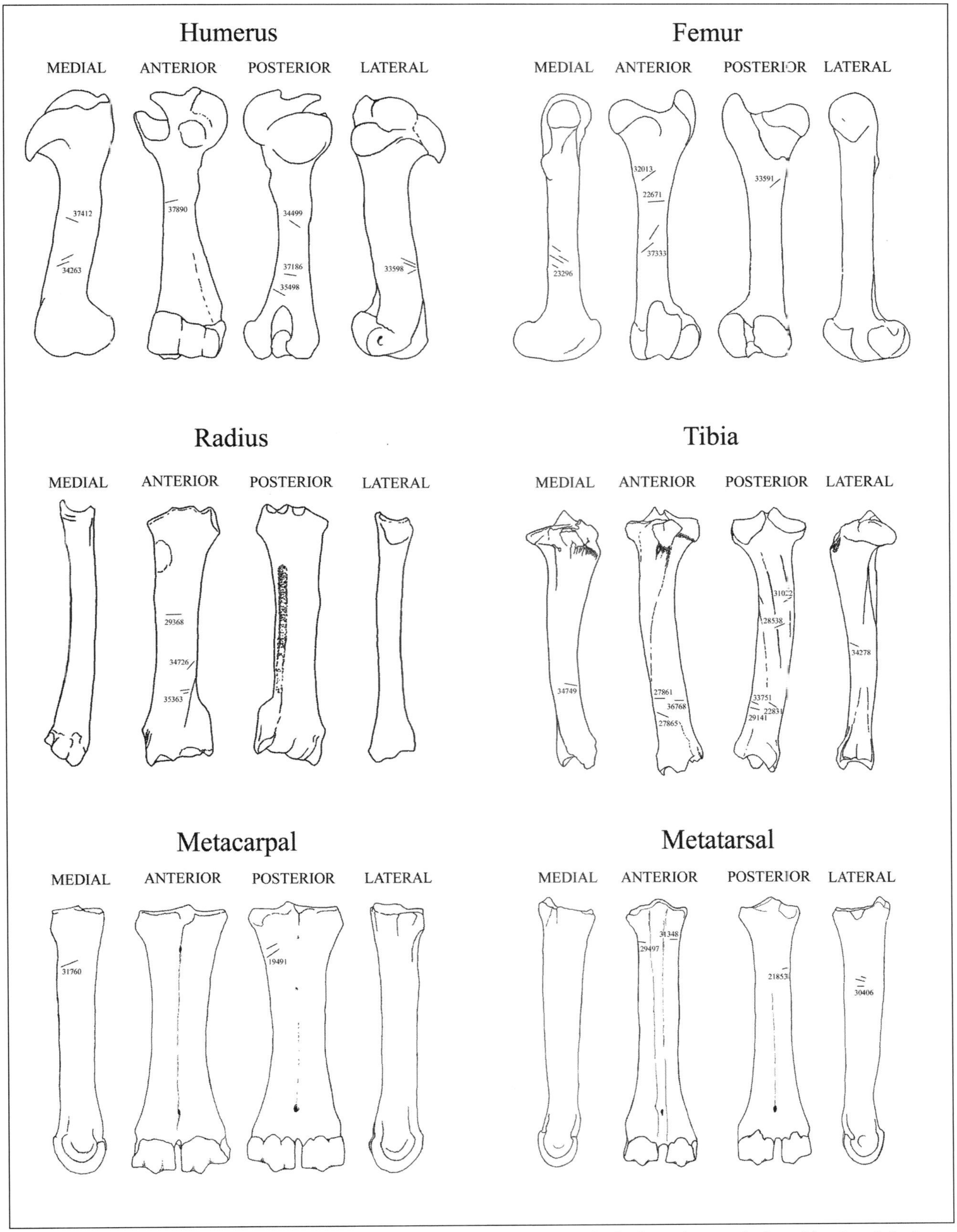


*Figure 3. Lines on limb bone templates showing the distribution of cutmarks in the sub-sample from Member 3 that was identified to skeletal element. Cutmarks occur on elements from both sides of the body, but left limb bones are used as the standard templates in this figure. Specimen catalog numbers are indicated next to the corresponding cutmarks; all catalog numbers are preceded by SKX prefixes, which are dropped in this figure. Several specimens with cutmarks were identifiable to element, but they could not be placed exactly in position on the element templates, so those cutmarks are not illustrated in this figure. Those unrepresented specimens are: SKX 24494 (radioulna); 25304 (metacarpal); 28786 (femur); 30429 (femur); 31396 (femur); 37424 (metatarsal); 37540 (radioulna); 45748 (metatarsal).*

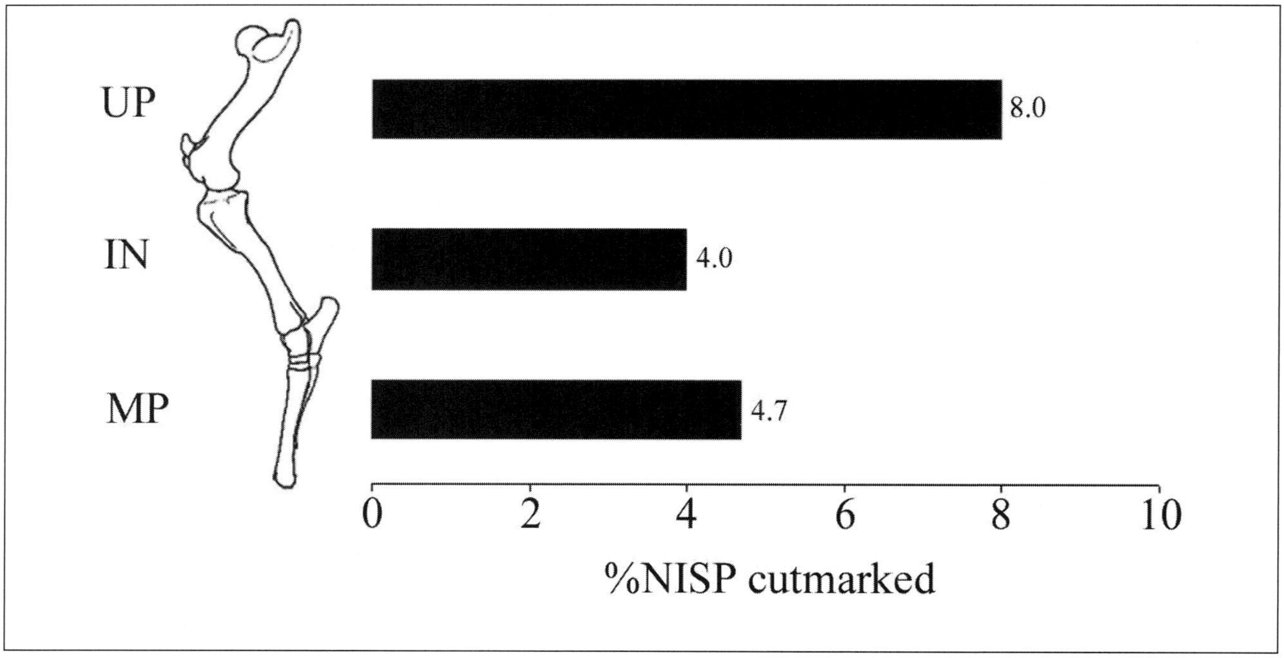


*Figure 4. Cutmark percentages in the Member 3 limb bone shaft assemblage for all animal body size classes combined (see Brain, 1974, 1981 for animal body size classes). Abbreviations: UP = upper limb bones (humerus plus femur); IN = intermediate limb bones (radioulna plus tibia); MP = metapodials; %NISP = percentage number of identified specimens.*

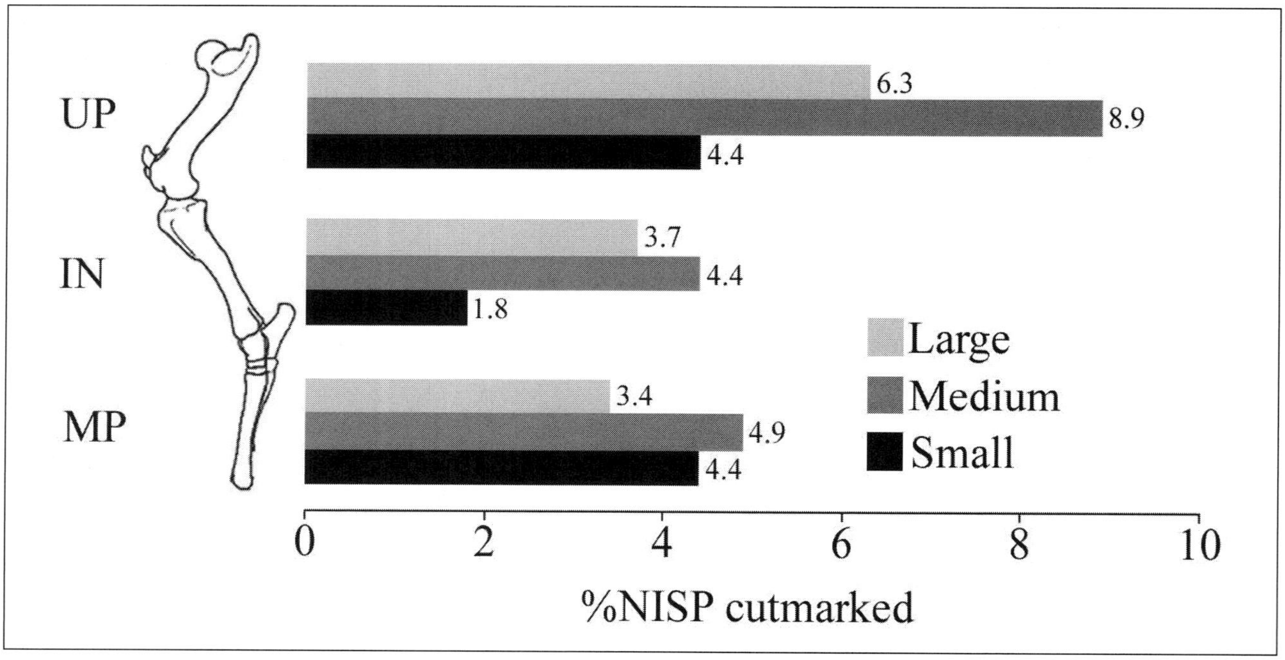


*Figure 5. Cutmark percentages in the Member 3 limb bone shaft assemblage broken down by animal body size: small = Size Class 1; medium = Size Classes 2 and 3; Large = Size Class 4 and above (see text for explanation and Brain, 1974, 1981 for animal body size classes). Abbreviations: UP = upper limb bones (humerus plus femur); IN = intermediate limb bones (radioulna plus tibia); MP = metapodials; %NISP = percentage number of identified specimens.*

limb segment are not significantly different from cutmark frequencies (e.g., for limb segment: $X^2 = 1.45$, 2 d.f., $p<0.5$). This seems to suggest fairly complete processing (i.e., both soft tissue removal and marrow extraction) of those limb bones that hominids acquired.

Less ambiguous for inferences of the carcass acquisition and utilization capabilities of Swartkrans hominids is the presence of cutmarked upper and intermediate midshaft limb bone specimens in the Member 3 sample. Even if the hypothesis of whole limb unit acquisition and deposition by hominids is false for Swartkrans Member 3, the fact that cutmarked upper and intermediate limb bone midshaft specimens have been identified still indicates early access by hominids to animal carcasses. The midshaft portion of upper and intermediate limb bones is a region defleshed early in the feeding sequence of a carnivore that has primary access to a carcass. For example, Domínguez-Rodrigo (1999b) observed that upper and intermediate limb bones from 28 ungulate carcasses displayed a paucity of adhering flesh after ravaging by lions; midshaft sections on upper limbs in this dataset displayed a *complete* lack of flesh scraps, while flesh scraps on the midshaft portions of intermediate limb bones were poorly represented after lion ravaging. Assuming that the prehistoric carnivores of the Sterkfontein Valley operated similarly, there would be no reason for hominids to have imparted cutmarks on upper and intermediate limb bone midshafts had they been relegated to scavenging passively (i.e., late access to carcasses) from the remains of picked-over carnivore kills (see, Bunn, 2001; Domínguez-Rodrigo, 2002; Domínguez-Rodrigo and Pickering, in press; Pickering and Domínguez-Rodrigo, 2004). No flesh would have been present on those bone portions in that scenario and thus there would be no reason for hominids to put a stone tool edge to upper and intermediate bone midshafts; in fact, there would be good reason *not* to do this because slicing into bone simply dulls the cutting edge of a tool (e.g., Bunn, 2001). Experimental butchery data corroborate the eloquent argument based on logic that cutmarks are unexpected on previously defleshed limb bone midshafts. For example, Domínguez-Rodrigo's (1997, 1999a) and Nilssen's (2000) large, modern datasets demonstrate convincingly that cutmarks from activities *other than defleshing* (i.e., skinning, disarticulation) *almost never* occur on upper and intermediate limb bone midshafts.

## Summary and conclusions

At face value, the Swartkrans Member 3 fauna would appear to be of fairly low integrity, and thus its potential for reconstructing early hominid carcass foraging minimal. However, the Member 3 assemblage preserves a much lower proportion of single bone specimens that have co-occuring hominid- and carnivore-derived surface modifications than do modern actualistic assemblages derived by the interdependent actions of both agents. This suggests instead that the fossil assemblage can actually be divided into two fairly independently formed and high integrity sub-assemblages—one created largely by the actions of hominids and the other created largely by the actions of carnivores (see also, Egeland et al., 2004; Pickering et al., 2004a, 2005). Overall, tooth-marked specimens are approximately three and a half times as common as hominid-modified specimens in the limb bone shaft subassemblage as a whole. However, when taking into account diagenetic breakage, cortical surface preservation and differential fragmentation, hominids and carnivores appear to have contributed similarly to the formation of the Member 3 limb bone shaft subassemblage.

Additionally important is the finding that evidence of hominid activity that *is* preserved is informative behaviorally. Cutmarks and percussion marks are distributed fairly evenly across all limb elements, suggesting fairly complete processing of whole limb units by hominids. By extrapolation, this might mean that hominids were acquiring whole carcasses for processing. Based on actualistic observations, cutmarks on intermediate and especially upper limb bone midshaft specimens indicate, at the very least, early access to carcass parts typically defleshed completely by primary carnivores early in their feeding sequences.

With addition of this new cutmark data from Swartkrans Member 3, the southern-most continental datum so far known, a pattern in the zooarchaeology of Early Stone Age Africa is confirmed. As with the Member 3 archaeofauna, cutmarks occur on upper and intermediate limb bone midshafts in the important assemblages from FLK *Zinj*, BK, FxJj 50 and the ST site complex (e.g., Bunn, 1982; Bunn and Kroll, 1986; Domínguez-Rodrigo, 2002; Domínguez-Rodrigo et al., 2002; Oliver, 1994; Monahan, 1996; see also Domínguez-Rodrigo et al., 2005)—suggesting hominid access at all sites to the largely fleshed carcasses of ungulates and contradicting predictions of passive scavenging models.

Given the extreme polarization of research groups working on the issue of early hominid access to large animal carcasses, it seems unlikely that our conclusions will be embraced by all. However, we will still be very gratified if this study accomplishes another broader goal of bringing important South African zooarchaeological data into the ongoing consideration by paleoanthropologists of this important topic.

## Acknowledgments

The first three authors owe a remarkable debt to Bob Brain for extending his unselfish encouragement to us to undertake this re-analysis of the Swartkrans archaeofauna. He is an inspiration and example to each of us, and it is a true honor to work with him on this project; his support, collaboration and friendship are appreciated always and gratefully acknowledged here. Thanks also go to Kathy Kuman and Ron Clarke for their continued friendship and hospitality, as well as to the Department of Archaeology at the University of the Witwatersrand. In addition, we thank the Northern Flagship Institution (formerly the Transvaal Museum) and especially Stephany Potze and Francis Thackeray for granting us permission to study the material and for facilitating its convenient accessibility. TRP was supported by a Summer Faculty Fellowship from the College of Arts and Science, Indiana University. He thanks Nick Toth and Kathy Schick for their steadfast encouragement in all things and especially for serving as co-organizers and co-editors on the *African Taphonomy* project. His thanks are also extended to the conference participants, contributors to this volume and to all the folks at the Stone

Age Institute who helped make this project an incredible success. Finally, he thanks his family for their continued support and understanding.

## References

Alcántara García, V., Barba Egido, R., Barral del Pino, J.M., Crespo Ruiz, A.B., Eiriz Vidal, A.I., Falquina Aparicio, Á., Herrero Calleja, S., Ibarra Jiménez, A., Megías González, M., Pérez Gil, M., Pérez Tello, V., Rolland Calvo, J., Yravedra Sáinz de los Terreros, J., Vidal, A., Domínguez-Rodrigo, M. in press. Determinación de procesos de fractura sobre huesos frescos: Un sistema de análisis de los ángulos de los planos de fracturación como discriminador de agentes bióticos. Complutum.

Barba Egido, R., Domínguez-Rodrigo, M. 2005. The taphonomic relevance of the analysis of long limb bone shaft features and their application to implement element identification: Study of bone thickness and the morphology of the medullary cavity. Journal of Taphonomy 3, 17-42.

Bartram, L.E. 1993. An ethnoarchaeological analysis of Kua San (Botswana) bone food refuse. Ph.D. Dissertation, University of Wisconsin-Madison.

Behrensmeyer, A.K. 1978. Taphonomic and ecologic information from bone weathering. Paleobiology 4, 150-162.

Behrensmeyer, A.K. Gordon, K.D., Yanagi, G.T. 1986. Trampling as a cause of bone surface damage and pseudo-cutmarks. Nature 319, 768-771.

Behrensmeyer, A.K., Gordon, K.D., Yanagi, G.T. 1989. Non-human bone modification in Miocene fossils from Pakistan. In: Bonnichsen, R., Sorg, M.H. (Eds.), Bone Modification. Center for the Study of the First Americans, Orono (ME), pp. 99-120.

Binford, L.R. 1978. Numamuit Ethnoarchaeology. Academic Press, New York

Binford, LR. 1981. Bones: Ancient Men and Modern Myths. Academic Press, New York.

Binford, L.R. 1985. Human ancestors: Changing views of their behavior. Journal of Anthropological Archaeology 4, 292-327.

Binford, LR. 1988. The hunting hypothesis, archaeological methods and the past. Yearbook of Physical Anthropology 30, 1-9.

Binford, L.R., Bertram, J.B. 1977. Bone frequencies—and attritional processes. In: Binford, L.R. (Ed.), For Theory Building in Archaeology. Academic Press, New York, pp. 77-153.

Bishop, J.E., Blumenschine, R.J. 1994. Surface marks on long bones from Makapansgat, Sterkfontein, and Swartkrans an their taphonomic implications. Paper presented at the 12th Biennial Conference of the Society of Africanist Archaeologists, April 28, 1994.

Blumenschine, R.J. 1986. Early hominid scavenging opportunities: Implications of carcass availability in the Serengeti and Ngorongoro ccosystems. British Archaeological International Series, 283, Oxford.

Blumenschine, R.J. 1987. Characteristics of an early hominid scavenging niche. Current Anthropology 28, 383-407.

Blumenschine, R.J. 1988. An experimental model of the timing of hominid and carnivore influence on archaeological bone assemblages. Journal of Archaeological Science 15, 483-502.

Blumenschine, R.J. 1991. Hominid carnivory and foraging strategies, and the socio-economic function of early archaeological sites. Philosophical Transactions of the Royal Society (London) 334, 211-221.

Blumenschine, R.J. 1995. Percussion marks, tooth marks and the experimental determinations of the timing of hominid and carnivore access to long bones at FLK *Zinjanthropus*, Olduvai Gorge, Tanzania. Journal of Human Evolution 29, 21-51.

Blumenschine, R.J., Marean, C.W. 1993. A carnivore's view of archaeological bone assemblages. In: Hudson, J. (Ed.), From Bones to Behavior: Ethnoarchaeological and Experimental Contributions to the Interpretation of Faunal Remains.Southern Illinois University Press, Carbondale (IL), pp. 273-300.

Blumenschine, R.J., Selvaggio, M.M. 1988. Percussion marks on bone surfaces as a new diagnostic of hominid behavior. Nature 333, 763-765.

Blumenschine, R. J., Selvaggio, M. M. 1991. On the marks of marrow bone processing by hammerstones and hyaenas: Their anatomical patterning and archaeological implications. In: Clark, J.D. (Ed.), Cultural Beginnings: Approaches to Understanding Early Hominid Life-ways in the African Savanna. Dr. Rudolf Habelt GMBH, Bonn, pp. 17-32.

Blumenschine, R.J., Cavallo, J.A., Capaldo, S.D. 1994. Competition for carcasses and early hominid behavioral ecology: A case study and conceptual framework. of early archaeological sites. Journal of Human Evolution 27, 197-213.

Blumenschine R.J., Marean C.W., Capaldo S.D. 1996. Blind tests of inter-analyst correspondence and accuracy in the identification of cut marks, percussion marks, and carnivore tooth marks. Journal of Archaeological Science 23, 493-507.

Brain, C.K. 1974. Some suggested procedures in the analysis of bone accumulations from southern African Quaternary sites. Annals of the Transvaal Museum 29, 1-8.

Brain, C.K. 1981. The Hunters or the Hunted? An Introduction to African Cave Taphonomy. University of Chicago Press, Chicago.

Brain, C.K. 1993. Taphonomic overview of the Swartkrans fossil assemblages. In: Brain, C.K. (Ed.), Swartkrans: A Cave's Chronicle of Early Man. Transvaal Museum, Pretoria, pp. 255-264.

Brain, C.K., Shipman, P. 1993. The Swartkrans bone tools. In: Brain, C.K. (Ed.), Swartkrans: A Cave's Chronicle of Early Man. Transvaal Museum, Pretoria, pp. 195-215.

Brain, C.K., Sillen, A. 1988. Evidence from the Swartkrans Cave for the earliest use of fire. Nature 336, 464-466.

Bunn, H.T. 1981. Archaeological evidence for meat-eating by Plio-Pleistocene hominids from Koobi Fora, Kenya. Nature 291, 574-577.

Bunn, H.T. 1982. Meat-eating and human evolution: Studies on the diet and subsistence patterns of Plio-Pleistocene hominids in East Africa. Ph.D. Dissertation, University of California, Berkeley.

Bunn, H.T. 1983. Evidence on the diet and subsistence patterns of Plio-Pleistocene hominids at Koobi Fora, Kenya, and at Olduvai Gorge, Tanzania. In: Clutton-Brock, J., Grigson, C. (Eds.), Animals and Archaeology, Volume 1: Hunters and Their Prey. British Archaeological Reports International Series, 163, Oxford, pp. 21-33.

Bunn, H.T. 1986. Patterns of skeletal representation and hominid subsistence activities at Olduvai Gorge, Tanzania, and Koobi Fora, Kenya. Journal of Human Evolution 15, 673-690.

Bunn, H.T. 1991. A taphonomic perspective on the archaeology of human origins. Annual Review of Anthropology 20, 433-467.

Bunn, H.T. 1994. Early Pleistocene hominid foraging strategies along the ancestral Omo River at Koobi Fora, Kenya. Journal of Human Evolution 27, 247-266.

Bunn, H.T. 1997. The bone assemblages from the excavated sites. In: Isaac G.Ll., Isaac, B. (Eds.), Koobi Fora Research Project, Volume 5: Plio-Pleistocene Archaeology. Claredon Press, Oxford, pp. 402-444.

Bunn, H.T. 2001. Hunting, power scavenging, and butchering by Hadza foragers and by Plio-Pleistocene *Homo*. In: Stanford, C.B., Bunn, H.T. (Eds), Meat-Eating and Human Evolution. Oxford University Press, New York, pp. 199-218.

Bunn, H.T., Ezzo, J.A.1993. Hunting and scavenging by Plio-Pleistocene hominids: nutritional constraints, archaeological patterns, and behavioural implications. Journal of Archaeological Science 20, 365-398.

Bunn, H.T., Kroll, E.M. 1986. Systematic butchery by Plio-Pleistocene hominids at Olduvai Gorge, Tanzania. Current Anthropology 27, 431-452.

Bunn, H.T., Harris, J.W.K., Isaac, G.Ll., Kaufulu, Z., Kroll, E.M., Schick, K.A., Toth, N., Behrensmeyer, K. 1980. FxJj 50: An early Pleistocene site in northern Kenya. World Archaeology 12, 109-136.

Capaldo, S.D. 1995. Inferring hominid and carnivore behavior from dual-patterned archaeological assemblages. Ph. D. Dissertation, Rutgers University, New Brunswick.

Capaldo, S.D. 1997. Experimental determinations of carcass processing by Plio-Pleistocene hominids and carnivores at FLK 22 (*Zinjanthropus*), Olduvai gorge, Tanzania. Journal of Human Evolution 33, 555-597.

Capaldo, S.D. 1998. Methods, marks and models for inferring hominid and carnivore behaviour. Journal of Human Evolution 35, 323-326.

Capaldo, S.D., Blumenschine, R.J. 1994. A quantitative diagnosis of notches made by hammerstone percussion and carnivore gnawing in bovid long bones. American Antiquity 59, 724-748.

Cleghorn, N., Marean, C.W. 2004. Identifying skeletal elements useful for behavioral analysis. Paper presented at the 69th Annual Meeting of the Society for American Archaeology, Montreal, March 31 – April 4, 2004.

Domínguez-Rodrigo, M. 1997. Meat-eating by early hominids at the FLK 22 *Zinjanthropus* site, Olduvai Gorge Tanzania: An experimental approach using cut mark data. Journal of Human Evolution 33, 669-690.

Domínguez-Rodrigo, M. 1999a. Meat-eating and carcass procurement at the FLK *Zinj* 22 site, Olduvai Gorge (Tanzania): A new experimental approach to the old hunting-versus-scavenging debate. In: Ullrich, H. (Ed.), Lifestyles and Survival Strategies in Pliocene and Pleistocene Hominids. Edition Archaea, Schwelm, pp. 89-111.

Domínguez-Rodrigo, M. 1999b. Flesh availability and bone modifications in carcasses consumed by lions: Palaeoecological relevance in hominid foraging patterns. Palaeogeography, Palaeoclimatology, Palaeoecology 149, 373-388.

Domínguez-Rodrigo, M. 2001. A study of carnivore competition in riparian and open habitats of modern savannas and its implications for hominid behavioral modeling. Journal of Human Evolution 40, 77-98.

Domínguez-Rodrigo, M. 2002. Hunting and scavenging in early hominids: The state of the debate. Journal of World Prehistory 16, 1-56.

Domínguez-Rodrigo, M., Pickering, T.R. 2003. Early hominid hunting and scavenging: A zooarchaeological review. Evolutionary Anthropology 12, 275-282.

Domínguez-Rodrigo M, de Luque L, Alcalá L, de la Torre Sainz I, Mora R, Serrallonga J, Medina V. 2002. The ST site complex at Peninj, West Lake Natron, Tanzania: Implications for early hominid behavioral models. Journal of Archaeological Science 29, 639-665.

Domínguez-Rodrigo, M., Pickering, T.R., Semaw, S., Rogers, M. 2005. Cutmarked bones from Pliocene archaeological sites at Gona, Afar, Ethiopia: Implications for the function of the world's oldest stone tools. Journal of Human Evolution 48, 109-121.

Egeland, C.P., Pickering, T.R., Domínguez-Rodrigo, M., Brain, C.K. 2004. Disentangling Early Stone Age palimpsests: A method to determine the functional independence of hominid-and carnivore-derived portions of archaeofaunas. Journal of Human Evolution 47, 343-357.

Fiorilla, A. 1989. An experimental study of trampling: Implications for the fossil record. In: Bonnichsen, R., Sorg, M.H. (Eds), Bone Modification. Center for the Study of the First Americans, Orono (ME), pp. 61-71.

Isaac, G.Ll. 1978. The food-sharing behavior of protohuman hominids. Scientific American 238, 90-108.

Isaac, G.Ll. 1981a. Archaeological tests of alternative models of early hominid behaviour: Excavation and experiments. Philosophical Transactions of the Royal Society of London 292, 177-188

Isaac, G.Ll. 1981b. Stone Age visiting cards: Approaches to the study of early land use patterns. In: Hodder, I., Isaac, G.Ll., Hammond, N. (Eds.), Patterns of the Past. Cambridge University Press, Cambridge, pp. 205-257.

Isaac, G.Ll. 1983. Bones in contention: Competing explanations for the juxtaposition of Early Pleistocene artifacts and faunal remains. In: Clutton-Brock, J., Grigson, C. (Eds.). Animals and Archaeology, Volume 1: Hunters and Their Prey. British Archaeological Reports International Series 163, Oxford, pp. 3-19.

Isaac, G.Ll. 1984. The archaeology of human origins: Studies of the Lower Pleistocene in East Africa, 1971-1981. In: Wendorf, F., Close, A.E. (Eds.). Advances in World Archaeology, Volume 3. Academic Press, Orlando (FL), pp. 1-87.

Keyser, A.W. 2000. Dawn of humans: New finds in South Africa. National Geographic 197, 76-83.

Marean, C.W., Cleghorn, N. 2003. Large mammal skeletal element transport: Applying foraging theory in a complex taphonomic system. Journal of Taphonomy 1, 15-42.

Marean, C.W., Kim, S.Y. 1998. Mousterian large-mammal remains from Kobeh Cave: Behavioral implications for Neanderthals and early modern humans. Current Anthropology 39 Supplement, S79-S113.

Marean, C.W., Domínguez-Rodrigo, M., Pickering, T.R. 2004. Skeletal element equifinality in zooarchaeology begins with method: The evolution of the "shaft critique." Journal of Taphonomy 2, 69-98.

Marean, C.W., Spencer, L.M., Blumenschine, R.J., Calpado, S.D. 1992. Captive hyaena bone choice and destruction, the schlepp effect and Olduvai archaeofaunas. Journal of Archaeological Science 19, 101-121.

Monahan, C.M. 1996. New zooarchaeological data from Bed II, Olduvai Gorge, Tanzania: Implications for hominid behavior in the Early Pleistocene. Journal of Human Evolution 31, 93-128.

Newman, R. 1993. The incidence of damage marks on Swartkrans fossil bones from the 1979-1986 excavations. In: Brain, C.K. (Ed.), Swartkrans: A Cave's Chronicle of Early Man. Transvaal Museum, Pretoria, pp. 217-228.

Nilssen, P.J. 2000. An actualistic butchery study in South Africa and its implications for reconstructing hominid strategies of carcass acquisition and butchery in the Upper Pleistocene and Plio-Pleistocene. Ph.D. Dissertation, University of Cape Town.

Oliver, J.S. 1989. Analogues and site context: Bone damage from Shield Trap Cave (24CB91), Carbon County, Montana, USA. In: Bonnichsen, R., Sorg, M.H. (Eds.). Bone Modification. Center for the Study of the First Americans, Orono (ME), pp. 73-98.

Oliver, J.S. 1994. Estimates of hominid and carnivore involvement in the FLK *Zinjanthropus* fossil assemblage: Some socioecological implications. Journal of Human Evolution 27, 267-294.

Pickering TR. 1999. Taphonomic interpretations of the Sterkfontein early hominid site (Gauteng, South Africa) reconsidered in light of recent evidence. Ph.D. Dissertation, University of Wisconsin-Madison.

Pickering, T.R., Domínguez-Rodrigo, M. in press. The acquisition and use of large mammal carcasses by Oldowan hominins in eastern and southern Africa: A selected review and assessment. In: Toth, N., Schick, K. (Eds.) The Oldowan: Studies into the Origins of Human Technology. Stone Age Institute Press, Bloomington (IN).

Pickering, T.R., Domínguez-Rodrigo, M., Egeland, C.P., Brain, C.K. 2004a. New data and ideas on the foraging behaviour of Early Stone Age hominids at Swartkrans Cave, South Africa. South African Journal of Science 100, 215-219.

Pickering, T.R., Domínguez-Rodrigo, M., Egeland, C.P., Brain, C.K. 2004b. Beyond leopards: Tooth marks and the contribution of multiple carnivore taxa to the accumulation of the Swartkrans Member 3 fossil assemblage. Journal of Human Evolution 46, 595-604.

Pickering, T.R., Domínguez-Rodrigo, M., Egeland, C.P., Brain, C.K. 2005. The contribution of limb bone fracture patterns to reconstructing early hominid behavior at Swartkrans Cave (South Africa): Archaeological application of a new analytical method. International Journal of Osteoarchaeology 15, 247-260.

Pickering, T.R., Egeland, C.P., Schnell, A., Osborne, D., Enk, J. 2006. Success in identification of experimentally fragmented limb bone shafts: Implications for estimates of skeletal element abundance in archaeofaunas. Journal of Taphonomy 4, 97-108.

Pickering, T.R., Marean, C.W., Domínguez-Rodrigo, M. 2003. Importance of limb bone shafts in zooarchaeology: A response to "On *in situ* attrition and vertebrate body part profiles (2002), by M.C. Stiner. Journal of Archaeological Science 30, 1469-1482.

Pickering, T.R., White, T.D., Toth, N. 2000. Cutmarks on a Plio-Pleistocene hominid from Sterkfontein, South Africa. American Journal of Physical Anthropology 111, 579-584.

Potts, R.B., Shipman, P. 1981. Cutmarks made by stone tools on bones from Olduvai Gorge, Tanzania. Nature 291, 577-580.

Robinson, J.T. 1959. A bone implement from Sterkfontein. Nature 184, 583-585.

Selvaggio, M.M. 1994. Identifying the timing and sequence of hominid and carnivore involvement with Plio-Pleistocene bone assemblages from carnivore tooth marks and stone tool butchery marks on bone surfaces. Ph.D. Dissertation, Rutgers University.

Selvaggio, M.M. 1998. Concerning the three stage model of carcass processing at the FLK *Zinjanthropus*: A reply to Capaldo. Journal of Human Evolution 35, 319-321.

Selvaggio, M.M., Wilder, J. 2001. Identifying the involvment of multiple carnivore taxa with archaeological bone assemblages. Jorunal of Archaeological Science 28, 465-470.

Shipman, P., Rose, J. 1983. Early hominid hunting, butchering, and carcass processing behaviors: Approaches to the fossil record. Journal of Anthropological Archaeology 2, 57-98.

Villa, P., Mahieu, E. 1991. Breakage patterns of human long bones. Journal of Human Evolution 21, 27-48.

Watson, V. 1993. Composition of the Swartkrans bone accumulations, in terms of skeletal parts and animals represented. In: Brain, C.K. (Ed.), Swartkrans: A Cave's Chronicle of Early Man. Transvaal Museum, Pretoria, pp. 35-73.

# Appendix

Butchered fossils identified in the Swartkrans Member 3 limb bone shaft archaeofauna. Animal size classes are based on Brain's (1981) well-known system for antelope. The catalog number of each specimen is preceded by a SKX prefix, which is dropped in the following Tables 1–3; specimens are listed here in numerical order by catalog number. Some listed specimens were recovered from screen bags, in which multiple specimens were originally assigned the same catalog number; in those cases we distinguished each modified specimen with a unique suffixed number after an added decimal point.

Following the tables, the hominid butchered bones from Swartkrans Member 3 are illustrated in two figures.

*Table 1. Cutmarked specimens (PM = percussion marks; TM = tooth marks)*

| Size Class | Specimen number | Element | Other damage |
|---|---|---|---|
| 1 | 29674.2 | Metapodial | |
| | 33575 | Upper | TM |
| | 33598 | Humerus | TM |
| | 33751 | Tibia | |
| | 37424 | Metatarsal | |
| 2 | 21853 | Metatarsal | |
| | 22068 | Limb bone | |
| | 22425 | Intermediate | |
| | 22671 | Femur | |
| | 24494 | Radius | |
| | 25304 | Metacarpal | TM |
| | 27861 | Tibia | |
| | 28538 | Tibia | PM |
| | 28786 | Femur | TM |
| | 29055 | Limb bone | |
| | 29141 | Tibia | |
| | 29156 | Upper | TM |
| | 29273 | Limb bone | |
| | 29674.3 | Limb bone | PM |
| | 30617.1 | Upper | TM |
| | 31022 | Tibia | |
| | 31348 | Metatarsal | |
| | 31396 | Femur | TM |
| | 31474 | Upper | |
| | 31760 | Metacarpal | |
| | 31765 | Upper | |
| | 32013 | Femur | TM |
| | 32905 | Metapodial | |
| | 33591 | Femur | |
| | 34278 | Tibia | |
| | 34315 | Limb bone | |
| | 34564 | Metapodial | |
| | 34749 | Tibia | TM |
| | 35363 | Radius | |
| | 36741 | Limb bone | |
| | 37186 | Humerus | |
| | 37890 | Humerus | |
| 3 | 19491 | Metacarpal | TM |
| | 22831 | Tibia | TM |
| | 23296 | Femur | PM, TM |
| | 27865 | Tibia | |
| | 28225 | Upper | |
| | 29368 | Radius | TM |
| | 29497 | Metatarsal | |
| | 30406 | Metatarsal | |
| | 30429 | Femur | TM |
| | 30631 | Upper | TM |
| | 34499 | Humerus | |
| | 34636 | Upper | |
| | 34726 | Radius | |
| | 36690 | Metapodial | |
| | 36768 | Tibia | |
| | 36805 | Limb bone | |
| | 37333 | Femur | |
| | 37412 | Humerus | |
| | 45758 | Metatarsal | |
| 4 | 34263 | Humerus | |
| | 35498 | Humerus | |
| | 36231 | Metapodial | TM |
| | 37540 | Radioulna | |

*Table 2. Percussion marked specimens (CM = cutmarks; TM = tooth marks)*

| Size Class | Specimen number | Element | Other damage |
|---|---|---|---|
| 1 | 94 | Femur | |
| | 24896 | Metacarpal | |
| | 29993.1 | Tibia | |
| | 31091 | Limb bone | |
| | 32182 | Metacarpal | |
| | 33540 | Limb bone | |
| | 36044 | Femur | |
| | 37013 | Metacarpal | |
| | 37409 | Tibia | |
| | 37863.1 | Femur | |
| 2 | 20057 | Metacarpal | |
| | 22658 | Limb bone | |
| | 26114 | Limb bone | |
| | 28095 | Radius | |
| | 28538 | Tibia | CM |
| | 28603 | Metatarsal | |
| | 28641 | Radius | |
| | 29674.3 | Limb bone | CM |
| | 29813 | Femur | |
| | 30081 | Tibia | |
| | 30917.1 | Tibia | |
| | 31040 | Radius | |
| | 33441 | Femur | |
| | 34844 | Femur | |
| | 35125 | Tibia | |
| | 35727 | Humerus | |
| | 36692 | Humerus | |
| | 37218 | Metacarpal | |
| | 37291 | Humerus | |
| | 37947 | Limb bone | |
| 3 | 19526 | Limb bone | |
| | 21563 | Metacarpal | |
| | 21858 | Tibia | |
| | 23296 | Femur | CM, TM |
| | 26726 | Tibia | |
| | 27348 | Humerus | |
| | 30638 | Metapodial | |
| | 32476 | Humerus | |
| | 33269 | Tibia | |
| | 33497 | Femur | TM |
| | 34639 | Tibia | |
| | 36776 | Metatarsal | TM |
| | 36806 | Humerus | TM |
| 4 | 19514 | Metapodial | |
| | 32532 | Metacarpal | |
| | 36231 | Metapodial | CM |

*Table 3. Impact flakes*

| Size Class | Specimen number | Element | Size Class | Specimen number | Element |
|---|---|---|---|---|---|
| 2 | 30035 | Humerus | ? | 30090 | Limb bone |
| ? | 22320 | Limb bone | | 30188 | Limb bone |
| | 22320.1 | Limb bone | | 30416 | Limb bone |
| | 22320.2 | Limb bone | | 30581 | Limb bone |
| | 22948 | Limb bone | | 30598 | Limb bone |
| | 23320 | Limb bone | | 30670 | Limb bone |
| | 24675.2 | Limb bone | | 30767 | Limb bone |
| | 24675.3 | Limb bone | | 30835 | Limb bone |
| | 29011.1 | Limb bone | | 30860.1 | Limb bone |
| | 29011.2 | Limb bone | | 30860.2 | Limb bone |
| | 29090 | Limb bone | | 31571 | Limb bone |
| | 29321.1 | Limb bone | | 32455 | Limb bone |
| | 29321.2 | Limb bone | | 32884 | Limb bone |
| | 29361 | Limb bone | | 33230 | Limb bone |
| | 29391 | Limb bone | | 33416 | Limb bone |
| | 29452.1 | Limb bone | | 33625 | Limb bone |
| | 29452.2 | Limb bone | | 33637 | Limb bone |
| | 29485 | Limb bone | | 33652 | Limb bone |
| | 29610 | Limb bone | | 34480 | Limb bone |
| | 29753 | Limb bone | | 34506 | Limb bone |
| | 29962 | Limb bone | | 34564 | Limb bone |
| | 30022 | Limb bone | | 34611 | Limb bone |
| | | | | 34675 | Limb bone |
| | | | | 34675.1 | Limb bone |
| | | | | 34675.2 | Limb bone |
| | | | | 34700 | Limb bone |
| | | | | 35057 | Limb bone |
| | | | | 35810 | Limb bone |
| | | | | 36967 | Limb bone |
| | | | | 37619 | Limb bone |
| | | | | 37929.1 | Limb bone |

*Figure 1. The sample of cutmarked fossils from Swartkrans Member 3. Note that two cutmarked specimens, SKX 24494 (radius) and 34315 (limb bone shaft fragment), identified in an earlier study (Brain, 1993) are not figured here.*

*Box (a), humerus specimens: top row = SKX 34263; second row (left to right) = SKX 33598, 35498; third row (left to right) = SKX 37890, 37186; fourth row (left to right) = SKX 34499, 37412.*

*Box (b), radioulna specimens (left to right) = SKX 35363, 34726, 29368, 37540.*

*Box (c), metacarpal specimens: top row = SKX 31760; second row (left to right) = SKX 19491, 25304.*

*Box (d), femur specimens: top row = SKX 22671; second row = SKX 28786; third row (left to right) = 32013, 23296; fourth row (left to right) = SKX 33591, 31396; fifth row (left to right) = SKX 37333, 30429.*

*Box (e), tibia specimens: top row (left to right) = SKX 31022, 36768; second row (left to right) = SKX 22831, 34278; third row (left to right) = SKX 27865, 33751, 34749; fourth row (left to right) = SKX 27861, 29141; fifth row = SKX 28538.*

*Box (f), metatarsal specimens (left to right) = SKX 37424, 21853, 45758 (top), 30406 (bottom), 29497, 31348.*

*Box (g), upper (humerus or femur) specimens: top row (left to right) = SKX 34636, 31765, 30617.1, 33575; second row (left to right) = SKX 29156, 31474, 30631, 28225.*

*Box (h), intermediate (radioulna or tibia) specimen = SKX 22425.*

*Box (i), metapodial specimens: top row (left to right) = SKX 36690, 34564; second row (left to right) = SKX 29674.2, 32905; third row = SKX 36231.*

*Box (j), limb bone shaft specimens: top row = SKX 36805; second row (left to right) = SKX 22068, 36741, 29055; third row (left to right) = SKX 29674.3, 29273.*

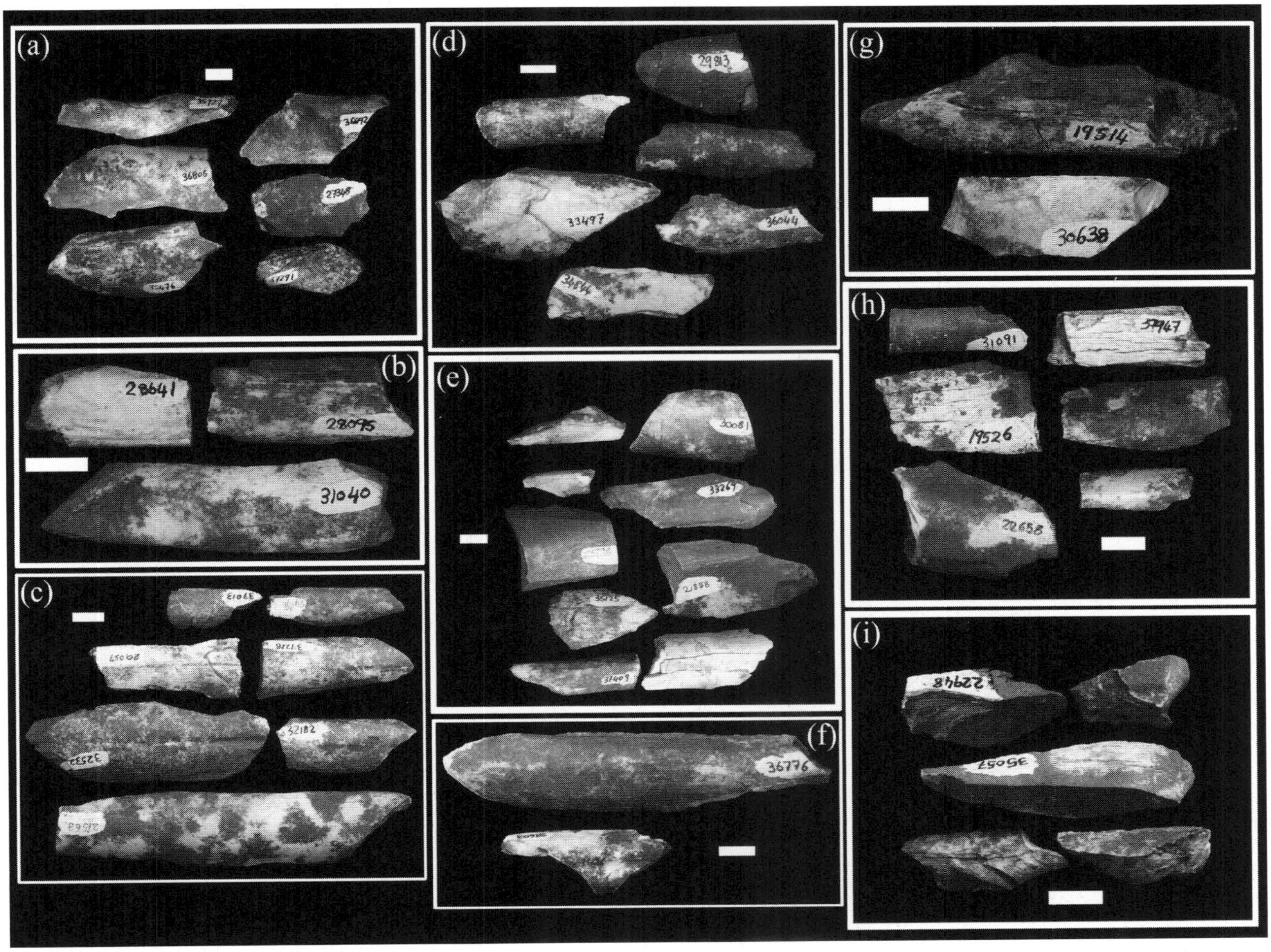


*Figure 2. The sample of percussion marked fossils from Swartkrans Member 3. Note that several percussion marked specimens are not figured below because they are pieces that also preserve cutmarks and are thus illustrated in the Figure 1 composites above. These specimens include: SKX 23296 (femur), 28538 (tibia), 29674.3 (limb bone shaft) and 36231 (metapodial). One additional percussed specimen, SKX 33441 (femur) is also absent in the figure above.*

*Box (a), humerus specimens: top row (left to right) =SKX 35727, 36692; second row (left to right) = SKX 36806, 27348; third row (left to right) = SKX 32476, 37291.*

*Box (b), radioulna specimens: top row (left to right) = SKX 28641, 28095; second row = SKX 31040.*

*Box (c), metacarpal specimens: top row (left to right) = SKX 37013, 24896; second row (left to right) = SKX 20057, 37218; third row (left to right) = SKX 32532, 32182; fourth row = SKX 21563.*

*Box (d), femur specimens: top row = SKX 29813; second row (left to right) = SKX 94, 37863.1; third row (left to right) = SKX 28359, 36044; fourth row = SKX 34844.*

*Box (e), tibia specimens: top row (left to right) = SKX 30917.1, 30081; second row (left to right) = SKX 29993.1, 33269; third row (left to right) SKX 26726, 21858; fourth row = SKX 35125; fifth row (left to right) = SKX 37409, 34639.*

*Box (f), metatarsal specimens: top row = SKX 36776; second row = SKX 28603.*

*Box (g), metapodial specimens: top row = SKX 19514; second row = SKX 30638.*

*Box (h), limb bone shaft specimens: top row (left to right) = SKX 31091, 37947; second row (left to right) = SKX 19526, 26114; third row (left to right) = SKX 22658, 33540.*

*Box (i), representative examples of impact flakes: top row (left to right) = SKX 22948, 30670; second row = SKX 35057; third row (left to right) = SKX 34506, 29485.*

# CHAPTER 14

# EQUIFINALITY IN CARNIVORE TOOTH MARKS AND THE EXTENDED CONCEPT OF ARCHAEOLOGICAL PALIMPSESTS: IMPLICATIONS FOR MODELS OF PASSIVE SCAVENGING BY EARLY HOMINIDS

MANUEL DOMÍNGUEZ-RODRIGO, CHARLES P. EGELAND AND TRAVIS RAYNE PICKERING

## ABSTRACT

The frequency and anatomical distribution of carnivore tooth marks figure prominently in models that envision Plio-Pleistocene hominids passively scavenging from felid kills. Some of these models assert that high percentages of tooth-marked limb bone midshaft fragments unambiguously reflect primary access to carcasses by carnivores (and, by extension, late access by hominids). This assertion emanates in part from analyses of tooth mark frequencies in modern "carnivore-first" carcass feeding experiments. However, because hyenas rather than felids are the predominant agent of bone modification in these actualistic controls samples, no study has yet provided the comparative tooth mark data required to accurately model hominid scavenging of abandoned felid kills. As a first step toward remedying this deficiency, we provide preliminary tooth mark data on carcasses consumed by leopards, lions and cheetahs. Our analysis demonstrates that tooth mark percentages on midshaft portions of felid-processed limb bones are (1) much lower than previously supposed and (2) even overlap in some cases with tooth mark frequencies produced by hyena ravaging of human food refuse. This potential equifinality in tooth mark frequencies and distribution highlights the palimpsest nature of many Stone Age sites and emphasizes the limited utility of tooth marks for inferring hominid foraging behavior.

## INTRODUCTION

Two landmark publications in taphonomy, C.K. Brain's (1981) *The Hunters or the Hunted?* and L.R. Binford's (1981) *Bones*, alerted zooarchaeologists to the fact that many Stone Age faunas are palimpsests; that is, assemblages accumulated and modified by more than one formational agent. Stimulated largely by these seminal works, archaeological taphonomy has spent the last 25 years constructing referential frameworks to decipher the integrity and resolution of Stone Age archaeofaunas. In the fallout of this innovative research, the foundations of the hunting hypothesis and the concomitant socio-economic models emphasizing home bases, delayed resource consumption, and food-sharing (e.g., Isaac, 1978, 1981, 1983, 1984) were largely undermined, while new models emphasizing various forms of scavenging emerged (e.g., Binford, 1981, 1985, 1988a, b; Blumenschine, 1988, 1991, 1995; Capaldo, 1995, 1997, 1998; Potts, 1982, 1988; Potts and Shipman, 1981; Selvaggio, 1994; Selvaggio and Wilder, 2001; Shipman, 1983, 1986; Shipman and Phillips, 1976).

Much of this work was focused on documenting faunal assemblages accumulated and modified by modern humans and carnivores, with special attention paid to skeletal part abundances (e.g., Brain, 1981; Binford, 1978, 1981; Bunn, 1982, 1983; Bunn et al., 1988, 1991; Cruz-Uribe, 1991; Hill, 1975; Klein, 1975; O'Connell et al., 1988, 1990, 1992; Potts, 1982, 1988; Pickering, 2001, 2002). Further studies led to the important realization that many bone-destroying carnivores can significantly bias skeletal part abundances in archaeofaunas by deleting less-dense axial bones and limb bone epiphyses (e.g., Bartram, 1993; Bartram and Marean, 1999; Blumenschine and Marean, 1993; Bunn, 1991, 1993; Bunn and Ezzo, 1993; Capaldo, 1995; Marean and Bertino, 1994; Marean and Frey, 1997; Marean and Spencer, 1991; Marean et al., 1992, 2004; Pickering et al., 2003; Cleghorn and Marean, this volume). However, rather than clarifying inferences of assemblage formation, this refined cognizance of carnivore bone destruction and its impact on skeletal part abundances led to disparate interpretations of early archaeological sites (e.g., Binford, 1981, 1984, 1985, 1988a,b; Blumenschine, 1991, 1995; Blumenschine and Marean, 1993; Bunn, 1981, 1982, 1983, 1986; Bunn and Kroll, 1986; Bunn and Ezzo, 1993; Isaac, 1983, 1984; Lupo, 1998; Potts, 1988; Shipman, 1986; O'Connell et al., 2002).

The research of Blumenschine and his colleagues (e.g., Blumenschine, 1988, 1995; Blumenschine and Marean, 1993; Capaldo, 1995; Selvaggio, 1994) utilized bone surface modifications to circumvent the problems of equifinality prevalent in skeletal part abundances and thus provided a powerful analytical tool for investigating the order in which carnivores and hominids intervened with carcasses. Based on actualistic assemblages of carcasses modified by humans and large carnivores in various combinations, it was argued that tooth mark frequency and anatomical distribution were the most important variables for assessing the order of hominid and carnivore access to carcasses. From these data the mode of hominid carcass acquisition was then inferred (hunting/aggressive scavenging versus passive scavenging).

This framework was elaborated by assuming "interaction" between hominids and carnivores in site formation, while criticizing previous interpretations that considered the contribution of each agent separately (Blumenschine, 1988, 1995; Blumenschine and Marean, 1993; Capaldo, 1995; Selvaggio, 1994). The application of this framework to the 1.75 million year old archaeofauna from FLK Level 22 (the *Zinjanthropus* Floor) at Olduvai Gorge, Tanzania led to a "multiple-pattern" interpretation of assemblage formation, in which defleshed and abandoned felid kills were scavenged by hominids and transported to "sites" for demarrowing with hammerstones. Hyenas intervened in the last stage of formation by consuming the grease-bearing portions of carcasses abandoned by hominids (Blumenschine, 1988, 1995; Capaldo, 1995; Selvaggio, 1994). Subsequent experiments emphasizing cutmark rather than tooth mark data led Domínguez-Rodrigo (1997a, b) to argue that hominids at FLK Level 22 were acquiring fully fleshed carcasses procured through hunting and/or aggressive scavenging. More specifically, and corroborating earlier studies (e.g., Bunn, 1981, 1982, 1986; Bunn and Kroll, 1986), Domínguez-Rodrigo (1997a, b) found that most cutmarks in the FLK Level 22 assemblage occurred on meat-bearing bone portions, indicating the presence of substantial flesh at the time hominids imparted them.

Such contradiction among analysts, this time focusing on bone surface modifications, led researchers to highlight the effects of equifinality on analyses of cutmarks (Blumenschine, 1991, 1995; Capaldo, 1995, 1998, Selvaggio, 1994, 1998), tooth marks (Domínguez-Rodrigo, 1999a) and both cutmarks and tooth marks (Lupo and O'Connell, 2002). Nevertheless, most analysts conform to the view that high tooth mark frequencies on midshaft portions of limb bones is an indicator of primary access to carcasses by carnivores, while low tooth mark frequencies on those bone portions are indicative of secondary access by carnivores (e.g., Blumenschine, 1988, 1995; Capaldo, 1995; Lupo and O'Connell, 2002; O'Connell and Lupo, 2003).

We argue here that although previous actualistic datasets are useful interpretive tools in specific cases, there are three factors that make them inappropriate analogues for modelling passive scavenging palimpsests. First, and most importantly, the extended concept of the archaeological palimpsest recognizes that bone assemblages are both accumulated *and* modified by more than one agent (terminology follows Binford, 1980, 1981; Egeland et al., 2004). However, many of the interpretive frameworks applied to Stone Age sites, especially in Africa, assume that assemblage accumulation is largely the result of a single agent, whether carnivore or hominid. For example, it has been suggested that many Plio-Pleistocene faunas are the result of redundant predation and carcass consumption by carnivores at the same sites (e.g., Binford, 1981). Other researchers insist that hominids were the primary bone accumulators, relegating carnivores to the marginal role of modifying what was abandoned at sites (e.g., Bunn and Kroll, 1986, 1988; Potts, 1988; Oliver, 1994). Even multiple-pattern models posit a single accumulating agent (hominids) at early sites, although the initial capture and consumption of carcasses is attributed to carnivores (e.g., Blumenschine 1995; Blumenschine et al., 1994; Capaldo, 1995; Selvaggio, 1994). Therefore, such models are only partially "multi-patterned": while recognizing multiple agents of on-site bone modification, they ignore the possibility of multiple accumulation agents that may (or may not) have acted independently. Ignoring this aspect of site formation will no doubt affect interpretations based on taphonomists' standard toolkit (e.g., skeletal part abundances, minimum number of individuals, mortality profiles, carcass size profiles, bone surface modification frequencies). For example, an assemblage created through the transport of carcasses by independently operating hominids and carnivores could generate skeletal part and carcass size profiles that mimic what would be expected if hominids were scavenging from carnivore kills or dens. This problem would be exacerbated if the carnivore contribution to assemblage accumulation greatly surpassed that of the hominids. We simulate such a scenario below and its impact on the interpretation of site formation.

The second factor that limits the utility of current

actualistic models concerns the application of tooth mark frequencies and distribution. Many scavenging interpretations use tooth marks to infer the order of carnivore access, but then proceed to reconstruct the hominid role in site formation using the same data. However, hominid behavior cannot be reconstructed directly from tooth mark data (Domínguez-Rodrigo and Pickering, 2003; Pickering and Domínguez-Rodrigo, in press). Tooth mark frequencies and distributions directly reflect only carnivore interaction with bones and can thus only be used to infer carnivore access to carcasses. We demonstrate below that in a fully realized palimpsest, if the carnivore contribution to site formation exceeds that of hominids the resultant frequencies of tooth marks can mask the original hominid contribution.

Finally, the type of carnivore used in actualistic studies impacts profoundly subsequent "tests" of passive scavenging models. We present data indicating that if felids are used as the "first" carnivore in multiple-pattern models (which, according to the models themselves, should be the case) rather than hyenas, tooth mark percentages drop drastically and can become non-diagnostic. In order to provide a framework for addressing these issues, this study provides preliminary tooth mark data on carcasses consumed by leopards, lions and cheetahs.

## Materials and Methods

The analytical sample used in this study is summarized in Table 1. Briefly, a total of four separate leopard kills from South Africa and one lion kill from Kenya were analyzed. The leopard kills, consisting of three impalas and a steenbok, were discovered and reported on by Brain (1981). The lion kill, a cow, was first reported on by Domínguez-Rodrigo (1997a). In addition, prey remains from two Namibian leopard lairs and an assemblage from a captive cheetah feeding experiment (Brain, 1981) are included in our sample.

*Table 1. Summary of the felid-ravaged assemblages analyzed in this study*

| Predator taxon | Prey taxon | Assemblage type |
|---|---|---|
| Leopard | Steenbok | Kill |
| Leopard | Impala | Kill |
| Leopard | Impala | Kill |
| Leopard | Impala | Kill |
| Leopard | Goat | Den |
| Leopard | Sheep | Den |
| Cheetah | Baboon | Experimental |
| Lion | Cow | Kill |

Following criteria summarized by Blumenschine et al. (1996), and with the aid of 10 x hand lenses, all conspicuous and inconspicuous tooth marks on limb bone specimens were identified. Tooth mark distributions, frequencies and densities per limb bone portion (epiphyseal, near-epiphyseal and midshaft; following Blumenschine, 1988) were also calculated (Figure 1).

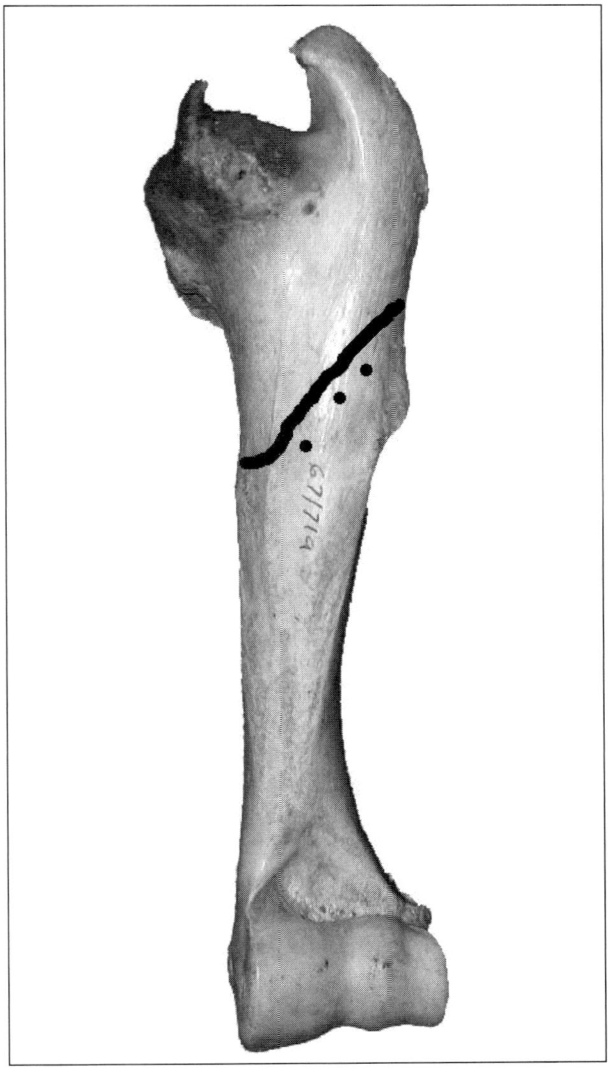

*Figure 1. Cranial view of a left humerus showing how tooth marks from the felid-modified bone assemblages were located on templates. Dots indicate tooth marks and the line indicates breakage plane.*

Most of the bones in the modern sample are complete (except those from the lion kill; discussed separately below). However, hammerstone breakage is an essential component of passive scavenging models. Because the complete bones in the modern sample could not actually be broken to replicate marrow extraction (after all, they comprise a valuable taphonomic study collection, created, not incidentally, by our honoree, Bob Brain), we chose to circumvent this problem by "virtually fragmenting" the comparative collection. Typical patterns of hammerstone breakage for each limb bone were established in reference to a sample of sheep carcasses that were de-

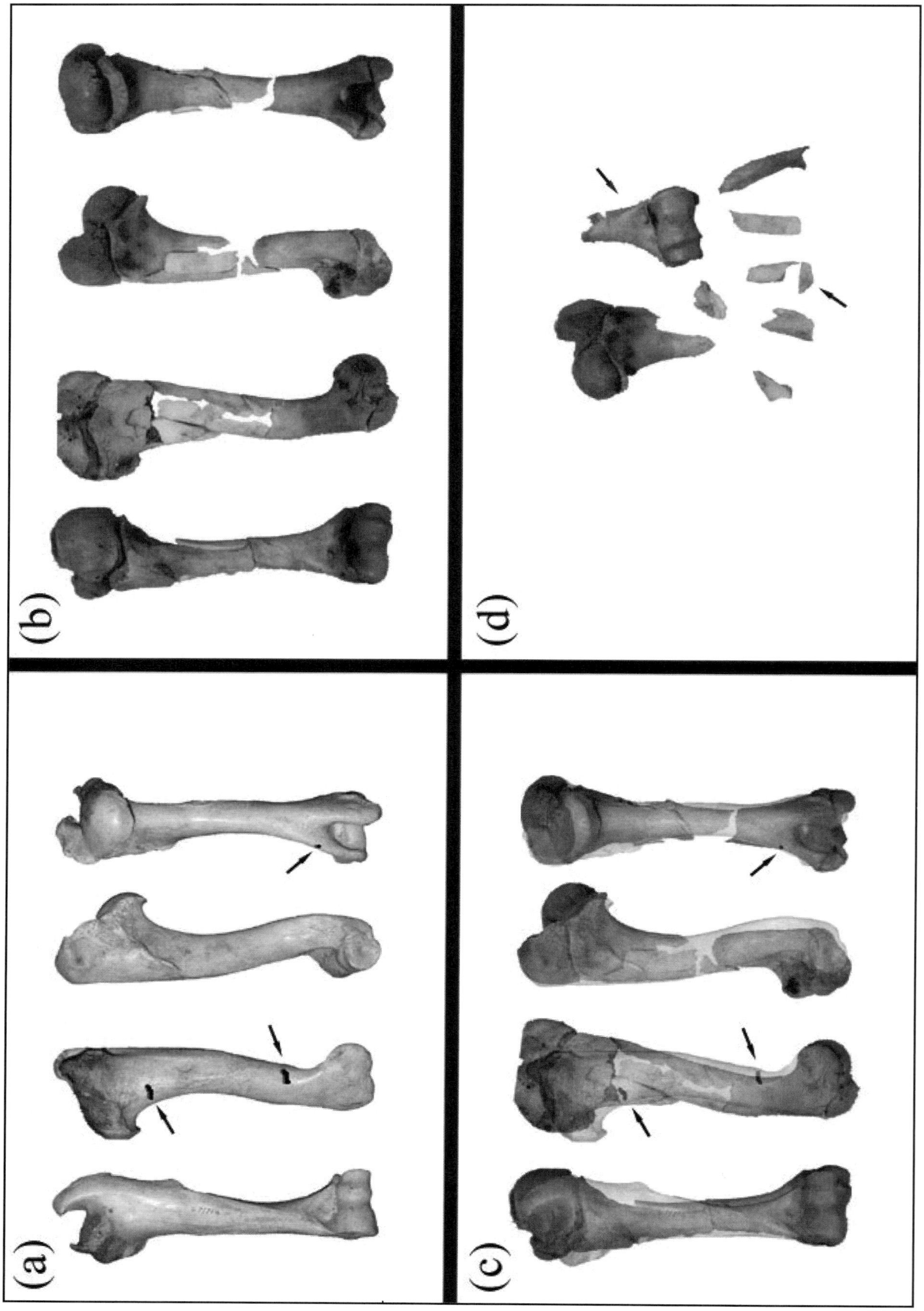


*Figure 2. Summary of the "virtual fragmentation" procedure, using a left humerus as an example. (a) Exact location of each tooth mark (indicated by arrows) on a template of the complete bone showing four anatomical aspects. (b) A left humerus broken through hammerstone percussion. (c) The fragmented humerus is superimposed on the tooth-marked template (tooth marks again indicated by arrows). (d) Virtual fragmentation of the tooth-marked humerus with arrows indicating the bone fragments that are tooth-marked.*

*Table 2. Frequency and distribution of individual tooth marks by bone section. Abbreviations: EP = epiphysis; PNEP = proximal near-epiphysis; DNEP = distal near-epiphysis; MSH = midshaft; HM = humerus; RD = radius; MC = metacarpal; FM = femur; TA = tibia; MT = metatarsal; L = left; R = right. Note: Given the low frequency of tooth marks on epiphyses, proximal and distal ends are combined.*

| Assemblage | Element | Side | EP | PNEP | DNEP | MSH | Total |
|---|---|---|---|---|---|---|---|
| **Steenbok** | HM | L | – | – | – | 1 | 1 |
| | HM | R | – | – | – | – | – |
| | RD | L | – | – | 4 | – | 4 |
| **Impala 1** | RD | L | – | 3 | – | – | 3 |
| | RD | R | – | – | – | – | – |
| | MC | L | – | – | – | – | – |
| | FM | R | – | – | 8 | 2 | 10 |
| | TA | L | – | 7 | – | 3 | 10 |
| | TA | R | – | 6 | – | – | 6 |
| | MT | L | – | 6 | – | – | 6 |
| | MT | R | – | 13 | – | – | 13 |
| **Impala 2** | MC | L | – | – | – | – | – |
| | FM | L | – | – | – | – | – |
| | TA | L | – | – | – | – | – |
| | TA | R | – | – | – | – | – |
| | MT | L | – | – | – | – | – |
| | MT | R | – | – | – | – | – |
| **Impala 3** | MC | R | – | – | – | – | – |
| | TA | L | – | 1 | – | – | 1 |
| | TA | R | – | 4 | – | 1 | 5 |
| | MT | L | – | – | – | – | – |
| | MT | R | – | – | – | – | – |
| **Goat** | HM | L | 2 | – | 6 | 2 | 10 |
| | HM | R | – | 1 | 1 | 1 | 3 |
| | RD | L | – | – | – | – | – |
| | RD | R | – | – | – | 5 | 5 |
| | MC | L | – | – | – | – | – |
| | MC | R | – | – | – | – | – |
| | FM | L | – | – | 5 | – | 5 |
| | FM | R | – | 2 | 5 | – | 7 |
| | TA | L | – | 7 | 1 | – | 8 |
| | TA | R | – | 4 | 1 | – | 5 |
| | MT | L | – | – | – | – | – |
| | MT | R | – | – | – | – | – |
| **Sheep** | HM | R | 2 | 3 | – | – | 5 |
| | RD | R | – | – | – | – | – |
| | MC | R | – | – | – | – | – |
| | FM | L | – | – | 6 | 4 | 10 |
| | FM | R | 4 | 2 | 1 | 1 | 8 |
| | TA | L | – | 1 | – | 2 | 3 |
| | TA | R | – | 2 | – | 1 | 3 |
| | MT | L | – | – | – | – | – |
| | MT | R | – | – | – | – | – |
| **Baboon** | HM | L | – | 4 | 7 | – | 11 |
| | HM | R | – | 2 | 1 | 1 | 4 |
| | RD | L | – | 5 | 1 | 2 | 8 |
| | RD | R | – | 1 | 1 | – | 2 |
| | FM | L | – | 2 | 6 | 10 | 18 |
| | FM | R | – | 3 | 2 | 7 | 12 |
| | TA | L | – | – | – | – | – |
| | TA | R | – | 9 | 2 | 1 | 12 |
| **Total** | | | **8** | **88** | **58** | **44** | **198** |

*Table 3. Tooth mark frequency and anatomical distribution in the virtually fragmented felid-ravaged assemblages. Numbers in the numerator are for the total number of tooth-marked specimens. Numbers in the denominator are for the total number of specimens in each category. Numbers in brackets show the percentage of tooth-marked specimens. Summary statistics: n = 7; mean %NISP tooth-marked = 15.5; standard deviation = 21.2; 95% confidence interval = 0.0 – 35.0. Abbreviations: HM = humerus; RD = radius; MC = metacarpal; FM = femur; TA = tibia; MT = metatarsal; L = left; R = right.*

| Assemblage | Side | HM | RD | MC | FM | TA | MT | Total |
|---|---|---|---|---|---|---|---|---|
| **Steenbok** | L | 1/12 (8.3) | 1/6 (16.7) | – | – | – | – | 2/18 (11.1) |
| | R | 0/10 (0.0) | – | – | – | – | – | 0/10 (0.0) |
| | Total | 1/22 (4.5) | 1/6 (16.7) | – | – | – | – | 2/28 (7.1) |
| **Impala 1** | L | – | 0/6 (0.0) | 0/7 (0.0) | – | 3/8 (37.5) | 2/6 (33.3) | 5/27 (18.5) |
| | R | – | 1/6 (16.7) | – | 3/9 (33.3) | 3/7 (42.8) | 1/7 (14.2) | 8/29 (27.6) |
| | Total | – | 1/12 (8.3) | 0/7 (0.0) | 3/9 (33.3) | 6/15 (40.0) | 3/13 (23.1) | 13/56 (23.2) |
| **Impala 2** | L | – | – | 0/7 (0.0) | 0/7 (0.0) | 0/8 (0.0) | 0/6 (0.0) | 0/28 (0.0) |
| | R | – | – | – | – | 0/7 (0.0) | 0/7 (0.0) | 0/14 (0.0) |
| | Total | – | – | 0/7 (0.0) | 0/7 (0.0) | 0/15 (0.0) | 0/13 (0.0) | 0/42 (0.0) |
| **Impala 3** | L | – | – | – | – | 1/8 (12.5) | 0/6 (0.0) | 1/14 (7.1) |
| | R | – | – | 0/7 (0.0) | – | 3/7 (42.8) | 0/7 (0.0) | 3/21 (14.3) |
| | Total | – | – | 0/7 (0.0) | – | 4/15 (26.7) | 0/13 (0.0) | 4/35 (11.4) |
| **Goat** | L | 4/12 (33.3) | 0/6 (0.0) | 0/7 (0.0) | 2/7 (28.5) | 2/8 (20.0) | 0/6 (0.0) | 8/46 (17.4) |
| | R | 2/10 (20.0) | 3/6 (50.0) | 0/7 (0.0) | 3/9 (33.3) | 2/7 (28.5) | 0/7 (0.0) | 10/46 (21.7) |
| | Total | 6/22 (27.3) | 3/12 (25.0) | 0/14 (0.0) | 5/16 (31.3) | 4/15 (26.7) | 0/13 (0.0) | 18/92 (19.5) |
| **Sheep** | L | – | – | – | 3/7 (42.9) | 3/8 (37.5) | 0/6 (0.0) | 6/21 (28.6) |
| | R | 1/10 (10.0) | 0/6 (0.0) | 0/7 (0.0) | 3/9 (33.3) | 2/7 (28.5) | 0/7 (0.0) | 6/46 (13.0) |
| | Total | 1/10 (10.0) | 0/6 (0.0) | 0/7 (0.0) | 6/16 (37.5) | 5/15 (33.3) | 0/13 (0.0) | 12/67 (17.9) |
| **Baboon** | L | 2/12 (16.7) | 4/6 (66.7) | – | 4/7 (57.1) | 0/8 (0.0) | – | 10/33 (30.3) |
| | R | 2/10 (20.0) | 1/6 (16.7) | – | 4/9 (44.4) | 3/7 (42.8) | – | 10/32 (31.3) |
| | Total | 4/22 (18.2) | 5/12 (41.7) | – | 8/16 (50.0) | 3/15 (20.0) | – | 20/65 (30.7) |

*Table 4. Tooth mark distribution by bone portion in the virtually fragmented felid-ravaged assemblages. Bone portion definition follows Blumenschine (1988). Numbers in the numerator are for the total number of tooth-marked specimens. Numbers in the denominator are for the total number of specimens in each category. Numbers in brackets show the percentage of tooth-marked specimens. Abbreviations: EP = epiphysis; NEP = near-epiphysis; MSH = midshaft. Summary statistics (EP): n = 7; mean %NISP tooth-marked = 15.9; standard deviation = 35.5; 95% confidence interval = 0.0 – 48.5. Summary statistics (NEP): n = 7; mean %NISP tooth-marked = 16.8; standard deviation = 19.6; 95% confidence interval = 0.0 – 34.8. Summary statistics (MSH): n = 7; mean %NISP tooth-marked = 8.2; standard deviation = 6.2; 95% confidence interval = 2.4 – 14.0.*

| Assemblage | EP | NEP | MSH | Total |
|---|---|---|---|---|
| Steenbok | 0/6 (0.0) | 1/9 (11.1) | 1/13 (7.6) | 2/28 (7.1) |
| Impala 1 | 3/12 (25.0) | 5/10 (50.0) | 5/30 (16.6) | 13/56 (23.2) |
| Impala 2 | 0/12 (0.0) | 0/8 (0.0) | 0/22 (0.0) | 0/42 (0.0) |
| Impala 3 | 1/12 (8.3) | 2/7 (28.5) | 1/21 (4.7) | 4/35 (11.4) |
| Goat | 9/24 (37.5) | 6/21 (28.5) | 3/47 (6.3) | 18/92 (19.5) |
| Sheep | 6/18 (33.3) | 4/14 (28.5) | 2/35 (5.7) | 12/67 (17.9) |
| Baboon | 12/16 (75.0) | 0/16 (0.0) | 7/42 (16.6) | 20/65 (30.7) |
| Total | 31/90 (34.4) | 18/85 (21.2) | 19/210 (9.0) | 69/385 (17.9) |

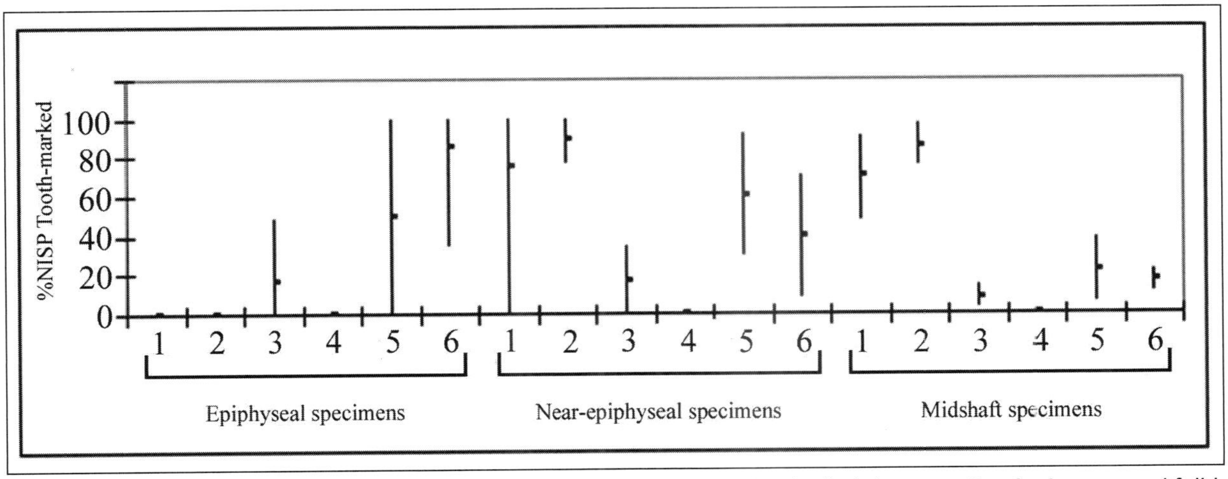


*Figure 3. Means and 95% confidence intervals of tooth mark frequencies by limb bone portion for hyena- and felid-modified assemblages. Numbers: 1 = small carcasses in "carnivore-first" assemblages (Blumenschine, 1995); 2 = large carcasses in "carnivore-first assemblages" (Blumenschine, 1995); 3 = leopard and cheetah assemblages reported on here; 4 = lion assemblage reported on here; 5 = small carcasses in "hammerstone-first" assemblages (Blumenschine, 1995); 6 = large carcasses in "hammerstone-first assemblages" (Blumenschine, 1995).*

fleshed and demarrowed at Complutense University as part of a butchery experiment.

These two modern datasets were combined by first drawing each individual tooth mark in the felid-ravaged sample on digital templates (displaying all four aspects) of each limb bone. Extreme care was taken to document the exact anatomical location of each tooth mark. Each limb bone was then "fractured" digitally by superimposing typical hammerstone-generated fragments onto the tooth mark templates. This procedure allows tooth mark frequencies to be reliably quantified in assemblages initially defleshed by felids and subsequently broken (virtually) by hammerstone percussion (see Figure 2).

## Results and Discussion

The analyzed sample of felid-ravaged limb bones from South Africa and Namibia is composed of 51 individual specimens (we discuss the lion-killed cow from Kenya below), 22 (43 %) of which preserve no tooth marks at all. This agrees with the results of Selvaggio (1994), and indicates that felids generally damage bones much less intensely than do hyenas. More specifically, a total of 198 tooth marks have been documented in the complete assemblage. Of these, only 44 (22 %) occur on midshaft sections (Table 2). A majority of the remaining tooth marks cluster on proximal and distal near-epiphyses. The relative intensity of tooth-marking on these sections is related to the breakage of bone and, in some cases, deletion of epiphyses. However, tooth mark density is low overall. Specimens that preserve >10 tooth marks are exceptional and when this density is reached it is again related to bone chewing on the limb bone ends, often coupled with partial or total deletion of epiphyses. Complete bones, which would have been the most attractive to marrow-scavenging hominids, rarely show more than 3 – 5 tooth marks.

In order to model hominid scavenging of felid kills, we include in the "virtual fragmentation" only bones with intact medullary cavities, as these are the only bones that hammerstone-wielding hominids would bother to break open for marrow. Superimposing the digital templates of tooth-marked limb bones over the modern hammerstone-broken sample yields a virtual assemblage from seven felid prey carcasses that is suitable for reconstructing tooth mark frequencies and distribution in a passive scavenging scenario. Table 3 indicates that overall tooth mark frequencies are very low in a fragmented limb bone assemblage initially consumed by felids. For all carcasses, tooth mark frequencies are <35 % (see Table 3 for summary statistics). This is in marked contrast to tooth mark frequencies in carcasses ravaged by hyenas, which show values between 80 – 100 % (Blumenschine, 1988, 1995).

Table 4 summarizes patterns of felid tooth mark frequency and distribution by bone portion in the virtually fragmented assemblages. Each portion (epiphyseal, near-epiphyseal and midshaft) displays lower tooth mark frequencies relative to assemblages of hyena-ravaged bones (Blumenschine, 1988, 1995). Most importantly, the mean (8.2%) and confidence intervals (2.4% –14.0%) for tooth-marked midshaft fragments are far below the >80% asserted to be a clear indication of "carnivore-first" assemblages (Blumenschine, 1988, 1995) (Figure 3). Furthermore, tooth mark frequencies in the felid-ravaged assemblages are indistinguishable from those reported for experiments that model secondary access of hyenas to hominid-demarrowed limb bones (Blumenschine, 1988, 1995). It is clear that felid defleshing of limb bones imparts few tooth marks; this finding corroborates both Selvaggio's (1994) earlier actualistic work and unpublished observations by T.R. Pickering and K. Kuman of a captive lion assemblage from South Africa. Higher frequencies of tooth-marked epiphyseal and near-epiphyseal sections relative to midshaft sections is likely related to the fact that felid gnawing concentrates on the less dense cancellous bone of limb ends.

Our analysis of an assemblage of cow bones from a lion kill compliment the findings discussed above for smaller prey animals. We provide these data separately because unlike the South African and Namibian assemblages, the left limb bones of the lion-killed cow were actually broken open with hammerstones subsequent to lion ravaging (see details in Domínguez-Rodrigo, 1997a). As the data from Table 5 indicate, tooth mark frequencies in the lion assemblage are also very low. As with the other felid-ravaged assemblages, epiphyseal fragments display the highest tooth mark frequency followed by near-epiphyses and midshafts.

*Table 5. Tooth mark distribution by bone portion in an assemblage modified by lions (Domínguez-Rodrigo, 1997a). Numbers in the numerator are for the total number of tooth-marked specimens. Numbers in the denominator are for the total number of specimens of each category. Numbers in brackets show the percentage of tooth-marked specimens. Abbreviations: EP = epiphysis; NEP = near-epiphysis; MSH = midshaft.*

| Assemblage | EP | NEP | MSH | Total |
|---|---|---|---|---|
| Cow | 4/12 (33.3) | 3/13 (23.1) | 2/17 (11.7) | 9/42 (21.4) |

## Summary and Conclusions

### Implications of results for current models of passive scavenging by hominids from felid-ravaged carcasses

A previous study that investigated tooth mark densities in leopard-created bone assemblages concluded that tooth mark frequencies similar to those documented in hyena-modified assemblages would be expected on prey limb bones (Cavallo, 1997). However, this overlooks the fact that most tooth marks occur on bone fragments and not on complete bones. As we mentioned above, if hominids were scavenging from felid kills (as most passive scavenging models posit), they would disregard broken, resource-depleted bone fragments and instead concentrate on the exploitation of marrow-containing whole bones. The preliminary data presented here indicate that tooth mark frequencies in such a scenario would be very low.

This conclusion clearly calls for a reconsideration of passive scavenging models and their application to actual archaeofaunas. A fundamental aspect of multiple-pattern passive scavenging models is that felids, not hyenas, initiated the exploitation of carcasses. The empirical evidence presented here supports the theoretical assertion that hyena-modified assemblages probably do not accurately simulate felid-ravaging. Our results also have more general implications for the use of carnivore tooth mark data as indicators of hominid foraging behavior. The fact that midshaft tooth mark frequencies generated by felids acting as primary agents of bone modification are very similar to those produced by hyenas acting as secondary agents of bone modification is particularly intriguing in this regard. This seriously diminishes the utility of tooth mark frequencies on limb bone midshafts for testing scenarios of passive scavenging, especially considering that lion and hyena tooth pit dimensions on limb bone diaphyses overlap (Domínguez-Rodrigo and Piqueras, 2003). Therefore, analytical manipulations of tooth mark frequency data are subject to equifinality if the taxonomic identity of the modifying agent(s) is not recognized and controlled for explicitly in the model-building process. We thus suggest that the term "carnivore-first" is too ambiguous a concept for modelling serial intertaxonomic processing of large animal carcasses by multiple consumers.

More fundamentally, we argue that only the use of hominid-imparted bone surface modifications such as cutmarks and hammerstone percussion marks can reliably inform zooarchaeologists about the order of hominid access to large animal carcasses. This runs counter to popular zooarchaeological opinion, which contends that carnivore tooth marks are less susceptible to equifinality and thus the most useful class of surface modification for reconstructing early hominid foraging behavior (e.g., Blumenschine, 1995; Capaldo, 1995; Lupo and O'Connell, 2002).

Recent tooth mark-based interpretations of hominid behavior at FLK Level 22 can now be examined in light of the results presented here. Blumenschine (1995) asserts that tooth mark frequencies on limb bone midshaft fragments at FLK Level 22 indicate early felid and late hominid access to carcasses. Tooth mark frequencies on midshaft portions at FLK Level 22 are lower than those imparted by hyenas with primary access to bones in Blumenschine's (1988, 1995) experiments but, as can now be seen, are also several times higher than frequencies reported here for primary felid access to carcasses. There are three reasons for this discordance in tooth mark frequencies. First, as Domínguez-Rodrigo (1999b) has suggested, it is likely that hominids did not break open all limb bones at FLK Level 22, which provided scavenging hyenas with some exploitable marrow. Second, a recent reanalysis of the FLK Level 22 archaeofauna (Domínguez-Rodrigo and Barba, 2006) found that previous estimates artificially inflated tooth mark frequencies because natural biochemical marks were mistaken for tooth marks. Finally, although FLK Level 22 largely reflects the interdependent contributions of both hominids and carnivores, it is likely that as a true archaeological palimpsest in the extended sense the site also represents the independent accumulation and modification of carcasses by these agents (even if in a minority of cases), much as Isaac (1983) originally suggested for Plio-Pleistocene sites in general. Given the likely ubiquity of such a situation, we conclude by constructing a theoretical model of a true archaeological palimpsest.

## Modelling an archaeological palimpsest

The important work of Blumenschine and his collaborators provides the fundamental datasets required to model an archaeological palimpsest. We restrict our simplified palimpsest to a two-agent system and begin by considering a single carcass modified by hominids and a single carcass modified by spotted hyenas. Capaldo's (1995) extensive actualistic dataset, which includes fairly complete carcasses, is probably the best source for gauging differences in bone fragmentation by hyenas and hominids. An average of 14 limb bone fragments survive hyena ravaging per complete carcass, while human-processed carcasses yield an average of 86 limb bone fragments (Capaldo, 1995). Capaldo (1995)

*Table 6. Tooth mark frequencies in a theoretical palimpsest. The initial palimpsest consists of one carcass accumulated only by hominids and one carcass accumulated only by hyenas. Each subsequent carcass represents one individual introduced to the palimpsest by hyenas, while hominids contribute no more carcasses (see text for full explanation). Total NISP represents the total number of limb bone specimens in the palimpsest. Hominid NISP represents the number of limb bone specimens introduced by hominids (this number remains the same because hominids only contribute to the initial stage of palimpsest formation). Hyena NISP represents the number of limb bone specimens introduced by hyenas (14 specimens per carcass). TM NISP represents the incremental increase in tooth-marked limb bone specimens as the number of carcasses introduced by hyenas increases. %TM represents the incremental increase in tooth mark percentages as the number of carcasses introduced by hyenas increases.*

| Carcass | Total NISP | Hominid NISP | Hyena NISP | TM NISP | %TM |
|---|---|---|---|---|---|
| 1 | 100 | 86 | 14 | 28 | 28.0 |
| 2 | 114 | 86 | 28 | 39 | 34.2 |
| 3 | 128 | 86 | 42 | 50 | 39.1 |
| 4 | 142 | 86 | 56 | 61 | 43.0 |
| 5 | 156 | 86 | 70 | 72 | 46.2 |
| 6 | 170 | 86 | 84 | 83 | 48.8 |
| 7 | 184 | 86 | 98 | 94 | 51.1 |
| 8 | 198 | 86 | 112 | 105 | 53.0 |
| 9 | 212 | 86 | 126 | 116 | 54.7 |
| 10 | 226 | 86 | 140 | 127 | 56.2 |

also reports a number of identified specimens (NISP) to minimum number of elements (MNE) ratio of 1.2 limb bone fragments for each complete bone recovered after thirty experiments. Human-created assemblages result in a much higher NISP:MNE ratio (7.6). The implication of these data for a theoretical palimpsest is that human processing of a single carcass will result in more limb bone fragments than a single carcass processed by hyenas at the same accumulation site.

Regarding bone surface modifications, Capaldo (1995) reports that ~70 % of limb bone fragments recovered from hyena-modified assemblages display tooth marks, which is only slightly lower than the 84% documented in Blumenschine's (1988, 1995) experiments. In addition, both researchers demonstrate that secondary access by hyenas to human-fractured bones results in low percentages of tooth-marked limb bone fragments (~20%) and, more specifically, very low frequencies of tooth-marked midshaft fragments (10 – 15%).

When these data are combined, a theoretical assemblage of 100 limb bone specimens (86 from the carcass accumulated only by hominids plus 14 from the carcass accumulated only by hyenas) is created. Of these 100 specimens, eleven (80%) of the 14 hyena-only fragments will bear tooth marks, while 17 (20%) of the hominid-created fragments will be tooth-marked as the result of hyena scavenging. This results in a total of 28 (28% of the original 100-fragment assemblage) tooth-marked fragments.

If hyenas were to deposit the limb bones of yet another carcass (without any subsequent hominid input of carcasses), tooth mark frequencies would increase to 34.2% (39 out of 114 specimens). In other words, the addition of one carcass by hyenas would increase the previous tooth mark percentage by 6%. Table 6 summarizes the progressive increase in tooth mark frequencies with the addition of up to ten hyena-accumulated carcasses to the theoretical palimpsest. After the addition of a seventh carcass tooth mark frequencies reach 50%.

This simplified model obviously ignores a number of important factors that contribute to tooth mark frequencies. Nevertheless, the important message is that although tooth mark frequencies can be very high in palimpsests to which hyenas have contributed significantly, these frequencies do not address even indirectly the timing of hominid access to the carcasses they are responsible for accumulating. This conclusion supports our earlier contention that only hominid-imparted bone surface modifications hold the potential to construct realistic inferences of hominid carcass foraging. Our well-published arguments on the analytical utility of especially cutmark placement and frequency suggest that hominids often gained early access to carcasses (Domínguez-Rodrigo, 1997a,b, 1999a,b, 2002; Domínguez-Rodrigo and Pickering, 2003; Domínguez-Rodrigo et al., 2002, 2005; Pickering and Domínguez-Rodrigo, in press), which agree with earlier conclusions drawn by other analysts using similar data (e.g., Bunn, 1982; Bunn and Kroll, 1986).

In closing, we re-emphasize that this study represents an initial (and simplified) step towards understanding Stone Age palimpsets. However, the important point remains that when reconstructing the formation of fossil faunas, zooarchaeologists need to recognize the prehistoric reality that hominids and carnivores operated both interdependently (as currently modelled in passive scavenging scenarios), *and* independently in site formation. We also wish to stress that our critiques inherently (and gratefully) acknowledge previous actualistic work as a catalyst for this expanded view of Stone Age assemblage formation. We have in particular the pioneering work of Bob Brain to thank for the nearly universal adoption of the actualistic approach that makes this research possible.

## Acknowledgements

First and foremost, we thank Bob Brain for constant inspiration and for initiating the florescence of the modern actualistic approach in zooarchaeology. We also thank the Northern Flagship Institution (formerly the Transvaal Museum) and especially Stephany Potze for access to their taphonomic collections. MDR thanks the Office of the President in Kenya and the Kenyan Wildlife Foundation for permits to conduct research, and expresses his gratitude to Travis Pickering, Nick Toth and Kathy Schick for having invited him to participate in this tribute to C.K. Brain. In addition, he thanks Rebeca Barba Egido and Elia Organista for their help in analyzing the butchery collections at the Complutense University. CPE thanks the Graduate and Professional Student Organization, Indiana University for partially funding his research in South Africa. TRP thanks Manuel Domínguez-Rodrigo and Charles Egeland for making our continuing collaborative work so instructive and fun. TRP also thanks his family and Nick Toth and Kathy Schick for all the support and the College of Arts and Sciences, Indiana University for a Summer Faculty Fellowship to conduct the research. As always, CPE and TRP are most appreciative to Kathy Kuman, Ron Clarke and the Brain family for the hospitality they extend during our time in South Africa.

## References

Bartram, L.E. 1993. An enthoarchaeologcial analysis of Kua San (Botswana) bone food refuse. Ph.D. Dissertation, University of Wisconsin, Madison.

Bartram, L.E., Marean, C.W. 1999. Explaining the "Klaises Pattern": Kua ethnoarchaeology, the Die Kelders Middle Stone Age archaeofauna, long bone fragmentation and carnivore ravaging. Journal of Archaeological Science 26, 9-29.

Binford, L.R. 1978. Nunamiut Ethnoarchaeology. Academic Press, New York.

Binford, L.R. 1980. Willow smoke and dog´s tails: hunter-gatherer settlement systems and archaeological site formation. American Antiquity 45, 4-20.

Binford, L.R. 1981. Bones: Ancient Men and Modern Myths. Academic Press, New York.

Binford, L.R. 1984. Faunal Remains from Klasies River Mouth. Academic Press, Orlando.

Binford, L.R. 1985. Human ancestors: changing views of their behavior. Journal of Anthropological Archaeology 4, 292-327.

Binford, L.R. 1986. Comments on Bunn and Kroll. Current Anthropology 27, 444-446.

Binford, L.R. 1988a. Fact and fiction about the Zinjanthropus Floor: data, arguments and interpretations. Current Anthropology 29, 123-135.

Binford, L.R. 1988b. The hunting hypothesis, archaeological methods and the past. Yearbook of Physical Anthropology 30, 1-9.

Blumenschine, R.J. 1988. An experimental model of the timing of hominid and carnivore influence on archaeological bone assemblages. Journal of Archaeological Science 15, 483-502.

Blumenschine, R.J. 1991. Hominid carnivory and foraging strategies, and the socio-economic function of early archaeological sites. Philosophical Transactions of the Royal Society (London) 334, 211-221.

Blumenschine, R.J. 1995. Percussion marks, tooth marks and experimental determinations of the timing of hominid and carnivore access to long bones at FLK *Zinjanthropus*, Olduvai Gorge, Tanzania. Journal of Human Evolution 29, 21-51.

Blumenschine, R.J., Marean, C.W. 1993. A carnivore´s view of archaeological bone assemblages. In: Hudson, J. (Ed), From Bones to Behavior: Ethnoarchaeological and Experimental Contributions to the Interpretation of Faunal Remains. Center for Archaeological Investigations, Southern Illinois University, Carbondale, Illinois, pp. 271-300.

Blumenschine, R.J., Selvaggio, M.M. 1991. On the marks of marrow bone processing by hammerstones and hyenas: their anatomical patterning and archaeological implications. In: Clark, J.D. (Ed), Cultural Beginnings: Approaches to Understanding Early Hominid Life–ways in the African Savanna, Dr. Rudolf Habelt GMBH Bonn, pp. 17-28.

Blumenschine, R.J., Cavallo, J.A., Capaldo, S.D. 1994. Competition for carcasses and early hominid behavioral ecology: a case study and conceptual framework. Journal of Human Evolution 27, 197-213.

Blumenschine, R.J., Marean, C.W., Capaldo, S.D. 1996. Blind tests of inter-analyst correspondence and accuracy in the identification of cut marks, percussion marks, and carnivore tooth marks on bone surfaces. Journal of Archaeological Science 23, 493-507.

Brain, C.K. 1974. Some suggested procedures in the analysis of bone accumulations from southern African Quaternary sites. Annals of the Transvaal Museum 14, 1-8.

Brain, C.K. 1981. The Hunters or the Hunted? An Introduction to African Cave Taphonomy. University of Chicago, Chicago.

Bunn, H.T. 1981. Archaeological evidence for meat-eating by Plio-Pleistocene hominids from Koobi Fora and Olduvai Gorge. Nature 291, 574-577.

Bunn, H.T. 1982. Meat-eating and human evolution: studies on the diet and subsistence patterns of Plio-Pleistocene hominids in East Africa. Ph.D. Dissertation, University of California, Berkeley.

Bunn, H.T. 1983. Evidence on the diet and subsistence patterns of Plio-Pleistocene hominids at Koobi Fora, Kenya, and at Olduvai Gorge, Tanzania. In: Clutton-Brock, J., Grigson, C. (Eds), Animals and Archaeology, Volume I: Hunters and their Prey. British Archaeological Reports International Series, 163, Oxford, pp. 21-33.

Bunn, H.T. 1986. Patterns of skeletal representation and hominid subsistence activities at Olduvai Gorge, Tanzania, and Koobi Fora, Kenya. Journal of Human Evolution 15, 673-690.

Bunn, H.T., Ezzo, J.A. 1993. Hunting and scavenging by Plio-Pleistocene hominids: nutritional constraints, archaeological patterns, and behavioural implications. Journal of Archaeological Science 20, 365-398.

Bunn, H.T., Kroll, E.M. 1986. Systematic butchery by Plio-Pleistocene hominids at Olduvai Gorge, Tanzania. Current Anthropology 27, 431-452.

Bunn, H.T., Kroll, E.M., Bartram, L.E. 1988. Variability in bone assemblage formation from Hadza hunting, scavenging and carcass processing. Journal of Anthropological Archaeology 7, 412-457.

Bunn, H.T., Kroll, E.M., Bartram, L.E. 1991. Bone distribution on a modern East African landscape and its archaeological implications. In: Clark, J.D. (Ed), Cultural Beginnings: Approaches to Understanding Early Hominid Life-ways in the African Savanna, Dr. Rudolf Habelt GMBH Bonn, pp. 33-54.

Capaldo, S.D. 1995. Inferring hominid and carnivore behavior from dual-patterned archaeological assemblages. Ph.D. Dissertation, Rutgers University, New Brunswick.

Capaldo, S.D. 1997. Experimental determinations of carcass processing by Plio-Pleistocene hominids and carnivores at FLK 22 (*Zinjanthropus*), Olduvai Gorge, Tanzania. Journal of Human Evolution 33, 555-597.

Capaldo, S.D. 1998. Methods, marks and models for inferring hominid and carnivore behavior. Journal of Human Evolution 35, 323-326.

Cavallo, J.A. 1997. A re-examination of Isaac´s central-place foraging hypothesis. Ph.D. Dissertation, Rutgers University, New Brunswick.

Cruz-Uribe, K. 1991. Distinguishing hyena from hominid bone accumulations. Journal of Field Archaeology 18, 467-486.

Domínguez-Rodrigo, M. 1997a. Meat-eating by early hominids at the FLK 22 *Zinjanthropus* site, Olduvai Gorge, Tanzania: an experimental approach using cut mark data. Journal of Human Evolution 33, 669-690.

Domínguez-Rodrigo, M. 1997b. A reassessment of the study of cut mark patterns to infer hominid manipulation of fleshed carcasses at the FLK *Zinj* 22 site, Olduvai Gorge, Tanzania. Trabajos de Prehistoria 54, 29-42.

Domínguez-Rodrigo, M. 1999a. Flesh availability and bone modification in carcasses consumed by lions. Palaeogeography, Palaeoclimatology, Palaeoecology 149, 373-388.

Domínguez-Rodrigo, M. 1999b. Meat-eating and carcass procurement by hominids at the FLK *Zinj* 22 site, Olduvai Gorge (Tanzania): a new experimental approach to the old hunting-versus-scavenging debate. In: Ullrich, H. (Ed), Lifestyles and Survival Strategies in Pliocene and Pleistocene Hominids, Edition Archaea, Schwelm, Germany, pp. 89-111.

Domínguez-Rodrigo, M. 2002. Hunting and scavenging in early humans: the state of the debate. Journal of World Prehistory 16, 1-56.

Domínguez-Rodrigo, M., Barba, R. 2006. New estimates of tooth mark and percussion mark frequencies at the FLK *Zinj* site: the carnivore-hominid-carnivore hypothesis falsified. Journal of Human Evolution 50, 170-194.

Domínguez-Rodrigo, M., Pickering, T.R. 2003. Early hominid hunting and scavenging: a zooarchaeological review. Evolutionary Anthropology 12, 275-282.

Domínguez-Rodrigo, M., Piqueras, A. 2003. The use of tooth pits to identify carnivore taxa in tooth-marked archaeofaunas and their relevance to reconstruct hominid carcass processing behaviours. Journal of Archaeological Science 30, 1385-1391.

Domínguez-Rodrigo, M., de la Torre, I., de Luque, L., Alcalá, L., Mora, R., Serrallonga, J., Medina, V. 2002. The ST Site Complex at Peninj, West Lake Natron, Tanzania: implications for hominid behavioural models. Journal of Archaeological Science 29, 639-665.

Domínguez-Rodrigo, M., Pickering, T.R., Semaw, S., Rogers, M.J. 2005. Cutmarked bones from Pliocene archaeological sites at Gona, Ethiopia: implications for the function of the world's oldest stone tools. Journal of Human Evolution 48, 109-121.

Egeland, C.P., Pickering, T.R., Domínguez-Rodrigo, M., Brain, C.K. 2004. Disentangling Early Stone Age palimpsests: determining the functional independence of hominid- and carnivore-derived portions of archaeofaunas. Journal of Human Evolution 47, 343-357.

Hill, A. 1975. Taphonomy of contemporary and late Cenozoic East African vertebrates. Ph.D. Dissertation, University of London, England.

Isaac, G.Ll. 1978. The food-sharing behavior of proto-human hominids. Scientific American 238, 90-108.

Isaac, G.Ll. 1981. Archaeological tests of alternative models of early hominid behaviour: excavation and experiments. Philosophical Transactions of the Royal Society (London) 292, 177-188.

Isaac, G. Ll. 1983. Bones in contention: competing explanations for the juxtaposition of Early Pleistocene artifacts and faunal remains. In: Clutton-Brock, J., Grigson, C. (Eds), Animals and Archaeology, Volume I: Hunters and their Prey. British Archaeological Reports International Series, 163, Oxford, pp. 3-19.

Isaac, G.Ll. 1984. The archaeology of human origins: studies of the Lower Pleistocene in East Africa, 1971-1981. Advances in World Archaeology 3, 1-87.

Klein, R.G. 1975. Paleoanthropological implications of the non-archaeological bone assemblage from Swartklip I, Southwestern Cape Province, South Africa. Quaternary Research 5, 275-288.

Lupo, K.D. 1998. Experimentally derived extraction rates for marrow: implications for body part exploitation strategies of Plio-Pleistocene hominid scavengers. Journal of Archaeological Science 25, 657-675.

Lupo, K.D., O´Connell, J.F. 2002. Cut and tooth mark distributions on large animal bones: ethnoarchaeological data from the Hadza and their implications for current ideas about early human carnivory. Journal of Archaeological Science 29, 85-109.

Marean, C.W., Bertino, L. 1994. Intrasite spatial analysis of bone: subtracting the effect of secondary consumers. American Antiquity 59, 748-768.

Marean, C.W., Frey, C.J. 1997. The animal bones from caves to cities: reverse utility curves as methodological artifacts. American Antiquity 62, 698-711.

Marean, C.W., Spencer, L.M. 1991. Impact of carnivore ravaging on zooarchaeological measures of element abundance. American Antiquity 56, 645-658.

Marean, C.W., Spencer, L.M., Blumenschine, R.J., Capaldo, S.D. 1992. Captive hyaena bone choice and destruction, the schlepp effect and Olduvai archaeofaunas. Journal of Archaeological Science 19, 101-121.

Marean, C.W., Domínguez-Rodrigo, M., Pickering, T.R. 2004. Skeletal element equifinality in zooarchaeology begins with method: the evolution and status of the "shaft critique". Journal of Taphonomy 2, 69-98.

O'Connell J.F., Lupo, K.D. 2003. Reply to Domínguez-Rodrigo. Journal of Archaeological Science 30, 387-390

O'Connell, J.F., Hawkes, K., Blurton Jones, N.G. 1988. Hadza hunting, butchering and bone transport and their archaeological implications. Journal of Anthropological Research 44, 113-161.

O'Connell, J.F., Hawkes, K., Blurton Jones, N.G. 1990. Reanalysis of large mammal body part transport among the Hadza. Journal of Archaeological Science 17, 301-16.

O'Connell, J.F., Hawkes, K., Blurton Jones, N.G. 1991. Distribution of refuse-producing activities at Hadza residential base camps: implications for analyses of archaeological site structure. In: Kroll, E. M., Price, T.D. (Eds), The Interpretation of Archaeological Spatial Patterning. Plenum Press, New York, pp. 61-76.

O'Connell, J.F., Hawkes, K., Blurton Jones, N.G. 1992. Patterns in the distribution, site structure and assemblage composition of Hadza kill-butchering sites. Journal of Archaeological Science 19, 319-345.

O'Connell, J.F., Hawkes, K., Lupo, K.D., Blurton Jones, N.G. 2002. Male strategies and Plio-Pleistocene archaeology. Journal of Human Evolution 43, 831-872.

Oliver, J.S. 1994. Estimates of hominid and carnivore involvement in the FLK *Zinjanthropus* fossil assemblage: some socioecological implications. Journal of Human Evolution 27, 267-294.

Pickering, T.R. 2001. Carnivore voiding: a taphonomic process with the potential for the deposition of forensic evidence. Journal of Forensic Sciences 46, 406-411.

Pickering, T.R. 2002. Reconsideration of criteria for differentiating faunal assemblages accumulated by hyenas and hominids. International Journal of Osteoarchaeology 12, 127-141.

Pickering T.R., Domínguez-Rodrigo, M. in press. The acquisition and use of large mammal carcasses by Oldowan hominids in Eastern and Southern Africa: a selected review and assessment. In: Toth, N., Schick, K.D. (Eds), The Oldowan: Studies into the Origins of Human Technology. Stone Age Institute Press, Bloomington, Indiana.

Pickering, T.R., Marean. C.W., Domínguez-Rodrigo, M. 2003. Importance of limb bone shafts in zooarchaeology: a response to "On situ attrition and vertebrate body part profiles" (2002) by M.C. Stiner. Journal of Archaeological Science 30, 1496-1482.

Potts, R. 1982. Lower Pleistocene site formation and hominid activities at Olduvai Gorge, Tanzania, Ph.D. Dissertation, Harvard University, Cambridge.

Potts, R. 1988. Early Hominid Activities at Olduvai. Aldine de Gruyter, New York.

Potts, R. and Shipman, P. 1981. Cutmarks made by stone tools from Olduvai Gorge, Tanzania. Nature 291, 577-580.

Schaller, G.B. 1972. The Serengeti Lion. Chicago University Press, Chicago.

Selvaggio, M.M. 1994. Identifying the timing and sequence of hominid and carnivore involvement with Plio-Pleistocene bone assemblages from carnivore tooth marks and stone-tool butchery marks on bone surfaces. Ph.D. Dissertation, Rutgers University, New Brunswick.

Selvaggio, M.M. 1998. Concerning the three stage model of carcass processing at FLK *Zinjanthropus*: a reply to Capaldo. Journal of Human Evolution 35, 319-321.

Selvaggio, M.M., Wilder, J. 2001. Identifying the involvement of multiple carnivore taxa with archaeological bone assemblages. Journal of Archaeological Science 28, 465-470.

Shipman, P. 1983. Early hominid lifestyles: hunting and gathering or foraging and scavenging? In: Clutton-Brock, J., Grigson, C. (Eds), Animals and Archaeology, Volume I: Hunters and their Prey. British Archaeological Reports International Series, 163, Oxford, pp. 21-30.

Shipman, P. 1986. Scavenging or hunting in early hominids: theoretical frameworks and tests. American Anthropologist 88, 27-43.

Shipman, P., Phillips, J.E. 1976. On scavenging by hominids and other carnivores. Current Anthropology 17, 170-172.

# CHAPTER 15

# Butchering Backstraps and Bearing Backbones: Insights from Hadza Foragers and Implications for Paleolithic Archaeology

Henry T. Bunn

## Abstract

This paper examines the dynamics of carcass transport by Hadza foragers near Lake Eyasi, Tanzania, and considers the archaeological implications for interpreting past human behavior from skeletal proportions in ancient bone assemblages. The Hadza possess a boiling technology (cooking fires and pots) for rendering edible fat from cancellous bones, which provides the incentive to transport defleshed backbones and some other axial and girdle elements to camps, where they are chopped into pot-sized pieces and boiled. Shifting temporal and technological contexts to the beginning of the Paleolithic, there is no evidence of a boiling technology, and, thus, no incentive for hominin transport of most axial and girdle elements from animal death sites to central locations ("home base" sites). An abundance of diverse large animals and an abundance of butchered limb elements in assemblages at sites formed over short time intervals, illustrated best at FLK Zinj, indicate that hominins selectively transported meaty portions of carcasses to such sites for further butchery, sharing, and consumption. The paucity of vertebrae, and consequently of the potential evidence of butchery damage to them, at such sites constrains consideration of how Plio-Pleistocene hominins may have used those carcass portions for food, but it is unnecessary to argue that hominins would have regularly transported many axial and girdle elements in the first place.

## Introduction

Known colloquially as backstraps to many in rural America, t-bone, loin, and tenderloin to steak lovers, longissimus and psoas muscles to anatomists, and as loins in this paper, these major, elongate muscles running along the dorsal surfaces of the backbone and ventral surfaces of the lower back, are a prime cut of meat. Attached as they are to the backbone (and dorsal rib shafts) of an intact animal, the loins present a challenge in butchery, particularly when the animal is large. Moving a large carcass by manual labor requires segmenting of it into transportable portions guided by consideration of animal size, size of labor force, animal anatomy, and so on. Basic division of a carcass into one or more axial and girdle portions and the four limb portions is typical, with decisions for further division and, if the entire carcass is not to be transported, for selective transport of only some portions, as options. The axial/girdle portions of a carcass are rich in edible meat and fat but more awkward and unwieldy to transport than the meaty limb portions. The question of how prehistoric foragers resolved such subsistence challenges is addressed through observation of modern-day foragers and through analysis of archaeological bone assemblages.

Skeletal proportions in archaeological bone assemblages offer a tantalizing yet challenging opportunity to reconstruct aspects of ancient foraging behavior, including carcass transport, site function, and many others. One of the commonest analytical measures employed since the late 1800s (Lartet and Christy, 1865-75), is the use of disproportions in limb and axial element representation to distinguish sites that are a product of selective transport of carcass portions by humans (i.e., base camps high in limb elements) from sites that are not (i.e., *in situ* death, kill, or butchery sites high in axial and girdle elements). One writer has rejected this interpretive principle stating repeatedly that it "lacks empirical support" which "places at risk" many archaeological reconstructions, including those at Plio-Pleistocene sites in East Africa (O'Connell et al., 1988, 1990). Consulting mainly his own short-term observations of carcass transport by Hadza hunter-gatherers and unique but inappropriate analytical approach, O'Connell has more recently alleged that the well-known home base (or central place) sites of the Plio-Pleistocene are actually *in situ* animal death sites exploited by hominins on the spot without transport of carcass portions or bones (O'Connell et al., 2002). But is that really so, or does empirical evidence in context (i.e., the real world) actually contradict what O'Connell has alleged? I address these questions herein by reviewing empirical evidence from African ethnography, particularly from the Hadza, and from natural and cultural landscape settings where carcass and bone distribution patterns have been measured.

## Taphonomy and African Foragers

During the past quarter century, taphonomic studies have documented some of the bone-related dynamics of carcass acquisition and use by several extant African forager societies. These include the !Kung (Yellen, 1977, 1991) and Kua (Bunn, 1983; Bartram et al., 1991; Bartram, 1993) in southern Africa, and the Hadza (Bunn et al., 1988, 1991; Bunn, 1993; O'Connell et al., 1988, 1990), Okiek (Marshall, 1993), and Aka Pygmies (Hudson, 1993) in eastern and central Africa. These studies have provided a wealth of empirical data and an informative array of insights for archaeologists. For present purposes, several salient points will suffice. Unsurprisingly, the dynamics of carcass transport are complex and often unknowable from ancient archaeological evidence. Factors influencing decision making by humans in field butchery and transport of carcasses include: (1) size and condition of carcass; (2) distance from death site to intended base camp (or other) destination; (3) number and strength of available carriers; (4) various logistical considerations, including time of day, weather, hunger level, and perceived competition from carnivores. Because most of these factors are hard to identify from archaeological data, fairly general statements about carcass transport are a prudent research objective.

The general tendency that stands out strongly in these studies is the intent to transport entire carcasses of all but the largest taxa and the successful realization or near realization of that objective most of the time. In other words, the food item for these forager societies is the entire, field-butchered carcass, not its constituent parts. Among the Hadza, this transport pattern is particularly strong and well documented (Bunn et al., 1988; Bunn, 1993), *contra* O'Connell et al., (1988, 1990), who employ an inappropriate analytical method and thereby mistake measurement of bone abandonment for measurement of the process of carcass transport. Bunn (1993) combined carcass transport data from the mid to late 1980s observations of Bunn (1986, 1988) and of O'Connell (1985, 1988) and presented skeletal profiles. Sample size from the independent research differed markedly, particularly for the commonest Hadza prey, impala (MNI: 24 in Bunn's research and 12 in O'Connell's) and zebra (MNI: 32 from Bunn and 10 from O'Connell). Although full discussion of these data is beyond the scope of this paper, Figure 1 illustrates the general outcome. The Hadza transport all skeletal elements more than 90 percent of the time, except crania (82%), mandibles (83%) and ribs (82%), which indicates a dominant pattern of whole-carcass transport.

How does the finding of whole-carcass transport among the Hadza and other African foragers impact the interpretive principle of measuring carcass transport archaeologically with limb/axial bone representation? Does it validate that analytical measure? Or, does it mean a lack of empirical support? To answer those questions, a closer look at the dynamics of carcass transport and processing by the Hadza is required.

## Dynamics of Carcass Transport by Hadza Foragers

Hadza foragers acquire a variety of large mammal prey by several hunting and scavenging strategies. Male hunters use powerful wooden long bows and an array of arrows in two principal contexts: (1) hunters construct blinds near water holes and wait in concealment to ambush water-dependent prey during the daytime and on moonlit nights during the dry season; (2) hunters encounter prey more opportunistically on daytime foraging walks from the base camp. Approximately 20 percent of carcasses are acquired by opportunistic scavenging during daily foraging activities either by small groups of hunters or by groups of women foraging for plants. Power scavenging is by far the most productive scavenging method, simply because it yields mostly intact carcasses and avoids the loss of most edible tissues to hyenas. Any form of late-access or passive scavenging from abandoned carcasses is generally unproductive, because in such contexts most edible tissues are consumed rapidly by scavenging hyenas.

Prey animal death sites thus occur in diverse locations on the landscape, and this poses diverse challenges regarding carcass transport. Prey animals may die within sight of a base camp. Given the occasional (but uncommon) paralyzing arrow shot to an animal's spine, animals may die adjacent to a hunting blind. Most commonly, however, prey animals run long distances after being shot, up to five kilometers or farther before falling dead, and that poses a recurring challenge to the Hadza: how to transport a large carcass a significant distance from the middle of nowhere in the bush to the larger group of individuals at the base camp.

As discussed above, the dynamics of carcass transport are complex, and although the degree of difficulty varies with each carcass, major trends do emerge. For example, with small to medium-sized carcasses (i.e., size

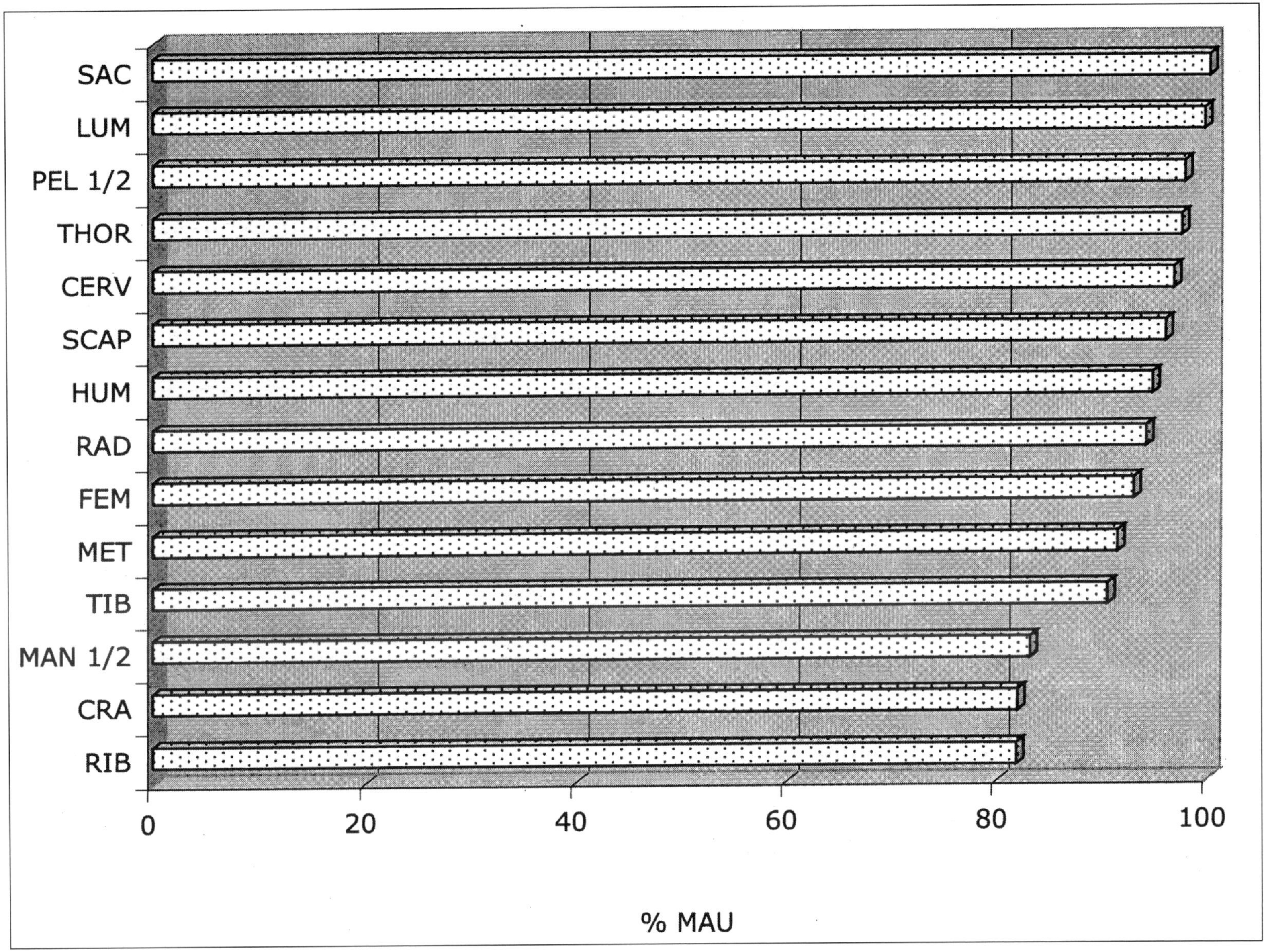


*Figure 1.* *Skeletal profile of Hadza carcass transport data combining all taxa, as observed by Bunn et al. (1988; Bunn, 1993). The Hadza transport all carcass portions from animal death sites to camps more than 90% of the time, except skulls and ribs, indicating a dominant pattern of whole-carcass transport. Adapted from Bunn (1993). Skeletal element abbreviations: SAC = sacrum; LUM = lumbar vertebra; PEL 1/2 = half of pelvis; THOR = thoracic vertebra; CERV = cervical vertebra; SCAP = scapula; HUM = humerus; RAD = radius; FEM = femur; MET = metapodial; TIB = tibia; MAN 1/2 = half of mandible; CRA = cranium; RIB = rib.*

group 1 = up to 50 lb live weight; size group 2 = 50-250 lb live weight; e.g., Brain, 1981; Bunn et al., 1988), transport from death site to base camp of entire carcasses, either intact or in field-butchered units, is routine. One to three carriers can easily achieve this even with significant transport distance. If necessary, one individual can transport an impala by splitting it longitudinally and mounting the halves onto a pole that can be carried on the shoulders. Similarly, with large carcasses (size group 3 = 250-750 lb), such as zebras and uncommon alcelaphine bovids (wildebeest, hartebeest), transport of entire field-butchered carcasses is a common and successfully realized objective of the Hadza. The Hadza successfully achieve this by enlisting additional carriers, with 7-10 individuals considered adequate. Deviations from this pattern have much more to do with particular Hadza attitudes about food quality (i.e., marrow) and about food sharing, than with concerns about reducing the energetic costs of food transport, *contra* O'Connell et al. (Bunn et al., 1988; Bunn, 1993; O'Connell et al., 1988, 1990). The Hadza's different handling of alcelaphine bovids (hartebeest and wildebeest) and zebra, all size group 3 taxa, illustrates this most simply. Some limb bones of alcelaphines (MNI = 4 from Bunn, 3 from O'Connell) are broken for marrow and abandoned at animal death sites, whereas axial and girdle elements, along with some limbs, are transported to camp. According to O'Connell et al., this means that transported bone assemblages at base camps should be dominated by scapulae, vertebrae, and pelves, and that the pattern should persist regardless of the mix of taxa in transported assemblages. According to O'Connell, this finding is what places at risk interpretations of Plio-Pleistocene sites high in limbs as transported assemblages. According to O'Connell et al., the partial abandonment of alcelaphine limb bones at kills results from a Hadza objective of reducing energetic transport costs, and they claim a similar pattern with even smaller impala carcasses.

According to the Hadza themselves, however, the partial abandonment of alcelaphine limb bones at kills results from the poor quality ranking of alcelaphine limb bone marrow by women and children who would be its likely recipients if transported to camp; the men compliantly eat some of the marrow and abandon some limb bones at kills.

These alternative explanations of cause and effect relationships governing transport dynamics can be restated as testable hypotheses and then tested with additional Hadza data. (1) If, as reasoned by O'Connell et al., the desire to reduce carcass transport cost leads the Hadza to abandon limbs at kills, then carcasses similar in size to the alcelaphines with correspondingly similar transport costs should be treated the same (i.e., abandonment of limbs). (2) If, as explained by Hadza informants, it is the poor quality of alcelaphine limb marrow that causes abandonment or non-transport of their limb bones, then carcasses similar in size to alcelaphines but with high quality marrow should be treated differently than alcelaphines (i.e., transport of limbs).

Zebra provide an ideal test case. They are abundant in the combined sample of Hadza transport data (MNI = 32 from Bunn et al., MNI= 10 from O'Connell et al.). Their meat is widely regarded as the favorite of the Hadza, and their limb marrow is ranked highly for its quality and sweetness. Like the alcelaphines, zebra are in size 3, but zebra are larger and heavier than either hartebeest or wildebeest; zebra, therefore, have higher transport costs. If a need to reduce transport costs of size 3 carcasses influences Hadza decisions in a meaningful way, at least as many, if not more, zebra limb bones should be abandoned at kills. The skeletal profile of zebra carcass transport shows that *all* zebra elements are transported by the Hadza more than 90 percent of the time from kill to camp except skulls and ribs (77-81 percent of time) (Bunn, 1993). Even though transport cost for zebra carcasses is higher than for alcelaphines, the Hadza routinely transport nearly entire zebra carcasses with only rare, negligible abandonment of limb bones. When it comes to transporting size group 3 carcasses, the Hadza could not care less about reducing transport costs.

With the selective, differential transport of very large, size group 4 and larger carcasses, a significant threshold is crossed in Hadza transport objectives. More carriers, up to 20-25 adults, may be enlisted, but unless transport distance is insignificant (i.e., less than one km. or so), many more skeletal elements are abandoned at kills. The progressive abandonment of more bones with carcass size increase is informative regarding transport decisions by the Hadza. With these largest of carcasses, there is a ranking of skeletal elements based on transport cost, further processing cost, and food yield. With Cape buffalo carcasses (size group 4), limbs are carried, but most axial elements and the pelvis are abandoned at the kill after being defleshed. Specifically, the loins are stripped from both sides of the backbone, gluteal muscles from the pelvis, and sheets of muscle from the ribs. With even larger giraffe carcasses (size group 5), most or nearly all skeletal elements are abandoned (again, unless distance to camp is short), including axial and girdle elements and now, limb elements, as well. Large packages of meat alone are the principal transported items.

Several salient points emerge from this review of the dynamics of carcass transport by the Hadza.

1. For small through large carcasses of zebra/wildebeest size (size group 3, up to 750 lbs), the Hadza regard the entire carcass, not its constituent parts, as the food item.

2. The dominant objective of transport (by far, unless poor food quality intervenes) is to move whole carcasses (minus a few snack items, such as skulls, ribs, and the occasional limb element) from kill to camp site.

3. Smaller carcasses are transported intact; larger carcasses are field-butchered into readily transportable units.

4. Some meat is separated from adjacent bones and the two transported independently. This involves some organ meat and superficial muscles from various elements but notably is extreme with the backbone.

5. The backstraps, or loins (certainly over 90 percent of attached meat), are routinely defleshed from the backbone at the kill. Yet backstraps and defleshed backbones are both transported from kill to camp sites (Figure 2).

6. Boiling technology (fires and large cooking pots) provides the incentive to transport defleshed backbones and other axial/girdle elements (pelves and ribs), which are chopped into pot-sized pieces and boiled to salvage contained fat and adhering bits of meat.

7. Even with the boiling technology, the Hadza rarely transport defleshed backbones of Cape buffalo, although, in principle, these could be carried to camp, chopped into pot-sized pieces, and rendered for fat. Evidently, the Hadza rank fully fleshed buffalo limbs for transport to camp but defleshed buffalo backbones for abandonment at kills.

The first question to consider from this review is: what would happen to the incentive to transport defleshed backbones if boiling technology were deleted from the Hadza's foraging adaptation? Given the obvious cause-effect relationship between boiling technology and the transport of defleshed axial elements for boiling among the Hadza, it is undeniable that a lack of boiling technology would greatly reduce the incentive to expend energy transporting defleshed backbones from kill to camp. Why carry defleshed backbones to a more secure or convenient location, such as a camp, if the contained fat cannot be readily used? Following that logic parsimoniously, the actual transport of defleshed axial and girdle elements, at least of medium and larger carcasses, would likely be reduced, and a relative abundance of limbs and elevated limb/axial ratio would result in bone assemblages produced by repeated carcass transport.

What O'Connell wishes to do is to take the effect (observed transport of defleshed backbones) out of the ethnographic context of the cause (boiling technology) in which it occurs. Hence, O'Connell's statements that

*Figure 2. Defleshing of loin meat from backbone at animal death site. (a) Backstrap meat from one side of zebra carcass being separated cleanly from backbone for transport to base camp as an independent meat unit. (b) Defleshed backbone and pelvis unit awaiting transport from death site to camp, where it will then be chopped into pot-sized fragments and boiled to extract fat and adhering meat scraps.*

interpretations of Plio-Pleistocene sites as locations to which carcass portions (abundant meaty limbs) were transported "lack empirical support" and are "at risk". O'Connell et al. (2002) consequently imagine Plio-Pleistocene sites abundant in limbs to be *in situ* death sites lacking significant transport, but that reveals a complete misunderstanding of the relationship between Plio-Pleistocene MNI values (ranging up to four dozen at FLK Zinj [Bunn and Kroll, 1986]), *in situ* carcass accumulation rates in modern analogue settings, and the restricted period of Plio-Pleistocene site formation (Bunn and Kroll, 1987). If, instead, a rudimentary attempt is made to contextualize transport dynamics appropriately for the Plio-Pleistocene boundary, then out go boiling technology and the incentive to transport defleshed backbones, and the interpretive problem regarding limb and meat transport that O'Connell wishes to construct simply collapses from a lack of any support.

## A bias against backbones ... and against backstraps

In addition to playing a pivotal role in reconstructions of site type relative to the transport dynamic, axial elements, and particularly backbones, provide the direct evidence of utilization of that meaty and fat-rich portion of carcasses. A proportional abundance of vertebrae provides evidence of access to backbones, and butchery damage on vertebrae provides evidence of how they were utilized for food.

The problem, of course, lies in the fragile structure of vertebrae relative to denser skeletal elements and in the fat content of cancellous portions. Many studies have shown that vertebrae, and other cancellous elements, are relatively vulnerable to destruction by a range of taphonomic and diagenetic processes (e.g., Klein, 1989; Marean et al., 1992; Lyman, 1994, and references therein; Lam et al., 1998). Prominent among these destructive processes are: human butchery and consumption; consumption by bone-crunching carnivores; differentially faster subaerial weathering of exposed cancellous portions; post-depositional crushing during sediment profile compaction; post-depositional chemical leaching of cancellous elements.

In short, the fossil and archaeological record has a high potential for bias against the preservation of backbones and other cancellous skeletal elements. Some researchers conclude pessimistically that these biasing effects are so pronounced and insurmountable that it may not be productive to use vertebrae (Marean and Cleghorn, 2003), or skeletal profiles in general (Klein, 1989; Klein et al., 1999), to reconstruct prehistoric human behavior from archaeological bone assemblages. If proven correct, that conclusion would mean that some of the following sorts of research questions may be unresolvable using archaeological skeletal profiles. Does a paucity of vertebrae in an assemblage mean that they were initially abundant but subsequently removed or destroyed *in situ*? Or, does a paucity of vertebrae mean an initial paucity at the site with behavioral significance? Does such a paucity necessarily become an analytical case of equifinality, or are there analytical approaches that enable a level of distinction between the alternatives? Similar questions could be posed for limb elements, which also have less dense, cancellous (epiphyses) and more dense (shafts) portions. For limbs, an optimistic resolution is well established: the use of denser limb shaft specimens to determine MNE values enables accurate reconstruction of initial limb proportions. Vertebrae in general may be less dense than limb shafts, but a similar analytical reliance on the densest vertebral portions might be productive. The most revealing experimental work on vertebral loss to hyenas has been conducted with fat-rich vertebrae of small to medium-sized domesticated bovids, yet, larger, size group 3 taxa predominate at Plio-Pleistocene archaeological sites. Larger vertebrae of size groups 3 and 4 taxa are more durable than small vertebrae, and they may have a significantly higher potential for preservation at a site. The reluctance of the Hadza to transport and boil Cape buffalo vertebrae illustrates this point; it also illustrates the obvious archaeological bias against transported and eaten backstraps, which may be archaeologically invisible from initial absence of transported backbones or from subsequent loss of them from a site.

For present purposes, my aim is simply to consider preliminarily how well vertebrae fare at Hadza sites and how butchery damage on vertebrae reflects the known intensive use of that carcass portion. Figure 3 shows the progressive loss of vertebrae as carcasses and bones are transported to base camps, processed for food and discarded, and abandoned to scavenging by hyenas. Based on a large sample of carcasses, it is evident, *contra* O'Connell et al., that essentially whole-carcass transport is the dominant Hadza pattern: axial, girdle, and limb elements are all very abundant in the transport data, (> 90 percent transported, except skulls and ribs, which are > 80 percent transported). Further processing by the Hadza, specifically chopping vertebrae, pelves, and ribs for boiling, followed by consumption of edible meat/fat and some gnawing and ingestion of fragmentary bones, decreases significantly the representation of those axial and girdle elements. Representation of vertebrae, for example, drops from the 96 to 100 percent range in the transport data, to values of 30 to 70 percent in the occupied camp sample (not vulnerable to scavenging carnivores). From an abandoned camp assemblage, which had additionally been scavenged by hyenas, the representation of vertebrae dropped further, to values in the 25-50 percent range. Notably, pelves and crania are quite abundant relative to other elements, because Hadza processing renders the denser portions of them unappealing to scavenging hyenas.

The timing of such skeletal changes helps to clarify Plio-Pleistocene site formation. Thus, at Plio-Pleistocene sites, crania, as the most likely skeletal markers of animal death sites, should have survived the effects of taphonomic biasing, if they were initially present. But at

a

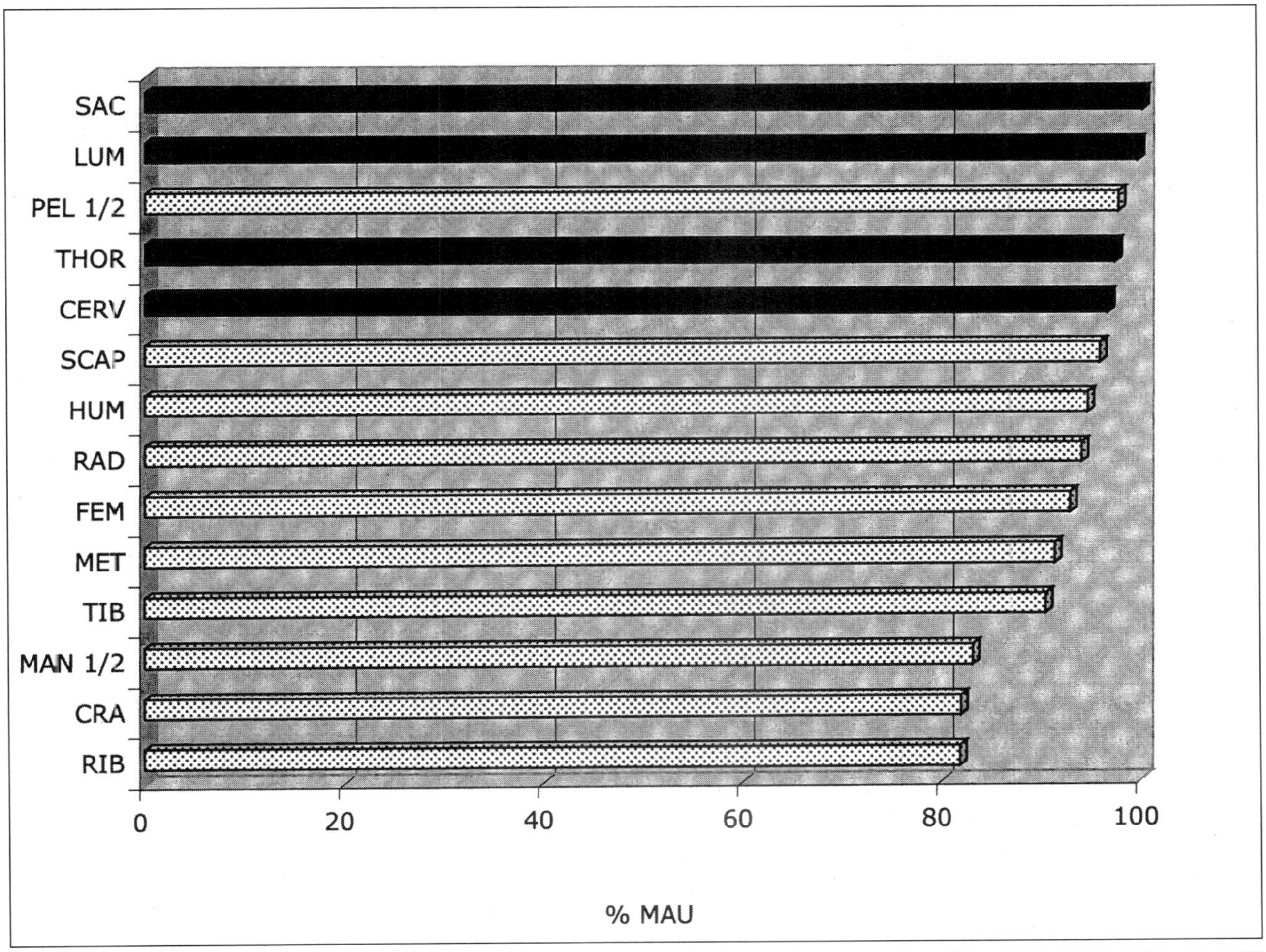
SAC
LUM
PEL 1/2
THOR
CERV
SCAP
HUM
RAD
FEM
MET
TIB
MAN 1/2
CRA
RIB
0
20
40
60
80
100
% MAU

b

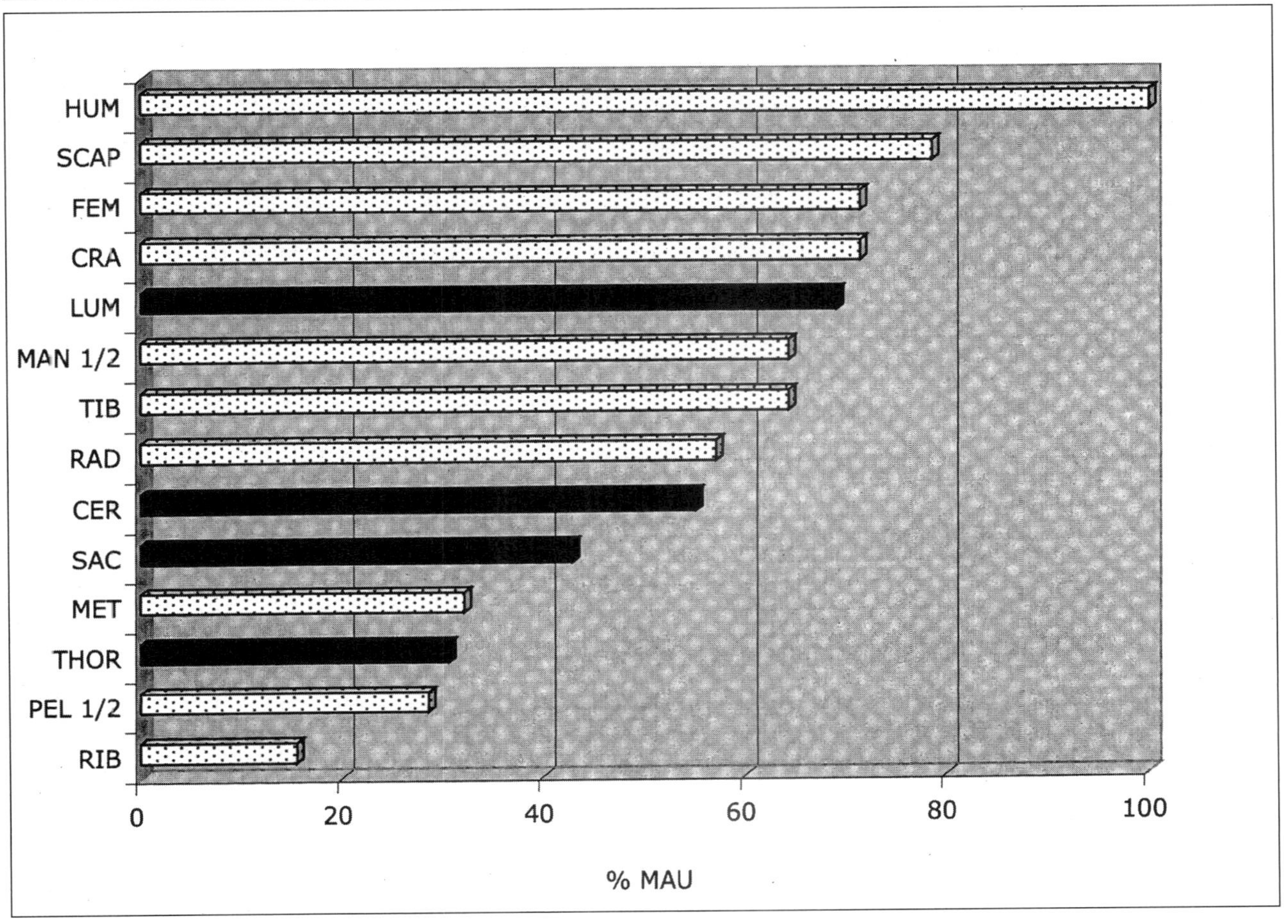
HUM
SCAP
FEM
CRA
LUM
MAN 1/2
TIB
RAD
CER
SAC
MET
THOR
PEL 1/2
RIB
0
20
40
60
80
100
% MAU

C

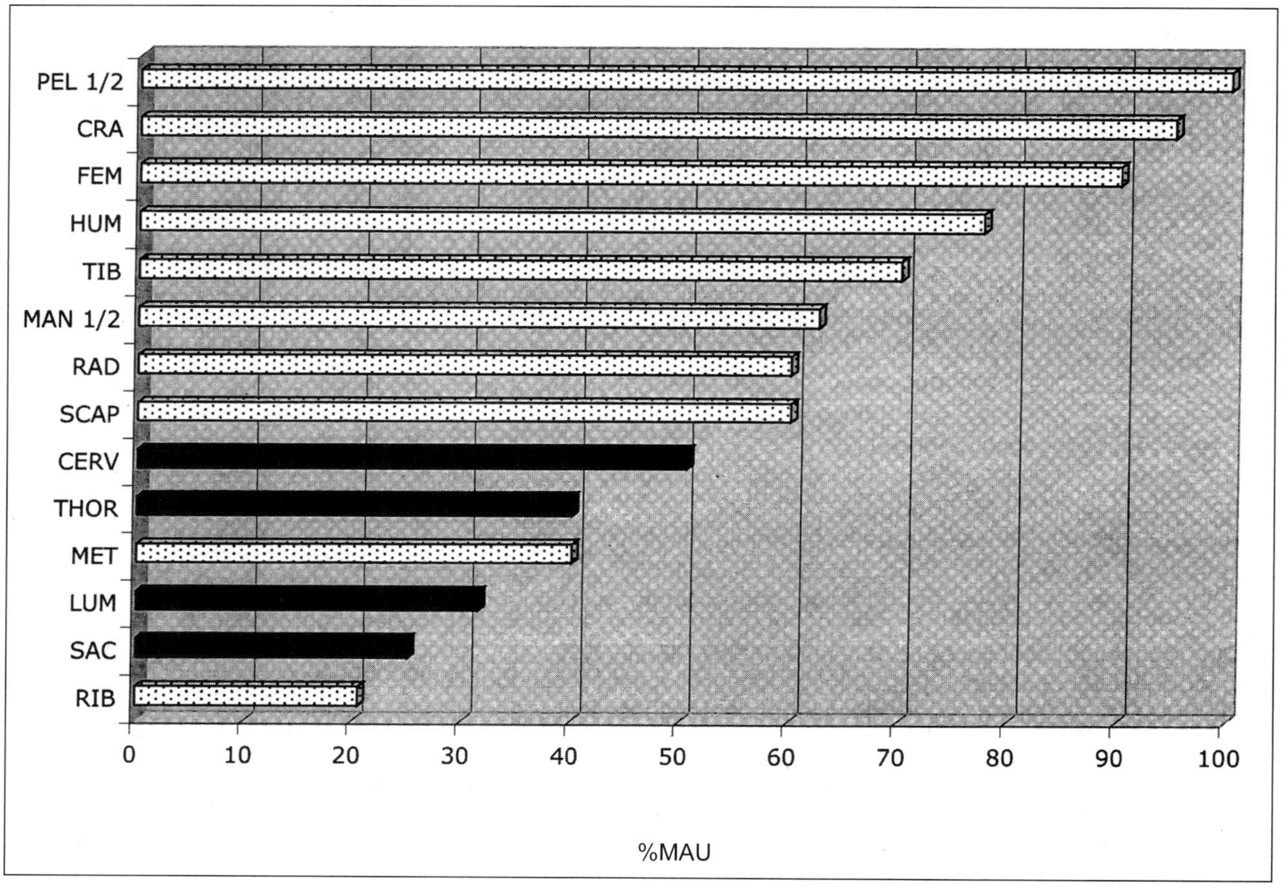


*Figure 3. Skeletal profiles showing progressive loss of vertebrae (black shading) as carcasses and bones pass through several stages of utilization by Hadza foragers. Data from Bunn et al. (1988, 1991; Bunn, 1993) with all taxa combined; skeletal element abbreviations as in Figure 1. (a) Skeletal profile of carcass transport illustrates that vertebrae and all other skeletal elements are very well represented, as a result of Hadza objective to transport essentially entire carcasses to camp. If all bones survived taphonomic processes at Hadza camps, this same profile would be reflected in bone assemblages following further processing by Hadza. (b) Skeletal profile of actual bone assemblage collected from Hadza base camp immediately following further processing i.e., secondary butchery, cooking, consumption of meat and fat, by Hadza. There is a marked reduction in vertebral representation relative to the transport profile, attributable to Hadza destruction of cancellous and other less durable bone portions. Data from Bunn et al. (1991). (c) Skeletal profile of bone assemblage from Hadza base camp, collected two years after bones discarded and camp abandoned by Hadza. A further, more modest reduction in vertebral representation is evident, which is attributable in this case to gnawing and removal by hyenas and possibly other scavengers attracted to still-greasy vertebrae following boiling and discard by Hadza. Well-represented portions of crania and acetabular portions of pelves survive these taphonomic processes because they are reasonably durable and nutritionally unappealing to scavengers following Hadza processing. Data from Bunn (1993).*

FLK *Zinj* and other Plio-Pleistocene assemblages, however, crania are poorly represented. Significant loss of vertebrae may readily result from differential destruction by human and carnivore processing for food and need not be attributed to post-depositional, density-mediated profile compaction or leaching in a buried context. As a pervasive geological force, profile compaction might be predicted to act uniformly across a buried site assemblage, thereby eliminating all, not just some, less dense portions. The presence of some undamaged and yet relatively fragile vertebrae at Plio-Pleistocene sites may indicate that it was more initial taphonomic processes, rather than much later diagenetic processes that shaped observed skeletal profiles in the bone assemblages.

Given the known intensive processing of vertebrae by the Hadza, what damage patterns result? From observations of Hadza butchery, cooking, and meat-eating, several types of damage to bones are predicted, including (1) defleshing cut marks oriented parallel to the backbone on neural spines and transverse processes from removal of the backstraps and tenderloins; (2) cut and chop marks from field butchery of vertebral portions and from pot-sizing of defleshed bones for boiling; (3) gnawing damage inflicted by the Hadza, and by scavenging hyenas. A sample of 100 vertebral specimens from an abandoned Hadza camp assemblage was examined to document how well these known taphonomic processes are reflected on the bones. Fifty specimens of each of the two commonest taxa used by the Hadza, zebra and impala, were examined for surface damage following standard procedures.

Several noteworthy patterns of bone damage were observed. For zebra vertebrae, the chopping of articulated portions into pot-sized pieces for boiling left deep gashes and sheared off, planar surfaces on most specimens. 80 percent of zebra vertebrae exhibited chop marks from the use of small steel axes by Hadza butchers. Cut marks from the use of steel knives to deflesh and disarticulate vertebrae were likewise abundant and occurred on 48 percent of the zebra vertebrae. Most of these were defleshing cuts oriented parallel to the backbone on neural spines, on the dorsal surfaces of transverse processes, and on the lateral surfaces of zygapophyses. A few were short, disarticulating cuts oriented transverse to the backbone on zygapophyses and near centrum epiphyses reflecting use of a knife to cut between bones rather than the overwhelming force of an ax to chop through them. 38 percent of zebra vertebrae exhibited both chop and cut marks, and 26 percent exhibited well-defined, large carnivore (hyena) tooth punctures (10 percent) or other gnawing damage. The gnawing damage illustrates that even after boiling by the Hadza, vertebrae retain sufficient grease and nutritional appeal to hyenas that some are gnawed and others probably removed from the site. For impala vertebrae, the results are similar qualitatively in the kinds of locations of damage, but the frequency of bone damage is lower (70 percent chopped; 43 percent cut; 33 percent both), and the relative completeness of individual vertebrae is higher. In other words, the Hadza are able to deflesh and boil the individually smaller impala vertebrae without inflicting as much damage to the bones as with larger zebra vertebrae. Although the frequencies of damage differ, both zebra and impala vertebrae provide strong evidence of Hadza defleshing of loins and chopping of vertebrae.

## Plio-Pleistocene Implications

As discussed elsewhere (Bunn 2001), applying Hada butchery patterns to reconstructions of the Plio-Pleistocene has many constraints. Although the anatomical locations and clusters of defleshing cut marks on limb elements from Hadza butchery show strong similarities to the Plio-Pleistocene FLK Zinj pattern, there is a marked contrast in the pattern of butchery damage to vertebrae between the two contexts. Chop marks and cut marks are abundant on vertebrae from Hadza butchery but rare on vertebrae from FLK Zinj. Although the frequency of chop marks produced during preparation to boil bones is not relevant to the Plio-Pleistocene example, the contrast remains strong even if restricted to defleshing cut marks resulting from separation of the loins from the backbone. This is an interesting contrast, and it emphasizes the need for improving the interpretive framework for reconstructing behavior from cut-mark data. A more comprehensive analysis of Hadza bones by vertebral portion, and more experimental work on butcheries by stone tools would help.

Given the known sequence by which large carnivores consume carcasses (e.g., Blumenschine, 1987), access to meaty limbs and particularly to hindlimbs, as documented at FLK Zinj, implies the availability of loins and backbones. Were loins butchered, transported, and then eaten by hominins at FLK Zinj? That is difficult to answer conclusively, although several alternatives for formation of the FLK Zinj site should be considered. First, the very low number of vertebrae and vertebral fragments may reflect an initial rarity of vertebrae at FLK Zinj; in other words, hominins preferentially transported limbs not backbones to the location from death sites elsewhere. Recall that crania, the last element to leave death sites in analogue studies, are poorly represented at FLK Zinj. The small sample of vertebral specimens precludes finding much direct evidence of loin defleshing and consumption.

Second, the number of vertebral specimens may be a small, biased remnant that survived removal by hyenas and/or other carnivores of most vertebrae. In other words, hominins transported limbs and backbones to FLK Zinj for further processing, followed by selective removal by scavengers of most vertebrae and other greasy portions. From this discussion of Hadza transport dynamics, particularly the cause-effect relationship between boiling technology and transporting of backbones, it is difficult to view this alternative as being equal in probability to the first alternative. In the Plio-Pleistocene context without the technology for boiling, and thus lacking also the incentive for transport and the means for chopping and boiling vertebrae for fat, where is the equifinality? Why expend energy transporting backbones that could not have been utilized efficiently?

Third, no meaningful transport of carcasses or portions thereof was required, because the FLK Zinj site was a repeated carnivore kill site, providing intact carcasses on-site, which were then scavenged by hominins (Binford 1981; O'Connell et al., 2002). According to O'Connell et al. (2002), scavenging of vertebrae by hyenas then yielded an FLK Zinj bone assemblage dominated by limbs and heads. This death-site alternative is neither an objective nor a parsimonious reading of available evidence, and it, thus, lacks merit. Even Binford disavowed it many years ago (Binford, 1988; Bunn and Kroll, 1988). Factual evidence from the bone assemblage and a substantial database from modern taphonomic studies contradict the death-site model. As reported several times, heads are not abundant in the FLK Zinj assemblage. Mandibles are abundant, but crania are poorly represented. To depict FLK Zinj as a death site, it will be necessary to conjure up a taphonomic agent responsible for removal of so many crania from the death site. To accumulate the four dozen or more (MNI = 48) large mammal carcasses documented in the bone assemblage from *in situ,* natural deaths at the FLK location within a conceivable time frame of site formation, which is several years (Bunn and Kroll 1987), not decades, centuries, or millennia (O'Connell et al., 2002), it would be necessary to invoke, without supporting evidence, a Plio-Pleistocene accumulation rate *many* times higher than observed in modern analogue studies (Behrensmeyer, 1983; Bunn et al., 1991; Sept, 1994; Tappen, 1995).

By tracing some of the dynamics of carcass transport by Hadza foragers and how these insights may impact understanding of ancient hominin behavior, I hope that the present study contributes in a small way to a long tradition of taphonomic studies in African paleoanthropology that was initiated in the 1960s by the pioneering research of Bob Brain, whose ongoing research continues to the present day.

## Acknowledgements

I thank Travis Pickering for developing the idea of the African taphonomy conference honoring the work of Bob Brain; Travis Pickering, Kathy Schick, and Nick Toth for hard work in making the conference happen in such a splendid setting as the Stone Age Institute; and Bob Brain for inspiring all of us.

## References

Bartram, L.E. 1993. Perspectives on skeletal part profiles and utility curves from eastern Kalahari ethnoarchaeology. In: Hudson, J. (Ed.), From Bones to Behavior: Ethnoarchaeological and Experimental Contributions to the Interpretation of Faunal Remains. Center for Archaeological Investigations, Southern Illinois University, Carbondale (IL), pp. 115-137.

Bartram, L.E., Kroll, E.M., Bunn, H.T. 1991. Variability in camp structure and bone food refuse patterning at Kua San camps. In: Kroll, E.M., Price, T.D. (Eds.), The Interpretation of Archaeological Spatial Patterning. Plenum Press, New York, pp. 77-148.

Behrensmeyer, A.K. 1983. Patterns of natural bone distribution on recent land surfaces: Implications for archaeological site formation. In: Clutton-Brock, J., Grigson, C. (Eds.), Animals and Archaeology, Volume 1: Hunters and Their Prey. British Archaeological Reports International Series, 163, Oxford, pp. 93-106.

Binford, L.R. 1981. Bones: Ancient Men and Modern Myths. Academic Press, New York.

Binford, L.R. 1983. Fact and fiction about the Zinjanthropus floor: Data, arguments, and interpretations. Current Anthropology 29, 123-135.

Blumenschine, R.J. 1987. Characteristics of an early hominid scavenging niche. Current Anthropology 28, 383-407.

Brain, C.K. 1981. The Hunters or the Hunted? An Introduction to African Cave Taphonomy. University of Chicago, Chicago.

Bunn, H.T. 1983. Comparative analysis of modern bone assemblages from a San hunter-gatherer camp in the Kalahari Desert, Botswana, and from a spotted hyena den near Nairobi, Kenya. In: Clutton-Brock, J., Grigson, C. (Eds.), Animals and Archaeology, Volume 1: Hunters and Their Prey. British Archaeological Reports International Series, 163, Oxford, pp. 143-148.

Bunn, H.T. 1993. Bone assemblages at base camps: A further consideration of carcass transport and bone destruction by the Hadza. In: Hudson, J. (Ed.), From Bones to Behavior: Ethnoarchaeological and Experimental Contributions to the Interpretation of Faunal Remains. Center for Archaeological Investigations, Southern Illinois University, Carbondale (IL), pp. 156-168.

Bunn, H.T. 2001. Hunting, power scavenging, and butchering by Hadza foragers and by Plio-Pleistocene *Homo*. In: Stanford, C.B., Bunn, H.T. (Eds.), Meat Eating and Human Evolution. Oxford University, New York, pp. 199-218.

Bunn, H.T., Kroll, E.M. 1986. Systematic butchery by Plio/Pleistocene hominids at Olduvai Gorge, Tanzania. Current Anthropology 27, 431-452.

Bunn, H.T., Kroll, E.M. 1987. On butchery by Olduvai hominids (Reply to R.B. Potts). Current Anthropology 28, 96-98.

Bunn, H.T., Kroll, E.M. 1988. Fact and fiction about the Zinjanthropus floor: Data, arguments, and interpretations (Reply to L.R. Binford). Current Anthropology 29, 135-149.

Bunn, H.T., Bartram, L.E., Kroll, E.M. 1988. Variability in bone assemblage formation from Hadza hunting, scavenging, and carcass processing. Journal of Anthropological Archaeology 7, 412-457.

Bunn, H.T., Kroll, E.M., Bartram, L.E. 1991. Bone distribution on a modern East African landscape and its archaeological implications. In: Clark, J.D. (Ed.), Cultural Beginnings: Approaches to Understanding Early Hominid Life-ways in the African Savanna. Dr. Rudolf Habelt GMBH, Bonn.

Hudson, J. 1993. The impacts of domestic dogs on bone in forager camps; or, the dog-gone bones. In: Hudson, J. (Ed.), From Bones to Behavior: Ethnoarchaeological and Experimental Contributions to the Interpretation of Faunal Remains. Center for Archaeological Investigations, Southern Illinois University, Carbondale (IL), pp. 301-323.

Klein, R.G. 1989. Why does skeletal part representation differ between smaller and larger bovids at Klasies River Mouth and other archaeological sites? Journal of Archaeological Science 16, 363-381.

Klein, R.G., Cruz-Uribe, K., Milo, R.G. 1999. Skeletal part representation in archaeofaunas: Comments on "Explaining the Klasies pattern: Kua ethnoarchaeology, the Die Kelders Middle Stone Age archaeofauna, long bone fragmentation and carnivore ravaging" by Bartram & Marean. Journal of Archaeological Science 26, 1225-1234.

Lam, Y.M., Chen, X., Marean, C.W., Frey, C.J. 1998. Bone density and long bone representation in archaeological faunas: Comparing results from CT and photon densitometry. Journal of Archaeological Science 25, 559-570.

Lartet, E., Christy, H. 1865-1875. Reliquiae Acquitanicae: Being Contributions to the Archaeology and Paleontology of Perigord and Adjoining Provinces of Southern France. Williams and Norgate, London.

Lyman, R.L. 1994. Vertebrate Taphonomy. Cambridge University, Cambridge.

Marean, C.W., Spencer, L.M., Blumenschine, R.J., Capaldo, S.D. 1992. Captive hyaena bone choice and destruction, the schlepp effect and Olduvai archaeofaunas. Journal of Archaeological Science 19, 101-121.

Marshall, F. 1994. Food sharing and body part representation in Okiek faunal assemblages. Journal of Archaeological Science 21, 65-77.

O'Connell, J.F., Hawkes, K., Blurton Jones, N. 1988. Hadza hunting, butchering, and bone transport and their archaeological implications. Journal of Anthropological Research 44, 113-161.

O'Connell, J.F., Hawkes, K., Blurton Jones, N. 1990. Reanalysis of large mammal body part transport among the Hadza. Journal of Archaeological Science 17, 301-316.

O'Connell, J.F., Hawkes, K., Lupo, K.D., Blurton Jones, N.G. 2002. Male strategies and Plio-Pleistocene archaeology. Journal of Human Evolution 43, 831-872.

Sept, J.M. 1994. Bone distribution in a semi-arid riverine habitat in eastern Zaire: Implications for the interpretation of faunal assemblages at early archaeological sites. Journal of Archaeological Science 21, 217-235.

Tappen, M. 1995. Savanna ecology and natural bone deposition. Current Anthropology 36, 223-260.

Yellen, J.E. 1977. Archaeological Approaches to the Present: Models for Reconstructing the Past. Academic Press, New York.

Yellen, J.E. 1991. Small mammals: !Kung San utilization and the production of faunal assemblages. Journal of Anthropological Archaeology 10, 1-26.

# CHAPTER 16

# CARNIVORA AND CARNIVORY: ASSESSING HOMINID TOOTHMARKS IN ZOOARCHAEOLOGY

TIM D. WHITE AND NICHOLAS TOTH


## ABSTRACT

The zooarchaeological implications of hominid bone chewing are relatively uninvestigated despite much progress in actualistic and archaeological research since C.K. Brain's classic 1960s studies of Hottentot/canid bone modification. Many investigations continue to make the unwarranted assumption that all evidence of chewing in zooarchaeological assemblages is attributable to nonhominid carnivores (the term "gnawing" is restricted here to rodents). In this contribution we evaluate some biases that may be responsible for the observation that hominid chewing continues to be underestimated or ignored in zooarchaeological contexts. We review the ethnoarchaeological and experimental evidence for hominoid chewing traces on bones. To illustrate issues surrounding the role of hominid teeth in patterning zooarchaeological assemblages we consider two cases of inferred cannibalism in the archaeological record. Fertile ground for continuing studies is identified, but given the morphological and anatomical parallels between the masticatory systems of hominids and carnivores, it appears that linking chewing damage with specific agents in archaeological contexts will be more difficult than previously imagined.


## INTRODUCTION

> If the student should ask me how the paleontologist tells the difference between hyaena and human teeth-marks on a bone, and particularly a bone that has been rotting in a cave since the everlasting hills were builded, I should answer that I don't know.
>
> Mark Twain 1871 (in Neider, 1961)

> Good science consists of strategically using prior knowledge to make projections from better-known domains to less well-known domains.
>
> Binford (2001)

By virtually any definition, Bob Brain has conducted exceptional science for nearly his entire life. His career as a paleobiologist is exemplary. There is hardly a topic in current human evolutionary studies untouched by his work.

During the two decades that followed Brain's 1981 classic *The Hunters or the Hunted? An Introduction to African Cave Taphonomy*, archaeological and actualistic studies of bone modification intensified and diversified. As a consequence, a routine component of any modern analysis of bones from archaeological contexts is the observation and recording of pits, scores, and fractures created by mammalian teeth. Such modifications are usually classified as "carnivore" damage. Standardization of the terms used to describe the modifications has proven elusive. Furthermore, a lack of inter-analyst consistency and replicability frequently plagues such descriptive and comparative studies.

Even the seemingly simple classificatory label "carnivore damage" is fraught with ambiguity. This is because the term "carnivore" has two meanings. One designates a mammalian order. The other is a term used to denote diet. The family Hominidae is precariously positioned relative to this ambiguity. Once the hominid niche was broadened by lithic technology during the Pliocene, consumption of large terrestrial mammals was no longer the exclusive realm of the Carnivora (see Stiner, 2002 for a review of what happened subsequently). This dietary shift is most recognizable by the stone artifacts with which it appears to have been associated—and by the signatures and patterns that the edges and surfaces of these implements leave on bones and bone assemblages. Ethnographic and primatological analogy both suggest that muscle and marrow-eating hominids would have chewed bones of the medium and large mammals they butchered.

Inspired by Bob Brain's work in South Africa, zooarchaeologists have made great progress in actualistic and archaeological studies that demonstrate the impact of both hominid and nonhominid carnivore behavior on bone assemblages. Attributes allowing the analyst to distinguish rodent gnawing, nonhominid carnivore, and even ungulate chewing damage on bones have been identified, described, and illustrated in both archaeological and actualistic contexts (see White, 1992 for a review). Analyses of these bone modifications play a central role in ongoing discussions about early hominid behavior and ecology.

Mammalian chewing alters individual bones and affects assemblages of bones. It modifies bone surfaces and deletes bone elements and element portions. Brain's classic work comprised both experiment and observation of these effects. He studied bone assemblages generated by traditional residents of the Kuiseb River's north bank. From the beginning, Brain recognized that these ethnoarchaeological bone assemblages bore the signature of two goat-eating agents—Hottentots and their dogs. Both agents were equipped with the masticatory means to modify bone surfaces and assemblages. Brain conducted experiments to *differentiate* the various signatures of bone modification and element representation patterns produced by these different taphonomic actors on the Hottentot ethnoarchaeological stage. He wrote:

> "It seemed advisable to separate the damage done to goat bones by the Hottentots themselves from that caused by their dogs...It was surprising to find that the Hottentots were capable of inflicting considerable damage on bones with their teeth...It is to be expected that Stone Age people would have done even greater damage to bones with their teeth than do Kuiseb River Hottentots" (Brain, 1981: 17-18).

Few of the hundreds of subsequent taphonomic and zooarchaeological studies inspired by Brain's work on South African cave assemblages have pursued research into bone modification caused by human chewing. As a consequence, our understanding of this phenomenon remains woefully inadequate. This is particularly surprising since many of these investigations are aimed at understanding the relationships between hominids and bone assemblages with which they are associated by their presence or archaeological droppings. The ability to discriminate bone chewing by hominids would be of central importance in archaeological studies ranging from the behavior of the earliest hominids to investigations of ethnohistorical cannibalism. Studies spanning wide cultural, spatial and temporal scales would benefit from the ability to discern and diagnose traces of chewing left by hominids as opposed to nonhominid carnivores. This has been clearly recognized for over a quarter of a century (see Binford's observations in 1981, described below), but the research remains unconducted.

In the sections that follow, we explore issues involving the subject of hominid bone chewing. Several examples illustrate how analysts have ignored the role that early hominid chewing might have played in generating modifications and patterns of representation in zooarchaeological assemblages. This lack of consideration is attributable to several factors, identified here as a series of biases that appear to have been imposed by modern human culture and anatomy. Anatomically, many researchers have seriously underestimated the potential of the early hominid masticatory complex to inflict osteological damage during chewing. Culturally, focus on hominids as technological creatures and modern western table manners may be co-conspirators in this analytical lapse. We discuss attempts to diagnose different bone chewers and address the likelihood that a substantial degree of equifinality will ultimately be demonstrated when the requisite studies have been conducted. Our report concludes with two presentations of case studies involving the possibility of human chewing of bone in contexts suggestive of cannibalism.

## Ignoring Hominid Chewing

The presence and activities of hominids are routinely inferred by paleoanthropologists on the basis of surface modifications to bones in zooarchaeological assemblages. Such inferences are today universally accepted. They are regularly grounded in solid actualistic research on cutmarks, hackmarks, percussion striae, and the like.

Quantification of bone surface modifications as well as skeletal element and element portion representations within assemblages is now standard practice in zooarchaeological research and reporting. However, debate persists on the standards to adopt in observation and recording of chewing modifications and extends to interpretation of the behavioral significance of modification to body part representation. For example, in a Lower Paleolithic context, the question of whether early hominids hunted, aggressively scavenged, or passively scavenged remains unresolved (Lupo and O'Connell, 2002; Domínguez-Rodrigo, 2003; O'Connell and Lupo, 2003; and references therein). The debate prominently features assessments of chewing damage to hominid-modified bone assemblages.

The residues left by primate and carnivore agents on African Plio-Pleistocene taphonomic landscapes have both spatial and physical components. For some occurrences, lithic assemblages have been subjected to repeated analysis and faunal remains have had their surfaces and proportions intensively and exhaustively investigated. The resultant data sets, derived from small windows excavated into vast paleolandscapes, comprise the scant evidence on which paleoanthropologists base key inferences regarding early hominid behaviors. Given the paucity of data, it is neither surprising that debates concerning these inferences continue, nor difficult to predict that studies of bone modification will continue to play a central role in such research far into the future.

Bone modification studies have a long and distinguished history in archaeology. Chewing-induced marks captured interest early, attested by Buckland's work in the early 1800s. By 1938 Pei was cautioning that errors of interpretation might befall investigators who neglected the role of mammalian carnivores in accumulating and modifying assemblages.

The FLK 22 "Zinj" excavation and its derivative contextual, artifactual, and zooarchaeological data sets have played central roles in discussions about the activities of Plio-Pleistocene hominids at Olduvai Gorge. As noted by numerous analysts of such assemblages, "actors" or "agents" active in the FLK taphonomic setting included wind, water, and ultraviolet light. Also present were nonhominid tramplers, nonhominid chewers and gnawers, plants with acidic bone-etching roots, and rock-wielding hominids (summarized by Capaldo, 1997).

The zoological agents of greatest behavioral importance in Plio-Pleistocene taphonomic settings such as FLK "Zinj" are nonhominid carnivores of various taxa. These animals shared the arena with at least two sympatric hominid species. Bone modification studies show unequivocally that at least one of these hominid species played active roles in influencing the recovered bone assemblages. The exact nature of such roles remains unclear. Furthermore, it is currently nearly impossible to infer the identity of the one or multiple contemporary hominid species involved with the lithic technology accompanying the bone assemblages.

Attribution of bone surface modifications documented on the Olduvai assemblages is most often accomplished by dichotomizing the bony trace evidence according to the agents inferred to have created it. The presence of "carnivores" is often inferred by chew marks. In contrast, the presence of hominids is inferred by signatures of tissue removal (cutmarks) and marrow acquisition (hammer and anvil striae; inner conchoidal scars on limb bone midshafts).

In Paleolithic archaeology, there is often the implicit assumption that any tooth marks discernable on a bone were made by Carnivora (hyaenids, felids, canids; but see contra, Brain, 1981; White, 1992; Pickering and Wallis, 1997). It is a short but dangerous step to then infer the order of access by the hominids and other carnivores in these Plio-Pleistocene settings. The superimposition (overprinting) of chewing traces and cutmarks/percussion damage are frequently the basis of such inferences, usually with the unwarranted assumption that the toothmarker belonged to the Carnivora. How valid is this assumption when we know that some early hominids even scarred their own *incisors* with sharp stone tools they used to slash tissue held between their front teeth (Fox and Frayer, 1998; Lozano-Ruiz, et al., 2004)? Did they not chew, too?

Paleoanthropologists can safely assume that nonhominid carnivores did not wield stone tools, and hence were not responsible for cutmarks or percussion processing evidenced by the faunal remains from such sites. But how safely can they assume that every toothmark on a bone was created by a nonhominid carnivore? And how might the relative abundance values of skeletal elements in an assemblage be apportioned to nonhominid carnivore chewing and transport as opposed to hominid defleshing, percussive marrow extraction, and chewing?

Negligible attention has been afforded to the possibility that hominids themselves could, and probably *did* chew bone portions. These were primates obviously attracted to meat and marrow. It is unrealistic to ignore the possibility that trabecular bone of the ungulate carcasses, often covered by only a thin bony cortex, was exploited by early hominids.

The mastication of bones by hominids would not have been limited to scarring of bone surfaces. It is predicted from Brain's Hottentot assemblages that these actions would also have deleted bones and bone portions from prehistoric sites. If both nonhominid carnivores and hominids themselves played roles in patterning the modifications and deletions that resulted in the recovered bone assemblages, how might their relative contributions be determined?

Actualistic studies of surface traces and fragmentation patterns have made it possible to set forth criteria that allow investigators to diagnose marrow processing via hammerstone versus marrow processing by nonhominid carnivore chewing (e.g., Blumenschine, Marean and Capaldo, 1996). Methodological problems persist (Lupo and O'Connell, 2002). Even in the apparently routine area of toothmark identification, wide inter-analyst disparities have arisen. As Capaldo (1997) notes, such disparities can be significant. Toothmark recognition for the Olduvai "Zinj" assemblage illustrates such methodological difficulty. Blumenschine (1995) reports tooth mark percentages four times higher than those reported by Bunn and Kroll (1986) for the same assemblage. This is clearly a serious problem but even *less* progress has been made in distinguishing potential hominid and nonhominid chewing on the *same* assemblages.

The magnitude and significance of this problem is manifested in Capaldo's (1997) dichotomization of "hominids" versus "carnivores" in the "Zinj" bone assemblage (see Capaldo, 1997 Table 1). To hominids he attributes "tool marks, hammerstone notches, spiral fractures, and bone fragmentation" (p. 559). To "carnivores" he attributes "tooth marks, tooth notches, spiral fractures, digestive etching, deletion of low density skeletal parts and portions, and bone fragmentation." But couldn't hominids have also created tooth marks? And couldn't they, by simple chewing, have effectively deleted low density skeletal parts and portions, *even without artifacts*?

In his book on the Olduvai evidence, Potts (1988: 145) states that "...one of the strongest effects of carnivore modification, which occurs at each Olduvai site, is the destruction of long bone ends." According to Potts' critique of colleagues (Bunn and Kroll, 1986) who interpreted the same Olduvai assemblages differently:

> "...the carnivore contribution to site formation is largely ignored and overshadowed by their attention to tool cut marks. As noted repeatedly here, carnivore damage to the bones indicates not only the presence of large and small scavenger/predators at the sites but also suggests that hominids were not responsible for consuming some portion of the edible tissues represented at these sites...the action of carnivores, as one element of site formation, is an important, consistent part of the context of hominid activities at Olduvai and, in fact, informs about the nature of hominid activities at these sites" (Potts, 1988: 306-307).

Perhaps, but it is obvious that *if* the "carnivore" responsible for a significant part of this "damage" and deletion were one of Olduvai's hominids, *then* the nature of the hominid/nonhominid carnivore activity would require a different interpretation. The dichotomization of the damage types according to the formulae: toothmarks = nonhominid carnivore, and cutmarks = hominid, while semantically and conceptually attractive, is illusory and naïve. It appears possible that advocates on either side of the debate about early hominid carnivory may have missed a significant component of behavior by employing such false dichotomization.

Arguments over the behavioral significance of the Olduvai zooarchaeological assemblages continue nearly unabated. The most recent incarnation is a debate between Lupo and O'Connell (2002, and O'Connell and Lupo, 2003) and Domínguez-Rodrigo (2003). At issue is the degree to which cutmark and toothmark frequencies may be indicative of early hominid carcass procurement and processing. Neither party explicitly considers the potential influence of hominid-generated toothmarks on their arguments. Indeed, this accurately reflects a very large literature on these assemblages—mere lip service has usually been paid to possible hominid participation in the creation of the toothmarks—bone modifications otherwise meticulously quantified, but almost universally attributed to "carnivores."

Oliver (1994: 270), at least, explicitly brings up the potential confounding effect of such participation: "Given the demonstrable hominid involvement with the fossil assemblage [references] it is likely that hominid teeth created at least some tooth marks." He goes on to explain his scoring of the Olduvai assemblages as follows:

> "Until actualistic data become available documenting differences between primate and carnivore-inflicted damage, I assume observed tooth pits and scores identify carnivores, but recognize that early *Homo* created at least some tooth marks."

The question of how many—and which ones—is patently important.

## Identifying Bias

The widespread inattention to the possibility that early hominids chewed bones and thereby produced bone modifications and deleted spongy bone parts in a manner that mimics patterns of damage usually attributed to Carnivora stems from at least three major causes. First is the modern practitioner's unfamiliarity with the masticatory apparatus and gnathic capabilities of hominids, particularly now-fossilized hominids. Second, some inattention seems predicated on the fact that despite Brain's call for ethnoarchaeological and experimental research in this area, there has been very little progress. We consider these aspects in subsequent sections. A third, less apparent, but perhaps even more important factor explaining why hominid chewing is relatively ignored involves *bias*.

A well-known bias in paleontology is what Raup (1979) has termed the "pull of the recent," the phenomenon by which species diversity appears greater in more recent deposits because of more complete sampling of younger units and misassignment of sampled fossils to recent taxa. A similarly well-known and parallel bias in archaeological research within Holocene and historic contexts is what Wobst has called the "tyranny of the ethnographic record." This is the phenomenon by which the richness of the ethnohistorical record in any geographic locale can skew interpretations of archaeologists working in the same area.

Parallel pitfalls pervade paleoanthropology. Recovery operations involving the Olduvai Hominid 62 specimen at the bottom of Olduvai Gorge in Tanzania, provided the excavators an opportunity to listen in on Serengeti tourists whose vehicles stopped on the adjacent road. A common theme of these overheard amateur conversations was incredulous wonderment: "why would anybody want to live in this dry gulley?" In assessing past environments, a fundamental challenge facing the student of paleoanthropology is to subdue perception of modern landscapes as a first step toward understanding the ancient precursors of these landscapes, thus escaping what might be called the "tyranny of modern topography."

Bias in the geographic realm is paralleled by another "pull of the recent" in the biological realm. Because paleoanthropologists are anatomically modern humans, ever since Darwin theirs has been a science beset by a "tyranny of evolutionary endpoints." In hominid paleontology, a classic example is the widespread but demonstrably false notion that the last common ancestor of African apes and humans was a chimpanzee. We know for certain that in anatomy and behavior early hominids were neither diminutive humans (although some have been given cute human names), nor upright chimpanzees (even though they are frequently depicted as such). If accurate interpretation of human origins and evolution is the goal, then using *either* extant organism as a proxy for morphologically and behaviorally extinct ones is fraught with epistemological peril. Early hominids were not humans, nor were they busy becoming humans. Exactly what they were, and what they were doing, is the research problem. Pretending to solve it with inappropriate analogies will not suffice.

In paleoanthropological assessments of bone assemblages associated with hominids of the Plio-Pleistocene, there may well be two other "tyrannies" at play. The first is the "tyranny of modern human mastication." The second is the "tyranny of table manners." Brain's actualistic and experimental studies nearly forty years ago allowed him to escape the confinement of these twin tyrannies. These studies allowed him to write that "Stone Age people" could have done even more damage to bones than what he had witnessed the Hottentots doing with their teeth.

White (1992: 155) noted the paucity of post-1981 work on human chewing of mammalian bone as follows: "This is an extremely underdeveloped, but important area of potential actualistic research." As Brain appreciated, the fact remains that hominid chewing of bone at archaeological sites has the potential to severely condition the assemblages available for interpretation across a wide sweep of archaeological circumstances.

Disproportions created by nonhominid carnivore chewing are well known and thoroughly documented from forensic (Haglund et al, 1988), ethnoarchaeological (Binford, 1978, 1981), and experimental work (Lupo and O'Connell, 2002) on chewing by nonhominid carnivores. Some important generalities have emerged from these studies regarding surface modifications and element preservation. For example, as summarized by Grayson (1988), canids prefer to attack the ends of major limb bones, whereas hominids tend to concentrate on midshaft portions by percussion. This generalization holds fairly well across a range of carnivore body sizes and bone destruction capabilities.

Both hominids and species of Carnivora can *delete* bones and bone portions through chewing. The *presence* and *activity* of hominids versus nonhominid carnivores may be demonstrated on the basis of accompanying bone modifications seen on assemblages. Cutmarks and percussion-related striae are only left by hominids. However, just because hominids had developed the percussive technology to access marrow and nutritive spongy bone, there is no reason to imagine that they simultaneously abandoned the tools of mastication that had served them and their ancestors so well over the previous hundreds of millions of years. Thus, Selvaggio's (2001: 469) inference that "it is unlikely that hominids frequently inflicted tooth marks on bones," and her contention that "[t]he development of stone tools is generally acknowledged to be a cultural innovation necessitated by the lack of shearing dentition in the hominid line" both appear to be entirely unwarranted.

Hominids seeking nutrition in the spongy ends of long bones can access it by chewing or pounding these portions between hammer and anvil. Oliver's (1994: 287) analysis of the FLK "Zinj" assemblage recognizes the possibility of both:

> "...the severity of the damages [sic] is more than required to remove marrow; the severity of hammerstone-induced damage demonstrates concern with, and consumption of the blood-rich cancellous ends. This intense processing may explain part of the loss of epiphyseal ends noted previously by Bunn."

Such processing, either for direct consumption or for grease rendering, can mimic carnivore activity in terms of the residual element and element portions. Whether the agent of destruction walked on two or four legs, the archaeological fallout of bone element and portion *survival* is conditioned by the structure of the bones themselves. When bone modification patterns and *traces* are taken into account, this ambiguity is reduced (see White, 1992). But what about hominid chewing of bone? Ethnographic and primatological observations suggest that hominids could have extracted nutrition in this manner. Such potential exploitation will not be recognized by preemptively denying its existence, by simple quantification of bone survival, or by assuming that hominid-induced bone modification is exclusively imposed by a stone edge. How might we recognize hominids who exploited the spongy parts of bone by chewing them? The required actualistic experimentation and observation advocated by Brain has yet to be conducted.

## Recognizing Carnivore Species

Substantial effort has gone into attempts to identify chewing modifications imposed by different *species* of Carnivora. Virtually none has been invested on attempting to differentiate these modifications from those left by hominid chewing. Haynes (1980) represents an early attempt to sort among carnivores based on damage patterns observed on spongy bone. He followed this (Haynes, 1983) with quantitative work comparing tooth puncture diameters with canine tooth profiles. Unsuprisingly, given the conical nature of many teeth, differentials in tissue penetrability, and inter-animal variation in bite force, Haynes found substantial ambiguity. Other investigators have attempted to identify carnivore species in more ancient contexts based on their tooth impressions. Oliver (1994: 285), faced with "carnivore damage" on 54% of the MNE from the FLK "Zinj" assemblage, argued:

> "...carnivore damage data also suggest the type or size of carnivore involved...rather a smaller carnivore that could make use of small meat scraps adhering to small bone fragments...these carnivore damage frequencies also corroborate the inference that carnivores were an important taphonomic agent at FLK Zinj."

Potts (1988), based on the same assemblages, implicates larger carnivores (hyaenids) in their generation.

Subsequent work in actualistic and archaeological contexts has explored bone modifications with the intent of identifying the nonhominid carnivore species involved in their creation. In her work with faunas from Late Pleistocene Italian caves, Stiner (1994) investigated the relationship between diameter of puncture marks and carnivore body size. She found that "The data separate carnivore agencies into two general body size groups, large and small, and do not distinguish hyaena from wolf, or wild cat from fox" (Stiner, 1994: 133). In her discussion, no consideration is given to the possibility that the smaller punctures were made by the deciduous teeth of larger species.

Selvaggio and Wilder (1995: 466) examined tooth marks "…selected by their resemblance to undamaged tooth cusps or complete crowns. Such marks are generally described as tooth pits." They cite Binford (1981) as the source for their adopted nomenclature, but his definition pertains to "pitting" rather than to individual pits. Given this and other ambiguities regarding how the marks that she measured were actually selected and defined, the results of Selvaggio and Wilder's study are difficult to interpret. They did conclude (not surprisingly), that pits on cancellous bone were larger than those on compact bone, and that it was difficult to distinguish between the known carnivore taxa used in her experiments.

Domínguez-Rodrigo and Piqueras (2003) performed another actualistic study that combined data from assemblages with elements chewed by captive and wild lions, jackals, bears, hyaenas, dogs, and baboons. They conclude that "conspicuous" (undefined) tooth marks can be used to distinguish three groups of carnivores, small, medium, and large. Like Stiner and Selvaggio and Wilder before them, these authors do not explicitly consider juvenile carnivores or the impact of their milk dentitions. In agreement with the latter authors, Domínguez-Rodrigo and Piqueras (2003: 1386) "are skeptical that specific carnivore taxa can be identified from tooth mark analysis alone."

Pickering et al. (2004) attempt to use these findings by concentrating, like Selvaggio, on "tooth pit" dimensions to evaluate the role of leopards in Swartkrans assemblage creation. A major advance of their work is its attempt to define what they mean by "tooth pit":

> "Tooth pits are bone surface modifications imparted by animal chewing and appear as discrete, roughly circular marks in plan view and result from scarring of bone without [significant] inward crushing of the bone cortex" (Pickering et al., 2004: 596).

## Identifying Toothmarks

In her extensive assessment of Pleistocene Italian cave faunas, Stiner (1994: 106) reported that "[n]o evidence of gnawing by human teeth was found," but did not suggest how such damage might have been recognized in the first place. As outlined previously, bone chewing by hominids might impact a bone assemblage in two ways, deletion or surface modification.

Despite decades of research, a diverse and often redundant vocabulary is currently employed by analysts to describe tooth scratches, punctures and pits made on bone surfaces by chewing mammals (Binford, 1981; Shipman, 1981; Cook, 1986; Lyman, 1987; Marshall, 1989; White, 1992; Blumschine et al., 1996). Investigations have failed to distinguish among extant species of Carnivora on the basis of these modifications except in the most general terms. Most of these studies have ignored the possible contribution of hominid chewing to this already difficult-to-disarticulate problem. In 1992 (p. 155) White noted:

> "It is evident from simple mechanical considerations that substantial overlap between human and carnivore chewing damage on bones will be shown by future research in this area…Meanwhile, tooth striae, punctures, and tooth pits should not be attributed to taxon in archaeological bone assemblages until further actualistic work is completed."

During the decade that has passed since that caution, substantial work has been done in attempts to discriminate between *nonhominid* carnivore *species* based on surface modifications to bones (see previous section). However, little further research has been conducted on hominid chewing and its effects.

Brain's actualistic studies of Hottentot modification of bones during the 1960s included an experiment in which he provided a small subadult goat to the local people. After the goat tissues had been consumed in the traditional manner, the remaining bones were collected prior to feeding to the village dogs. Brain (1981) provided a summary of damage to the skeleton. Maguire, Pemberton and Collett (1980: 88) echoed Brain's observations about the extent of bone damage by human chewing, and elaborated as follows:

> "It was surprising to note that the Hottentots were capable of inflicting a considerable amount of damage on the goat bones with their teeth. Ragged-edged chewing, practically indistinguishable from that produced by hyaenas on the more frail skeletal elements, was observed in particular on the scapulae and pelvic bones…Splintering was also common to both the carnivore and hominid samples."

These authors note that the five categories of damage encountered on their comparative hyaena-modified actualistic sample (striations, pitting, grooves, scooping of cancellous bone, and etching by stomach acids) were *not* observed on the Hottentot sample. They went on to report that crushing by human teeth represents a category of damage that:

> "...can be reproduced by repeated crunching with the molars and premolars on a fairly soft bone, such as a chicken limb or immature goat or sheep bone, after the articular epiphyses have been removed so as to leave a splintery, inwardly depressed margin to the shaft." (p. 88)

No further work with the Brain Hottentot sample has been conducted, but Pickering and Egeland (personal communication, 2004) are currently re-analyzing it.

Additional studies of chewing of mammal bone by modern people have been ethnoarchaeological and archaeological in nature rather than experimental. For example, in her 1989 work on northern Kenyan pastoralists, Gifford-Gonzalez decries the lack of needed experimental and observational research on the production of bone assemblages. In particular, she notes that much of the chewing damage she observed on faunal remains in a Dassanetch bone assemblage could not be attributed specifically to either human or nonhominid chewing (she cites Solomon's unpublished 1985 bachelor's thesis as noting the possibility that human teeth might mimic carnivore marks). The ethnoarchaeological literature contains many accounts of modern people chewing off the ends of long bones (e.g., Jones, 1983). Even extant hunter-gatherers continue to provide ethnoarchaeological evidence of hominid chewing and consumption of spongy bone:

"On some long bones, cancellous tissue in articular ends may be gouged out with a knife or some other pointed object and eaten. Ribs are sometimes snapped or chopped into sections, and the broken ends gnawed and sucked" (Lupo and O'Connell, 2002: 87).

The necessity for additional experimental work on hominid chewing of bone is neatly illustrated by Binford, whose 1981 book echoed Brain's concern regarding experimentation:

> "This is an area where diagnostic properties might well be developed through direct experimentation with modern subjects. However, the gnawing would have to be directed and controlled for the strength of the subjects and they could not be allowed to select what they wanted to gnaw. Instead, the experimenter would have to instruct his subjects to gnaw assigned anatomical parts in specified ways so as to obtain as complete a picture as possible and likely under different conditions...we need data on this problem before tooth modifications on bones can be assigned to nonhuman agents in a totally reliable manner" (Binford, 1981: 148).

Three years later, Binford illustrated and described damage to a Middle Stone Age Klasies bushbuck metatarsal. He attributed it to hominid chewing (Binford, 1984).

Given the constraints of human subjects committees at institutions of higher learning, to say nothing of the expense of cosmetic and restorative dentistry, the lack of post-Brain bone chewing experimentation in humans is at least partly understandable. However, surprisingly little ethnoarchaeological research has been conducted on this topic. Meanwhile, bone chewing by captive and wild chimpanzees has proceeded in productive directions.

Three different studies have noted that modern chimpanzees can and do modify the bones of prey carcasses they consume, and that various aspects of that modification can mimic modifications documented for Carnivora. Plummer and Stanford (2000) report on analysis of a small bone assemblage made by chimpanzees at Gombe, and Tappen and Wrangham (2000) report on another taphonomic study of bones from chimpanzee dung at Kibale. Inspired by Brain, Pickering and Wallis (1997) undertook a captive study, finding that chimpanzees were capable of producing mastication damage similar to that produced by nonprimate carnivores.

## Recognizing Hominid Toothmarks

Hominid teeth, both fossil and modern, are often mistakenly presumed to be so excessively low-crowned as to be incapable of inflicting the kind of damage that is routinely attributed to Carnivora in zooarchaeological assemblages. Maguire, Pemberton and Collett (1980: 88-89) consider the masticatory apparatus of early hominids as follows:

> "Although the teeth of [*Australopithecus*] *africanus* were larger and more robust than those of living Hottentots and were thus potentially capable of inflicting a greater degree of damage, it is unlikely that the bunodont teeth of this species were capable of producing pitting, striations, grooves or scooping damage. There is certainly no basis for attributing such damage to the teeth of hominids when it can be demonstrated conclusively that carnivores can and do produce such damage."

Pickering and Wallis (1997: 1116) follow at least some of the same questionable logic:

> "While modern chimpanzees are not necessarily the best models for study of all aspects of Pliocene hominid behaviour, we feel that they serve as good models for deriving predictions of early hominid mastication damage on archaeological bones, due to the close similarities in chimpanzee and Pliocene hominid morphology, technological grade and bite force. While the chimpanzee dental arcade is different from that of the gracile australopithecines in shape and in having more procumbent incisors, larger canines, sectorial lower third premolars and smaller cheekteeth (see Swindler, 1976), we contend that the posterior dentitions of the two taxa are more similar to each other than either is to the Carnivora.

This is especially true regarding the bunodont configuration of both hominoids' cheekteeth."

There are two kinds of effects that hominid chewing might have on bones; the crushing/deletion of spongy portions (usually near-epiphyseal), and the surface marking of more resistant portions (mostly diaphyseal) of bone elements. In an effort to better visualize the kind of damage that early hominid teeth might impose on a bone surface, we performed some simple comparisons.

Figure 1 illustrates the general crushing potential of the early hominid dentition compared to modern humans and chimpanzees. Note that the A.L. 288-1 specimen ("Lucy") of *Australopithecus afarensis* (one of the earliest and least specialized species of its genus) had diminutive body size coupled with large postcanine teeth. Note also how large the postcanine platform of its sister species, *A. africanus*, can be. The dentitions of all the Plio-Pleistocene hominids were occluded under the force of a powerful masticatory apparatus (Kimbel et al., 2004). Crushing of nutritious spongy bone under a thin cortex would have been easy for any of these creatures.

There is abundant evidence that individuals of many species of early hominid loaded their teeth forcefully against hard objects. Figure 2 illustrates the kind of damage that results from this activity. Here, massive step-fractures on the labial surfaces of the canines of a Swartkrans adult hominid show that the teeth were pressed (or impacted) against a hard object with such force that the enamel failed. Subsequent wear polish superimposed on the fracture surfaces shows that this biting happened well before the death of the individual. The illustrated teeth are from *A. robustus* of Swartkrans, but similar polished fractures indicative of forceful biting against hard objects are found throughout the hominid fossil record. It is not possible to tell whether this damage was incurred during bone chewing, or the chewing of some other hard material. Similar fractures are also occasionally observed among hyaenids and other carnivores that crush bones with their teeth. Figure 2 includes such a fractured carnivore tooth, a wild spotted hyaena—a species thought to have specific histological adaptations to mitigate against such enamel failure.

Figure 3 compares the dental "business ends" of hominid and hyaenid teeth. The term "bunodont" is adequate to describe gross morphological differences between such morphologically disparate species as hominids and horses. However, this term should *not* conceal the fact that hominid tooth cusps, particularly when little worn or deciduous, can be sharp, salient, and capable of inflicting the kinds of pits, scores, and even punctures that are often attributed exclusively to Carnivora. Modern human and modern hyaenid deciduous and permanent dentitions bear an array of cusp morphologies capable of inflicting a wide range of damage types often ascribed exclusively to different species of Carnivora.

Figure 3b shows a comparison of cusp tip morphology in the two hominid species known to have been present on the Olduvai FLK "Zinj" floor, *Homo habilis* and *Australopithecus boisei*. Both have cusps comparable in their potential to modify bone surfaces during chewing activity, and it is difficult to imagine how such surface signatures might be differentiated from those left by the wide suite of nonhominid carnivore species.

Several of the studies cited in the previous section have attempted to make taxonomic attributions from bone surface modifications through the use of dimensions of surface pits and scores. None of them measure the depths of the pits analyzed. However, it seems fair to observe that most of the surface defects measured in these studies are less than a millimeter or two deep, particularly on the cortical bone of limb bone shafts. Thus, it is not the overall morphology of the tooth, or even its cusp that is important in such comparisons. Only the cusp's tip is responsible for interfacing with the bone surface to produce the observed modification. A simple consideration of the physical parameters of bone surfaces (tissue cover, maturity, texture, density) and the factors that interface with them (sharpness, deciduous or permanent, chewing force, taxon, tooth category) reveals a complexity not easily resolved into specific alternative taxa, or even grossly different body sizes within Carnivora, let alone discriminate them from the hominids with which they shared their prehistoric taphonomic arenas.

## Equifinality Predicted

Given the anatomical and physical illustrations provided in the last section, an unavoidable and unfortunately pessimistic prediction follows: except in very rare instances (such as a clean puncture by a taxon-specific, anatomically diagnostic tooth) no single morphological characteristic of a mammalian tooth mark on an ancient bone will allow the taxonomic identity of its maker to be unequivocally established. Such pessimism is only a slight extension of the findings of other studies already conducted and cited above.

Given such apparent equifinality (different causes producing the same end or result) involving the agents of modification responsible for scores, pits, and punctures on bones, Pickering and Wallis have followed others in advocating a configurational approach (attempting to focus on the anatomical context of the marks rather than the marks themselves). But given the variability in the species of Carnivora that have access to most archaeological bone assemblages, and these investigators' findings that bone damage caused by chimpanzee chewing is "nearly identical to carnivore gnawing damage" (Pickering and Wallis, 1997: 1125), how much room is there for optimism?

Optimism is probably only warranted at the assemblage level. Work on bone assemblages conducted with the goal of identifying the main agent of modification has taken place in a variety of archaeological contexts mentioned in previous sections. Perhaps no context has been so controversial as the one involving the question

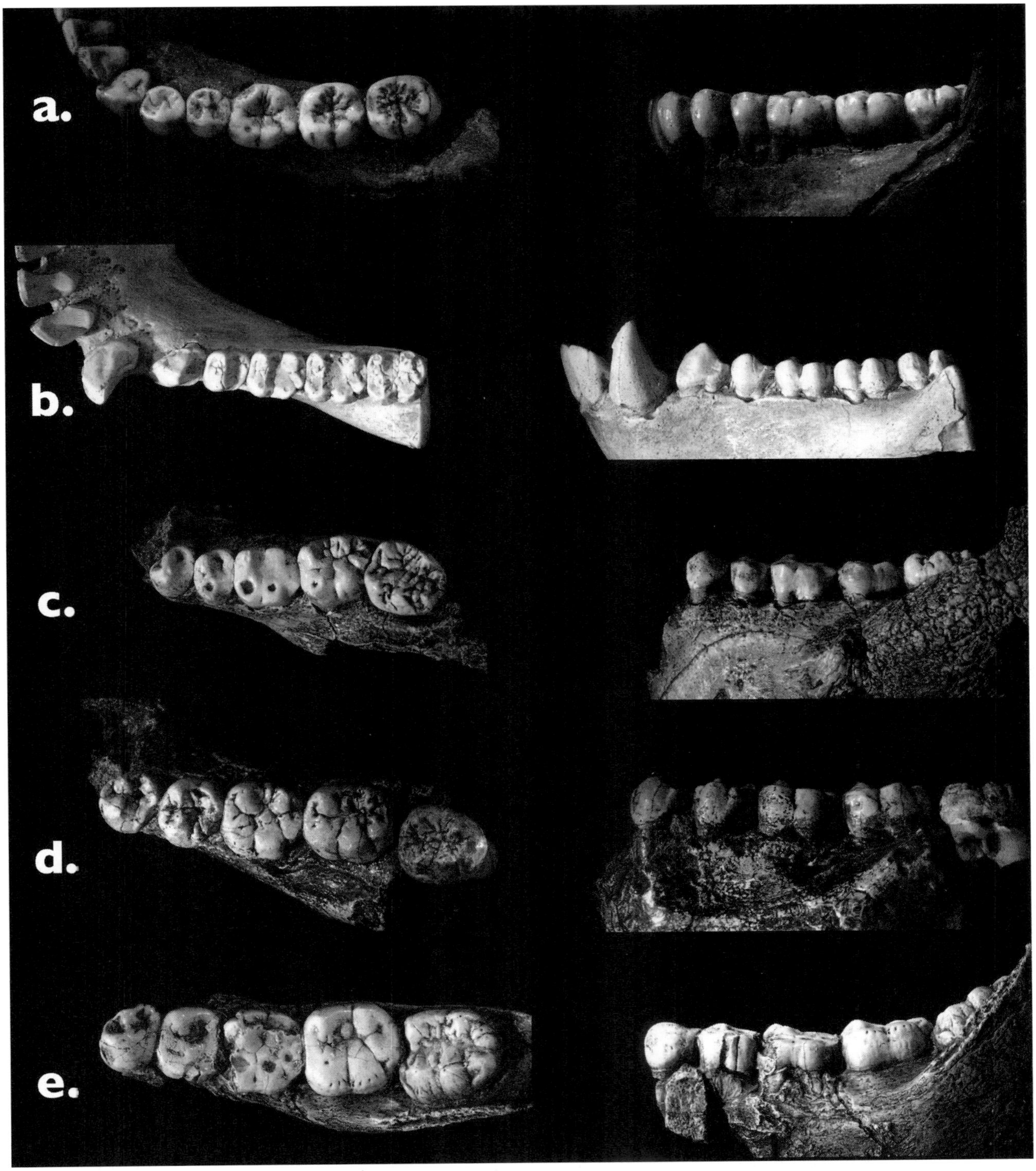


*Figure 1. The masticatory apparatus of early hominids featured robust jaws and teeth fully capable of crushing many vertebrate bones, particularly spongy ends. Seen here via casts in side and occlusal views, it is evident that the dentitions of any of the fossil species had the potential to inflict bone surface modifications that would easily surpass the dramatic effects already documented for the modern human chewing. Even the primitive* Australopithecus afarensis *(c; A.L. 288-1 "Lucy," reversed) was a megadont species compared to anatomically modern humans (a; Qafzeh 9) or modern chimpanzees (b; CMNH B-3412). The utterly inappropriate term "gracile" has been applied to taxa such as* Homo habilis *(d; KNM ER-1802, here with OH 16 M3, reversed) and* Australopithecus africanus *(e; STW 498), but it is clear that these megadont species also had bone chewing potentials far exceeding that seen in modern humans or chimpanzees.*

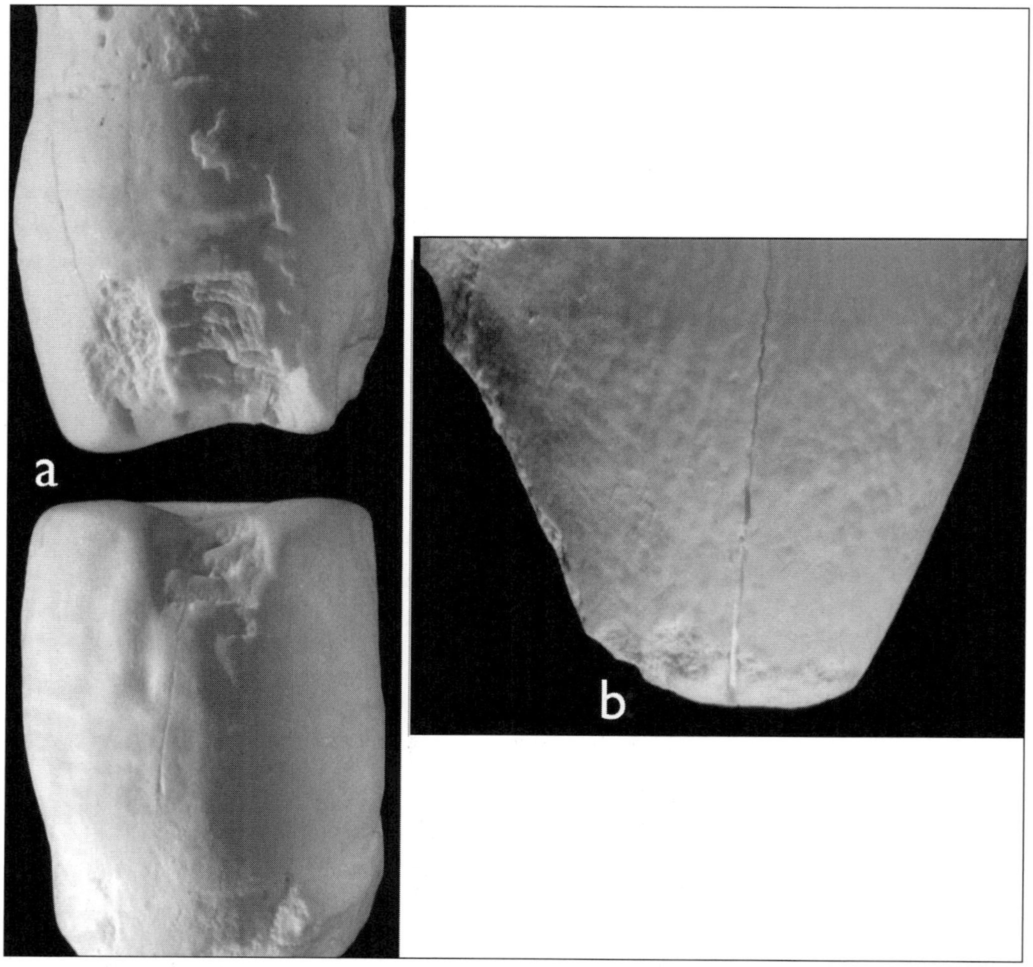


*Figure 2. The probably female Swartkrans* Australopithecus robustus *maxilla (SK 65+67+74) features prominent step fractures on the labial surfaces of the canines. Such damage, with post-fracture wear polish and striae superimposed to show that the trauma was incurred* in vivo, *indicates forceful biting of a hard surface. Such damage is frequently encountered in different early hominid taxa. When found in hyaenids (U.C. Berkeley MVZ specimen 173762, wild, Narok, Kenya 1981) such damage is usually interpreted as evidence of bone chewing.*

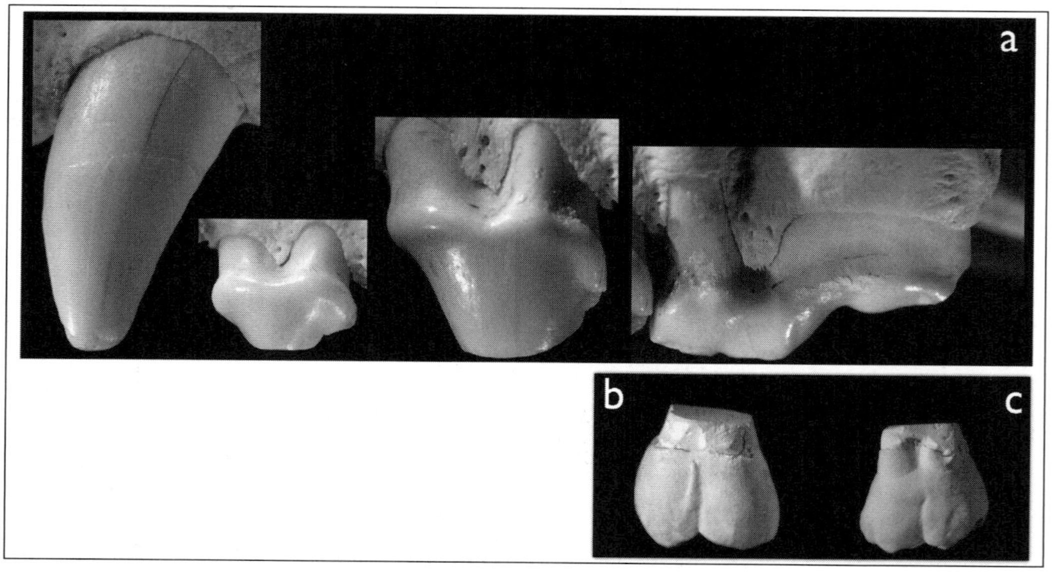


*Figure 3. Tooth profiles in different fossil and modern mammals. The Figure illustrates a) an adult modern spotted hyena (U.C. Berkeley MVZ specimen 173762, wild, Narok, Kenya 1981), b)* A. boisei *(OH 30), and c) early* Homo *(OH 45) cusp morphologies. It is evident that all three taxa are capable of inducing bone surface modifications that include punctures, pits, and striae such as seen on zooarchaeological remains from localities such as FLK "Zinj."*

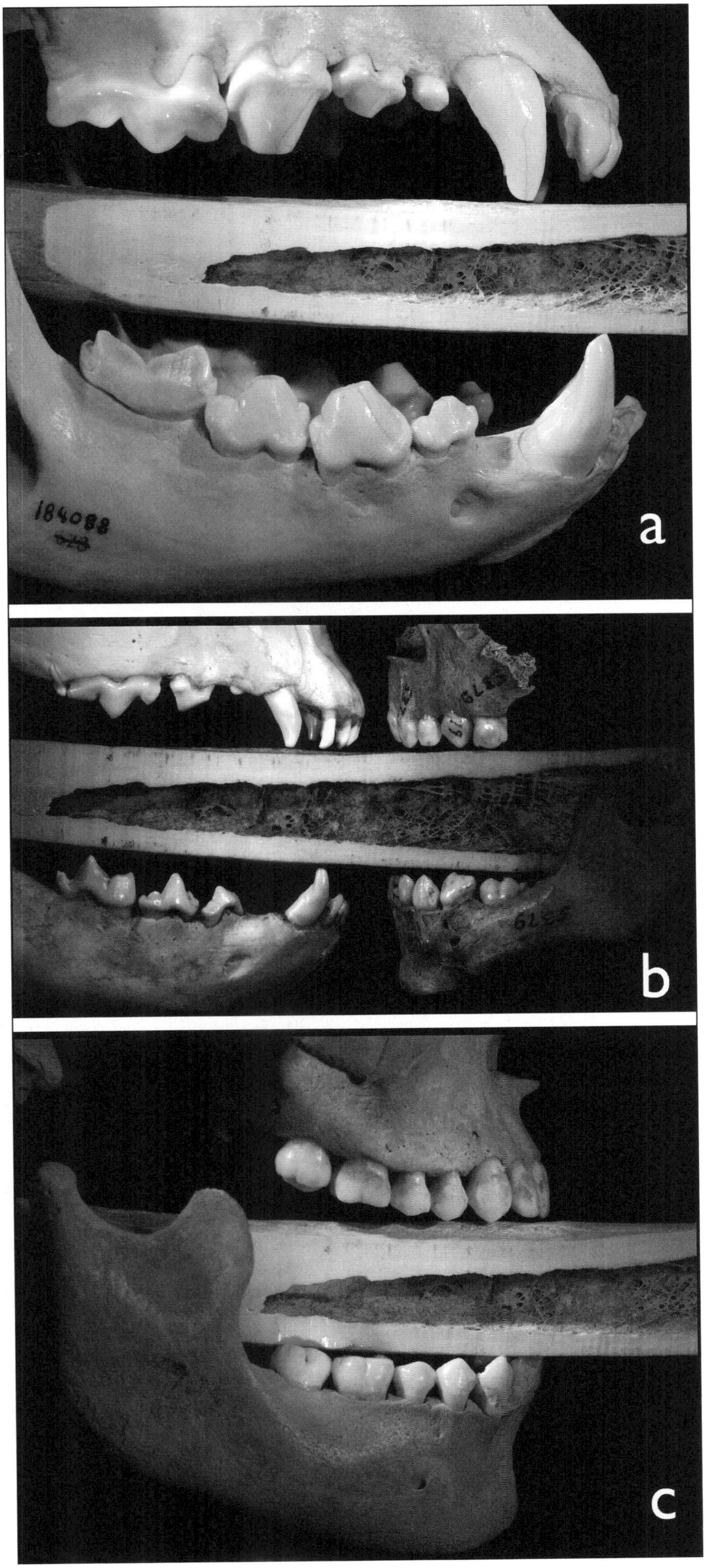


*Figure 4. This illustrates the wide range in cusp morphology that would be responsible for bone modifications made by either* Crocuta crocuta *(U.C. Berkeley MVZ specimens 173762 and #173772, wild, Narok, Kenya 1981) or Homo sapiens (U.C. Berkeley PAHM specimens, Native American). When both permanent (a, c) and deciduous (b) teeth, in worn and unworn conditions are taken into account, it seems highly unlikely that any particular cusp "signature" will be shown to differentiate these taxa, thus introducing equifinality into the identification of the "chewers" of any zooarchaeological bone assemblage. We do not mean to imply that hominids and hyaenas would both have crushed these midshafts. They are used here only to illustrate the relationship between tooth and bone surfaces.*

of prehistoric cannibalism. White (1992) concludes that it will only be possible in extremely rare cases to identify a toothmark signature that is uniquely attributable to a human agent. Employing a configurational approach, however, he illustrates bone modification that is likely, but not certain, to be the result of human chewing at a prehistoric site in the American Southwest.

## Cannibalism

A substantial literature featuring various interpretations of human bone assemblages in the American Southwest has been reviewed by White (1992) and Turner and Turner (1999). For decades, Turner argued that cannibalism is indicated for many of these assemblages. In support of this thesis he has developed what he refers to as "the minimal taphonomic signature of cannibalism." White advocates a more functional, anatomical, configurational approach opposed to the checklist approach favored by Turner and Turner. Both authors conclude that the dozens of bone assemblages they review are indicative of cannibalism among the Anasazi, a conclusion bolstered by more recent biochemical work (Marler et al., 2000). After detailed analysis of bone modification and element representation, both Turner and White emphasize the difficulty of distinguishing canid and hominid chewing damage on the human bone assemblages they analyzed.

Turner and Turner (1999) list as one of the items on their cannibalism checklist the following: "Animal gnawing and chewing occurs on only a small proportion of all elements, usually less than 5%. Some gnawing may have been done by humans and/or their dogs." They specify (p. 14): "…we have often been unable to decide whether finger or toe bones were broken by carnivores, humans, or natural mechanical-physical agencies." White's (1992) treatment recognized a pattern of damage to these manual and pedal elements that he concluded, in the context of that assemblage, was strongly suggestive of chewing by the human inhabitants of the Mancos Pueblo.

Using insights into hominid processing of spongy bone that were afforded by the Mancos analysis, we undertook a still unpublished study of the bone assemblages from the Neanderthal site of Krapina in Croatia. This Mousterian site and its contents have been the subject of numerous and varied studies (summarized in Radovčić, 1988, and a more recent review by Patou-Mathis, 1997). A debate over whether cannibalism was practiced among the Neanderthals there and elsewhere has been entertained for over a century (see Defleur et al., 1999 and references therein). We shall not review those arguments, even the ones specific to the Krapina Neanderthal assemblage and its composition. Rather, we wish to draw attention to bone modifications encountered in our study that may represent the kind of unusual evidence which, in rare cases, might make it possible to link actor with effect in a paleoanthropological setting.

The Krapina Neanderthal remains were recovered over a century ago. Unfortunately, the excavation and curatorial techniques employed compromised the behavioral value of the collection. Much of the associated fauna was discarded at the excavation and the smaller component of the bone assemblage was not recovered by sieving. Many of the recovered hominids were extracted in a rough fashion and with coarse provenience. There is abundant evidence of bone damage during recovery and transport. This damage was exacerbated by the fragility of the Krapina osseous material. Better-preserved fragments of Neanderthals have been recovered from cannibal-generated assemblages of limestone caves such as Vindija and Moula-Guercy (Defleur et al., 1999). The sandstone rockshelter of Krapina represented a different depositional and fossilization environment, leading to retention of exquisite surface detail on the bones, but preserving the bone tissue itself as chalky and soft, hence extremely fragile and subject to post-recovery damage. Without preservative, this bone was so soft that pencils used to number it have indented the bone surface. Due to these factors, most of the Krapina hominid bones were coated with variably deep layers of preservatives. This treatment protected the fragile specimens, but left a thick translucent cover that now serves to obscure important details of bone modification on many Krapina specimens.

It is evident from a comprehensive study of the Krapina hominid collection that these Neanderthals were processed intensively with stone tools. Clear stone tool cutmarks and hammer/anvil percussion damage is manifested throughout the collection. In addition, element and element portion preservation resemble those seen in the Mancos collection described by White (1992) (Figure 5). Notions that rock fall might be responsible for this trauma (Trinkaus, 1985) are obviated by the fact that large diameter limb bone shafts are virtually entirely absent (except for percussion-marked splinters), whereas the much more fragile fibular midshafts are intact, but missing their spongy ends.

Figure 6 shows a Krapina Neanderthal fibula shaft, specimen number 230. This specimen lacks a proximal end, but the broken distal end includes a sliver of the articular facet. A series of shallow, paired marks approximately 1.2 mm-wide, cross the shaft perpendicular to its long axis. There are three sets of these marks. The first set, the more proximal one, is 11.5 cm distal to the broken proximal end of the bone. Its short parallel grooves are separated from each other by 5.5 mm (measured between groove centers). The second pair, another 16 cm downshaft, is similar in orientation, morphology, and depth. Its groove centers are separated by 7.5 mm. The third pair, another 26 cm downshaft, is a similar set of shallow grooves with centers separated by 8.5 mm.

There is little or no evidence of diagnostically non-hominid carnivore chewing on the entire Krapina Neanderthal assemblage of over 800 specimens. There is ample evidence of cutmarks made by stone tools, and

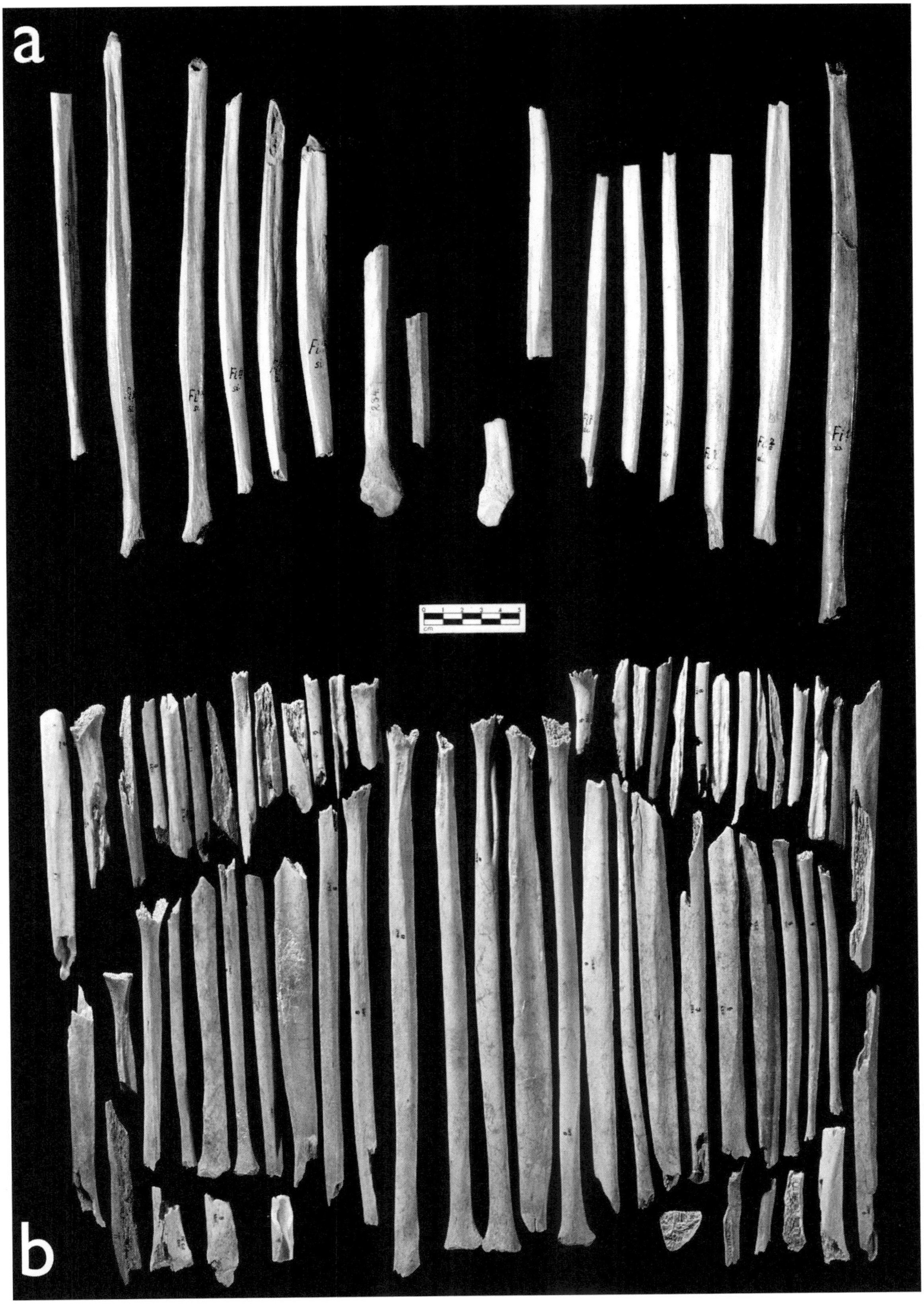


*Figure 5. The fibular samples from two cannibalized assemblages of* Homo *bone. Note that the Neanderthal subassemblage (top, from Krapina, Pleistocene, Croatia) and the Anasazi subassemblage (bottom, from Mancos SMTUMR-2346, Holocene, Colorado) share a pattern whereby the more friable spongy bone ends are missing whereas the midshafts tend to be preserved intact. This pattern is entirely different for larger-diameter limb bones from this occurrence, such as the tibia and the femur. The latter, larger-marrow capacity bones exhibit clear modification evidence associated with percussion by hammerstone that is inferred to have been directed at marrow procurement. Krapina 230 is third from the left in Figure 5a.*

marrow processing by hammer/anvil technique. Is it possible that the shallow marks seen on the 230 fibula and other specimens like the Krapina 217 tibia of this collection (with similar paired marks as well as cutmarks, peeling, and percussion damage) were made by hominid teeth? Provocatively, the Krapina 49 maxilla, also illustrated in Figure 6b, has a bi-central incisor occlusal edge breadth of 9.0 mm (and an equivalent central-to-lateral incisor breadth), producing an irregular incisal cutting edge that conforms remarkably well to the marks preserved on the fibular shaft. We are not contending that this particular individual used its incisors to remove muscle and/or periosteum from the fibula. We do, however, suggest that this kind of patterning is difficult to explain any other way. How much of the trauma to the Krapina hominid assemblage, or to the hundreds of other collections of zooarchaeological remains, is attributable to chewing by hominid carnivores? This is a research problem that is likely to be difficult to solve given the equifinality predicted above. It may prove that unusual specimens like the Krapina 230 partial fibula will eventually help meet the challenge of identifying hominids among the several carnivorous chewers that might have co-created these assemblages and their characteristics.

## Conclusion

From the glacial deposits at Moula-Guercy in the Ardeche to the desert environments of the Afar depression, studies of bone modification yield insight into the prehistoric past. A review of the evidence of hominid chewing as a modifier of zooarchaeological assemblages shows us that Bob Brain's prediction about the teeth of "Stone Age people" (and their ancestors) must no longer be ignored. Brain's early appreciation that early hominid teeth might be expected to impact bone assemblages is but one of many incisive observations in a body of work that has established him as the world's pre-eminent vertebrate taphonomist. May his good science continue to serve as our example, and his curiosity continue to inspire our pursuit of the past.

## Acknowledgements

The research on fossil hominid remains was conducted with the support of the H.F. Guggenheim Foundation. Thanks to the dozens of museum and university personnel on three continents who generously shared expertise and collections. Henry Gilbert is gratefully ac-

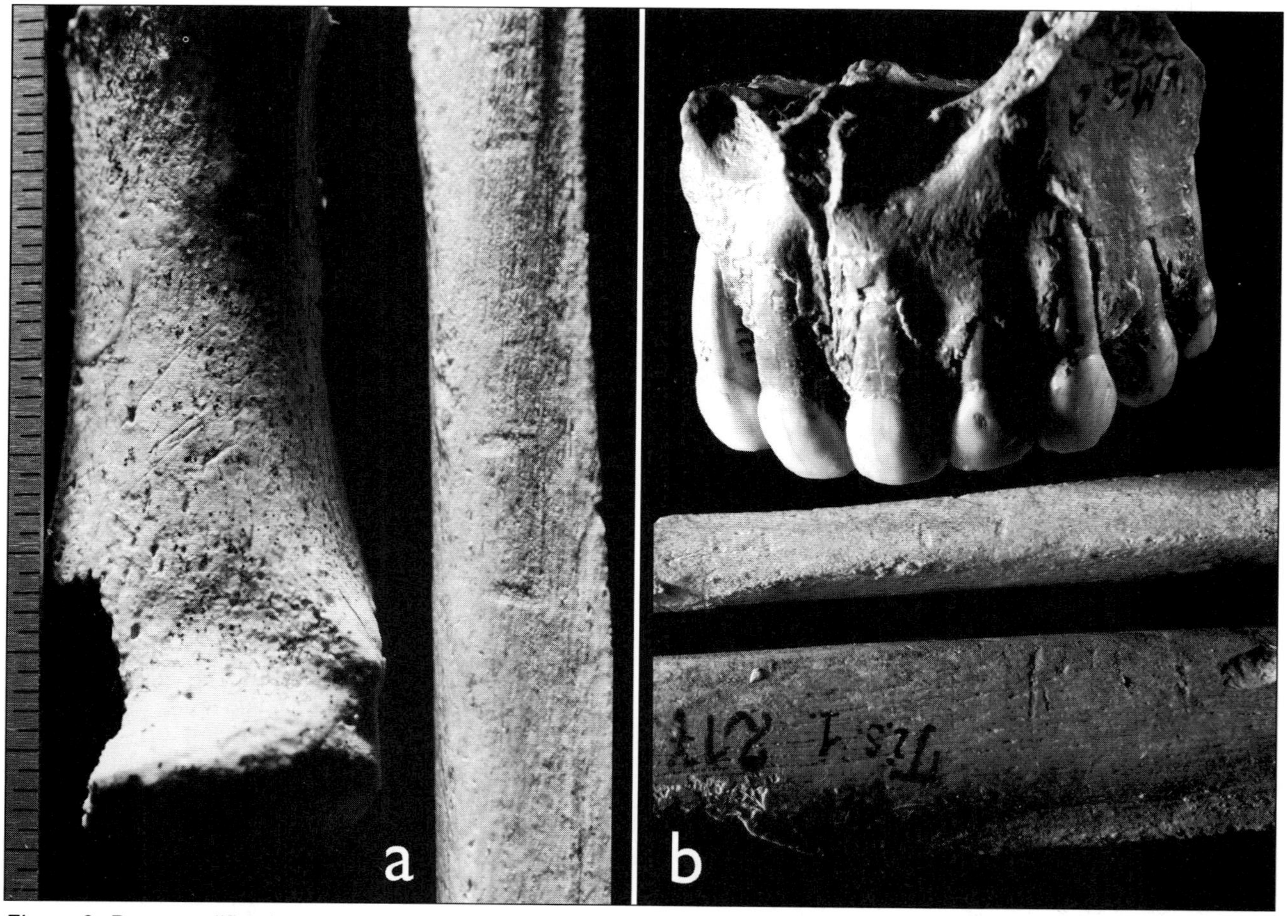


*Figure 6. Bone modifications on the Krapina Neanderthals. Figure 6a is a closeup of two specimens, the Krapina 246 first metatarsal with cutmarks on its dorsal surface; and the Krapina 230 fibula with two of the sets of shallow, short, parallel striae described in the text. Figure 6b is a closeup of the latter marks compared to the incisal edge of the Krapina 49 maxillary specimen, which shows that the inter-crown center measures for canines and incisors on this biting edge approximate the features seen on the fibula. This suggests, but does not demonstrate, that these bone modifications are the result of stripping flesh with the front teeth.*

knowledged for the illustrations. He and Nuria Garcia provided valuable comments on a draft of the manuscript. Thanks to the Stone Age Institute, CRAFT, Indiana University, and Travis Pickering for organizing an exceptional conference to honor an exceptional colleague.

## References

Binford, L.R. 1978. Nunamiut Ethnoarchaeology. Academic Press, New York.

Binford, L.R. 1981. Bones: Ancient Men and Modern Myths. Academic Press, New York.

Binford, L.R. 1984. Faunal Remains from Klasies River Mouth. Academic Press, Orlando.

Binford, L.R. 2001. Constructing Frames of Reference. University of California, Berkeley.

Blumenschine, R. J. 1995. Percussion marks, tooth marks, and experimental determinations of the timing of hominid and carnivore access to long bones at FLK *Zinjanthropus*, Olduvai Gorge, Tanzania. Journal of Human Evolution 29, 21–51.

Blumenschine, R. J., Marean, C. W., Capaldo, S. D. 1996. Blind tests of interanalyst correspondence and accuracy in the identification of cut marks, percussion marks, and carnivore tooth marks on bone surfaces. Journal of Archaeological Science 23, 493–507.

Brain, C.K. 1981. The Hunters or the Hunted? University of Chicago, Chicago.

Bunn, H.T., Kroll, E.M. 1986. Systematic butchery by Plio-Pleistocene hominids at Olduvai Gorge, Tanzania. Current Anthropology 27, 431-452.

Capaldo, S.D. 1997. Experimental determinations of carcass processing by Plio-Pleistocene hominids and carnivores at FLK 22 (*Zinjanthropus*), Olduvai Gorge, Tanzania. Journal of Human Evolution 33, 555-597.

Cook, J. 1986. The application of scanning electron microscopy to taphonomic and archaeological problems. British Archaeological Reports International Series 296, 143-163.

Defleur, A., White, T.D., Valensi, P., Slimak, L., Crégut-Bonnoure, E. 1999. Neanderthal cannibalism at Moula-Guercy, Ardèche, France. Science 286, 128-131.

Domínguez-Rodrigo, M. 2003. On cut marks and statistical inferences: Methodological comments on Lupo and O'Connell (2002). Journal of Archaeological Science 30, 381-386.

Domínguez-Rodrigo, M. and Piqueras, A. 2003. The use of tooth pits to identify carnivore taxa in tooth-marked archaeofaunas and their relevance to reconstruct hominid carcass processing behaviours. Journal of Archaeological Science 30, 1385-1391.

Fox, C.L., Frayer, D.W. (1998) Non-dietary marks in the anterior dentition of the Krapina Neanderthals. Int. J. Osteoarch. 7:133-149.

Gifford-Gonzalez, D. 1989. Ethnographic analogues for interpreting modified bones: Some cases from East Africa. In: Bonnichsen, R., Sorg, M. (Eds.), Bone Modification. Center for the Study of Early Man, Orono (ME), pp. 179-246.

Grayson, D.K. 1988. Danger Cave, Last Supper Cave, and Hanging Rock Shelter: The faunas. American Museum of Natural History Papers 66, 1-130.

Haglund, W.D., Reay, D.T., Swindler, D.R. 1988. Tooth mark artifacts and survival of bones in animal scavenged human skeletons. Journal of Forensic Science 33, 985-997.

Haynes, G. A. 1980. Evidence of carnivore gnawing on Pleistocene and recent mammalian bones. Paleobiology 6, 341–351.

Haynes, G. A. 1983. A guide for differentiating mammalian carnivore taxa responsible for gnaw damage to herbivore limb bones. Paleobiology 9, 164–172.

Jones, K.T. 1983. Forager archaeology: The Ache of eastern Paraguay. In: LeMoine, G.M., MacEachern, A.S. (Eds.), A Question of Bone Technology. University of Calgary Archaeological Association, Calgary, pp. 171-191.

Kimbel, W.H., Rak, Y., Johanson, D.C. Holloway, R.L., Yuan, M. 2004. The Skull of *Australopithecus afarensis*. Oxford University, Oxford.

Lozano-Ruiz, M., Bermúdez de Castro, J.M., Martinón-Torres, M., Sarmiento, S. 2004. Cutmarks on fossil human anterior teeth of the Sima de los Huesos Site (Atapuerca, Spain). Journal of Archaeological Science 31, 1127-1135.

Lupo, K.D., O'Connell, J.F. 2002. Cut and tooth mark distributions on large animal bones: Ethnoarchaeological data from the Hadza and their implications for current ideas about human carnivory. Journal of Archaeological Science 29, 85-109.

Lyman, R.L. 1987. Archaeofaunas and butchery studies: A taphonomic perspective. Advances in Archaeological Method and Theory 10, 249-337.

Maguire, J.M., Pemberton, D., Collett, M.H. 1980. The Makapansgat Limeworks Grey Breccia: Hominids, hyaenas, hystricids, or hillwash? Palaeontologia Africana 23, 75-98.

Marler, R.A., Banks, L. L., Billman, B.R., Lambert, P.M., Marlar, J.E. 2000. Biochemical evidence of cannibalism at a prehistoric Puebloan site in southwestern Colorado. Nature 407, 74-78.

Marshall, L.G. 1989. Bone modification and "The Laws of Burial." In: Bonnichsen, R., Sorg, M. (Eds.), Bone Modification. Center for the Study of Early Man, Orono (ME), pp. 7-24.

O'Connell, J.F., Lupo, K.D. 2003. Reply to Dominguez-Rodrigo. Journal of Archaeological Science 30, 387-390.

Oliver, J. S. 1994. Estimates of hominid and carnivore involvement in the FLK *Zinjanthropus* fossil assemblage: Some socioecological implications. Journal of Human Evolution 27, 267–294.

Patou-Mathis, M. 1997. Analyses taphonomique et palethnographique du materiel osseux de Krapina (Croatie): Nouvelles donnes sur la faune et les restes humains. Préhist. Europ. 10, 63-90.

Pickering, T.R., Wallis, J. 1997. Bone modifications resulting from captive chimpanzee mastication: Implications for the interpretation of pliocene archaeological faunas. Journal of Archaeological Science 24, 1115-1127.

Pickering, T.R., Domínguez-Rodrigo, M., Egeland, C.P., Brain, C.K. 2004. Beyond leopards: tooth marks and the contribution of multiple carnivore taxa to the accumulation of the Swartkrans Member 3 fossil assemblage. Journal of Human Evolution 46, 595-604.

Plummer, T.W., Stanford, C.B. 2000. Analysis of a bone assemblage made by chimpanzees at Gombe National Park, Tanzania. Journal of Human Evolution 39, 345-365.

Potts, R. 1988. Early Hominid Activities at Olduvai. Aldine de Gruyter, New York.

Radovčić, J. 1988. Dragutin Gorjanovi'c Kramberger and Krapina Early Man Hrvatski Prirodoslovni Muzej, Zagreb.

Raup, D.M. 1979. Biases in the fossil record of species and genera. Carnegie Museum Natural History Bulletin 13, 85-91.

Selvaggio, M.M. 2001. Identifying the involvement of multiple carnivore taxa with archaeological bone assemblages. Journal of Archaeological Science 28, 465-470.

Shipman, P. 1981. Applications of scanning electron microscopy to taphonomic problems. Annals of the New York Academy of Science 276, 357-385.

Stiner, M.C. 1994. Honor Among Thieves. Princeton University, Princeton.

Stiner, M.C. 2002. Carnivory, coevolution, and the geographic spread of the genus *Homo*. Journal of Archeological Research 10, 1-63.

Tappen, M., Wrangham, R. 2000. Recognizing hominoid-modified bones: The taphonomy of colobus bones partially digested by free-ranging chimpanzees in the Kibale Forest, Uganda. American Journal of Physical Anthropology 113, 217-234.

Trinkaus, E. 1985. Cannibalism and burial at Krapina. Journal of Human Evolution 14, 203-216.

Turner, C.G., Turner, J.A. 1999. Man Corn: Cannibalism and Violence in the Prehistoric Southwest. University of Utah, Salt Lake City.

Twain, M. 1871. A brace of brief lectures on science. In: Neider, C. (Ed), Life as I Find It. Harper and Row, New York.

White, T.D. 1992. Prehistoric Cannibalism at Mancos 5MT-UMR-2346. Princeton University, Princeton.

Wobst, H.M. 1978. The archaeo-ethnology of hunter-gatherers or the tyranny of the ethnographic record in archaeology. American Antiquity 43, 302-309.